The Primal Code

RONALD SIMONAR

A THOUGHT IN PROCESS

EVENTHOR

To Anna

This book is a work of plain logic. Most of its material systems and energy fields exist far below the radar of the Planck constant.

The first twenty pages are used to set out the principles of the Primal Code. It has not been clothed in the fashionable garbs of mathematics. The Code has the attributes of a theory in terms of precise principles without representation of complex equations. There is therefore no way to hide any shortcomings if the proposed understanding is lacking.

The remainder of this book offers carefree conceptual models based on the template of inviolable principles that reach us sifted through many layers of material systems that warp them like funny mirrors reflecting our faces.

This is a different world from yours.

ISBN 978-91-979667-4-0

Foreword

The Primal Code is based on the behavior of a corporeal primal atom built of nothing but itself with no inherent internal qualities except its relation to zero out of which it is split along with an adverse twin atom. The Primal Code lays out a handful of principles concerning internal relationships and why each atom can cause identical random splits.

The primal atom has no energy in any sense of the word and, as the smallest Universal unit of mass (and energy), it cannot project external effects across a void. It does have a unique relation, a coordinate system relative to linear Universal movement and rotation of all other atoms. Its blend of parental and newborn coordinates is not really a property of the atoms but their relativity is the only thing keeping the Universe going.

The conceptual Primal Code explains the nature of material systems and different faces of primal energy with duality of mass and movement, generated by a collective and incredibly vast primal interaction. The Code takes you from the smallest Big Bang ever conceived through formation of gargantuan globular Mattrons built of 10^{912} primal atoms in equal adverse numbers. The Mattrons gather in globular Hydrons build of 10^{228} Mattrons that exchange the rare emission from primal splits. Neutrons, Protons, Electrons and Neutrinos are all Hydrons.

This Primal Code shows you Gravity and the Gravitational Constant, the Strong force and the why's and how of electromagnetic wavelength and frequency, the photoelectric effect and photon absorption and more.

Having been told for a lifetime to shut up and calculate, the hardest thing is to lose that baggage and travel light; to build solely on the principles of a Primal Code. The theory is falsifiable and open for attack, but should its premise be flawed, then why is the world it creates when it peeks above the parapet of the Planck constant, exactly like our Universe and its laws.

The Primal Code is not an accessory to the Standard Model. It is a self-contained theory of many material layers of which the Standard Model knows nothing; a template waiting for you to fill in the correct numbers. Perhaps the Primal Code can provide a foundation for the painstakingly constructed mathematical Standard Model, an irrefutable edifice of human ingenuity, so regrettably built on quantum sand.

Ronald Simonar

Table of Contents

Part 12 - <u>Photon Properties</u>

Part 1: The Primal Code

Introduction

Imagine if you can, a world void of matter and energy, and consequently void of all properties ascribed to these.

Having made a stab at this impossible task, I ask you to accept a proposal that with a handful of qualifying assumptions will in due course explain the evolution of our Universe from its beginning to the present; along with all the properties of matter and energy. My arbitrary proposal is therefore certainly expedient and its underpinnings run broadly as follows;

Accept that the Universal void you have imagined is in part and parcel composed of two adverse 'components' that in a combined state form the entity we call 'nothing' (in the way we view zero in mathematics; an entity made up partly of logically related properties described as +1 and -1).

The proposal is that that a part-volume of 'nothing' can actually split into two identical 'real' entities geometrically defined by the original part-volume. When the two adverse entities have separated spatially out of 'nothing', they are 'different' from nothing. Should the two ever occupy the same 'space' again they return to being zero.

Secondly; make this arbitrary part-volume split up (of nothing) repeatable. The + and - entities (primal atoms) have no 'properties' that work across the emptiness of 'space' (nothing). According to the Primal Code, all that exists in our Universe, then as well as now, are such simple entities along with a zero void where +1 always 'equals' another +1 and -1 always equals another -1.

Please forget all properties of mass or energy which are merely collective descriptions of the austere primal attributes (of +1 and-1) when they reach us heavily garbled through multiple layers of relatively immensely vast mass structures (relatively speaking as some of these structures are so small that we did not know they existed).

It is impossible to carry the excess baggage of our conventional views so let us travel light; the ball is mine to throw and yours to catch. You can make up the math as you go along, if you must, but my council is to grasp the big picture before you go down that road. Before I can begin to trace Universal evolution from a first austere split, I must minutely define the existential facts that are present at that barren moment.

These qualifying assumptions, the fundaments of the Primal Code, fall woefully short of the exotic character displayed by 'matter and energy'. Even so, these simple attributes must stand as the unbroken chain of command that governs every event in the Universe. They are all there is.

There are no attributes other than those of primal atoms, forthwith to be presented. In this, the laws of the Primal Code as a theory of principle, are inviolable. As we progress from this handful of austere assumptions, which takes but a few pages, one cannot manipulate an incorrect outcome of a heavily deformed Universe by using *'what if'*. If the correct outcome does not flow logically from the Primal Code, then the genetic strain of the code is flawed or not fully understood.

I ask you therefore to bear with me through these deceivingly simple and repetitive descriptions. Their complicated heritage will tax your mind soon enough. For convenience, I have arranged the handful of assumptions of the Primal Code under seven headings and adding three more to describe the aftermath of this very miniature bang.

Creation

You have now tried to imagine a world void of all matter and all energy and all the properties ascribed to these. You have found that such an unbroken universal state does not enter into tangible existence without the formation of some independent reference point.

Contriving an independent reference point within the framework of this imaginable void where nothing is the only existing quantity, it is from the 'nothing' that a reference point must be extracted.

Significantly such an event cannot alter the sum of the Universe, which is and will always remain logically equal to '0'.

Accept therefore my arbitrary request to bring into being two entities or 'primal atoms' so possessed that they united equal '0' or specifically a part thereof. These two primal reference 'atoms' that are 'split' out of 'part of nothing' will now populate and therefore define the otherwise empty 'Universe'.

Refer to these two entities as adverse primal atoms; which is which is unimportant. They are exactly the same but 'adverse' in the same way as plus and minus one. Give them the symbols + and - but try to avoid thinking of them as mathematical entities. They are real enough.

Furthermore accept that these two solitary 'newborn' atoms that originated from the same spatial reference point after the split are by necessity receding along a 'universal line' away from the spatial point where the split occurred. It is of course not much of a Universe, or much of a line.

The two newborn primal atoms move on their axis in relation to each other and the line of their common recession runs along that rotation axes. In our minds eye, we would 'see' that both atoms 'spin' about the 'straight' polar line along which they recede and note that they by necessity spin in 'opposite' direction; else there would be no spin.

To soften the pangs of logical conscience, you may allow the rotation and linear movement to stand as the cause of the arbitrary split of the part zero that formerly contained both primal atoms. Accept therefore that not only is each atom surrounded by the zero void 'o' but that the spatial volume that is occupied by each atom after the split also contains its original 'part-o' as an integral part of the material manifestation of each primal atom.

In other words, if we could snap out fingers and make either primal atom disappear, we are not faced with an empty bubble into which the engulfing 'o' subsequently 'flows'. The reason is that before this disappearing act, the 'part-o' was already occupying that part volume along with its materialized components of either + or -.

As the 'empty' atom sized volume 'o' can also be described as + and - in a bonded state and its part volume is also occupied by the materialized volume of either one + atom or one – atom, a more correct description of a primal atom would be ++- or --+.

Now hold onto your academic hat and accept that there exists an internal discord within each primal atom where two identical components alternate to form its none-materialized manifestation 'o'. As the ++- atom has two + components that alternate with - to form the 'o' combination, the material spin of + and + would in that case be identical.

Accept that this internal discord 'from time to time', at no specific interval, sometimes or never, causes a split up of the internal 'o' into two independent newborn primal atoms that are of identical shape and volume and only 'materialize' as a pair of 'opposites'.

There is no logical 'reason' to be found for the 'timing' of such splits and no apparent cause for the relative movement of the primal atoms but we shall soon see that the relative movement is not the property of the atoms, which may call into question the necessity for such a cause.

Before a split is consummated, the + and - atom are in the unified state of 'o' while moving out of the parental volume. It is first on the outside of their parent volume that they 'split' into a pair of independent adverse primal atoms that float away from each other (and a parent) as they separate across 'empty space' in a birth that causes no change in the 'material sum' of the Universe.

Logically, should two newborns split up already within their parental volume, there would be a period after 'birth' where two material manifestations in the same sign (+ or -) occupy and move simultaneously through the same spatial volume (+++-- and ---++). I reject this possibility because I can see no provision for such internal 'material fusion', no matter how fleeting or partial.

Each primal atom is an indivisible 'real' entity without any internal material properties that can affect it or shape it.

The Primal Code holds that in relation to the surrounding 'o', each material digit is made of only itself (and the internal 'part-o' that cohabits with its material manifestation within its spatial volume at each time).

Aside from the opposite linear displacement after birth, each of the two newborn primal atoms is endowed with relative opposite spin on a common axes about their line of mutual recession; the spin is simply a different form of relative displacement.

With no apparent logical element setting the direction taken away from a parent; the two primal but adverse virgin atoms leave the parent in any 'Universal direction' along a common axis of recession about which both 'rotate'; also in opposite 'Universal directions'.

At the birth of all subsequent atoms that will be split by these first two primogenitor atoms and later by their descendents, one must differentiate between two distinct sets of inherent coordinate for both types of relative displacement; linear and spin.

The Universe that came into being with the split of the first two primal atoms will in time likely see each of these two atoms split primal a pair of virgin atoms in their own image. This happens in total accord with a parents coordinate system; aside from the offspring's individual inherent coordinates of linear displacement and spin.

The parents coordinate system is a unique and inviolate universe to itself. Aside from its private birth coordinates, a newborn primal atom in its virgin state moves in total adherence to its parents spatial coordinate system. In regard to spin and linear motion, the spatial coordinate system within which it exists relative to the rest of the Universe are those of its parent and it has no logical reason to move differently at any distance.

There is also the second set of inherent coordinates that are independent of those of the primogenitor parent but identical, albeit opposite, for the newborn virgin pair. This second set of spatial coordinates is a new birthspin and the linear displacement away from its adverse twin atom and these independent coordinates are totally new to the parent.

Aside from the independent coordinates of newborn spin and that both newborns are moving away from a parent, they also move in perfect step with inherent parental coordinates relative to the rest of the Universe.

When I talk of newborns or virgin particles being emitted by a parent, it is always in this fashion.

An obvious consequence of all the aforesaid is that there exist in this primeval Universe neither mass (as we know it), nor any energy or gravity fields; internal or external. A primal atom does not and can never have the slightest influence upon other primal atoms across 'empty' space. There is nothing in that equation but the actual atom itself.

According to the Primal Code, all later generations of primal atoms are for the above reasons identical to the primogenitor atom in size and shape and their birth does not and cannot cause a change in the zero sum of the Universe. The universal zero-sum applies to both the numbers of 'adverse' particles and the two types of relatively opposite local 'movement'.

It would be a mistake to imagine a pair of primal atoms as matter and antimatter. That would be to associate them with ordinary physics which is premature. My advice is again to travel light.

And so far, the correspondence to mathematics is in perfect order.

The Conceptual Universe

After the reformation of the big zero with the creation of the first two opposite primal atoms, the Universal void has become more conceptual.

I do not concern myself with the shape of these identical atoms. It suffices to state that each atom has exactly the same diameter in three-dimensional space and there consequently exists a definite unit of length.

I have given the primal measurement unit the name 'ãnna', expressed by the symbol 'ã'. A measure of length is then expressed as 'ãr' where the term 'r' stands for the atom radius ('r' actually describes the diameter of a primal atom but his makes my relative size discussions easier to grasp).

This unit of length can in our mind's eye be applied to the increasing gap between two receding atoms, thus defining the distance between them. With nothing else in the Universe but these two adverse primal atoms, this is the only distance there is.

In the fog of mathematics, the idea of time as a physical entity has taken root. At the dawn of creation we can clearly see that time has no physical aspect. Time does not exist. Time is a conceptual crutch, the division of the duration of one event into the duration of another event. It is mathematics.

Thus the number of 'ãr' by which the gap between the two newborn atoms widens can be applied as a measure of the primal time (expressed as 'ãt') it takes for each primal atom to complete a single rotation around its axes.

In our minds eye in a barren Universe of two primogenitor atoms, we can work out a divisible distance and divisible time that gives us the means to define the splitting speed of the first two adverse newborns.

Such a mental exercise in defining 'speed' cannot in any way affect 'reality' on the scene and gives us no leeway in 'warping' space or suchlike.

The concept of linear 'universal directions' away from a primogenitor atom is at the outset restricted to a pair of opposite directions. In 'time', with a growing number of primal atoms in the early Universe and growing distances between them, the number of definable universal directions increases. If the Universe were an expanding balloon, the number of definable directions from its center increases in tune with additional points on its imagined 'sphere'.

Virgin newborns stand an equal chance of receding away from a parent along a 'straight' line in two opposite, out of all possible, directions.

I would like the speed at which all newly split primal atoms move apart to be equal but I can find no logical reason to support that stance so my only recourse is to settle for an 'average' speed of recession.

The average addition to the 'ãr' distance between the two newborns during one average 'ãt' rotation can be used to measure the average 'primal' splitting speed that I have come to call 'ãc' or more simply ¢.

In the same vein, I see no logical reason for a constant splitting frequency or splitting rate at which a new pair of two opposite virgin atoms is emitted by each existing parent atom of either sign. The average splitting frequency could then be quantified by an 'average' creation constant 'ḱ'.

The internal discord within a primal atom where two identical components alternate to form its internal part void of zero, a primal atom can from time to time (sometimes or never) facilitate a split of its identical part volume of zero. I set a preliminary value for the creation constant ḱ as 10^{-1192}ãt where each atom causes a split of a pair of adverse atoms on average once every 10^{-1192}ãt.

I caution you to take this 'average constant' on faith because nothing can be calculated at this point; the only way to adjust 'ḱ' to reality is by reverse engineering against the Universe that leads from this set of circumstance. In this respect, the Primal Code is an all or nothing enterprise.

For the displacement speed, spin or splitting frequency to be constants would require these elements to be a function of some presently unknown (and most likely unknowable) existential factors.

The values of splitting frequency, linear speed and relative rotation of any newborn pair can therefore vary dramatically but as the number of primal atoms that play part in our calculations grows, the value of all these 'average constants' becomes ever more accurate.

Even so, one must accept a great deal of individual uncertainty and settle for sudden unexpected jumps in any of these relative effects at this 'lowest' material level. When I refer to constant birthspeed, constant birthspin or a constant rate of new primal atomic splits, I am always using average values.

As the Universal population of primal atoms increases at an accelerating rate, the average constants within this soon gargantuan expanding cloud of primal atoms can with increasing accuracy be regarded as constant.

Linear Motion

There is an important point to be noted concerning each newborn primal atom; besides being in no way externally affected across empty space by any other primal atom, its coordinate system is independent in its relation to the rest of the primal Universal mass.

Each primal atom is a universe to itself. It moves within a totally private and universally valid coordinate system; a hard coded fusion of its inherent birth coordinates and the also inherent coordinate system of its parent. *One of the most important changes in our logical perception is therefore that any primal atom is best observed from its private reference point.*

The inherent coordinate system of a primal atom dictates how its spatial position alters in relation to other equally independent coordinate systems. This is especially relevant should its coordinate position be shifted.

You will find this apparently humdrum fact to be one of the greatest sources of 'power' in the Universe.

The only correct way to judge the relative 'motion' of the first two totally independent primal atoms in the Universe is by necessity using the medium coordinates of both participants as there is no other reference to be made except a parents *observation* which confirms this identical speed.

The two-point reference illustrates a cardinal truth; two atoms approach and recede (or otherwise move in relation to each other) at speeds that under all circumstances must be ascribed to both atoms equally.

We seen this most clearly on studying the first two primogenitor atoms as they enter into 'being' in an otherwise materially empty Universe. Asserting that one of these two atoms is moving while the other is standing still is an absurdity in the context of a two-point coordinate system that incorporates the inherent coordinates of both atoms.

You cannot change this fact by adding a third independent observation point, or by adding enough primal mass to fill a Universe. It is a fact that will not change and that forces us to always look at the primal scene from this only relevant observation post.

Any two primal atoms at a changing distance always move at equal linear speed in opposite direction when their relative motion is measured. Note that the inherent birthspin of the first newborn split by either of the primogenitor twins is by half the orbital movement of the primogenitor atom around it and increasingly so its primogenitor twin that is distancing itself with its orbital movement therefore speeding up.

The two-point reference also illustrates perfectly that the relative speed of a primal atom need not be 'constant' or, more precisely, such perceived speed can change erratically for no apparent reason when viewed from a third reference point.

For obvious reasons, linear displacement is rarely along straight lines and there is really no such thing. If we shape linear displacement of two atoms as a mental exercise where they move along two vast circular orbits placed side by side in the same plane, we would, using the Universe as a reference point, see these two orbital paths as two circles that almost intersect.

By 'wiping out' the remaining Universe, we can look at the two atoms from a proper two-point reference, which in no way affects them as there exist no forces to insidiously affect these independent primal atoms.

If we let these two atoms head in opposite directions on their orbital paths, their relative linear motion in a two-point reference system is *observed* to be back and forth along a straight line. The relative speed grows from standstill at the farthest point towards the middle distance where it starts to slow until it reaches dead stop at the closest point. From there it starts to back off; and so on ad infinitum.

Using our minds eye as an *observer* situated on either atom in a one-point reference system, the other atom is seen to behave in this fashion while the observer atom is of course standing still.

All this comes about without any forces being exercised and without any inertia of a Newtonian world. We are in the realm of spatial coordinates where everything is relative and movement is rarely in the eye of the beholder. Primal motion has little to do with primal atoms or their 'mass'.

Had the two atoms above been moving along the two even sized circles in the same direction on the same orbital plane at equal speed, the two point reference system would show them to be completely stationary; that the two atoms have no relative linear motion at all.

As the perceived movement of these two primal atoms is not the property of the particles, there can be no 'energy' implied as a property of their 'mass' by way of such perceived movement.

The observed one-point motion is governed by the independent coordinate system into which each atom is born and within which it exists. The relative motion first becomes apparent upon introducing a 'foreign' reference point. Without that distant reference point with no effect upon the primal atoms, that specific interpretation of their relative motion is quite irrelevant.

The two-point reference illustrates the fundamental fact that atoms would always approach each other and recede along 'straight' lines were it not for their inherent coordinates of relative spin; the next basic assumption to be fleshed out in the Primal Code.

The Spin

The first two primal atoms split out of 'part-o' were set as rotating around the axes of their linear recession in relation to each other (it was suggested that this internal out of sync movement might cause the splits).

In relation to the remaining Universe this observed spin is just as inherent part of a virgin atoms coordinate system as its linear motion. And like linear motion, its inherent coordinate system extends to the outermost edge of the Universe.

As the ratio between the average birthspin of a newborn virgin atom and its average birthspeed is not within my grasp, I simply set the values of these two average 'constants' as being equal.

This says that each newborn pair on average covers the distance of one primal diameter 'ār' away from the parent atom in relation to an independent reference point while each completes one whole axis rotation in opposite direction to its adverse twin.

To understand the relative 'motion' of spin and judge its manifestation for the atoms involved, we must study it in a two-point coordinate system where spin may complicate the picture but not the principles.

Picture in your mind two virgin atoms some distance apart. Pick a pair of atoms that from our selected universal *observation* point would be described as one spinning and one stationary atom situated in the others plane of spin (instead of the common rotation axis of a split pair).

Having picked these two atoms, snap your fingers and make the rest of the Universe disappear along with our observation point. This disappearing act does not in the least affect the atoms as no insidious forces work across empty space.

With no other reference points than the coordinates of the two left behind atoms, we can in our minds eye ascertain the actual positional changes that take place and ascribe them to each atom.

When we study the situation from the one-point vantage of the observed 'stationary' atom, we accept that that the other left behind atom is indeed spinning on its axis while we remain at complete standstill.

When we study the situation from the one-point vantage of the observed 'spinning' atom, we accept that the other atom is orbiting it. From that perspective, the atom that was earlier termed 'stationary' is responsible for all of the relative motion of both atoms by virtue of its orbital movement.

Just as is the case with linear motion, all observed displacement must be divided equally between the two atoms; there is no other way that reflects reality within their inherent two-point coordinate systems.

That the entire Universe may be spinning around one of the atoms or that our former observation point had the same atom at standstill does not enter into that equation at all.

The only logical conclusion to be drawn from this mental exercise is that the atom we initially perceived to be spinning does spin at half the observed rate whereas *the atom we perceived as stationary is orbiting it in an opposite direction at a fast speed that accounts for half of the observed speed of spin noted at our earlier observation point.*

Place the 'non-spinning' atom (at our universal vantage point) somewhere out along the spin axis of the 'spinning' atom (not in its spin plane) and observe that from the 'spinning atom' it is seen to spin at the same rate that we noted earlier at our observation point.

It is patently impossible to ascribe any relative coordinate movement to one of the involved atoms. It is absurd to proffer upon any primal atom the responsibility for relative movement, let alone to ascribe to that atom any change in mass or suchlike by the virtue of that arbitrary choice.

The relative motion we ascribe to things is purely a consequence of the setting we select to calculate from; it is not an inherent quality of the atom itself. This is what true relativity is all about. You may think this is a small difference but it will be seen to shake your physics to the core.

The above results are valid at whatever extreme distance we select with unsettling complications that we are forced to accept; and it is not easy.

If the rest of the Universe that we made to disappear materializes again, it too must be judged to orbit the single spinning atom in an opposite direction at a speed that can account for half the observed rotation speed.

And the further you place your observation post away from the 'center' of such a coordinate system, the faster the entire Universe orbits the primal atom. It is important to stress that there is nothing logically abnormal in observing the Universe orbit a single primal atom. That we will have serious difficulty in accepting this has to do with deeply rooted mental coordinates, not reality.

The two-point coordinate system gives the narrowest but also the most relevant interpretation and this will be seen to beget tangible real effects in collisions between allied primal atoms.

Let us first look at a strictly governed world from the inherent viewpoint of a newborn atoms spinning coordinate system. The nature of a spinning coordinate center is that the further you go from its center, the faster the apparent orbital motion of other atoms becomes.

Such increase in the relative motion of an unrelated atom is a consequence of its inherent coordinate system and has nothing to do with the involved atoms, at any distance. The paramount fact that flows from this bit of relativity is that all atoms endowed with the same primal coordinate system are stationary within that system, even when not stationed at its center. The coordinate system does not belong to the materially manifested atom; it is a system of coordinates that the manifested atom happens to be completely at home within.

As a single primal atom would seldom appear stationary to the world at large, it is beneficial to look at a curious phenomenon of relativity that has to do with the movement of atoms governed wholly or partly by one inherent coordinate system.

Governance by a coordinate system can be achieved either through birth of a virgin atom into that specific system, or through collisions with an allied atom; an important aspect that will be treated under a special heading. But let us for now look at the perceived changes caused by a single coordinate system during the birth of an atom.

Assume the role of a newborn leaving a virgin parent at the center of its original inherent parental reference system. As a newborn virgin you are yet untouched by any other particle of other values than our own and, of course, that of our left behind parent.

By a strange coincidence someone has just made the remaining Universe disappear. The Universe had up until then been spinning around our parent but of this we know nothing.

To simplify this observation, our virgin atom happened to be born without any independent birthspin (it would cause our parent to orbit us) but our birthspeed (of which we know nothing) is causing our parent to recede. Other than that, our parent is perfectly stationary in relation to us.

From our parents point of view we are moving away from it in a straight line further into the parental coordinate system into which we were delivered that is also inherently ours (aside from moving apart) at whatever distance we reach from its center that is occupied by our parent.

Our position changes in no other way than with 'straight' linear departure in relation to that center of our common coordinate system. Born without our own birthspin, we are endowed with the same spin coordinates as our parent (as perceived by a Universe that we make to disappear).

The following now becomes clear; two newborns (without birthspin) move in an opposite direction along a 'straight' line within a common coordinate system away from a stationary parent at its center. Without independent birthspin in relation to our parent we have no possible logical reason to undertake to orbit our parent within our coordinate system (identical aside from the recession).

Consequently, if our parent was observed to be spinning in relation to a universal reference point, there is only one type of relative movement that allows our newly split atom to fit both observations; every newborn atom is seen to 'follow' its parental coordinates of rotation at whatever distance.

If the newborn heads onto a parent's single plane of spin, it makes the circumference in exactly the same time as the parent. Its orbital speed in that plane of spin increases in direct relationship with the distance between the parent and its newborn (the radius of its orbit) as perceived from an outside reference point.

This is a natural effect of relativity and there are no 'forces' involved. The observed 'boost' in relative motion is not the property of the primal atoms but a consequence of each being a universe to itself endowed with that unique frame of reference that is its system of coordinates.

A newborn atom can leave a 'spinning' parent in any direction and from the parents point of view the outcome is identical. However, from a universal *observation* point, the relation of its directional path to a parental coordinate

18

system determines the orbital movement of the virgin atom as perceived by a universal *observer*.

A newborn that is emitted into a parent's plane of spin becomes associated with rapidly increasing speed as it 'orbits' away from its parent. Studied by the same universal *observer*, a newborn that heads out along the parental rotation axis is seen to retain the correct coordinate relation to its parent by spinning on its own axis at the same rate as its parent.

Particle emission at other observed angles between the single plane of spin and a single rotation axis are seen to carry relevant variations of these two extremes.

A virgin atom receding out across a single plane of parental spin increases its relative orbital 'sweep' by the distance traveled (the orbital radius). This sweep also increases the newborn atoms 'collision potential' by sweeping ever more spatial points per 'ăt' after leaving the parent.

The spatial presence of a virgin is set by the spatial volume (1) that its material manifestation fills multiplied by the number of equally sized spatial points (ã) that it sweeps through or occupies during a primal time unit (ăt).

An atom emitted into a parent's single plane of spin reaches orbital speed 'r¢' (relative to a universal *observer*) at distance 'r'. As every 'ãr' thick sphere away from its parent is represented by r^2 spatial points (losing all constants), the spatial presence or sweep of a newborn virgin on each sphere in the parents plane of spin is depicted by its spatial bulk (1) multiplied by r.

However, when a newborn leaves a parent along the parents rotation axis it is not perceived by a universal *observer* to gain any relative boost in orbital speed; it conforms to its parental coordinates by spinning on its axis in sync with its parent. On whatever r^2 sphere is reaches out at right angle to the parental plane of spin along a perfect axis line, the sweep of its spatial bulk (1) is multiplied by 1 as its collision potential on every sphere.

The chance of the two relative directions away from a parent atom that has a single plane of spin lies between one polar line and 'r' number of equatorial directional lines. Every newborn pair leaves a parent along an imaginary line that has a relation to the parental surface arc somewhere between its plane of spin and its single axis line.

The definition of each specific linear direction taken by the virgins spatial bulk increases with distance. Drawing a centerline to any distant point on an imaginary outside sphere does not alter the definition but does enhance its accuracy.

A line close to the parental pole would therefore pass for a non-orbital line at a small distance but become distinctly orbital at larger distances. One line on any imaginary outside sphere is the perfect polar axis line; only one line on a sphere of universal reach exhibits no trace of orbital sweep while r number of lines on that sphere depict the number of equatorial lines.

For all imagined distance spheres, all the possible paths of emitted primal atoms would display an average increase in orbital sweep between the polar boost (1) and the plane of spin boost (r). For all universal creation of primal mass, one can infer that the average orbital sweep of a non-spinning virgin as it moves away from its parent is the √r.

The average collision potential of all newborn atoms that leave a parental coordinate center is increased by √r. On each r^2 sphere reached, the average virgin suffers merely r√r loss in collision potential.

I end this by reiterating that on top of the parental coordinate boost, there is also the independent spin of every newborn atom about its axis. Taking this into consideration complicates an illustration but does in no way alter the relative values described above.

I point out that from a universal perspective, the spin of a primal atom is not restricted to a single plane. It will soon turn highly convoluted through collisions when the growing universal cloud of primal atoms becomes large and dense enough to allow such collisions.

So let's now pass on to the basic assumptions of the Primal Code that deal with the actual collisions of primal atoms.

Adverse Linear Collisions

A collision of two adverse atoms, (+ and -), would in accordance with the already covered principles of the Primal Code, render both again into "o".

This union is effective to the extent of the two meeting volumes. There is a tiny chance that the volumes meet so perfectly that they completely cancel out but there exist no 'forces' here that can fit them more perfectly together. Furthermore, such a merger is only effective during the fleeting moment in time while the two adverse bodies occupy the same spatial volume.

One can arrive at two perfectly logical results from such an encounter; a situation not to be appreciated when the consequences can radically affect the Universe and the case can even be made that a benign blend of two logical possibilities exists.

The likely result, the one I have promoted to the power of an assumption, is that the inherent instability of such a 'collision merger' does not (or rarely) result in a permanent change of the involved adverse atoms. The two atoms float through each other without any effect other than the momentary creation of part zero during the overlap of their volumes.

In the next 'instance', the adverse atoms have passed and become separate entities again. The Primal Code accepted earlier that the internal discord of + and – that alternated in forming the material manifestation of a primal atom, succeeded at extreme intervals to cause a split of new atoms. I can therefore accept that there is no logical reason for such a part unification during an infinitely more fleeting moment to be effective in uniting the two entities back into zero.

The less logical result is that the fleeting sharing of the two volumes will permanently change the size and shapes of the two adverse particles that are then left without whatever part of their atomic mass volume was rendered into 'o' during passage.

Although I discard this option, partly annihilated atoms behave exactly as whole particles after the passage. Their spatial presence or collision sweep is set by the remaining volume left untouched and these smaller atoms would thereafter split particles out of their internal 'o' in their diminished image.

The main assumption of the Primal Code is that adverse atoms simply float through each other with no effect but a fleeting non-permanent creation of 'o' during the moment of their partly overlapping volumes. Which concludes this short discussion on the pristine consequences of collisions between adverse primal atoms.

Allied Linear Collisions

Allied collisions between primal atoms carry vastly more complicated consequences and demand a more detailed discussion.

When two allied atoms collide, their surfaces meet. There is no inertia and no attraction. There is no charge and no repulsion, - no 'forces' of any kind. All that exists are the atoms, each an indivisible digit built of nothing but itself and unable to alter size or shape.

Significantly, as two allied atoms collide, two different inherent coordinate systems come to govern every relation of the two colliding particles relative to all foreign *observers*.

At the point of no distance between two colliding allied atoms, their relative and perfectly equal movement changes to nil in every aspect.

Even at what we would consider incredibly fast speeds, a change of shape or some kind of a 'fusion' of two atoms is a logical impossibility according to the

Primal Code.

It must at no time be assumed that an atom strives in its direction, thus implying force or inertia. This would attribute properties to the primal atom that it does not possess.

After a collision, with two touching surfaces, with no distance between the two, the allied atoms remain without motion in relation to each other. This tells us that the two formally independent coordinate systems have become one and are now identical.

That is another way of saying that the unique coordinate system of each atom still exists but the relation of each atom has changed in relation to its original coordinate system. Reality has not changed but the relationship to the two coordinate systems is altered and this has tangible effects.

For the individual virgin atoms, being kicked out of their old coordinate center does not really change the old systems; it just gives each atom a new and different relation to those earlier coordinates. One can of course view this blend of two independent coordinate systems as a wholly new coordinate system that is inherent to both atoms but that may be confusing.

After this coordinate-marriage through collision, the common identity of the two allied atoms has become greater than that between a parent and its allied newborn; after the collision, neither of the atoms has any movement at all in relation to the other.

The collision causes a fundamental shift in relation to their two coordinate systems as neither atom is any longer stationed at its respective coordinate center. Although now with a single set of coordinates, a useful description is that each virgin has been kicked out of the center of the unique inherent coordinate system it gained at birth and has by so doing a different relation to the other atoms coordinate system.

The outcome of a coordinate adjustment is an identical set of coordinates for both atoms which is not the same as a new unique coordinate system but as it function like that, it can even be considered like that. In allied collisions the relation to all coordinate systems in the Universe undergoes fundamental changes although the atoms are in no way affected across distance.

If we from the vantage point of the two colliding atoms *observe* a nearby third unaffected atom that was earlier *observed* to be stationary to one of colliding atoms before the collision, we would see this third atom float away after the collision.

From the vantage point of the nearby atom that was an independent fellow traveler of one of the colliding atoms, the collision between the other two gives the impression that an approaching atom on the move kicked a nearby stationary atom away at half its approach velocity.

With no transfer or annihilation of force in this collision, we are confronted with the illusion of force when the third uninvolved atom reveals its latent relative motion of a coordinate system that remains unaltered.

This also illustrates quite well that such an illusionary 'kick' is in no way affected by the 'mass quantity' of the atoms partaking in the collision.

We can for example look at an atom annihilated by half in a collision with an adverse atom (if possible) and ask whether the spatially twice-larger mass is twice as effective in 'dislocating' a smaller atom in the above collision. The answer is an empathic no!

In neither case did a perceived motion have any connection to mass quantity. The latent relative motion of a third unaffected reference atom did not change. It cannot have changed. And it is that atom that signaled to us the appearance of altered linear motion when the spatial coordinate systems of the two colliding atoms became mutually adjusted.

From this we draw a logical consequence; when a pair of touching atoms, endowed with a single coordinate system, undergoes a new allied collision, the pair experiences the same dislocation effect as would a single atom.

For two close atoms with surfaces that are not touching; where a 'distance' exists between them, however small, the outcomes is altogether different. If a close non-touching pair is on a collision course with an incoming atom, the result is affected by all three involved independent coordinate systems. In this case the outcome is a double gear-down. Viewing the collision sequence from an independent observer would apparently shows it twice halving the incoming atoms original speed of approach.

And nowhere is a 'force' lost or gained.

The effect of linear dislocation of primal atoms in collisions is directly linked to the number of distances closed; not to the bulk number of the participating particles.

Spin in Allied Collisions

The same principles that applied for linear motion at the moment of collision are brought to bear on colliding atoms whose coordinates signal spin or orbital movement relative to independent *observers*.

Relative spin motion and therefore also its mirrored orbital movement must be evenly divided upon two atoms participating in an allied collision.

If a third point reference tells us that one atom spins and another does not, the coordinate system of the spinning atom tells us that the stationary atom is indeed orbiting it and accounts for all the relative movement. The fact that the orbiting atom may seem to be doing this in the company of the entire Universe cannot in any way alter the outcome.

The only logical conclusion in a two-point reference system is that each of the two atoms involved in a collision accounts for half the relative spin and orbital movement incumbent upon these coordinates.

The spinning atom rotates at half the observed speed and the stationary atom (along with our observation post and the rest of the Universe) has an opposite orbital motion that, even though well hidden by its universal conformity, accounts for half the observed speed.

In a collision between two atoms, all relative movement of both ceases to exist at the point of contact. Of course, all the other parts of an orbiting Universe remain totally unaffected.

Imagine two atoms that we *observe* to be stationary in a linear mode some distance apart. Call them 'A' and 'B' and observe that atom 'A' appears to be spinning on a single axis while the apparently non-spinning atom 'B' appears to be stationary some distance away in 'A's single plane of spin.

Looking at the transfer of relative motion, linear and spin, between two primal atoms that are governed by two independent and different coordinate systems, it may help to mark these particles down on paper to be better able to picture them in your mind.

We now confirm that *observed* from the apparently non-spinning atom 'B', atom 'A' appears to be spinning and an independent *observer* further away that is also stationary in 'A's plane of spin confirms that 'A' is spinning in the same way as was perceived from 'B'. However, *observed* from atom 'A' the atom 'B' appears to be orbiting it in the same time and manner as would be the independent *observer* at greater distance.

With this setting in place, imagine that the apparently spinning atom 'A' splits a pair of primal atoms of which 'C' is allied to the non-spinning atom 'B' in 'A's' plane of spin.

When the newborn twin atom 'C' is observed from its parent 'A', it is seen to recede its parent in a straight line without orbital movement.

Now, assume that the outgoing newborn atom 'C' collides with atom 'B'.

To obtain a net result, we must ascertain how all relative linear and spin coordinates are transferred in relation to an independent *observer* stationed at a greater distance. In this case the relative movement has three aspects.

The first is the orbital or spin of the two atoms, 'A' and 'B'. Although we saw 'A' spin and 'B' at standstill, we have already concluded that each of these two atoms does account for half of the relative speed.

The second aspect is the relative linear movement of the newborn twin atom 'C' away from its parent atom 'A'.

The third aspect is the new independent birthspin of the twin atom 'C'.

There is no way around the logical fact that although the newborn atom 'C' comes at atom 'B' away from a stationary parent atom 'A', then 'B' accounts for half the speed of approach. It changes nothing that atom 'B' is at the time *observed* by the Universe to preserve its distance to the parent 'A'.

This is not a curious sort of a mind-warp but a simple confirmation that the coordinate systems of both atoms 'A' and 'B' had been moving exactly in the same way in relation to the coordinates that later became the inherent coordinate system of the newborn twin 'C'.

Upon reaching the distance where 'B' is stationed, the newborn atom 'C' must therefore 'in reality' be orbiting its parent 'A' at exactly half the apparent parental rate of spin. This while the atom 'B' (and the independent far off *observer* in the parents plane of spin) accounts for the other half of this motion by orbiting 'A' in the opposite direction.

There is no logical conflict in these observations although the massive conformity of Universal coordinate motion serves to mislead us. When the collision occurs between the two atoms 'C' and 'B', the resulting standstill between them brings to the fore the latent half-speed orbital motion of the remaining Universe and latent half-speed rotation of the parent atom 'A'.

From the parents 'A's point of view, the orbital motion of the now collided 'B' has been cut in half and an identical orbital movement has also been assumed by its formerly straight line offspring 'C'.

The speed of recession by atom 'C' away from 'A' has also been cut in half and the same rate of recession has by no coincidence been assumed by the orbiting atom 'B'.

Observed from parent 'A', both 'C' and 'B' are now orbiting but only at half the speed of the far off observer (and of the remaining Universe).

The parent 'A' and the universal *observer* in its plane of spin are unaffected by any coordinate change. As it should, the Universe viewed from the parent atom 'A' appears to orbit it at the same rate as before. Conversely from the vantage of the universal *observer* in its plane of spin, the parent atom 'A' appears to spin at exactly the same rate as before.

What has changes is from the universal vantage in 'A's plane of spin, the formerly stationary 'B' and newborn 'C' (that is no longer a virgin atom) are receding in opening orbit away from the spinning 'parent A' in its plane of spin. The touching side by side atom pair completes one widening orbit in the half time of 'A's spin; *at whatever distance these atoms reach.*

If atom 'B' had instead been stationed out along the single rotation axis of parent 'A' and the newborn twin 'C' had collided with it out along that single axis line, the 'BC' particle pair undertakes no orbital sweep. The touching pair would instead be seen to confirm the transfer of half the axis spin of the parent atom 'A' and half the recession speed of the newborn twin 'C'.

The linear dislocation of atom 'B' in the company of twin 'C' at half the (average) birthspeed '¢' does not change; it is effective away from parent 'A' in whatever universal direction the orbit of 'BC' makes actual for the pair at any given moment. Its coordinate system is simply moving away from 'A'.

In other words, the repelling effect of 'A's emission that is aimed into its plane of spin is not directional in any one specific universal line (like out from a perfect rotation axis). It is effective away from 'A' along all universal directions incumbent on the resulting orbit. Only emission along the precise polar axis line (from a parent that spins in one plane) is repulsive in one specific universal direction (or rather in opposite directions for each sign in a primal split).

I have no wish to complicate matters for the reader, but there is now a third coordinate aspect that must enter the picture as the newborn twin 'C' had a unique birthspin of its own. It is at the center of a unique coordinate system that is itself 'moving' in accord with the parental coordinates of 'A' until 'C' collides with atom 'B' and transfers and adjusts to its coordinates.

Before the collision of 'B' and 'C', half the new virgin birthspin of 'C' must thus be ascribed to atom 'B'. When atom 'C' hits 'B', 'C' is moved out of its unique virgin coordinate center by the relative retro motion of 'B'. If this were the only orbital motion, the pairs path away from the collision point would be along an opening orbit that widens away from that newborn coordinate center.

This newborn coordinate center is still orbiting 'A' at half the speed that 'C' orbited 'A' earlier when 'C' occupied the center of that coordinate system. The orbiting coordinate center is left empty as 'C' moves 'B' at half orbital speed away from its coordinate center. The pair in this opening local orbit also orbits 'A', making one circumference at half the originally perceived speed at whatever distance. Apparent relative motion in all the adjusted coordinate systems govern part of the resulting unified motion.

The earlier coordinate center of 'C' is left behind by its physical newborn that is kicked out of it by the relative motion of 'B'. The vacated coordinate center continues its orbital and linear departure away from parent 'A' in accordance with the coordinate input formerly described.

The physical pair 'BC' is united in a common coordinate 'relation' to their original coordinates while 'BC' also recedes away from the unique coordinate center of 'C' that is now common to both atoms. The pairs widening orbital departure from 'A' is therefore affected by a second simultaneous widening orbit as the pair orbits away from the coordinate center of 'C'.

The 'dislocation' input in the collision between 'B' and 'C' is therefore not merely half the birthspeed of 'C' but locally also half the birthspin of 'C'.

Its inherent birth coordinates had 'C' rotating once while it moved linearly across one diameter (r) away from its parent 'A'. If the collision path with 'B' was in its plane of spin of 'C', this orbital speed is halved but it increases in perfect accordance with the widening of that orbit in the plane of spin. The orbital speed at any distance is 'r' times greater.

As the pair 'BC' moves away from its collision point at orbital speed that vastly exceeds the speed by which it moves linearly (away from 'A'), the speed input into any eventual next collision will be that much greater.

A new collision orbit away from such a new common center will open up faster and gain orbital sweep far more quickly than could be achieved by 'CB' that receded more slowly from atom 'A'. This means that 'CB' and some third atom that the pair hits move away from that next common coordinate center after a new collision in an opening orbit at a speed that vastly exceeds that by which 'CB' moved away from 'A'.

This subsequent speed input into an eventual next collision will be that much greater and a new larger orbit away from a new common center opens up faster. Occupying more spatial points during each unit of time immensely boosts spatial presence of primal atoms and thereby their collision potential among the increasing universal population of primal atoms.

The pair can even come back at 'A' in faster far more penetrating widening orbits out from that new center or any other new coordinate center.

This tells us that a parents relative repulsion by way of emitting a pair of newborns is not merely linear repulsion. Proliferating primal mass cannot as easily expand away from its original birthplace as it multiplies in the wider universal field. It will also frequently return to shore up the central regions of a consolidating primal mist that is part and parcel of the Universe itself.

Linear and orbital speed governed by myriad coordinate centers continues to slow relatively in collisions as assimilate coordinates adjust. The common coordinates hold an increasing collective sway over all the atoms that do not remain in a virgin state, locally and at large. Many are still marked by part of some parents now widely spread coordinates while the coordinate heritage of the primogenitor atom is common to all universal atoms.

Let me finally put to you an example that takes these points out of the realm of primal particle creation and brings independent collisions on to a firmer ground using atoms have undergone multiple collisions and have then subsequently been kicked asunder in similar collisions.

Look at any touching pair of primal atoms like 'C' and 'B' that now orbit a common coordinate center. This is a common situation to most primal atoms in the progressed Universe. It is interesting to see what happens when either 'C' or 'B' are collided apart by a foreign 'straight-line' strayparticle 'D' in a way that for example leaves 'C' behind. The 'trajectory' of the new particle pair 'DB' serves well to illuminate the principles at work.

For the sake of simplicity we decide that the approaching atom 'D' does not spin on its axis as it approaches from deep space along a 'straight' universal line and closes in on the 'CB' pair that is locally orbiting in a single plane of spin. With that we have the relative coordinate movement of all the three participants in a coming collision or an eventual near miss.

It is of paramount to get the correct picture from the vantage of the single atom that will be left behind untouched after a collision, namely atom 'C'.

We do this best by imagining the approach path of 'D' in relation to 'C' as if its ultimate collision partner 'B' (now coordinate twin of 'C') was not there. This allows us to visualize the real relative approach path of 'D' during either a hit or a near miss. Half of all perceived movement, linear and orbital, belongs to the coordinate system of the 'CB' pair.

Observing the approach path of 'D' from the locally orbiting atom 'C' if it were single, we note that it is not in a straight line but in ever-closer orbits that ultimately almost touch the surface of 'C' (where 'B' would be stationed with a 'CB' pair in case of a hit) and 'D' then passes by 'C'.

After the near miss, its receding phase is also in widening orbits where each orbit is completed in the same 'time' no matter what the distance. Had we not opted for simplicity, this smooth motion would be made erratic by the to and fro modes inherent in the pairs multiple local orbits and eventually also the possible virgin spin of the incoming 'straight-line' atom 'D'.

As half of the observed movement is ascribed to the left behind atom 'C' (a coordinate twin of 'B'), half of the observed movement would be annulled in 'D's eventual collision with the 'B' of the 'CB' pair.

After an actual collision at an opportune angle, observing from a left behind 'C', the collision pair of now touching and joined 'DB' recedes in a widening orbit at half the orbital and half the approach speed of the former incoming strayparticle 'D'.

The cardinal point is that while the straying atom 'D' came at the particle pair 'CB' along a straight universal line, the Universe was orbiting and is still orbiting the left behind atom 'C' in a totally unaffected fashion.

The result is heralds dramatic consequences of relativity.

When any primal atom travels along a universally straight line and hits a stationary spinning or locally orbiting atom, the resulting path of the collision participants becomes a widening orbit away from a new common coordinate center, i.e. not a straight line at all.

Viewed from a universal reference point, all terminated linear and spinning motion has been transformed into half speed orbital movement around and away from the now vacated collision point.

On the other hand, had atom 'B' been single, there would be no left behind atom with twin coordinates to aid our understanding of what just happened; its absence however does not affect the outcome.

A newborn atoms path through the Universe is altered at its first collision point from being linear motion at 1¢ birthspeed into a new average $\sqrt{r}$¢ orbital sweep away from a local coordinate center with its collision potential dramatically increased throughout the local spaces.

Eventually most primal atoms in the Universe, except those most recently born will have collided myriad times and been partly assimilated. This means that most parent atoms have long lost their virginity and activated primal spin with $\sqrt{r}$ boost has assumed a slow collective movement within an immense local coordinate system governed largely by the continuous input of new and influential newborn birthspin.

Relatively early in the proliferation of the Universal mist of primal atoms, most newborns do not carry the primeval coordinates of a single plane of spin about a single axis. Instead they move away from a parent that is already inherently endowed with a differential spin in many shifting planes of spin and therefore in full or part possession of the average $\sqrt{r}$ parental coordinate boost in all universal directions for any new primal split.

Part 2: The Primal Universe

The Primal Universe

I have now to the best of my understanding laid bare the assumptions of the Primal Code. By building on this logic alone, there remains the daunting task of unraveling the chain of Universal causality.

The natural point of departure is the original part split of 'o' that started the upheaval of a new Universe. This first event brought the 'Universe' into a conceptual framework is followed by endless repetitions of identical primal splits.

We have before us a zero Universe where harmonious 'breeding' of primal atoms is taking place. Using the mathematical crutch of 'time', these rare splitting events in the infant Universe continue to weave an unchanging process through almost eternity.

Eventually we can in our minds eye observe an exceedingly tenuous but enormous mist expanding ever unhindered into the encompassing void at vast speed. Deep within this procreating primal mist, occasional atoms of both signs (+ and -) pass and orbit by an observer from every conceivable 'direction'.

Later, this universal mist starts also to consolidate inwards, shored up by atoms in Omni-directional returning orbits that would help bolster the atom population of central areas after being re-directed out of their virgin collision centers. Take this 'consolidate' in the correct sense; the universal mist will always remain incredibly tenuous.

My equal ratio of newborn birthspin and birthspeed is an arbitrary choice and it affects the average number of orbits a newborn must make before colliding with another primal atom (we could tell this if we knew the density and size of the particle mist).

A newborn split into a parent's single plane of spin completes one orbit around its parent while it covers the distance of one primal diameter further away. On the other hand, given the unfathomable size of the primal mist in relation to a primal atom and the remarkably tenuous nature of the Universal mist, the number of non-populated orbits is overwhelming.

Imagine a newborn split into a parent's single plane of spin at the center of our present Universe. Allow it, as some virgins would even today, to reach the outskirts of our observable Universe without a single collision. Accept then that this virgin newborn orbits the entire Universe in the same time it takes its parent to complete a single rotation on its axis. A primal rotation takes place in an incredibly trivial fraction of the smallest measure of time that has ever been mentally conceived by a human being.

So much for the speed of light!

The composition of this enormous universal particle mist is that the number of straying atoms increases towards its center where most mass is also created. From that birthplace of advanced structures, atoms can only split new atoms with an outward motion in relation to the Universal mist and their coordinates are 'expansive' throughout this area.

Is there any reason to assume that any of this, even with all the time in the world, can ever assemble into anything remotely like the clockwork Universe we have become accustomed to.

The Background Strayfield

This enormous and incredibly sparsely populated universal primal mist is composed exactly half and half of allied and adverse atoms in a blend that is for the most part evenly mixed.

Within this great expanding primal mist, much of the population is now orbiting in local parts of the mist although less geared down atoms are also crossing in every conceivable direction on orbital paths of every conceivable kind. These utterly eccentric gyrating paths of immense variety and speeds derive from activated virgin spin; soon extensively slowed in collisions.

Each and every newborn in its virgin state moves rigidly within parental coordinates as if the rest of the world did not exist; any assimilated collision coordinates are now inherent.

This growing field of straying atoms can be considered a normal condition of otherwise 'empty' space that is now conceivable within the boundaries of this slowly expanding and proliferating universal primal mist.

Although material fields are complex, they must be treated as reality; the material that defines them is real and relative to all other particles so that we can now take independent mental snapshots to provide a picture of this new reality and its ongoing interactions.

The Primal Code applies to everything in nature as manifestation of this principle of fundamental primal evolution. Any descriptions of the gathering subclouds within the universal mist must fit comfortably under its umbrella. Nothing in the Universe can stand aloof from it.

The average experienced frequency and speed at which the stray particles occupy or pass through any given point in space can differ due to local mass concentrations. The deep background strayfield of the universal mist would offer a smooth more constant average value of the frequency and speed at which atoms pass through every point in space.

Every primal atom in space is part and parcel of the generating strayfield; no matter whether this is a single atom alone in emptier reaches of space or part of some immense local particle collective.

All these primal atoms have the same potential to emit newborns at an average relative speed of 1¢ in opposite sign and directions to be boosted by the average √r travelled, parentally or primally, on their continuing journey through space.

This √r boost of newborn atoms stands for most of the observed speed of straying atoms and their spatial presence within the universal mist.

As boosted atoms collide, they are most frequently captured in fast local orbits around new coordinate centers. This brings increased spatial presence in local areas that contain most atoms. Eventually, they are more easily assimilated within the general area of their birth where the probability of their capture is greatly enhanced by many subsequent collisions before vast orbital speed is reached.

Extremely great orbital speed, once achieved, makes the curvature of a first collision orbit too small for capture within a more tenuous local area; even for a virgin with birthspin when it collides with a spinning atom.

For such fast atoms it takes continued collision passages through large well populates areas, producing spiraling showers of particle pairs of successively new smaller local curvatures until the fast moving spreading particle herd that was stampeded by the original passing atom can be downgraded in speed enough to be effectively absorbed with all its inherent Universal coordinates by a larger or eventually smaller primal submist.

All spatial coordinates inherent in the coordinate system of a primal strayparticle are eventually passed on in myriad bits and pieces as these are divided among the entire particle collections at the receiving end where the original fast atom is trapped within a submist after causing these multiple collisions along with some of the vast primal heard it set moving.

A slow moving atom has a better chance of multiple local collisions after a first collision in a tenuously populated area. In collision with locally spinning or orbiting atom, the curvature of the resulting orbits is far greater and the resulting particle shower is more likely to be trapped within that local area.

Within the central regions of the increasingly vast universal mist, we would see faster incoming atoms arrive from further from the universal center where the population is more tenuous. These collide on average into these central areas of the universal mist from all directions equally.

Although the background mass field does not have much directional or linear 'push' on the larger universal scale, atoms in fast vast orbits can be linearly compressive closer to the slowly expanding universal center when implanted locally directional movement is on average annulled by collisions into the central primal collection from the opposite direction.

If a vast particle collection in is not constantly confronted by atoms from a directly opposite direction, then the absorbing collective will be directionally moved (relatively speaking) by the field effect. We can then regard this as one sided linear pressure, although it is of course nothing of the sort.

During immense swaths of time, it is fair to assume the formation of more structured clouds or collectives of primal atoms in the central regions of the Universal mist. These central regions are first to experience the momentous event of gathering mass into denser structured submists.

The Condensation Effect

Any spatial volume that contains a large collection of primal atoms experiences the birth of additional primal atoms. Initially there is little to block the flight of these into the vast surrounding spaces and the empty outside zero space where they continuously spread and breed.

Given the tenuous environment, the field of spin plays the most important role in increasing the local spatial presence of primal mass; enormously increasing local traveling paths of emitted newborns that follow a parent's relative rotation at whatever distance within that a mist, greatly enhance their collision potential.

The new atoms collide with other moving and spinning atoms, constantly undergoing changes and adjusting their coordinate systems, only to collide again with other atoms. And not only are they collided together, they are also collided apart, if not as often.

Relative speed sets the curvature and opening up speed of a resulting orbit taken by a particle pair away from a collision point. High speed collisions open the new orbit up at half the incoming speed while the orbital movement around the collision center is usually much slower.

For this reason the average $\sqrt{r}$ orbital sweep boost away from a collision center grows faster than when an orbit is opening at half birthspeed. A result of rapidly lengthening orbits per 'ăt' is that a particle pair comes to sweep an immensely greater number of spatial points in the same period of time.

Consequently, a boosted collision pair entertains another collision sooner than would a less boosted single atom. Due to a larger orbital size achieved more quickly, the input speed in a second collision is also greater. In this way the orbital speed is successively jacked up in each forthcoming collision.

There is however a distinct upper limit to orbital sweep set by distance. Each single orbit must be traveled without a new collision during the orbital period.

When the equation of volume divided by the primal population within a larger mist equals the r√r spatial points swept by an atom that heads out to that full orbital circumference after a collision, the atom will on average collide already within its first completed orbit.

There exists by this rule a clear upper limit of average orbital size and thus also a maximum average orbital speed to be gained. Both are directly linked to the congestion of primal atoms within a given volume.

The average speed input in a new collision can increase in value until it reaches this upper limit. When that point is reached, there can be no further increase in the average speed or orbital size before another collision comes about to slow the atoms linear penetration.

For every successive collision, on average once during a complete orbit, the almost equal components of orbital sweep and outward movement are halved until all relative movement has been annulled and the coordinate input assimilated into all atoms of that local primal collection.

The density of a primal mist determines the average radius of orbits likely to be traveled by its population of primal atoms.

Within the universal mist and especially its more central areas, in the aftermath of untold eons of collision showers, there remain few virgin residents with an intact birth coordinates.

Within a large local mist of spatially contained atoms, the opposing spin of activated newborns will in time be annulled. Every parent splits newborns in both sign. If a virgin escapes a local mist while its adverse twin does not, its unopposed spin coordinates retained by a local mass collective will in time become common to all its primal atoms as inherent collective coordinates.

As communication of spin is exclusively transferred between atoms of the same sign in direct collisions, such a spin message, when transferred to the opposite signs, lags behind. Each primal atom of the adverse but cohabiting mass collections must wait to be brought in line with the other sign by split newborns from its already orbiting allied mass. This disparity may cause two local mass signs that cohabit in one volume to spin in opposite directions, albeit extremely slowly.

Most primal atoms become veterans with extensively aligned and slow and locally closed orbits about many now common coordinate centers at different rate about each. Should they ever be collided out of otherwise non-populated and therefore non-collision orbits within a local collective by a strayparticle, the new orbital path affords them varying spatial presence whereas if hit by a virgin newborn would ensure swift collision and local capture.

It is important to note that though an increase in radius away from their coordinate center boosts their speed, atoms are not exclusively kicked away from the center of a coordinate system but even towards such a center. If the resulting direction after a collision is towards a particles coordinate center, its average speed in inherent orbit decreases by $\sqrt{r}$.

There is less potential for spatial boost when two veterans collide; their universal coordinates are already so aligned that most local collisions simply work to keep them within their local spaces.

Condensation is therefore always more effective in areas that have become slightly more populated than surrounding areas and it is within such spaces that the greatest number of collisions has already taken place.

There is within every large atom mist a point that can claim the smallest average distance to all of the atoms. The vast spaces around this point in a mist become its natural mass center that has the smallest average distance to the greatest number of primal atoms. It is in such local spaces that incoming strays or outgoing virgins entertain the greatest collision potential.

The natural center(s) of our Universal mist will therefore in time become the area(s) where most atoms of both signs are captured and detained in inherent local non-collision orbits.

These expanding central regions are always under assault from a relatively constant high speed stray bombardment from the background universal field that supplies robust orbital pressure more or less equally from all directions. This slows the expansion of central areas that collect such primal atoms awhile they (initially) procreate at an accelerating rate.

Before I turn my attention to any potential subclouds built up of primal atoms within the universal mist, I want you to feel the atmosphere within the unstructured primal nebula before it evolves functioning submits.

After describing mass being kicked about at bizarre speeds in every conceivable direction, one might get the impression that the inside of a forming submit is a boiling inferno of activity.

Not so!

The average splitting constant k that sets the creation of newborn primal atoms describes an extremely rare occurrence but atoms within a forming submit that wait on this rare event are not about to go away.

After all, nothing much is happening.

The primal atoms simply exist, a function they can keep up for all eternity without any effort. Observing a nascent submist is therefore with benefit best done on the primal time scale. In our minds eye we *observe* enormous regions, so vastly spread that it beggars belief and extremely sparsely populated by primal atoms that are neither standing still nor dispersing at great speeds. The atoms are mostly locally circling or diversely spinning but few would be seen to be rapidly escaping out of this enormous mist.

It is this conformity, gained though collisions and sharing of coordinates, linear and spin, that earns a local nebulous collection of primal atoms within the universal mist the name 'submist'.

So take a peek into a gargantuan submist on the primal time scale.

On the primal scale we observe an event deep within the mist. Two new adverse primally spinning virgin atoms move away from a locally orbiting parent and away from each other. Accented by the vastness of the mist, we find their speed insignificant on the primal time scale but we notice that they follow a slow orbit around a parent atom and observe that this orbital speed is gradually picking up.

What happens depends on the size of the submist. If we assume that this submist is first and largest within the universal cosmos that will become a galactic super cluster, each of the adverse twin atoms eventually acquires great speed in opening orbit which input was set by the local orbital speed of a parent. The orbital speed dwarfs the unchanged insignificant linear speed away from their common parent.

After a first allied collision by both atoms, four affected atoms are on the move away from the parent, inwards or outwards within the mist around two collision adjusted coordinate centers by coordinate assimilation.

The curvature of the newly activated virgin spin in orbits out from the collision point is set by the large input speed of the two former virgins that had by then gained highly boosted parental orbits.

And again nothing much is happening.

After a much shorter period at greater distances and orbital speed, we see four atoms undergo further collisions within the vast submist and observe how they rearrange ever more particle positions in other coordinate systems.

During each and every one of these multiplying collisions, different linear directions and orbital coordinates are being aligned and partially cancelled as opposite directions demand.

After a long time and uncountable collisions later, all that remains of the original movement of the two newborns is the local circling of an enormous number of atoms and an almost undetectable collective movement that highlights the coordinated conformity of all atoms within the submist.

Then again, almost for an eternity, nothing else happens.

Things on the primal scale are not be quite this calm, but due to the vast number of atoms and the almost imaginable relative vastness of the empty spaces involved, things are certainly far removed from a boiling inferno.

I am confident that this broad illustration follows irrefutable logic and is therefore an inevitable consequence of the Primal Code.

Primal Submist Fields

With this handful of qualifying assumption of the Primal Code laid bare, there can be no acceptance of other principles that do outside of those flowing logically from its theoretical base.

Primal uncertainty and unclear constants aside, the Primal Code is an all inclusive hardwired program. Questions remain on the present number of primal atoms in the Universe or in our corner of the Universal mist and on how many atoms the various structural submists can support under present conditions and what would be their relative sizes. The numbers with which to prime this theory are breathtakingly wide.

With time as a relative concept, fundamental questions arise and must be narrowed down; such as the average value of the splitting frequency 'ќ'.

Although the Primal Code itself has now been clearly stated, there remains the task of tracing its consequences. I face the difficulty of picking values out of a wide range of possibilities and while only one represents the present day Universe, I must take my shots in the dark in order to show the system at work; leaving it to others to work out the finer details and the mathematical framework.

To come to grips with the effects that individual primal atoms collectively yield through collisions within a vast submist, it is helpful to use and focus the discussion on 'fields of influence'. The fields are vast collective tendencies that pool individual and partially contradictory primal coordinate effects that can easily build into powerful collective surges.

The fields of influence within the universal mist can fashion locally denser parts of the universal mist into submists where our observable Universe is likely just a part of the biggest submist. Eventually, the collective effects of these fields that work different according to scale will be seen to mold the apparently incoherent universal mist into local areas of individual proactive structural submists that affect and respond to other nearby submists.

Although all submists are made up of primal atoms, the smallest submist is made up of primal atoms as its only sub-unit. Other submist types will eventually form from the smallest submist in the food chain which then gather as a building blocks in superstructures where they are prominent (but not exclusive) sub-unit.

We are used to this way of thinking, as we consider galaxies as submists made up of suns but not exclusively so because of their subatomic particles. We will see how a composite submist functions as a dynamically responsive structure in a way that the primal atom cannot function.

The influence of a small submist upon a gargantuan super cloud of similar submists is similar in nature to a photon within a sun but is incredibly less influential upon that submist than the photon is upon a sun.

Every link in the chain from the smallest submist up through successively larger submist structures responds to the same primal fields of influence. The primal atoms are however being differently channeled between various internal structural units. Every link in the chain responds to identical but differently channeled primal fields of influence.

Although the primal influence becomes ever more heavily modified with each step up the chain of submist systems, the character of each link in that chain is always and exclusively based on the apparently insignificant primal atom that with incessant splits is the Primus Motor of all submists, up to and including the vast Universal nebula of primal atoms.

The coordinate system of the primal atom, as set out in the Primal Code, is the only motor that keeps the Universe going.

As the first stepping stone in an evolutionary chain, a firm understanding of the inner workings of the smallest sustainable submist of primal particle is paramount before continuing up the submist chain.

I have set out six collective fields of influence that govern the main inner workings of any submist and I count them here;

- EMISSION FIELD
- SPIN FIELD
- MASSFIELD
- CHARGE FIELD
- STRAYFIELD
- GRAVITY FIELD

All these six fields are interdependent variables that form through primal interactions and govern a variety of submists. Some of these are stable and others ephemeral but it does not serve the purpose of this presentation to deal more than fleetingly with fringe structures.

To discuss these six conceptual fields of primal influence in succession is ill suited for clarity as the division of primal coordinates into six separate fields is for clarity. The underlying truth is that only the entwined effects of all interacting influences makes any sense of a functioning submist.

To avoid becoming hopelessly mired down by incessant cross-references, I have set up this general philosophical study of submists in a succession that affords the clearest insight into the functioning universal chain.

The Universal Nursery

The expanding universal mist is not always perfectly evenly distributed. Bound by the logic of the Primal Code, small temporary submists can form by chance within bigger submist. While these chance gatherings cannot yet hold on to their own newborns and much less faster traversing strays, they are quickly dispersed; evening out the density within the greater submist.

Eventually as the primal population gains density, there comes a time when the smallest submist becomes sustainable. First in line of these proactive structures is the MATTRON.

This smallest permanent submist is exclusively built of primal atoms and holds no other internal structures. This diffuse grid of primal atoms counts the greatest number of internal building blocks of any submist and its small size has the greatest relative difference to the size of its largest structural building block (primal atoms) of any submist.

Long before that, at the dawn of creation, more than $\sqrt{}$ of all newborns went off along minimally curved lines in all direction; rarely to return. Their departure at birthspeed gave them diminished spatial presence and collision potential within the central and faster breeding universal mass.

Through eons of time, this straighter minority has and is still traveling outwards from the more populated universal center across distances beyond our comprehension, constantly expanding the material Universe.

In time, secondary strands by these relatively fewer penetrating newborns sprout and consolidate, just as they did closer to the center after the first split.

These descendents cannot make any real headway backtracking in the opposite direction; their lineage leaves a vague trail of newly split material in the in the universal void, reaching out with primal shots of their own. Such primal trails ultimately intertwine with others and consolidate to condense into immense sub-clouds, most at distances far beyond our comprehension.

In the vast center of this universal mist with immense primal population, the less penetrating orbiting newborns of greater spatial presence have bred enough primal atoms locally to cause occasional collisions among their kind while all continue ever more slowly 'outwards'.

The incoming strays are mostly activated newborns in primal and parental orbits with a side sweep dictated by the average $\sqrt{r}$ spatial boost. They arrive from the surrounding universal mist in vast orbits more or less equally from all directions and their contradictory collisions exert a slight but uniform pressure upon the denser central submist.

Eventually the original universal center, perhaps a submist greater than our observable Universe, is opaque enough to block almost all newborns created within its borders and most atoms that orbit into that area from the outside mist. The center is marginally denser and its locally orbiting atoms become an ever greater spatial obstacle for traversing primal atoms but no reformation of this central nebula can take place before sufficient particle density is at hand to set it off and to sustain the reaction.

This universal center is the nursery of all submists.

The Submist Chain

Before embarking on the trail of universal evolution, it is fair to put you in the picture of how I see these largest and smallest sustainable submists in relation to our everyday world, both as regards particle population and size.

Ultimately, as smaller primal submist start to form within the 'borders' of the universal center and are at first only temporarily sustained, all of them are unstructured submists barely denser than the atom mist that surrounds them. It is premature to call them proactive structures but to some extent they becoming just that. And with increased density, the larger submists will in time sustain vastly smaller primal submist within them.

My original thought was to follow what I perceived to be an actual chain of $(unit^2)^2$ links visible in the sky. You can follow the chain links of dwindling number of building blocks as you move down from galactic super groups to galaxy groups and from there down to the largest galaxies to $(unit^2)^2$ more nuclei making up a sun.

The dropping unit numbers of all these submists followed a rough rule of thumb that seemed prophetic in the possible construal analyses of nuclei.

This rule of thumb is part and parcel of the Primal Code except that the cosmic chain does not stop there; it does not provide a sun with building blocks of Protons, Neutrons or Electrons that are almost indivisible.

Reverse engineering shows that the number of submist units within a Proton is set by the $(unit^2)^2$ rule. And it does not stop there. There is one more last step to take before the ultimate realm of the primal atom. This last $(unit^2)^2$ jump brings us to the smallest of all submists; the Mattron.

A sturdy $(unit^2)^2$ link chain from a maximum $3,16 \times 10^3$ galaxy group built of 10^{14} suns built of 10^{56} subatomic particles, each of these built of 10^{224} mattrons with each Mattron built of 10^{896} primal atoms. The Milky Way and the Local Group fall short of these mass numbers but our Sun with its 10^{57} subatomic particles (Hydrons) does not.

The chain relies on conventional count of suns in average or maximum galactic groups and likewise nuclei in an average or maximum sized suns. The lack or excess of primal mass in a higher system must affect the unit numbers in its lower submist systems and it is hard to cast the chain in stone. My simplified strategy in illustrating the lower regions of the Primal Code is by using Our Sun as its centerpiece.

Our Sun contains roughly 10^{57} Protons which indicates that Protons are built of 10^{228} Mattrons that are made up of 10^{912} primal atoms each.

Whatever the real value, it suffices to say that a Mattron at the end of the structural chain contains an almost inconceivably number of 10^{912} primal atoms in each sign compared to the paltry 10^{80} atoms of what is termed to be 'Our Universe'.

Furthermore and importantly, this mind-boggling primal population of atoms is so tenuously dispersed within the boundaries of a single Mattron that it appears to make a mockery of the Mattron as a structural entity.

It is healthy to keep in mind the inconceivably vastness of a Mattron in relation to a primal particle (its exclusive building block). Initially, before the primal population of a stable Mattron condenses from the evenly distributed primal mist, its volume could be in excess of 10^{2328} primal volume.

My initial take is that a fully formed Mattron holds 10^{912} primal atoms in each sign and has a stable volume of 10^{1896}ã points which offers each of its atoms roughly 10^{984} private spatial points to cavort in. This suffices to make the Mattron a very empty place.

To gain a practical insight into just how empty, take the volume of 10^{744} Universes (as you know it), chuck out all galaxies and other mass and keep all that extensive void. Then toss into that entire combined empty void one single Neutron and you got the density just about right; the relative private space that each primal atom has to itself within a Mattron.

And yet, as the mattron is compared to its primal population, the size of this smallest sustainable submist relates feebly to our world. In the model I use, it takes 10^{228} of these gargantuan structures to form a single Proton, that keeps a collective primal population of 10^{1140} atoms.

It is therefore, for the time being, wise to drop any direct reference to our everyday world and make this a purely mental exercise, keeping only with the logic of the Primal Code. My reference to a Mattrons stupendous size or population is valid only on the primal scale. The Mattron scale highlights the insignificant influence that a single atom can wield upon the whole Mattron; an effect that should not be underestimated.

Let me finally stress that the level of primal containment that a submist must achieve to form a Mattron is facilitated by its size in relation to its primal building block. With great relative size, even a slightly bent path out from a collision point will end as an opening orbit weaving across immense primal distances within the Mattron. Given a smaller size, a Mattron must contain more mass to wield the same spatial presence.

The Mattrons survival as a stable entity hangs on the ability of its primal mass to bend the paths of internally split newborns and the paths of faster moving strays. Relative distance is therefore even more important than the physical blocking ability of its primal population. And if a primal submist is too small in both these respects, it will be ripped apart by penetrating strays that tear it asunder and expand its mass instead of compressing it.

The loosely described ongoing general condensation does not cause primal atoms to clutter together in constricted areas. Fast strays are too disruptive to allow that to happen. The environmental uncertainty can chance to allow a denser local population to gather for a while but such peripheral groupings are inevitably dispersed until the circumstances allow a different outcome.

It is first when this necessary large-scale density has been reached for the relatively immense Mattron in its homogeneous nursery that gathered primal atoms can by virtue of their genetic code launch a monumental upheaval.

This is the birth of the first fully-fledged proactive structure in the primal universal mist, - and the quantum split that follows it.

Part 3: The Mattron

The Mattron Embryo

Let us (in our mind-eye) place an imaginary globular sphere into the central universal mist. This imaginary sphere is a logical exercise that serves exclusively to define a volume within the even distributed mist. For the sake of argument, set its radius as 10^{800}ã primal points.

Although our imaginary sphere is remarkably tiny in relation to the central universal mist, it is incredibly vast in relation to the primal atom and large enough to serve our purpose with a population of 10^{912} primal atoms.

We see occasional straying atoms arrive more or less evenly from all directions to pass into and through the central universal mist that engulfs our imaginary globular sphere.

Let us assume that we have the condition necessary for the formation of a mass collective within our imaginary sphere. This requires that the density of the particle mist within the defining sphere is such that a majority of slow moving strays or newborns that collide within are ultimately blocked and captured by the mist.

In the long run-up to this density, the capture rate within out imaginary sphere would need to be greater than for strays that collide through similar neighboring volumes of the universal our sphere. Before this general density is present, there is no local tendency to consolidate the surrounding primal mist into the imaginary sphere.

An average boosted collision where birthspin is activated covers roughly r√r spatial points on average within our imaginary sphere. This leaves r√r untouched spatial points for every point swept within the defined sphere.

The eventual entrapment of a virgin atom passing at birthspeed through the imaginary sphere volume from outside can only be guaranteed if the presence of primal mass within our sphere is sufficient to guarantee a collision that activates the newborns birthspin. This would quickly lead to multiple other local collisions and likely capture.

Such a density is reached within the sphere when the number of spatial points occupied by its atoms exceeds the radius of the imaginary sphere 'r' as measured in primal diameters and especially after the density edges closer to r√r.

As this relation is the same in any sphere, one might be tempted to conclude that sphere size does not matter; that the smallest densest central area within the universal mist is the first to collapse into a more defined submist, can simply grow from there.

Because of the immensely boosted and disruptive speeds of penetrating strays, submist size is of utmost importance and the general condensation starts with the largest entities; the universal center being first in line. The relative vastness has the advantage that straying atoms take long in primal time to traverse these immense widths where even slightly crooked paths become ever more orbital and therefore more collision prone.

Although areas of more than galactic super-cluster size have started to consolidate at the Universal center and other submists within them, this diffuse gathering of primal mass at the top of the structural chain cannot yet be termed functioning material structures.

The cumbersome gathering of mass into large entities takes longer as mass gathering are not easily herded linearly. On the other hand, the collapse into a Mattron at the center of a vast primal gathering is relatively rapid when the density allows. Were we to guess the birthplace of the first Mattron embryo in its nursery, we could select the greatest submist; the universal center that is on par with our observable Universe. Nonetheless, the spatial dimensions of all entities at that time would have vastly exceeded their ultimate size.

From the vantage of our imaginary sphere that defines a Mattron embryo, this Mattron nursery is a diffuse submist within even less structured larger submists in this vaguely condensing central area of the universal mist. At this beginning, the smallest and densest submist, precisely because of its density making up for the relatively shorter distances to be closed, is first to gain a more structured interior.

It is important to keep in mind that although relative distances have been shrinking between the repulsive primal atoms, these average distances are still gargantuan on the primal scale. During the initial condensation, spin and atom orbits have become more aligned in collisions and new coordinates can only be communicated by procreation to atoms of the opposite sign.

To a Mattron embryo (the mass and space within our imaginary sphere) at the center of a consolidating primal nursery, the immense outside mist can be viewed as an all-engulfing Universe. This smallest stable submist does not form from scratch in empty and open spaces; the scattering by fast strays would effectively prevent consolidation under these circumstances.

After the smallest confinable embryonic mass is eventually able to spatially contain its own primal production and to trap slower strays that enter and collide within it, its density is set to grow at an accelerated pace due to both procreating mass and a shrinking volume. Perhaps a more fitting phrase for this growth is 'less excruciatingly slowly'. So let us trace the first steps of the embryo towards Mattron-hood.

Envision again the smallest possible imaginary sphere that we earlier placed at the center of the universal mist. Assume that its internal volume has reached the required primal particle population to enables its capture of most newborns split within it and even some faster strays that pass through it from the outside mist as described above.

Very little about this imaginary sphere deep within the even mist of primal atoms would remind us of a primal structure. Within the sphere and on the outside it, we would only find almost the same incredibly scarcity of primal particles and would see that mist continue away in all directions.

The envisioned boundaries of our imaginary sphere are deep within the evenly distributed central area of the great universal mist. The proximity of atoms within and outside the imaginary sphere is almost more or less the same in every direction for immense distances.

In this homogeneous environment, given the orbital nature of primal collision centers, the transfer of collision effect and alignment of spatial coordinates is not very effective in a straight-line sense. Newborn 'pressure' travels the vast mist 'slowly' as it blooms like Omni-directional bubbles all over the place due to birthspin activation while ever more particles become more aligned through collision-transfer of coordinates.

Such continuous redistribution of a primal population cannot be completely homogeneous all of the time. Some areas will always temporarily contain marginally more primal atoms than surrounding areas and it just so happens that we have placed our imaginary sphere in such a way that an incidental primal gathering of marginally less tenuous nature has become its natural center.

By definition this center has the shortest average distance to all the atoms within our imaginary sphere. Thus the marginally denser central areas within our imaginary sphere now become the focal point where particles entertain the greatest collision rate.

Aside from this focal point, should we pass out through the boundaries of our imaginary sphere, we would fail to see any falling off in the density of the evenly populated universal mist that surrounds it and continues away from our sphere across unconceivable distances.

Although the number of primal atoms within the imaginary border of our envisioned sphere, as well as around it, has been slowly increasing to gain a marginally better general density, only its imaginary center has passed the point of required opacity. The atom number within the sphere may be just a tad more than that the sphere radius and still far below the square root of its volume.

With its marginally greater density and therefore slightly more effective blocking of newborns and straying particles, the volume within our sphere will inexorably through eons of time become the smallest local area within the universal mist where the greatest number of particles is procreated and trapped.

As the forming Mattron passes across this threshold, it may seem like an isolated event but it is not. The formation of this single embryo launches upon the Universal mist a massive upheaval that shifts its foundation.

I call it the 'quantum split into Mattrons'.

Concentrated around the early incidental focal point of its center, the growth of primal mass within our imaginary sphere is facilitated on two fronts. With moderately more mass, the slow creation frequency (mk) leads to faster growing mass in that area but even more entrapment of newborn atoms and slower strays from outside this vast imaginary sphere.

With eternity on its hands, the less dispersed primal population deep within our envisioned sphere continues to grow in numbers but, as we shall see, not in size.

As the tenuous primal population of this first smallest stable submist rises within its outer boundaries, one might be tempted to argue that the inward circling particles from the surrounding primal mist will keep adding to its mass and that the spiral of self-generated growth can continue ad infinitum to add mass with no upper limits to the attainable atom number.

There are serious flaws with this argument.

Start with the spatial facts where the 'surface' of our imaginary sphere encloses the volume of what becomes a Mattron embryo. Give the imaginary sphere the width of a primal particle to make it more tangible. If the primal population were evenly distributed, this imaginary surface sphere would sport the greatest number of primal atoms of all other internal spheres we could fit into its globular volume from the center outwards.

Our globular 10^{2400}ã volume embryo has a sphere that is 'r' times less or 10^{1600}ã surface points. Its marginally denser 10^{912} primal population is merely a 10^{-688} fraction of the spatial points available on that sphere and by far the largest number of atoms populates its relatively immense outer regions.

The size of the Mattron sphere and its extensive areas of outer mass is necessary to allow rotation axes newborns from within to collide more frequently and to slow and trap strays that collide within its denser central areas. Collisions turn strays more locally orbital with increased collision potential within larger slightly less populated volume that insures multiple collisions on traversing the vast outer regions of the Mattron embryo.

Having achieved more local density, the center helps this extensive main mass to capture a growing number of newborns in relatively fast parental orbits that circle in from the engulfing mist. A colliding activated incoming newborn causes a great shower of atoms to issue out from the relatively aligned mass of the slightly more opaque embryonic central space in tighter primal orbits. Having already served to slow down the traversing atom, the orbits will increase its primal presence within the embryos extensive outer spaces and ultimately enable its capture within it.

The growing primal bulk within our vast imaginary sphere, itself a small lump in this primal soup, is increasingly being achieved at the expense of the slightly less populous primal mist that surrounds our imaginary sphere; a chipping out effect that reaches outwards across many sphere diameters.

It is self evident with this development that primal particles in the vicinity of our imaginary sphere stand a statistically greater chance to be collided into the denser globular cloud by orbital atoms and to be captured within it than to be collected by the outside mist. The majority of orbital newborns that arrive into the globular sphere are also frequently accompanied by particle showers carried from the surrounding primal mass.

The surrounding areas in the immediate vicinity of our imaginary sphere are through this process destined to grow increasingly transparent. Atoms that arrive in widening orbits from outside the imaginary sphere and are trapped by its marginally greater density are now gradually stripping these outside areas of primal atoms.

For a long time thereafter, on the primal time scale, the primal population of the mist is fortified at an accelerating, not only by procreation but by local exterior atoms being continually chipped out of the gradually emptying shell that surrounds the globular embryo.

On the other hand, the primal population outside the imaginary sphere has started to taper off into the more empty void for a distance of least one embryonic radii into the surrounding mist.

The Hollow

In the other direction, newborns issue out of our imaginary sphere, split by atoms in the outermost voluminous regions and when activated there, many peep out of the imaginary sphere in opening orbits myriad times before their collisions slow the swarm enough for local capture.

These peeping orbits pass repeatedly through the more depleted areas outside the imaginary sphere where they are less prone to collide than when they orbit back into and through parts of the denser embryo. This situation prevents the embryo from thinning gradually into the original density of the standard mist; it thwarts the formation of a seamless structure.

The orbits of such outgoing activated newborns must reach a radius distance that exceeds the width of the embryo before the opening orbits of the resulting swarm can move free of the globular embryo. Chances are that the atom swarm is overwhelmingly harvested by the vast embryo before it can cross the more transparent hollow.

After some time, the primal population of the Mattron embryo is no longer fortified by local exterior atoms chipped out of the surrounding void as these are no longer present while and ever fewer orbiting atoms can reach the embryo from the original mist across the void.

Across the emptier void for a distance one embryonic radii, the emptier hollow that the Mattron embryo has carved out of the original mist gives way to the original density as the universal mist that stretches out in all directions for incredible distances.

The hollow exposes the flaw in arguing that the upper limit of mattronal mass depends solely on the availability of surrounding mass. The fallacy is in ignoring the loss of uniform nearness.

After the formation of a hollow, a newborn primal particle within the inner lining of the universal mist bordering the hollow (outside the depleted spatial belt that surrounds the embryo) can still orbit across this emptier barrier to join the embryo. Whether in parental orbit or activated in that new lining of the universal mist, the atom must first orbit enumerable times into and out of the hollow in the form of a slowing swarm of atoms before reaching the embryo.

As the swarm orbit in and out of an original outer mist that is now much denser than the depleted hollow, the effect is a two pronged depletion of the ever rarer and more transparent outside belt.

The hollow has torn the uniform fabric of material space.

Setting aside the enormous relative distances, particle movement within a primal cloud is affected by uniform nearness of the atoms. Overall medium closeness makes for step by step advancement of newborns in any direction. It is not their universally linear speed that gives them most local presence but rather the activated orbital sweep that immensely enhances their collision potential. With increasing density in any submist, the boost ensures their entrapment at relatively close quarters.

The further a newborn must travel away from its spinning or orbiting parent before it entertains a collision, the faster the (apparent) orbital speed within its coordinate system becomes around its parent. The further an activated newborn must travel away from its first collision point before it has encountered enough primal mass slows or stops its linear away movement, the faster the orbital speed of the resulting swarm becomes around that birthspin coordinate center.

The vast unchanged mist that still engulfs our mattron embryo is now richer in orbiting newborns than ever before. When a split occurs at the closest distance to the new mattron embryo, it is outside the new hollow that separates the mist from the embryo and the newborns must traverse a great linear distance on the primal scale through a more rarified hollow in order to reach the mattron embryo.

Let's keep the earlier estimate that the initial imaginary sphere we placed into the universal mist is the initial size of a Mattron embryo with a birth radius of 10^{800} primal radii. The hollow would therefore be roughly the same distance outside that no longer imaginary sphere. Activated newborn do not easily penetrate through a mist and most of those hitting the embryo earlier came from the hollowed out part of the mist.

Now, in order to collide within the embryo, each split newborn from surrounding mist must move inwards across the vast empty distance of the hollow. From the inner lining of the universal mist, an activated orbit across from the hollow opens at half 1¢ linear birthspeed and can theoretically gain 10^{400}¢ average orbital speed. This affords the atom swarm that is rapidly slowing in collision (both linearly and orbitally, immense spatial presence within the bordering mist, before it can hit the embryo.

More to the point, each mist newborn must approach the Mattron embryo across a vast hollow while it orbits away from its collision center within the engulfing mist where it entertains most collision. Each atom in that swarm has a widening path away from a collision point where inherited coordinates of the parent that started it all off causes it to orbit both out of the engulfing mist into the depleted hollow and then deeper back into a mist lining that is denser than the hollow; all this while advancing ever more slowly towards the mattron embryo.

This circling is repeated an incredible number of times by every atom involved and coordinates of an activated mist newborn have no direct linear path towards the distant mattron embryo. It is orbiting a collision center, advancing on the distant embryo covering on average $\sqrt{r}$ points on each new circumference while moving one point further from the first collision center and closer to the embryo.

In each swept completed orbit, more than half of the primal points swept by the newborn as it advances towards the Mattron embryo are within the more populated surrounding mist. A mist newborn is therefore statistically much more likely to collide myriad times within the denser mist than to cross the great divide and thereafter start to collide within the forming mattron embryo for a smaller part of each orbit.

The Hollows Inner Lining

Take this reasoning one step further. For each parental orbit that partly cuts the original mist and partly the more depleted hollow, less mass on average blocks the traversed points of any orbital distance. Because of this, newborns within the outer hollow lining achieve greater parental distance than they would within the mist and collide therefore ever further from the mist lining and deeper within the engulfing mist.

With that gain in penetration and linear reach, collisions by newborns take on average place deeper within the engulfing mist; further from the mattron embryo than for ordinary splits that happen deeper in the mist where orbits never reach into the hollow.

Collision swarms from shallower splits where the orbital showers reach into the hollow, cause entrapment further from the lining of the mist wall bordering the hollow and deeper within that local mist. This initially moves affected atoms, swarms of already adjusted mist atoms) further away from the lining of the hollow and at the same time away from the Mattron embryo.

The transformation of the universal mist does not stop there. The breaking of the symmetry of material distribution means that the universe passes into a new state after the overall uniform average primal distances are lost.

With the universal mists seamless homogeneity broken, the primal mass that lines the wall where the hollow meets the universal mist is soon seen to be moving slowly away from the embryo. This also means that the lining is building up a greater density than the density of the original universal mist across 10^{800}år distance beyond the lining away from the hollow.

The denser lining of the mist has undertaken a collective spatial movement away from the hollow that has started to actively widen away from the mattron embryo. Both the double hollow and increased density of the lining make it harder for a newborn to escape the mist and hit the mattron.

A similar phenomenon is active at the 'surface' of the now better defined material embryo. Newborns on its original 'surface' cross in and out of the condensing embryonic mass and are more likely to become trapped deeper within it than to escape that 'surface' by crossing the widening hollow.

The mattron embryo is contracting as the hollow outside it widens.

Non-activated newborns in marginally bent parental orbits that escape the forming mattron to collide within the engulfing mist have little chance of returning to the embryo and certainly not under their own coordinate motor. Their linear coordinates are moving away and when they collide within the hollow lining and are activated, they aid in herding the engulfing material shell further outward as it recedes away from the hollow in all directions.

This is a reciprocated effect and another change is taking screw as the incoming collisions from the surrounding primal mist are undergoing a gradual change in character upon the shrinking mattron embryo.

Look at the repulsion character from a newborn that is split deep within the embryo. Neither linear birthspeed or the vastly greater orbital phase can move local mass closer to the embryos center, except to the degree that mass within that tiny interaction span is denser inwards compared with outwards. Due to the immense size of any mattron on the primal scale, and more so a vast embryo, this is a minuscule fraction indeed.

While the width of the mattron embryo is the width of the hollow, a virgin or an activated newborn that hits its spherical 'surface' away from the mist passes tangentially by the embryo.

At this early stage in adjusted coordinate systems, I assume that relative collective coordinate systems are not as adjusted through collisions and that both mist and nascent embryo have yet to slow parental coordinates from full radius boost to ultimate $\sqrt{r}$ and that the spatial presence of parental mass and its opaqueness is thereby massively facilitated.

As a virgins speed in parental orbit can be great and then boosted, it will often have to be slowed down before it can be ultimately captured. Showers of resident particles are torn out of a mattron embryo in orbits that may or may not pass repeatedly through the embryo; just as for activated newborns.

If primal mass within a mattron embryo were stationary, there would be a large number of straight unblocked paths through the mattron sphere for every path blocked by a primal resident. A primal population in local internal non-collision orbits will immensely increase the collision potential of traversing atoms.

Particles that have crossed the hollow in primal orbits and gained immense speeds may after colliding within the embryo, allow too little time for local mass presence to develop. Nevertheless, atom that traverse an embryo at extreme speeds do so without collisions or have their stride marginally broken; or they leave it in the company of a great shower of the residential atoms they moving.

On the other hand the high speed atoms are likely return on another passage after a short period, thus enhancing the collision frequency with this double sided effect.

Even a relatively slow collision can take place on the mattron surface *on the way out of it*, thus working to temporarily expand it if captured and it is incorrect to assume that all pressure blocked from outside compresses the primal mass within a Mattron sphere towards its center.

Incoming newborns in universal orbits that pass only once through the center of any particular mist volume of embryonic size have more likelihood of being trapped if they have been slowed. For entrapment to succeed, their linear speed must hold such proportions that enough subsequent collision take place after the atom collides within that volume.

A Mattrons size and average level of locally moving mass are inter-linked qualities that affect its stability, size and external emission. Compression leads to adjusted coordinate systems that turns a Mattron more transparent with dropping primal presence, which can increase its expansive function instead of its compression through blocking ability.

We come to these aspects later but I wanted here to emphasize the change in character of incoming atoms that enter and collide within the embryo. In the earlier mist and within our imaginary sphere, orbital pressure from split atoms was almost equally in and out upon the collective mass, embodied in the same particle as it circled within. Because of uniform nearness, such orbits supplied contradictory local 'pressure' while marginally compressing the collective mass within the imaginary sphere towards its center.

After the hollow, this former state of affairs changes for incoming strays. Newborns that succeed in crossing the great hollow from the receding wall of the universal mist pass through the mattron embryo at orbital speeds along lines are relatively straight and spatially directional in relation to the body of the globular mattron embryo.

If any atom from the mist is not captured within the mist lining, which is statistically easier, it slices into and out of the embryos top 'layers' in vast orbits that take it back into the outside mist for each of the between 10^{400} and 10^{800} primal steps it has moved closer to the embryo for each orbit in contrary directions.

If an orbital collision takes place within the embryo, it can constitute compression by all inwards aim of the remaining birthspeed upon local mass within the embryo; i.e. it is no longer dependent on a minuscule fraction of increasing inwards density. Compression comes about when similar opposite pressure bears upon the embryo from 'all' directions and turn it vastly more center oriented.

The mattron embryo is condensing faster that before. As the hollow widens and the embryo continues to shrink, reclaiming primal mass from the universal mist is increasingly difficult. After its conception, the number of newborns able to cross the widening hollow is rapidly curtailed and its collective mass enjoys marginal accumulation of primal mass from outside sources; its continued growth is hereafter supplied by internal procreation.

The birth of the first Mattron does not pass without the Universe paying heed. This is a starting point of a monumental structural upheaval on a universal scale; the first quantum split of the universal mist into Mattrons.

Rapidly shrinking at the center of its initial hollow, with increasing center wise compression and increasing internal opacity by mass and movement, the Mattron has become a functioning responsive entity.

After its conception, the embryo enjoys little accumulation of primal mass from outside. The ever denser lining of the depleted hollow is now gradually receding and, as its inner condensing wall moves outwards with its primal mass being moved ever closer, the embryos main source of additional mass is effectively withdrawn. The embryos continued growth is hereafter supplied by its own internal procreation that continues at a slower rate.

I put the average mass number in this primal cloud as 10^{912} atoms when fully compressed. The embryos formation mass is probably bigger and stable due to its great initial size and internal mass presence that helps to block all strays and newborns that pass through it at well above birthspeed. Later, as the embryo continues to be compressed, its mass continues to procreate and build up its mass to the average atom number.

The birth of this first mattron does not pass without the Universe paying heed as this is the starting point of a monumental structural upheaval; the quantum split of the universal mist (or at least a mist the size of a galactic super group) into functioning mattrons.

The Quantum Split into Mattrons

As a watershed in structural conception, the slowly consolidating outer lining of the hollow is an important new factor on the universal scene, albeit a tiny local event.

The material wall (or wave) moves away from the first Mattron embryo, herding the already existing primal particle mass of the local mist before it in all directions with that mass growing gradually denser. As the lining of the hollow gains in density, that mist lining grows ever more proficient in blocking newborns and strays.

The primal population within this wave crest has a more uniform nearness among itself than to the rarified primal population within the hollow and is therefore less likely to be collided into the emptier areas in the direction of the first Mattron embryo.

However, due to the same laws, the original mist outside this denser material wave is having atoms more frequently chipped out to join the denser wall that is condensing outwards in all directions. These bordering areas outside of the new material crest are becoming increasingly tenuous.

Gradually, the recession or rather consolidation of this material wall away from the first hollow into the uniform mist is stymied by the increasing density of the building material wave itself.

Although the outward moving crest takes the visual appearance of a slowly expanding globular material shell that is consolidating and growing denser than the original universal mist that engulfs and dwarfs it, this is a truth of averages. The outward distribution of the Universal mist's primal population away from the first mattron embryo is simply local relocation of primal mass. Most of this primal mass will collide and ultimately be brought to relative standstill where the density of the wave is greatest. As an entity, the wave is not outward moving at all.

There is a self-governing effect that governs the process of this quantum split. The movement of primal particles is locally orbital and not universally linear. When a wave has grown dense enough to block their orbital paths, it puts a stop to the recently gained linear outward movement. The ultimate relocation and capture by the lining population gives the visual appearance that the consolidating shell has slowed in space and stopped expanding.

It is helpful to understand that this quantum split spans eons of primal time. Every single resident particle must collide independently and in most cases a great number of times to join the initial apparent movement of the expanding globular wave that gathers primal particles propagating out through the original local part of the Universal mist.

Grasping what makes this primal mass wave recede from the hollow, i.e. the orbital relocation of primal particles affords us a fine understanding of what follows. Namely, that the effect that caused the quantum split in the first place must eventually also be felt from the other side of the same new material crest.

As the extensive spaces within the crest have gained greater material presence than the original mist outside it in every direction, newborns that circle away from parents within the outer lining of that broad material crest stand a greater chance of colliding on their orbital way back into it.

The mist on the outside of the crest is thus also becoming more tenuous than the density it had achieved earlier. For a distance of at least the crest thickness that is of a similar size and opacity as the earlier hollow, the close outside spaces that were of the same density as the central universal mist (whatever size that is) are slowly but surely being stripped of primal atoms. The particle population of the mist near the wave crests outer lining is being moved closer to the wave's average center in the same fashion that the population of its inner lining was moved earlier.

As this new hollow shell starts to be chipped clean from the outside of the forming quantum wave, the primal mass of this material crest begins to consolidate from both sides. We see how the greater orbital pressure from the hollow building on the outside of the crest is of the same character as it was upon the first mattron embryo.

And just like then, the apparent propagation of the wave stops and seizes as a universally linear movement to become spatial relocation of primal mass towards a more opaque area; the center of the wave wall. The originally perceived outward movement of the wave is shown to be a condensation that gathers the evenly distributed primal atoms towards a thinner but denser globular material belt that now surrounds the new embryo that is situated between two hollows.

The quantum wave of primal material is now at a relative standstill and by the same principles, yet another crest of primal atoms outside the second hollow is starting to relocate mass outward away from that second hollow.

This second globular quantum wave of primal mass will consolidate and eventually be formed at a certain distance outside of the first material wave by the same laws; the distance being dictated by the mist density.

This quantum transformation of a local universal mist carries on arranging the originally evenly distributed primal atoms into 10^{800}är broad onion-skins of denser primal mass which are separated by 10^{800}är broad hollows that are soon largely empty of primal atoms.

Wave after hollow after wave continues to be materialize out through the embryonic mist for as long as there is enough material presence to allow such primal condensation to take place outside the last material wave.

During eons of primal time, this wavelike condensation ripples outward through the layers of the diffuse submist that up until this point had been slowly condensing in a more general primal fashion. Finally, mass presence outside of the last wave will prove too tenuous to be consolidated and the local relocation of the central universal submist stops.

The modification of this denser center of the immense universal mist is driven solely by primal coordinates; especially birthspin coordinates.

Left behind in space are an immense number of globular quantum shells built of primal atoms in both signs; the overwhelming majority of the shells of the original quantum split probably has sizes we would consider on the comic scale. These quantum shells are interspersed with hollows away from the long lost point where the first mattron embryo ripped the seamless homogeneity of the vast submist that provided the first required local density to start off a quantum split.

The evenly spread universal central mist becomes gradually more tenuous outwards as the quantum separation into shells progresses further from the first embryo that ripped its seamless homogeneity. Lesser density causes the hollow width between the material skins to grow. One can say that the 'frequency' of the material shells decreases while the shells are less affected as they call for a standard combination of mass number and size to bend the path of traversing primal atoms. The shells must exhibit the required opacity to prevent its dispersal and sustain the shell structure.

Similarly, if the propagation of quantum waves passes through local areas of greater density, the width of the hollows between material skins shortens correspondingly. If the propagation passes through local areas where primal mass is for some reason a bit more tenuous, the width of hollows lengthens. This is a marginal effect; or else, any abnormal density on such a large scale would already have ripped the vast submists seamless homogeneity.

The density and dimensions inside each material quantum wave is such that it can withstand dispersal by strays and newborns, just as a mattron embryo. As was the case with the original mattron embryo, these are the characteristics that made a Mattron the smallest stable structural entity built of primal particles that can be sustained.

Roughly a cube of this material wall, the <u>*thickness*</u> *of a quantum wave, would ultimately qualify for the formation of one independent mattron.*

By the laws of chance, and more so due to the orbital orientation of primal coordinates that are constantly narrowing the width of each material shell, there will inevitably form myriad denser spots within each vast shell.

As denser focal points continue to consolidate and absence of uniform nearness is already experienced inward and outward from the quantum shell on which they sit, sideways depletion around these denser focal points will inevitably reform the primal mass into myriad independent embryos.

This process is only marginally aided by the curvature of each quantum shell and it will quickly cause the formation of an immense number of new mattron embryos. On each walls of every quantum wave on an unfathomable number of quantum shells, mattron embryos form like so many tiny dots on the skin of an immense balloon.

Most primal atoms in the original mist will therefore ultimately withdraw into spherical and sustainable mattrons embryos, each vast enough to withstand the scattering by relatively fast passing strays. The distances between embryos across a quantum wave hollow are largely the same as are the sideways distance between embryos on each narrowing shell.

After this transformation of the immense local or universal mist, all these mattrons are securely locked onto shells, except the shells are soon gone. In reality, the primal mist has become an immense and even grid of Mattrons held tight in bonds of mutual repulsion and attraction (fields to be dealt with later). Whatever unsubstantial primal particle clouds that lace this material grid will be continually and extensively depleted, driven by collisions by fast strays that deposit primal atoms into some Mattron embryo or another.

We can also readily understand from the coordinate principles how orbital stray particles tend more effectively to chip away primal atoms situated in the vast outermost regions of a Mattron where population density is insufficient to trap them locally.

Any irregular shapes or tenuous bolsters that protrude away from the main bulk of a forming Mattron stand a far greater chance of dispersal than do its central regions. Such outer formations would tend to be continually chipped away by passing strays and newborns in activated opening orbits. No bolsters of primal surface clouds are opaque enough to trap such atoms locally and it is the main mass of the Mattron that benefits from their depletion.

Of all primal atoms that originally border a mattron embryo, only those most broadly shielded from passing strays by its main bulk will eventually form its outer limits. This is a fancy way of saying that a sufficiently massive Mattron will in time be chiseled into a perfect spherical shape.

After the quantum upheaval, an incredible number of perfectly neutral mattrons that contain primal atoms of both signs in equal numbers can continue to increase their internal mass from two sources; mainly by their own breeding of mass and marginally from the entrapment of strays.

Just as there is a minimum level of mass density or size that sets the formation of a stable Mattron, the question is whether there is an upper limit to the growth of its primal population. Can a fully formed stable Mattron continue to grow ad infinitum or is there some maximum level of mass and density that it cannot exceed without its inner structure or spatial balance being in some way compromised.

And so there is, - but the question cannot really be dealt with rationally before we have looked generally and in some rough detail at the inner workings of this smallest sustainable submist.

The Standard Mattron

Within each collapsing embryonic submist on its way to mattron-hood, every primal atom carries the key to the inherent disturbance that can cause it to split a pair of two independent and adverse newborn atoms of the atoms zero volume.

I earlier set the primal time unit called ãnna (ãt) as the 'period' it takes for a newborn at average linear birthspeed (¢) to traverse across the distance of a primal diameter (ãr).

My preliminary arbitrary value of the creation constant 'ḱ' was 10^{-1192}ãt which states that every primal atom splits a new newborn pair on average once every 10^{1192} ãnnas. I caution you to take this 'average constant' on faith as I am just testing the waters here.

Once more, I stress that newborn 'emission' by no means implies the existence of any force in our traditional sense.

Within a spherical mattron, every primal atom of its mass (m) adds its own average 'emission' coordinates with the average relative linear birthspeed movement (¢) where the mattrons internal mass field generates a collective if contradictory linear movement (mḱ¢ãt).

The average straight path through any sphere is the distance of its radius. To have some grounds upon which to stand for this illustration, I set the radius distance of a *fully developed and balanced standard mattron* model as 10^{632} primal diameters (that I call the radius of an indivisible atom). Although vastly less than the radius of its original sustainable embryo, this is no mean distance.

For example, it takes a newborn on average a period of 10^{632}ãt to pass from its place of birth within the Mattron and exit its surface along a straight line. I call this traversing period for its 'radius time'. The time is the same whether a virgin atom orbits out of the Mattron in a parental orbit or has a more linear path out along the parental axes.

Merely during the relatively short 'radius time' it takes an untouched virgin newborn to leave its Mattron, the Mattron itself gains an immense number of additional newborns in both signs through its own procreation. Some (10^{912}m$\times10^{-1192}$ḱ$\times10^{632}$ãt) newborn pairs are split by each mass sign during this period of passage; adding 10^{352} additional pairs of adverse atoms.

As primal particles are on average born at birthspeed, it stands to reason that their linear movement can in principle aim in all possible directions and this 'push' must be largely eliminated in internal collisions before primal mass can be collectively contained within the Mattron sphere.

It is first when a newborn virgin atom in parental orbit hits another atom that its linear aim relative to the rest of the Universe is realized out of all possible linear directions away from its parent caused by parental spin. This is not complete realization as one collision partner cannot brake its parental coordinate stride (being responsible for half of it). After myriad collisions with the parental coordinates divided upon different and extensively cross collided primal mass, there remains some small linear residue that has not been neutralized.

Whatever directional excess remains, i.e. whatever unopposed movement remains within a Mattron when all the opposing linear directions have been annihilated is the Mattrons collective linear movement in relation to the rest of the Universe. A Mattron can easily set itself on the move, albeit extremely slowly, without any outside influence.

Such directional excess is especially 'clear' when coordinates of unpaired newborns or unpaired strays are assimilated by the Mattron mass while the twin atom born with exactly opposing directional coordinates, linear or spin, does not collide within the Mattron; these twins counter coordinates do not become part of its internal mass field.

This illustration can be made clearer if we for that purpose assume that all virgin newborns move in a straight line. As mentioned, it is misleading to describe a collision effect within a Mattron in terms of inward or outward. A parent on a Mattron periphery may emit a newborn on a linear inward path that can pass through the Mattron and collide on its opposite surface with a decidedly outward effect.

If one for illustration freezes all of its ever ongoing local mass movement, my Mattron model becomes highly transparent. It has a transparency ratio of 10^{352} where only one out of 10^{352} radius sized paths are occupied by a primal atom. Any linear stray that hits a primal resident does so along a universal line that holds that single atom. Furthermore, out of the 10^{352} possible radius sized stray paths through the Mattron, only a single path delivers perfect compression aim towards its center.

It is therefore wise to avoid using the directional terminology 'inward or outward' for any pressure within a Mattron where both orientations are embodied in orbital and linear coordinates of a single particle; now on its way inward and now on its way outward.

It is nevertheless clear that a parent close to the Mattrons perfect center can only emit newborns endowed with a coordinate system that has outward direction in relation to the Mattron body. Such coordinate systems have the same orientation as the Mattron center. It does not matter how extensively rearranged outwards path of such virgins become through other coordinate influences; the linear aim of their parental coordinates remain entirely outward in relation to the Mattron body.

From the Mattrons perfect center and to a lessening degree further from it, none of the orbital spin input can be turned inwards by virtue of inherent coordinates. Even if an emitted central newborn, after colliding with atoms further out, comes back at central mass by courtesy of a different coordinate system, the coordinate of that incoming swarm will have assimilated all of its outwards coordinates. As there are no perfectly opposing collisions that can neutralize linear movement of coordinate centers away from the Mattron center, these coordinates will carry through in the long run.

Despite the central condensation brought on by the orbital drive during the creation process and thereafter; evening out the primal inter-atom distances within the Mattron, its outward linear expansion away from its center keeps a small upper hand.

Most newborns are activated within a standard Mattron. Opening primal orbits in convoluted planes of spin boost cause collisions that rapidly gear down the linear phase in the immediate locality of a split. Contradictory aim of the linear phase that is tiny compared to the orbital phase is rapidly and extensively neutralized by contrary coordinates that push atoms in partly opposite linear directions. This allows the mass to remain in non-collision local obits at substantial speed where the greater spatial presence are part of all parental coordinates.

The Mattrons expansion central excess that penetrates across the width of the Mattron and works to expand it is a function of its mass procreation and is governed by the creation constant 'ҝ'. There exists within the Mattron no natural influence to counter this slow long-term outwards 'pressure' that remains after all the other contradicting linear movements have run the local short-distance collision course and been annihilated.

In relation to the Mattron body the outwards potential of internal emission is determined by the parental position. The further away from the mattron center a parent atoms coordinate center is situated, the less of its linear emission 'thrust' counts as outwards and this diminishes with each primal diameter added to the parental distance from the Mattron center.

The net average outward excess in a newborn atoms linear phase is found in the relation of its primal size to the widely different sizes of (conceptual) internal spheres upon which its parent sits relative to the mattron center.

At the perfect Mattron center, the sphere size upon which the parental atom sits is that of a primal particle (1ã) and the outward excess function for its eventual emission from that 'sphere' is (1); its entire linear emission effect within the Mattron.

The number of particle sized spatial points on the sphere that defines my standard 10^{1264}ã mattron 'surface' area has 10^{352} more primal points than the Mattrons gargantuan number of 10^{912} primal atoms. To obtain a net outwards in excess of inwards potential for the linear thrust of a primal surface particle; divide the effect of all emissions that take place on its sphere by the Mattron radius. This while the total expansion excess of its central mass is any coordinate direction remains outwards and unpaired as linear movement.

Split newborns that collide within the Mattron surface and have their virgin spin activated can exert a good deal of orbital compression upon lower layers. Until spent and countered from an opposite orbital direction, this push is harvested deeper by virtue of the *initially* greater density. This helps to gather and keep the main outer mass to quarters at more even density.

For every newborn coordinate center that moves inwards along a line that cuts the mattron center, an adverse twin heads outwards along the same axes (convoluted as it may become in parental orbits). Clearly the net outward push ascribed to a surface newborn is not great but even parental particles near the surface can move mass outward but they are in the end almost as likely to do so inward.

The mattron can be said to carry 'r' layers of internal spheres (each being 1ãr thick). Primal atoms have a natural tendency to collide in more densely populated areas due to greater physical particle presence there. This gives particle distribution within a forming mattron the impetus to start averaging upon the available (r) number of internal conceptual particle thick spheres as we move away from the mattron center.

A population averaged on the conceptual onion-skins within the Mattron leads to greater central mass concentration, which increases the net outward potential of the generated internal emissions. I deal with this generally by stating that the 'radius time' expansion function has the same relation to all internal emission push as the 'average sphere' has to the mattrons outermost sphere; namely the square root.

Central particles are extensively cross-collided and each is therefore endowed with less spatial presence. They are at greater rest but there are more of them per volume unit. Because of the lesser movement, the center would be more transparent, were it not for the greater number of particles. Atoms in the outer regions are fewer per volume but endowed with faster orbital movement that boosts their spatial presence and collision propensity.

The remaining average linear expansion of the mattron by its newborns is set against two 'forces' that serve to spatially balance the mattron. The first is the more temporary orbital compression by the same newborns after they are activated. The second is the compression by the harvest of incoming strays.

Again, the most potent effect that causes mass to gather towards denser central area is the Omni-directional circling of activated newborns that reach out of the submist surface. This temporary bridging pressure would not be inward effective at all, even temporarily, were it not by virtue of the greater mass blocking opposing paths as they cut deeper within.

As the central mass grows and its size shrinks, the Omni-directional circling of newborns in fast orbits from the outermost mass grow fewer and these miss the shrinking central mass more often. This would tend to ease the pressure upon the central mass. The smaller achievable boost of a greater number of newborns split in central areas cannot much compress the soon more transparent center. The stray input into the transparent outer sphere is still blocked because of great spatial presence there but it cannot affect the small massive central mass in the same way as it did earlier.

The mathematics that would help in accuracy seems to come in the way of my quest for understanding the mattrons expansion excess. I therefore deal with it simply by saying generally that this part of the expansion function has the same relation to all internal emission as the 'average sphere' has to the mattrons outermost sphere; namely the square root.

If we had enough time at our disposal, linearly opposing newborn would collide along all the available directions, whereupon the 'average sphere' of parental placement signals its outward potential. I must however assume that this tiny square root expansion function of all the internal opposing directions is but a tiny part of the mattron actual expansion effect.

This is to say that if emissions in all available directions were supplied in time, they would neutralize but they are not. Arrange all spatial points to give the mattron 10^{632} self-regulating two-dimensional planes. During what I call 'disc time', one newborn pair is then split on each plane or 10^{632} in all.

Each of the two adverse newborns split on each plane has linear coordinate motion in relation to allied mass within that plane in whatever direction it goes. Imposing a single direction, this movement will eventually turn outward and equal to 1pu (primal pressure unit). This expansion pressure is far more disruptive than the small square root expansion function that would unravel mattron into space if the neutralizing newborns were available in time.

Although the mattron gains 4×10^{352} additional newborns in the time it takes an untouched newborn to leave its interior, the emission in 'disc time' cannot supply newborns in all available directions to neutralize the imposed movement. In my neutral mattron model, only 10^{632} newborn in each sign are emitted in the 5×10^{911}ät disc time and so whichever way each pushes linearly upon residential disc atoms, there is no chance that another newborn is split on the same plane to head in an opposite direction that neutralizes that movement. On these terms, the mattrons expansion function is close to the linear effect of all its newborn.

With no antidote supplied to this net outward movement within the mattron in time, this total expansion function must consequently be balanced against the compression yielded by incoming strays that are gathered by the mattron sphere, i.e. if the mattron is to be spatially balanced.

The mattrons stray harvest is not set directly by mattron size as the straypressure is mostly local and highly orbital with its linearly stopped flux returning in orbit repeatedly until it is blocked by the nearest mattron of whatever size and all its primal atoms have come home to roost.

As not all newborns within a mattron collide within and some of them escape, the mattrons collective expansion function derives from internally split newborns that are captured within it. As relatively few newborns escape the mattron, this loss which affects the expansion function is marginal.

The ability of the mattron to contract or expand (if not at maximum size) affords great leverage in tapping into the abundance of roaming long distance strays. But whereas the gathering of such strays crisscrossing the strayfield is determined by the variable mattron sphere, its internal procreation is a 'constant' function of time and mass.

Few incoming strays aim at the mattron center. Their paths are tangential to the curvature of its surface sphere and here the mattron size has a direct influence in compression by linear strays. Once within a mattron, a stray that collides and is blocked sets mass on the move and can when opposed by other strays from the opposite direction compress this mass along some straight line that is on average equal the length of the mattron radius.

It should be noted that (aside from local mass movement) my mattron model is highly transparent with a transparency ratio of 10^{352} where only one out of the 10^{352} average possible radius sized paths through it is occupied by a primal atom. A linear stray that hits a primal resident does so along a universal line that contains only that atom. Of the 10^{352} stray paths within the mattron, only one can therefore signal complete compression towards the mattron center.

Although individual strays, once captured and successfully resisted, can only collide effectively with half of the neutral mass, their linear movement or compression effect is in time shared by other mass by secondary emissions.

The same is true of newborns within a mattron. Each can only collide with half its neutral mass and once activated and captured, is slowed by contrary motion induced by incoming strays. This increases the time it takes to move mass across the radius and time allows more newborns in both signs to split. Newborn emission, pushing allied mass, communicates this expansion as well as compression movement by strays even to adverse mass. Thus both expansion by procreation and compression by random straypressure, when eventually balanced is shared and communicated to all the mass.

The balancing act between constant expansion and variable straypressure (number of strays that enter a mattron times their average speed) allows the mattrons to automatically adjust their sphere sizes (and average density) to allow their compression to match their expansion at some size. Shrinkage cannot adjust straypressure blockage, as reigning local flux is highly orbital and returns repeatedly again and again.

The dislocation effect of a single high speed stray can be said to represent linear 'pressure units' that equal its 'speed'. The primal pressure unit (pu) equals the dislocation effect of a primal atom at birthspeed. In this fashion, 10^{400} blocked strays at boosted speed 10^{100}¢ are equal in their dislocation effect to 10^{500} strays at 1¢ and both constitute 10^{500}pu. Similarly, 10^{400} strays at 10^{-100}¢ are equal to 10^{300} strays at 1¢ that constitute 10^{300}pu.

Newborns are split in adverse pairs and neutral mattrons are equally composed of allied and adverse particles. Both parental types contribute movement to the mattronal fields in both signs. As most of its newborns are captured within a balanced neutral mattron, expansion and compression are eventually communicated to mattron mass in both signs although the effect of compression will prove vastly less efficient.

Each individual atom has only a fifty-fifty chance to entertain an effective collision for each encounter with another particle. In half of all cases, the primal atom passes right through its potential collision partner with no permanent effect at all.

For the sake of simplicity, without calculation; not to become mired down in details, I set up my tentative 'standard' mattron model with a 10^{912} primal population in each sign and give it a maximum 10^{632}år sustainable radius in a balanced standard mattron grid. Please view this as an illustration.

For this mattron to balance, the compression by the straypressure must equal its internal procreation of newborns in radius time; the time it takes the average virgin to leave the mattron without collision. In this tentative mattron model, the sphere is in principle the instrument that allows it to find spatial balance by harvesting exchanged local pressure.

If straypressure in a standard 10^{632}år mattron must be harvested at the rate of its expansion squared, that would not be a balancing factor as mass compressed into a smaller volume becomes more transparent; not less. This jacks up the compression ability of harvested strays as the number of lines for possible passage diminishes.

Excessive stray pressure distributed upon the mattrons 'surface', taken at face value under heavy pressure, would therefore not bring balance and my neutral mattron model would simply collapse.

But the straight linear balancing act of pressure within this simple model is not all there is; main ingredients in its balancing act derive from various other properties that wait in the wings to modify its apparent inability to find footing under pressure.

All of that will further muddle my illustrations but first I have a few things about the mattron model to clear up.

Spin Coordinates

The first primal particle and its descendant to pass through our empty Universal space gave birth to newborns and from the generations upon generations that followed, eventually great swathes of primal mass were born until eventually galaxies were brought to condense in the sky.

For this reason, all primal mass in our part of the Universe and elsewhere has a common primal ancestor, a progenitor particle from which all the rest has eventually split. The spin coordinates of this progenitor are still firmly carried and passed on to all newborn particles, imbedded in the relationship that each single atom has to its parental co-ordinates.

However, the mattrons collective field of spin has very little to do with this inherent common ancestry. But it is interesting to note that we are still under its spell and we can neither shake nor can we conceive this orbital heritage as every particle we see is endowed with these same common coordinates.

This is in many ways similar to the heritage in a unified mattron coordinate center, the part forged from 10^{912} individual different coordinate centers in a much shorter span of time. Like the old progenitor coordinates, the movement has become common to all residents within the mattron. All particles within a mattron heed its inherent call, whatever eccentric individual movement they can be caused to make temporarily in local orbits.

If we could magically make all the individual parental or activated motion disappear within a mattron (making it a perfect mattron), the inherent spin that is common to all the particles would not be perceived as movement at all from the vantage of its atoms but rather as the absence of it.

It would also be a flawed conception to imagine a primal particle within the mattron having a perfect orbit around a single fixed axis in a single plane that would then also represent the mattron spin. Instead, each single fixed axis along which all the individual atoms are moving around the mattron center at various depths is most likely simultaneously orbiting the mattrons center.

Allow me to illustrate. Imagine two allied particles in an otherwise empty space moving in two independent and closed circular orbits at exactly the same 'speed' at the same distance around one common center. The orbit of each atom shares the same sphere with the other atom (as defined by their common center, - let us say this is the mattron center).

Imagine that these two closed orbits around two fixed axis are such that one atom completes a whole orbit around what we arbitrarily term to be the equatorial plane of their common sphere in the same time that the other atom completes its orbit by crossing over the likewise arbitrarily designated poles of their common sphere.

On their common sphere, the two particles therefore share only two spatial points where their perpendicular orbits cut. If the two allied particles collide in either of these two points, their coordinates become totally aligned.

Neither particle would be seen by an independent universal observer to continue traveling along its initial orbit. From a universal reference point, the pair is seen to make two convoluted circumferences around their common center (here the mattron center) before arriving at the same spatial point from which the pair started.

Put it another way, each primal particle completes its former orbit around the mattron center at twice the time it took earlier and it does so in the same time it takes the circumference of that orbit to complete one perpendicular orbit of its own around the mattron center.

The time it took earlier for both particles to complete one entire orbit has indeed been halved but the spatial distance the particle pair covers on a new convoluted path has twice the number of spatial points. Thus both relative movements have been preserved in the alignment. (If one of the two atoms had been on an orbital path while the other 'sat still', the orbital time for the pair would have been halved in the original orbit in a collision. The same orbit has equal number of spatial points with both relative movements preserved in the alignment).

The above phenomenon is well suited to illustrate the mattrons collective field of spin. The difference for a mattron is that the orbiting speed of all its individual atoms and the aligned orbital displacement is cut by 10^{-912}.

I emphasize that the two particles in the above illustration are positioned in the plane of spin of their individual coordinate systems. A particle that is not in that plane of spin will still complete the same spin as its coordinate center and in the same time. Out from the axes point of that coordinate system it keeps to its coordinates by simply spinning in the same way as the center and need not sweep any extra spatial points to do so.

The term I have chosen to describe this phenomenon of half-and-half spin of particle and its orbital displacement for all types and gatherings of primal submists is 'differential spin'.

The Collective Spinfield

Long before the birth of the first mattron, its population to be had become linearly at rest and was locally orbiting in differential orbits. A new spinning center within a submist larger than the mattron would eventually aid in its separation and help define it as an individually structured submist within the somewhat less coherent primal soup.

Within a mattron embryo, myriad collisions have cancelled out universally linear modes of past emission inputs and closed the increasingly aligned local orbits that by now allow the primal population to be spatially contained.

The first unpaired primal spinfields were eventually passed by proxy to all the embryos other mass and turned into a coordinated internal effect. This coordinate system is common to all its mass at any given time and this is the mattrons original collective spinfield but it can be altered.

Although a mattrons collective spinfield is firmly rooted in the 'odd man out' spinfields culled from all its newborns, the first cohesive spin to carry the day is difficult to reverse from within the mattron, given the number of newborns that orbit in contradictory directions. All newborns from within the mattron carry these same common coordinates and cannot therefore contradict them, but given half a chance, they can certainly embellish its reigning spin.

Every atom within the boundaries of the mattron initially started off as a newborn spinning at birthspin in some single or convoluted plane of primal spin, traveling in a 'straight' line at birthspeed.

My mattron model has a mass number far below r$\sqrt{}$r and even an activated newborn would stand a small chance of collision and consequent entrapment if the mattronal residents were at standstill within it. Here the incessant local movement of primal particles within the mattron facilitates collisions and the newborns consequent entrapment is therefore all but assured.

Among inter-colliding newborns, rearranged orbits open with increasing swiftness as each new sub-collision orbit is opened up faster by the earlier boosted orbital input. The activating input is also orbital, and cannot make the resulting path more directional in the bigger picture within the mattron.

As in the early stages of creation, linear birthspeed is not only gradually geared down in collisions but metamorphosed into vast orbits, some widening away from coordinate-centers or tightening towards them.

As changing orbits reap further collisions, the linear birthspeed motion of newborns is eliminated in endless shifting collisions before its input through the mass it sets on the move can be contained by the mattron. Whatever unpaired excess of universally linear directions remains when all adversely paired directions have been mutually annihilated becomes the linear speed of the mattron as a collective body.

A mattron can therefore easily change its linear course for no apparent reason but such non-boosted linear speed is obviously slow as it has to carry all the mattronal mass along through the input of a single newborn and the effect is therefore not apparent.

The linear birthspeed is accompanied by a corresponding birthspin. This orbital movement can never become universally linear (even if it can appear to be so when boosted newborns pass through local mattrons after they have been activated at far off distances).

Within a mattron the movement away from the newborns coordinate center after its first collision turns the virgin spin into an open differential orbit and makes its linear movement effective in many universal directions 'away' from whatever common coordinate center 'governs' its orbital phase.

When a highly boosted particle collides into another in an opening orbit, its linear input causes the new common orbit to open faster. Such a collision can (conceivably in rare cases) result in a straight linear movement where all spin is cancelled to leave only the 'away' coordinates of two systems intact (such continued motion would not be boosted). In other rare cases, linear inputs may exactly contradict and the resulting orbit closed, thus neither expanding nor contracting. But generally after a first collision, orbital movement quickly becomes the dominating movement of the affected primal atoms.

Most foreign incoming strays are long since activated, moving in fast primal orbits and few are in possession of virgin spin. Being relatively few compared to the mattrons own production of unpaired newborns they do not constitute a decisive source of odd coordinates to build its future collective spin.

The mattrons greatest source of odd collective spin is the many newborns that escape the mattron while leaving an activated identical opposite twin within. Such unpaired deposits without conflict continue to spread and be aligned until all available directions within a mattron have been filled.

Ultimately, all the involved coordinates of the unpaired strayparticle are transferred to all the mattrons primal population in both signs and all linear or orbital displacement matched by another atom within it is annulled. Any unpaired (unopposed) universal direction becomes a movement that has become common to all its primal population.

Differential orbital boost within a mattron easily reaches an average speed that is the product of average birthspin (1¢) and distance (r). The great orbital speed enhances a newborn's collision potential with the rest of the residential mass. It covers on average 'r' spatial points within the mattron for every point it moves outward without being carried out of its coordinate system (although it may be carried away from the plane of spin of that coordinate system).

The average spatial presence of an activated newborn within a mattron is r√r, which must not be confused with actual movement within its coordinate system. The particles every position within the mattron allows it to move (and push) in perfect accordance with its spin coordinates where its orbital time is in perfect alignment with its coordinates and distance is irrelevant.

When it collides with another newborn it will (much like the particle pair in my earlier illustration) make two convoluted circumferences in twice the time around a common center (a mattron center and perhaps a longer distance by half) before arriving at the same spatial point from which it started.

The alignment is not about the present position of the two primal particles in relation to a common coordinate system. The two independent systems of coordinates are instead entirely aligned into one single system that pertains to all particles within it. The planes of spin become the defining elements of the unified system by courtesy of their greatest relative movement.

In this differential coordinate system, the plane of spin is dominant. The individual orbit of a particle speeds up away from the mattron center in full accordance with its distance from this convoluted plane. The fastest particle movement in relation to the Universe is at its surface where the differential planes of all individual aligned spins are manifested in orbits around the mattron center.

The particles heed the dictated speed of their present position wherever they find themselves within the aligned coordinate system. The end speed in orbit is on average determined by the distance from the mattron center and the number of other particles that have become aligned through collisions, either direct or by proxy.

The mattron is a gargantuan collection of extremely sparsely distributed and cross-collided particles and its primal mass has just barely achieved the broad coordinate similarity necessary to prevent incessant collisions of its internal primal mass. The limited number of rotational orientations affords each collision a choice of 10^{632} spin orientations of which any given collision involves only two. The chance of a total annulment collision is small and a marginal adjustment would be par for the course in each internal collision.

The extra drag by the adverse half of the mass slows the collective spin in a single input by exactly half as it is eventually communicated to both signs. The unpaired spin of a single primal newborn in an (unopposed) single plane orbit will therefore in time translate into the identical movement for the entire 2×10^{912} mattron mass and become common to all its resident particles. Every particle will eventually be aligned in a coordinate system with central speed coordinates of $5 \times 10^{-913}¢$ that flows from this single input in one sign.

The actual speed of any atom in its orbit around the mattron center is then found by multiplying this $5 \times 10^{-913}¢$ central speed with that particles distance from the mattron center. 10^{632}ãr away from the perfect mattron center at its surface, the speed of any atom would then be $5 \times 10^{-281}¢$.

This collective movement in perfect concert is of course differential through the enormous number of single opposite inputs within the mattron. Setting aside all other individual particle movement, every residential atom passes at whatever depth within the mattron for one full orbit around its center in a non-collision direction at geared-down speeds in the same time that it takes the circumference of that same orbit.

An individual surface atom 10^{632}ãr away from the perfect mattron center at its surface must then travel twice that distance to return to the point in space whence we saw it 'start' its differential orbit. It would therefore take 4×10^{912}ãt to complete one differential rotation around the mattron center at its speed of $5 \times 10^{-281}¢$ (instead of 2×10^{912}ãt).

The spin coordinates of newborn atoms have nothing in common with the rest of the (perfect) mattron mass except those of this common collective spin. It takes a long time for all the primal mass to become this closely related by communicating spin through collisions; the only way it is possible. Different newborn or captured spin inputs may in the distant future become common to its entire particle mass but if they were a part of the perfect mattron, they cannot shake this inherent orbital heritage.

At the surface of a perfect mattron, atoms reach maximum orbital size and speed, after which they can reach no further boost within the mattron. For a surface atom that leaves such a peripheral surface position, the likelihood of a shrinking orbit that affords it less spatial movement within the mattron is equal to the likelihood of a larger opening orbit that brings faster movement but also eventually its escape from the mattron.

Compressed 10^{12} fold in radius from 10^{632}ãr to 10^{620}ãr with 2×10^{912} mass, the collective $5 \times 10^{-281}¢$ surface spin-speed slows to $5 \times 10^{-293}¢$ as the smaller mattron radius yields 10^{12}-times slower surface orbits. However, due to the shorter distance to be traveled to complete an orbit, whether in one plane or in a differential plane, the mattron completes one collective rotation in exactly the same period as before.

As actual communication of excess spin is in collisions between atoms of the same sign, the message of spin coordinates in one sign to the collective mass in the other sign lags behind by the difference of 'ќ' (10^{-1192}) creation constant and the 10^{912}āt consummation period of allied spin. A neutral new communication exceeds the 10^{912}āt new allied consummation 10^{280} times and would take a more extensive 10^{1192}āt period.

This means that a neutral mattrons old collective spin is identical for both its signs but that at every time there is another not fully consummated spin composed of two adverse collections of cohabiting mass that each must wait to be brought in line with the others spin direction and speed by secondary emissions. This causes the two adverse mass collectives to rotate slowly in different (not necessarily opposite) directions. And when collective spin out of past eccentricities has been consummated, new inputs have already affected the allied mass and are waiting to be collectively consummated.

Something we are not used to in spinning bodies is that the propagation of inherent collective rotation in a unified coordinate system increases in direct relation to its radius as the mattron expands. If compressed, the speed of its inherent internal rotation (not exterior communication of these coordinates) decreases at any depth in direct relation to the mattrons shrinking radius.

As a mattron starts to display its inherent spin, the Universe that has undergone no such alignment has gained a different coordinate relation to the mattron. Viewed from within the mattron, the Universe has taken to orbiting it (each internal sign can perceive this to be in different direction). From our universal vantage point, the entire mattron would be seen rotate about its center, most likely with a slow differential movement that appears inherent to all its mass (whatever the local movement of its primal particles).

We can view collective spin in each sign most clearly in a perfect (imaginary) mattron where all primal or parental spin has been annulled and a tiny excess of a great variety of unpaired spins in unopposed directions has transmuted into one single collective spin that is common to all its primal population.

Within a perfect (and imaginary) mattron without any internal movement, allied newborns experience no spatial boost in relation to other allied atoms. Along the single opposite trajectory followed by newborn twins, each virgin moves along a straight line in relation to the mattrons entire aligned population in its sign.

On each opposite trajectory, the adverse virgins find the mattrons entire population in the adverse sign to be moving slowly in relation to it is path but the virgin can have no direct contact with this part of the primal mass.

A newborn in a perfect 10^{632}ār radius mattron is emitted along one of 10^{1264}

possible straight paths where only one in 10^{352} paths is blocked by a single atom out of the 10^{912} allied particles in its sign. Without any internal mass movement to enhance its collision potential, the chances of an allied newborn to escape the perfect mattron without collision are excellent. Only 10^{-352} of them would have their virgin spin activated in a collision while the majority would escape the perfect mattron unimpeded.

We can now discard the perfect mattron model, as the incubation period that transforms a condensing mattron from a large embryo into a full-fledged slowly rotating mattron is constantly hampered or changed by the addition of newborns that implant new and potent spinfields in one of the two signs. The orbital offspring of such primal spin keeps all the prospective parents of a mattron population more or less in a constant whirl.

However, the fact that new and exotic primal spinfields constantly affect the mattron mass does not alter the fact that the excesses of its past are an inherent part of its synchronized mass, - that the common denominator of the past is the collective spin of the moment.

Although the net coordinates of individual spin, whether in each or both signs may be zero when eventually paired in collisions (born of secondary emissions), these have not been communicated between the signs. And where the numbers have yet to be added up, there is this lot of movement afoot.

New coordinates are constantly added but no atom can forever maintain a closed non-collision orbit and remain untouched. Pushed into widening or tightening orbits, a primal atom encroaches on the orbits of other atoms and becomes vulnerable to collisions and further alignment.

In time a great number of unsynchronized individual spinfields are relayed to all allied mass through collisions and an immense multitude of different eccentricities and spin directions result in extensive canceling of independent orbits. Slowly but surely, the process enters all the actual spin denominations into this field of conflict and only unpaired spin is eventually relayed into common coordinates for the allied population and ultimately for the entire mattron or a greater collection of mattrons.

Lastly, from the hip, I would like to set some usable illustration of the time required for new eccentric spin inputs to be consummated as collective spin.

As collective spin depends on mass M and is affected by $\sqrt{r}$ primal boost within a mattron and input by the creation constant k, I set consummation for a standard mattron as $M \times \sqrt{r} \times (M \times \acute{\kappa})$ or $(M^2 \times \sqrt{r} \times \acute{\kappa})$. For consummation in a larger group or grid of mattrons, I simply multiply by the mattron number; which may of course land me in trouble later on.

This reckless approximation gives me a certain sense of time where the standard mattrons mass of 10^{912} primal particles and 10^{632}ãr radius along with the 10^{-1192}ãt creation constant yields a consummation period for new collective spin of 10^{948}ãt.

The Parental Spinfield

While collective spin is being forged, there is orbital movement at widely divergent speeds ever ongoing and newly added in small or wide local orbits and in opposing directions on ultimate collision courses. Such mass will eventually collide and its relative motion is then ultimately annihilated by mass moving in the opposite direction.

Even so, the closed partly aligned local orbits allow a good deal of average orbital movement. At any given time, there is a lot of unresolved and yet to be eradicated movement afoot. Of this individual local orbital movement there is an ever ongoing average within the mattron, which rate of neutralization is in balance against the rate of further agitation by new spin inputs.

The available number of primal paths the size of mattron circumference is greatest in the mattrons surface 'layer'. This surface layer determines the greatest number of available directions in which a particle can orbit a full circle around the mattron center without encroaching on other such paths.

The number of the available directions in which a primal particle can move about a mattron center is therefore the mattron radius.

In collisions, primal spin and its resulting orbital movement is affected at the same rate as linear directional movement. The fully matured and boosted orbital phase is however a far greater motion than the linear component but faces identical levels of contrary movement. It too is eventually completely cancelled out where any two spin inputs are exactly contradictory.

In the 'Primal Code' a newborns 'single plane' spin direction was adversely matched in a pair of adverse twins in equal opposite directions around their 'single' axis of linear regression.

With all primal mass thoroughly cross-collided, parental spin coordinates within the mattron have long since become differential. Consequently, there are not many parental single axis orbits left in relation to the mattron. For this reason, 'emission' of newborns is to some extent differential right out of the starting gate while their activated and more potent primal spin is not.

The differential collective parental orbit around a mattron center moves at 5×10^{-281}¢ and is immensely slower that the single plane orbit of an activated primal newborn that has reached a distance of 10^{632}ãr where the two involved atoms can complete that mattron sized orbit at 5×10^{631}¢.

The immensely faster primal spin can be in a 'single' plane of spin, albeit convoluted. Inputs into partly aligned closed particle orbits turn into opening local orbits that encompass the average differential 'parental' motion that will be boosted out through the mattron if a resident atom is lodged from its less aligned local spin. When brought out of such a local orbit to a position by up to 90 degree to that individual plane, its boost dwindles on that account. The mattrons affords each of its 10^{912} primal residents some 10^{632} possible planes of spin to cavort in.

Again, the likelihood of a single linear collision that cancels out universally linear motion 'at one go' is equally rare as one that exactly cancels orbital motion at one go. The likelihood of a single collision that exactly cancels both linear motion and spin is extremely small.

Do not confuse the spatial presence of a communicated message with the coordinate message that is delivered and then collectively assumed by mass at the receiving end.

The spatial presence of a primal messenger is on average diluted by rvr but the delivered message in each instance depends solely on the position of the recipient that blocks the messenger in relation to that coordinate center.

While collective spin is really the absence of relative motion in a submist, most primal particles are at the same time orbiting wildly at various speeds and directions. Even if such ever-ongoing simultaneous movement might well be in orbits around the mattron center, much of it is in direction opposite to other such atoms and none of this is part of the mattrons collective spin.

The direction of such individual spin is relative to a mattron body that must be perceived at standstill in the respective sign. Its new effect can thus in time accelerate or decelerate the inherent mattron rotation, although on average it does neither but is annihilated instead.

Having looked at primal spin and collective spin, we must take a closer look at the man in the middle, the average orbital speed of every primal resident within a mattron. I have chosen to call it the 'parental spin' and will now give a rough illustration of the effect that is ever-ongoing within every mattron and every submist.

Using the 5×10^{-913}¢ collective central spin speed as a foundation, I set the much faster average parental spin of primal residents (the average individual unresolved movement of primal residents within a mattron) as a product of two additional values. These two values derive from the constant supply and eventual assimilation of virgin coordinates in an ever-ongoing cycle.

The first value is the time-related number of newborns split in each sign during 10^{632}ãt radius time (the time it takes a newborn to pass out of the mattron). This average addition of 2×10^{352} newborns in each sign is activated and captured and the resulting motion comes about in contradictory pares and is eventually geared down by opposing orbits.

The second value is the boost in the average local orbit. The conflicting, as yet non-cancelled local orbits of primal, movement are sized between $1\tilde{a}r$ and $10^{632}\tilde{a}r$. This means that a few atoms rotate in individual coordinate centers while a few reach orbits that are $10^{632}\tilde{a}r$ wide. I set the average free local orbit achieved by internal particles (all of them eventually being parents) as having a width of $10^{316}\tilde{a}r$, whereupon their actual orbital movement in relation to their coordinate center has potentially been boosted on average 10^{158} fold.

To approximate the collective input of the individual parental coordinate system at its center, I increase the central input of the mattrons collective spin by the product of these two above values, making the central parental spin input $(2\times10^{352})\times(10^{158}¢) = 2\times10^{510}$ times greater than the central collective spin of the entire neutral mattron.

Thus, while collective coordinates cause a particle at the mattrons collective center to spin at $5\times10^{-913}¢$, the atom at the heart of each parental coordinate system spins on average 2×10^{510} times faster and its individual spin at that center is consequently at $10^{-402}¢$.

Some residential particles would then be spinning at $10^{-402}¢$ when situated in their coordinate center. A few would be found in large $10^{632}\tilde{a}r$ sized orbits about that individual coordinate center and boosted on average 10^{316} times from their central $10^{-402}¢$ speed to an orbital speed of $10^{-86}¢$.

To set the average and ever ongoing orbital movement of primal residents in individual local orbits, I simply assume that they have reached an average $10^{316}\tilde{a}r$ distance from their individual coordinate center. At that point they would on average have been boosted 10^{158} times away from that center, i.e. from $10^{-402}¢$ to $10^{-244}¢$, the average ever-ongoing primal movement within my standard mattron model.

The value changes with size and mass. A simpler estimate says that each of the 10^{352} newborns split in radius time gain on average $10^{316}¢$ boost within the mattron and can move all its 10^{912} atoms about at $10^{-244}¢$. This constant orbital motion of residential atoms is further boosted or diminished if brought out of local orbits. If collided into a widening orbit, the additional boost they can receive within a mattron is at the $10^{632}\tilde{a}r$ periphery with extra 10^{158}-fold boost that increases the individual $10^{-244}¢$ average speed to $10^{-86}¢$ on exit.

If a resident is moved inwards within its coordinate system, the parental movement diminishes on average by the square of the shrinking radius. The internal emission of atoms tends to be straighter for mass near the mattron center where parental motion is most hampered by its density.

Each two-dimensional mattron plane contains 10^{280} primal atoms. A stray can traverse along 10^{632} different paths in 10^{596}ãt at 10^{36}¢ with its 10^{-352} chance of hitting a resident atom. Here the 10^{-244}¢ parental spin turns the table as the spatial motionless presence of 10^{280} atoms increases by 10^{352} or (10^{596}ãt÷10^{-244}¢) to a presence of $10^{280} \times 10^{352} = 10^{632}$ã during its 10^{596}ãt time of passage along one of 10^{632} possible paths.

Strays at 10^{36}¢ will therefore collide once within the mattron; the parental movement enables a mattron to block passing strays at 10^{36}¢.

As the numbers are exponential, atoms set moving by a stray collide twice again in such a period, then four times and so on. The real stoppable speed might be somewhere between 10^{35}¢ and 10^{36}¢ but I let the latter stand.

Every primal atom within the mattron is a potential parent endowed with constant local movement in relation to the mattron structure while separately carrying its slow collective coordinates. Primal atoms in the open strayfield are also potential parents endowed with local movement, albeit different.

The average central 'parental spin' is of less interest than the ever-ongoing orbital movement it gives rise to. When I use this name ' parental spin', I am also generally referring to this average orbital movement of primal mass.

When an emitted newborn escapes the mattron endowed with this boosted parental spin, the 10^{-86}¢ average exit sweep will be further boosted outside the mattron by the square root of the traveled distance as measured by the yardstick of its 10^{632}ãr parental radii.

In explicit contrast to the random parental motion by newborns that escape a mattron and generate an ultimately conflicting self-neutralizing spinfield, the slow synchronized 10^{-280}¢ collective rotation of the mattron population represents an allied and directionally non-opposed spinfield. Whereas faster parental spin supplies its own antidote inside and outside a mattron, there is no antidote in the external field that works against a collective spinfield save its own adverse input (if mass at the receiving end is neutral).

I must caution that the picture of synchronized parental motion is painted here with clear colors as a substitute for a rather more disorderly reality. This concludes my discussion on parental spin.

Escape Rate of Newborns

Most of the mattrons internal emission is contained as 'pressure' and only a tiny part of it escapes. The rudimentary conceptual estimates that I have used to narrow down the escape rate of newborns can be set up lucidly.

The passage of a newborn out of a mattron takes 10^{632}ãt and during that period, each of the 10^{912} primal residents in its sign moves about in local orbits at the average speed of 10^{-244}¢ per ãnna.

During a newborn's passage, this parentally moving mass therefore sweeps through (10^{632}ãt×10^{-244}¢) or 10^{388} times its physical presence; a 10^{1300} spatial points swept by all its primal mass. With its 10^{1896}ã volume, 10^{596} spatial points within the mattron remain untouched by the primal population that is allied to a virgin during the 10^{632}ãt it takes to pass out of the mattron.

This blocking ability of moving mass faces every newborn that aims to escape the mattron as it follows its parental coordinates receiving an average exit boost at 10^{-86}¢ for every point it moves forward at 1¢. A virgin sweeps 10^{-86} times the 10^{632} spatial points it must travel to exit the mattron, which adds marginally to the 10^{632} linear points of its passage.

To escape, a newborn must avoid collision as this activates its virgin spin that would secure its capture within the mattron but needs 10^{632} spatial points to avoid collision during passage. The orbitally active mattron mass leaves only 10^{596}ã spatial points untouched during the 10^{632}ãt exit period.

The average chance for internal newborn emission moving through neutral mattron mass to escape out of that mattron is therefore 10^{-36}.

The escape rate is closely related to parental spin that decreases upon compression and increases upon expansion. Parental spin is tied to whatever mass that can actively split newborns and circulate freely in parental orbits. A compressed mattron will respond to pressure not only by turning repulsive by emitting newborns at a higher rate but these would be less orbital and be more effective as repulsion in fewer universal directions.

There is a point to be made about the chances of escape by the newborns activated within the outer layers of a mattron, the place where primal mass is most tenuous. Such newborns move outwards in activated orbits while their coordinate centers are within the mattron. The newborn must after leaving the mattron 'surface', cross back into and through the mattron surface myriad times in orbits as a resulting flux shower.

Thus with residential mass averaged upon 10^{632} primal skins within the mattron, the splitting rate within the mattrons outmost layers is extremely tiny in comparison with its entire primal collective.

External Emission and the Spinfield

Not to overly complicate things, I assume that an average 10^{-36} chance of escape represents the average bulk of the mattrons freely escaping emission. I also assume that the majority of newborn escapees are not made up from splits that originate near the mattron surface but rather deep within the mattron, specifically and on average at its center.

An important aspect of mattrons internal fields of spin is the influence carried by freely escaping emission out past the mattron boundaries. One of these aspects is the mattrons collective field of spin.

Let me first put in a word on 'straight-line' emission from each 'stationary' parent in a perfect differential collective orbit within a perfect mattron. The newborns outgoing path may be apparently straight in relation to the mattron mass but it is certainly not directional in one universal line only.

The mattrons excruciatingly slow 10^{-912}¢ central collective spin does not sound like much of a movement. The split of a central newborn moving at 1¢ birthspeed would for most purposed appear to be in a relatively straight line, even to the Universe at large. As the newborn passes out of the mattron with these slow coordinates, it is already also moving marginally in a sideways orbit at 10^{-280}¢.

The definition of a straight line is redefined by distance. What sets the standard for a 'straight' line in relation to the mattron is the mattron radius (the average distance a newborn must travel to exit it). To be on a straight trajectory within the mattron, the exiting newborn must not even orbit single parental axes of recession. It must in other words gain nothing in spatial presence. Any primal emission that escapes a mattron will of course sooner or later run afoul of this definition of a straight line. Eventually the gradual enlargement of its crooked path bends even a lightly bent path into a regular open orbit.

Beyond the mattron boundaries, linear push increasingly deviates from its initially straighter line as primal, parental or collective mattronal coordinates define its path. In this fashion, even the relatively slow but externally boosted collective motion about the mattron center increasingly corrupts its originally 'universally' straighter line.

The much faster parental orbit of a newborn that passes out of a mattron without collisions to start a long journey through space is already moving in some orbital direction at 10^{-280}¢ and after being activated it gains on average a speed of 10^{316}¢ across the same distance of one mattron radius.

Linear repulsion imbedded in a single newborn particle endowed with the mattrons fully consummated inherent spin is ultimately no longer effective in any one specific direction but has become effective in many universal directions away from a parent and from the mattron.

However, while the linear 'away' phase of a parental coordinate center and eventually of activated newborns coordinate centers is not opposed by other emission from a mattron center, the orbital phase of primal and parental spin mostly in direct conflict within the mattron and out in the strayfield. As the orbital phase has been perfectly annulled in the outside strayfield upon collision, the opposites leave the linear phase that becomes effective in one universal line only; away from the coordinate center of these newborns.

Aside from that linear 'away' phase, there is only one orbital phase that is in no conflict, namely that of the mattrons collective spin that is inherent to all its mass and therefore identical in all newborns split of that mass.

Upon leaving the confines of the mattron, all three types of orbital input gain in value, the vast but contradictory primal boost of a newborn, the fast but contradictory input of individual parental spin and the slow synchronized orbital input of the collective spin of the entire mattron.

To understand the influence of this emission field, picture in your mind a model of a perfect imaginary mattron that for our convenience rotates in a single plane (not a differential one). This perfect mattron is built of totally aligned primal atoms with no movement in relation to the Universe other than each has a circular path around its mattrons single axis of rotation.

Focus upon one primal atom. Let this surface atom have its orbit around the mattron in its single equatorial plane of spin. Focus then on another atom positioned at its axis surface point where it rotates but does not orbit. All aspects of this perfect imaginary mattron are governed by a single coordinate system.

Picture in your mind's eye (looking in from outside the mattron) all the orbital paths of its primal population. They range between a slow rotation of the single atom on the mattrons axis (10^{-352} atoms per radius line) to faster orbits at its equatorial 'surface' point, both combined in a unified coordinate system, the backbone of the mattrons collective spinfield.

If you were a primal atom inside this perfect mattron, all other primal atoms would be linearly at rest, each and every one of them perfectly still and non-spinning. The only thing moving would be the outside world, which you would see orbiting this mattron in a single plane of spin.

In a perfect imaginary mattron with no other movement, the collective spin is totally consequent (in a real mattron, collective spin is just as consequent but offers all the younger inputs of additional unresolved movement).

The external collective spinfield of the perfect mattron is however at the peak of its power as 'all' its internal emission escapes the mattron to become external emission of great field penetration. Imagine that a resident atom in the single plane of spin of this imaginary mattrons splits two adverse newborns into that plane of spin. If we ignore their birthspin, the twins, like their parent, have no orbital movement in relation to the rest of the mattron mass as they pass out of it in whatever direction that is their lot at birth. They move away from the mattron in a rigid linear relationship to the single parent they have left behind.

We realize that as the distance of both newborns from their parent (and by proxy from the mattron center) increases - so does their slow collective orbital movements in relation to the Universe.

The newborns that have escaped into the perfect mattrons single plane of spin still complete the entire orbit around the mattron in harmony with the orbit that their parent completes within the mattron. In this single plane of spin (really the many planes of spin in a mattron with differential spin), the distance simply does not enter the equation.

The collective coordinates constitute unopposed movement in the outside field with each newborn adhering to the universal coordinates of the mattron. There is no natural antidote in the outside strayfield to neutralize the effect of this synchronized orbital movement that through collisions is brought to bear upon outside mass by newborns escaping into the mattrons exterior fields.

When an escaping newborn collides with a universally stationary atom in the vicinity of the mattron, only half the resulting orbital speed can be ascribed to the newborn. The primal atom it collides into will be carried along at half the input speed (in relation to the Universe) both 'in orbit' and 'away'. This confirms that the Universe with all and sundry was already orbiting the mattron in the opposite direction at half the observed speed.

In the aftermath of the newborns first collision, the spin input easily allows the collision pair to pass in orbit over the single rotation axis of this perfect mattron and then through its single plane of spin.

In this simplified example, although the fast primal movement is dominant, it carries the mattrons collective spinfield. The pair speeds up in the orbital direction of the mattrons spin whenever it passes through its single plane of spin. And it receives a lesser boost when it is absent from that plane of spin while continuing further away (if its linear aim is not corrupted), affirming the average $\sqrt{r}$ presence rule in the emission spinfield.

The delivered collision effect is different from the field distribution that is enhanced by $\sqrt{r}$ for every linear point of distance. In a collision, the newborn simply delivers the spatial coordinates dictated by its relation to its parent at the collision point. It is boosted by the unadulterated distance it has traveled in the mattrons plane of spin. If as a collision pair it passes over the axis of the perfect mattron, its only orbital movement in relation to it as it crossing the axis (aside from new primal orbit) is half the mattrons extremely slow central axis rotation.

As a newborn moves outward at its primal birthspeed away from its parent and perfect mattron, the slow collective and faster parental orbital movements are the underdogs from the start. This facilitates great penetration of escaping newborns into the outside domains of a mattron.

The collective and parental exit sweeps of a newborn gain an 'r' boost (when situated in a plane of spin) measured in mattronal or parental radii (whereas the linear outward birthspeed gains nothing).

This applies also for the faster independent primal movement after activation a newborns birthspin; a result measured in primal radii and also on average boosted by $\sqrt{r}$. After the first collision, the activated birthspin swiftly becomes the overwhelming influence on any collision pair's continued path, which in no way alters the above.

The collective spinfield is an extremely slow orbital generator but except for the different input of rotational versus linear, the behavior of a newborn on leaving a perfect mattron is perfectly parallel to that of a newborn leaving a spinning or locally orbiting parent, - which is the basic reality anyway.

There is therefore a point of distance where the slow collective sweep of the mattronal coordinates (10^{-280}¢ at the mattron surface) boosted by 'r' in the plane of spin grows equal to the linear 1¢ birthspeed, namely 10^{280} mattron radii away. From that point on (as the fields of emission and spin continue to fade), the collective sweep continues to outgrow the primal birthspeed.

In a real mattron, its entire bodies spins twice before completing one unit of spin that returns each primal resident to the spatial orientation it started from (after summing up all individual movement). Consequently, its exterior spinfield also rotates around one axis in the time it takes that circumference to circle the perpendicular axis.

As all the mattrons rotation 'planes' are manifestations of individual primal planes of spin, any spin (including collective spin) boosted in one plane by distance 'r' is communicated though collisions with outside mass by $\sqrt{r}$. The communicated effect is accumulative and outside mass will be gradually accelerated in an internal spinfield that stays the same over time.

In the empty domains outside a mattron, endowed with differential spin, there are fewer blind spots in the field to which a mattron cannot transmit its emission and communicate its spin over time. It is especially aided in its collective distribution by the virtue of an external emission field carried by a multitude of more powerful and more eccentric parental and primal orbits.

The individual parental coordinates of an escaping newborn must not to be confused with the slow collective spin coordinates that are carried by the same newborn. Parental orbits have no orbital coherence in the exterior field; for almost every orbital direction delivered, some other escaping newborn is likely to carry an opposite parental spin direction.

After a first allied collision occurs, the newborn's virgin spin kicks in and its spatial presence undergoes a vast increase but none of its powerful and new conflicting orbits in the exterior field are endowed with collective coherence. The conflicting orbital motion brings immense collision pressure out into the mattrons exterior fields but gives no coherent directional spin orientation.

There is an important point to be made on the inherent 'repulsion' effect of an emission field boosted by a spinfield. The orbital boost of each outgoing newborn enhances its presence in the exterior field and in turn increases its collision potential but it does not increase its finite linear mode away from its parent. Increased collision potential cannot increase its movement away from a coordinate center. While the boost allows each outward kick to be projected over a larger part of an outside sphere, the mattrons total finite cumulative repulsion is in no way increased by that orbital sweep.

Collectively, the linear 'away' repulsion in the exterior field remains equal to the birthspeed effect of the finite number of atoms that escaped the mattron. To the extent that another nearby mattron can absorb virgin newborns emitted in its direction (or parts of such coordinates delivered after collisions), the parental and collective spinfield have been greatly boosted.

However, in spite of the greatly increased and scatter-brained orbital flux pressure that is delivered (and is equally disruptive as compressive), the total linear away motion imbedded in the actual number of harvested newborns (or parts thereof) will bring into the outside mass exactly the same collective linear motion as was originally embedded in that newborn number.

While the parental and primal orbital effects broadly cancel out, the more insignificant part of the message that is the mattrons collective spinfield is also identical in every single emitted atom (or parts thereof).

Another 'mass' in the vicinity will be gradually nudged into the first inklings of a differential orbit around the emitting mattron by an effect that is vastly smaller than the unsynchronized primal orbital pushes signaled by the same emission atoms or the parts of them harvested as flux.

Although the 1¢ outward kick by a virgin newborn outguns the collective sweep, the continued nudges over time of slow coherent orbits work to corrupt a straight-line approach of outside mass towards a differentially spinning mattron and this sideways push continues while its collective spin remains unchanged.

Directions of External Emission

Like every other aspect of mass at the primal level, the direction of emission that escapes a mattron plays a role in the collective harvest and behavior of nearby material systems.

All the emission atoms have four traits in common; individual birthspin, outward birthspeed, local parental movement and the slower rotational coordinates of the mattron itself.

The diversity of new primal spinfields, realized in collisions outside the mattron, offer no coherent direction. To an opaque neutral outside mass it comes across as just so much Omni-directional straypressure which boost serves mostly to distribute newborn coordinates widely around the mattron.

But through this static the collective voice of mattron emission and of its collective field of spin will eventually carry.

In the vicinity of a mattron, the newborns that escape it are repulsive in whatever universal direction incumbent upon their individual split away from the center of their originating coordinate system. Given time, newborn virgins leave a mattron more or less equally in all universal directions.

Let us therefore for a moment put aside the effect of other fields and look only at the effects of the emission field. It delivers a one-sided push away from the mattron. Taken on its own, this induces other nearby submits to move away from the mattron. If mattronal positions change, this push works in whatever universal direction an individual newborn aims at that point.

Opaque mass at rest in the vicinity of a mattron would be repelled in one universal direction by each emission atom (or parts thereof) it assimilates and the opaque mass starts slowly to 'fall' away from the repulsive mattron along a universal line that on average cuts the center of both masses.

When we take into account the increasing orbital effect of the collective spinfield carried by the same newborns (or parts thereof), the same linear repulsion starts slowly to be affected with the contrary difference that this effect is more effective the further the opaque mass is from the mattron.

The only synchronized orbital phase absorbed by the opaque outside mass is communicated by the slow spin of the entire mattron (while this remains unaltered) and this relative side-sweep increases with distance although the delivered push has been diminished.

The transmitted parental and virgin spins are neutralized as their orbital effect is almost entirely contradictory but as the mattrons collective spinfield takes screw, it becomes increasingly difficult for linear emission to repel distant mass in any single universal direction.

If the distant opaque mass is moving (its position shifting sideways in relation to the emitting mattron) it will soon be pushed away along a new universal line by each received emission atom (or parts thereof) while the carrier flux is orbitally neutralized. It is also being pushed slowly sideways in one and the same orbit and therefore accelerated by newly received newborn emission from the 'rotating' mattron.

At far distances where the field effects are rapidly decreasing, the opaque mass is increasingly being pushed around the mattron and relatively less away from it. There are other fields at work here that affect this interaction but, as I said, we are ignoring these fields for now.

It is unlikely that parental sweep of a mattrons virgin emission can eclipse its linear birthspeed within distances where its repulsion is still a force to contend with and the field effects we are dealing with are more like slowly cumulating nudges that edge this faraway mass sideways.

With time to work its magic, a faded collective spinfield works relatively more effectively in pushing distant mass sideways, as the inherently stronger directional linear push that is harvested through part deposits by flux is diluted $\sqrt{ar}$ faster than orbital push.

If allowed time, the small coherent orbital movement becomes part and parcel of any faraway mass that starts to gradually fall around the rotating mattron, as well as away from it (ignoring other effects).

The first dose of newborns that collided into a faraway mass at standstill and started to push it away from the emitting mattron traveled along a universally straight line that cut the center of both masses. When opposing parental or primal orbits had been annihilated, the effect of these premiere newborns was universally directional and caused the opaque mass to start falling away in that single universal direction.

As time passes and the collective spin of the emitting mattron remains unchanged, this first line is slowly corrupted by a collective orbital influence of the mattrons collective spinfield. Each new orbitally neutralized kick at linear speed still pushes the faraway mass away along a line through both masses and infuses its inherent repulsion in that single universal direction.

Except now, this much later, the distant mass has gradually been induced to shift its position sideways in tune with its first inklings of an orbital path and the line of repulsion has changed focus. Having been shifted sideways, the faraway mass experiences a net repulsion that is directional along a different universal line.

The directional motion infused by earlier emission into the same opaque mass carried coordinates along the first line and in that direction only; this former directional push is no longer being reaffirmed or accelerated.

Gradually each new linear implant comes to find itself more out of line with the first emission push, which linear coordinates are still (albeit partly) moving the distant mass in that original direction. Soon the new repulsion has therefore come to aim in a slightly different direction than the original repulsion.

In this fashion (ignoring other fields) the sideways development, given time, continues with the mass slowly accelerating in orbital free fall away from and eventually around the emitting mattron.

Building the slower sideways sweep into an open orbit takes a lot of primal time. The point may however come (often through the aid of the independent linear movement of the involved masses) when the foreign outside mass has moved 180 degrees around the emitting mattron.

All the new active directional implants that the foreign mass receives and absorbs at this point are effective along the very first line of repulsion we discussed earlier, except now the harvested push is in an opposite universal direction. The total net universal repulsion along that first line is at 180° confronted with its own antidote.

The repulsive movement prompted in the original direction that is still part of the mass coordinates is thereby totally cancelled (although this may be at a different rate due to changes in distance). From the point that a full orbit is realized, the directional repulsion by mattron emission becomes a perpetual victim of changing vectors. At this point an orbital bond can in principle be formed as the faraway mass can no longer be repelled by the mattron while it is adjusts to the possible acceleration against the collective orbit.

In a similar fashion, universal movement of a distant approaching mass may be partly thwarted by the collective repulsion while it advances on the mattron. If after the deceleration phase during approach has been brought by this collective side-shift into an elliptical orbit, the recession impacts with an acceleration away that exactly matches its deceleration during approach; prodded away by emission implants along a contrary Universal line.

Under certain circumstances this could also cause very distant mass to enter into a repulsion-bond with the mattron if acceleration away from the mattron that would otherwise be the result of active emission cannot be realized in a closed orbit where the source of emission is at regular intervals pitted against its own former effect.

In such weak and far-off bonds, the collective spinfield carried by each newborn emission atom is unopposed in the field and though the induced orbital movement needs serious time to grow in stature, it can continue to accumulate within the mass.

The acceleration in an orbital bond forged by the effect of a collective spinfield has an upper limit set by that collective field of spin. It is clear that the orbital speed of the faraway mass can never exceed the average point speed of the orbital message that every newborn (or parts thereof delivered by flux) carries when harvested by that mass.

The 5×10^{-281}¢ collective sweep at a mattrons surface is spread upon the distance squared but boosted by 'r' as measured in mattron radii (this is a single inherent but collective coordinate system) where the inverse product caps the upper speed limit in the differential plane of spin.

The communication of this collective field away from a mattron is often in activated primal orbits that reach outside mass with $\sqrt{a}r$ boost (that stymies a straight line shadow) but the small fraction of every newborn coordinates that moves in concert and represents a mattrons inherent spin, is deposited part and parcel of every newborn or part thereof when harvested.

Considerable distance is called for to allow the orbital message of a collective spinfield, mostly brought across by activated flux, to mature into an effective exterior orbit. It is likely that the strength of an emission field has become extensively exhausted at the far distances where the orbital message becomes strongest as a relative effect.

A return to the more chaotic reality of all combined fields throws somewhat of a wet blanket over this clear simplified picture of active repulsion. Linear emission between two mattrons is mostly harvested as flux after being met in inter-mattronal space by the other mattrons contrary flux where it is linearly slowed and neutralized before the flux is consummated.

As mattrons have 'more or less' similar masses and sizes, I assume that the flux exchanged between mattrons has its linear push overwhelmingly geared down in this fashion. The virgins that cross an inter-mattronal distance aim locally in one direction and remains of this linear message from the original driving newborn is deposited in tiny bits of coordinates into the two mattrons along with opposing direction that have suffered little 'linear' slowdown.

The linearly stymied activated flux will still be able to carry the collective spin coordinates of both mattrons almost intact across and deposit them respectively in any two mattrons when harvested.

The Mattrons Gravity Shadow

The field of gravity acquires influence by way of (apparently) straight-line push deposited by stray particles. Later the overwhelming number of strays arrives in the form of flux ripped out of a great number of grid mattrons. As the individual mattrons cannot stop them, flux is mainly active between the larger systems eventually built of mattrons. For the individual mattrons, the local gravity shadow is also fuelled by virgin atoms.

Each atom eventually collides through some mattron at slow enough speed to be blocked by its enormous particle population along with all coordinates inherent to the harvested atom; a single side sweep built of more than 10^{352} independent side sweeps that (on a larger scale) aims in various directions and is better adept at causing mattron spin that casting a linear shadow. As the orbital drive is enclosed within the mattron, coordinates that drove to widen that orbit linearly are also embedded as the drive.

The relatively few newborn virgins (only 10^{-288} on their first inter-mattronal run) that survive the inter-mattronal trek untouched along relatively straight lines rarely hit a mattron. On the other hand, when they hit, these virgins are purely driven and deposit their uncut linear push coordinates.

The mutual direct exchanges by mattrons in the local strayfield generate massive flux but as no mattrons are positioned on those flux orbs, no gravity shadow is realized before the flux is harvested. To study the gravity field in simply terms, I must select a setting that can best describe the mass field interaction of two adjacent mattrons within a larger grid of mattrons.

The larger background of the grid strayfield supplies straying atoms that pass through the larger local grid in diminishing numbers that from time to time cut through a volume of space that houses a neutral mattron with its potential obstruction field of internally active primal mass.

The stable strayfield in the mattron grid is mostly composed of activated primal particles but even a few virgins. The further off these strays originate, the straighter their path across local spaces from far off but their composite side sweep is also more powerful while the finite drive to open that orbit is divided upon more flux atoms. After a journey from a distant coordinate center, the newborn drive slowed in repeated collisions through different mattrons moves in large orbits composed of myriad bits of coordinates from a large number of atoms and few display locally boosted spatial presence.

Rarely a driving newborn can be directional along a universal line where it covers a great many extra spatial points around a widening 'polar' cone away from a parent in a single plane of spin. By orbiting its axis of progression, the newborn can in theory cover up to 'ãr' extra points locally (ãr being its distance to that linear axis) while advancing one point in its linear direction. Unlikely to penetrate far in mattron grid, its spatial presence could in theory cover (up to) $ãr^2$ points within a mattron but although this increases ability to transmit coordinates locally, it cannot increase its linear 'push'.

Looking at the gravity shadow cast by a single mattrons, the background strayfield in the mattron grid presents a great variety of straying atoms in large orbits that appear linear to a local setting. Of the enormous multitude of the slowest moving primal strays that enter a mattron along a universally directional line, a number experiences internal collisions with residential atoms. Caught up in parental coordinate centers and geared down in speed, they orbit ever more tightly out of every mattron until they are ultimately trapped within one mattron.

For the purpose of studying a mattrons gravity shadow, it is best to ignore the mattrons emission field for a clearer view. We then see plainly that the mattrons entrapment of strays with apparent universally directional motion causes fewer of them to issue out of the mattron than enter into it and the mattron mass casts a definite linear shadow set by its sphere.

For the mattron mass the absent straypressure (on exiting the mattron surface on the other side) is the product of all internally trapped incoming strays and their average speed. This absent 'backhand' is not effective as a linear shadow in any one universal direction as the orbital drive of every flux particle is endowed with a massive composite side sweep that is brought into tighter orbits within the mattron instead of working linearly.

Only as virgin has no imbedded coordinates other than its own inherent coordinates and can cast a full out linear gravity shadow according to its boosted parental sweep (few virgins gain speeds that cannot be blocked by the first hit mattron) and a few activated once-hit newborns are effective.

The mattrons are highly transparent to fast moving straight line strays and if they are not brought to a full stop, the part of residential mass set on the move heads into the open as a shower of atoms at geared down speeds that collectively constitute the same linear and orbital effect. It is such slowdown that continues in increasingly tight orbits out of mattrons at diminishing speeds until the more even flux showers are ready for ultimate capture.

The mattrons opacity derives from the spatial presence of its mass during the exponential time it takes a stray to traverse the mattron (as mass at rest has little presence). Even a perfect (imaginary) mattron with tiny blocking ability where the vast escaping emission widely outguns and obliterates its tiny gravity shadow does not alter the fact that the gravity shadow exists.

The consequence of a gravity shadow is that primal mass in its vicinity stands a statistically greater chance of being collided into the mattron than away from it (when ignoring a mattrons external emission field). The gravity shadow is an integral part of a mattron structure and that emission dilutes that shadow does not prevent it from affecting surrounding submits.

A mattron experiences the shadow of another nearby mattron as a 'black spot' in the sky out of which fewer straying atoms issue. The space occupied by the mattrons was earlier exposed to directional bombardment by stray from all directions equally. From the direction of the other mattron, this bombardment of linear strays is now lesser upon the other mattron.

From the opposite directions to mutual incoming shadows between two adjacent mattrons, strays still bombard them at the same rate, i.e. linearly against the black spot in the sky out of which fewer stray particles issue.

A mattron receiving unaltered apparent linear pressure experiences an apparent linear displacement that now aims linearly in that direction and gives the appearance of active attraction. This fractional linear push in one broad universal line is mostly effective according to the drive that widens the orbital flux and is delivered as tiny parts of the coordinates of each flux atom and, it should be noted can also be in opposite linear directions.

This apparent attraction is caused by a greater number of strays that push each mattron linearly, albeit fractionally, in a straight line (from two opposite directions) against mutual shadows; the dislocation being greater towards a gravity shadow than away from it.

A smaller submit mass casting a gravity shadow blocks fewer available linear directions of passing strays with more erratic 'backhand' dislocation; especially given a short period of time.

Perhaps the best way to measure a gravity shadow is to 'count' the primal flux that would have been deposited in mattron if it weren't there; multiply by its average speed and downsize accumulated orbital push by the average primal boost affecting the flux.

The residual inter-mattronal backhand of the apparent linear push is a distinct gravity shadow with a perfectly linear deficiency in all directions away from the 'enormous' mattron mass. With these constrictions, the residual shadow is shared equally by all points on every sphere away from it; tapering off with the inverse of the distance squared.

If the 10^{316} newborns emitted in 10^{632}ãt between adjacent mattrons that are boosted to 10^{388}¢ and harvested as 10^{704} pressure units were effective as a gravity shadow (they are not), the linear backhand residue would count the none-boosted linear widening drive and yield a 10^{316}pu linear shadow; the same as the number of virgins emitted in radius time.

When we count in the mattrons emission field, the field balance outside a mattron can be either net attraction or repulsion. As emitted virgins gain activated presence, the point repulsion is more effective with distance while both fields fade in strength; a mattrons gravity shadow drops off faster than its activated Omni-directional repulsion (by square root of distance). Two mattrons may well be mutually attractive but it would still be rash to write off the potential repulsion delivered by the external emission fields.

Internal Origins of Gravity shadows

We have seen that mutual repulsion between two primal particles is made more effective if the two are situated well apart with emission-boosting space between them increasing their presence. And that within the dictates of the primal uncertainty, the potential primal shadow of two such nearby atoms is outgunned by their mutual repulsion.

In a large primal mattron submist the point has been reached when local availability of orbiting newborns supplies enough collective center-wise and general pressure to allow them to overcome this individual repulsion.

The shadows within a primal cloud are what collectively make its eventual central condensation possible. To resist the unraveling by its own emission in 'disc time', the mattron must get a helping hand from outside.

Between two neutral mattrons, the power of attraction over repulsion has gained significantly in comparison to the same effect between two primal particles but the messenger is also more easily opposed. The gravity shadow between two atoms is transmitted by the entire particle and with great field strength versus the mass to be moved, - when once in a blue moon the field has something to communicate with.

Aided by the internal movement of that residential mass in local orbits, a blocked strayparticle at 10^{36}¢ transfers all its directional coordinates, first to a part of and eventually to all the mattrons residential atoms. As each stray collision is multi-directional, the linear part push is transferred to mass in any of the 10^{1264} possible universal directions incumbent on its sphere.

If stray-push at 10^{36}¢ goes unopposed, it initially moves mass in its own sign along in its orbital direction and eventually all 10^{912} allied residents at the slow linear speed of 10^{-876}¢. As it moves the adverse mass by proxy, the resulting speed slows to half to 5×10^{-877}¢. Before that happens, the mattron experiences a collision at the average 10^{36}¢ stoppable speed in a broadly opposite direction and when that new linear movement has been transferred to the rest of its mass, the initial directional push is cancelled and the mattrons is compressed instead.

When a stray collision within a mattron encounters no opposing motion, eventually that unpaired linear push delivers a garbled collective directional movement to all its mass. The fractional dislocation by a stray without a counterpart is transferred to all its mass in accord with the imbedded boost and downgraded to a slow linear motion of the mattron.

We have earlier looked at how internally emitted newborns are harnessed by a mattrons collective mass while a small minority escapes. The way the mattron mass harnesses straight-line straypressure is similar. During the time of stray-passage at birthspeed (radius time), all but 10^{-36} of all strays that traverse a 10^{632}ãr-sized mattron collide within it. During passage at 10^{36}¢ half of all strays collide to be orbitally brought off course by internal parental movement and are thereafter destined for ever more gear-down through further collisions and eventual capture within the mattron.

In one ãnna, every residential particle occupies or sweeps through 10^{-244} additional spatial points. The mattrons 10^{912} mass in each sign can therefore block as many spatial points as set by the time it takes a stray to traverse the mattrons interior along the average distance of its radius. The effect may look like some curious sort of space-time continuum, but it is not.

It takes a stray at 10^{200}¢ only 10^{432}ãt to traverse the mattron on average and one of 10^{164} such strays has a statistical chance to meet a slowly moving residential particle to collide with. The transparency is due to lesser spatial presence that mass can build in a short period; great stray speed increases chances of free passage.

Fast strays issue out of a mattron on the other side, accompanied by one or more collision particle at slower speeds. The greater the speed, the faster is the more meager particle shower that issues out of the mattron and that flux is no longer part of the mattron mass. This gear-down continues in repeated passes through other mattrons until slow enough for capture.

When a strayparticle crosses a mattron at slow speed, it prolongs its stay within the mattron. By allowing the sweep of residential atoms time to work its magic, its more leisurely pace dictates fewer available untouched points during its passage and beds for greater internal collision potential.

As set out earlier the mattron must block enough straypressure to gain compression equal to its internal expansion; its only way to resist dispersal. If harvested straypressure falls short of balance, the emitting mass issuing out of an unraveling mattron may eventually provide the lacking pressure if or when its secondary emission circles back into the issuing mattron.

If the effective pressure in the strayfield cannot match mattron expansion, a greater size is needed to more effectively harvest compression. The orbital emission from an unraveling corona may also help to shepherd a temporary balance for the central mattron.

The gravity shadow consumed within the mattron in order to allow it to find spatial balance is tied to the expansion function of its internal emission and therefore directly based on its mass number. There is a firm relationship between the intensity of the strayfield and the size of the mattron (also the distance that a coronal newborn can penetrate into the main mattron to compress it more effectively).

The smaller the mattron, the fewer of its outermost coronal newborns gain large enough orbits (after activation) to effectively compress its main mass. I have chosen to ignore this gradual effect in my mattron model. To lay these concepts bare, I set a pivotal break instead of a gradual thinning coronal development in my mattron model that (in a balanced mattron grid) accepts no atoms beyond the 10^{632}ãr radius distance as residential atoms.

When a mattron is subjected to greater pressure by strays, it is forced to a balance at a smaller sphere. Most of a mattrons expansion is already tapped to resist the strayfield pressure. As the internal emission field has no more to give, there is danger of central collapse if inward pressure increases if the harvest of linear strays does not dwindle at the same rate as the sphere.

As the mattrons transparency decreases when it shrinks, its compression becomes easier for incoming strays. At the same time the orbital pressure from its outermost mass diminishes due to smaller boosts and because of greater density that blocks newborn before they can develop greater boost. There is therefore a limit to the shrinkage a mattron can handle before it collapses and this is governed by various fields that will be dealt with later.

A shrinking sphere does not mean fewer blocked orbital strays are as these are mostly local inter-mattronal strays that return repeatedly until blocked (whatever the mattron size).

After a stray with exact contrary linear movement hits somewhere within a mattron and its orbital opposition has affected the multitude of coordinate systems within; first then can its linear coordinates constitute compression towards its center. Until that happens, the directional movement remains a local or collective linear movement of some (or all) of the mattrons mass.

Even without any straypressure, excess expansion does not simply cause the mattron to disintegrate outwards unchecked as a coronal response fights a rearguard action to slow the process. But even if equilibrium can be found at some size, the mattronal balance is at best fluctuating and it inevitably becomes more erratic and uncertain the fewer spatial points are involved.

A drop in straypressure causes a mattron to expand when its expansion proves the more robust of the opposing influences. Up to a point, its larger sphere may then block more strays and also do it more effectively. A greater than 10^{632}ãr corona would also become more effective in keeping mattronal mass to quarters. Even collective mattron spin may also the situation before a new balance is reached.

We have therefore two sets of balances. The mattrons internal balance that sooner or later finds itself - and external field balance that is inevitably lost in the process. The external unbalance is the collective gravity shadow that materializes when straypressure is 'consumed' by primal mass.

Whereas the harvest of linear strays from across further than the average grid distance is a small fraction of a mattrons consumed straypressure, it is the overwhelming part of the mattrons effective gravity shadow. And while a mattrons total gravity shadow is on average based on mass number, it varies according to size and internal blocking ability in its local strayfield.

But as the dynamic mattron finds internal balance, the consumed strays become an external unbalance represented by a field deficit, the 'force of attraction' brought on by a one-sided stray push that flies in the face of the gravity shadow (the fraction that represents the opening drive across far distances in activated orbits).

Directional Pull of Gravity Shadows

At the best of times, the mattron is a volatile system with a tendency to oscillate between parameters of extremes. Its spatial equilibrium would tend to fluctuate in a never-ending balancing act between uncertain compression and expansion that I have narrowed down to a balance between internally created newborns and the cohesive pressure part from incoming strays.

I want to emphasize that according to the Primal Code, strays are orbital even when approaching a rotating mattron along what may appear a straight line. When a 'straight-line' stray approaches a spinning mattron, half of that relative movement belongs to the stray.

Half of the relative approach speed and spins of mattron and strayparticle set the 'actual' orbital course of both. The collision effect of the orbiting stray atom works partly as linear pressure against expansion and as resistance pressure against the collective spin direction.

If we use a two-point coordinate system when viewing a strayparticle that approaches a spinning mattron, we appreciate it does not have its effect of compression upon the mattron diluted by its collective spin but rather strengthened. The incoming atom is on orbital course toward the mattron center, no matter if it is carried off course over to the mattrons other side by internal mass.

This aspect of random straypressure can slightly diminish the pressures required for balance against expansion. The faster a mattron spins the more precisely strays aim at its center and wield a bit more compression. As the spinning mattron is compressed by the much faster linear push of strays, this drag by the opposite orbital movement is not very effective. With great orbital component speed involved in most stray compression, the great speed downsizes mattron rotation relative to a stray's linear phase. And yet, pitted against 10^{912}

primal residents in its sign orbiting a common mattron center, the rotational 'drag' is not that effective.

To focus on the directional aspects of a gravity shadow; imagine a neutral mattron with differential collective spin that traps all linear strays that pass though it at stoppable speeds. For any opaque mass in its neighborhood (let's make this another neutral mattron), its collective gravity shadow is therefore a 'black spot' in the sky.

When a strayparticle collides into a distant mattron along a line that runs against the black spot the shadow represents, a linear dislodging effect is transferred to its mass and realized as apparent attraction towards the shadow source.

The next strayparticle to arrive along the same line adds its directional dislodging effect onto whatever cumulative effect already implanted into the mattron mass. This continuously builds up a directional movement along a broad universal line (according to the principle of original newborn drive) as mass under the influence of a constant gravity field receives acceleration that is a function of time.

Mass that is pushed closer towards a distant gravity source finds that its rate of acceleration in the gravity field increases in phase with the distance squared. But this increased rate of acceleration along a relatively straight line may not be realized due to the aberrations of other directional aspects than the gravity shadow itself.

For example, a mattron falling towards a distant spinning gravity source that is another mattron also receives repulsive emission that dilutes this shadow. Such activated and boosted repulsion is orbitally incoherent and delivers repulsion away from the source of the gravity shadow to counter the straight shadow (rarely to obliterate it).

The mass of the 'falling' outside mattron heeds the call of collective and primal spinfields imbedded in a boosted differential emission field (without many blind spots). Nonetheless, only the collective spinfield phase imbedded in that emission delivers a concerted sideways orbital push that is steadily in the same orbital direction (while mattron spin remains unaltered).

A great distance allows great collective boost and the repulsive messengers resist 'attraction' and infuse a collective spinfield into distant mass where the coherent orbital motion becomes part and parcel of its coordinates. With that, the distant mass begins ever so slowly to 'fall' around a mattron at an accelerated rate, - as well as towards it.

Strays that move a mattron towards another's gravity shadow travel along universal lines that must cut both masses and the residual push is therefore (more or less) universally directional, as is the shadow that the blockage of the strays creates. Each kick of gravity pushes the distant mass along one universal line drawn through the center of both the masses, implanting a linear motion in the specific direction embedded in its coordinates.

After a while, however, the opaque far off mattron has experienced many small implants of collective orbital push and has been induced to shift its faraway position a bit sideways. Now the same distant mattron experiences a net attraction that is directional along a different universal line.

Directional motion along the first line was already implanted into each mattrons collective coordinate system by the independent earlier strays (or absence thereof from deep opposite space) and this coordinate motion is still at work within their distant mass as they slowly accelerate in a free fall towards a faraway shadow source. At the same time, the development of its continuous sideways movement increases.

Even at great distances, the total orbital input cannot become as great as the 'away' phase carried by every emitted newborn. But contrary to that away push, the orbital phase is 'always' in the same direction. We can ignore the individual parental and primal emission sweep; in spite of wielding far greater pressure than the collective sweep or the potential repulsion, both are directionally quite incoherent.

With passage of time, each new linear implant by strays becomes more out of line with the first linear stray implant that is still pushing mass in that 'original' universal direction as an inherent part of the mattrons collective coordinates. Increasingly, these old implanted directions have come to point beside the line that presently designates the distant source of attraction.

This development continues while the distant mass finds its acceleration towards the faraway source of gravity shadow increasingly thwarted by this constant shifting of focus.

The point may come for that distant mass in a collective field of spin, aided by a side shifting gravity field (or independent movement of the mass itself), that the mattron has been moved 180 degrees in its elliptical path around the other mattron. At that point, a new directional implant is again received along the 'original' line but now in exactly opposite universal direction. The net universal attraction is (like repulsion) confronted with its own antidote at 180° (although a field strength may be dictated by another distance).

If the distant mass were only guided by the collective spinfield and not its own speed, it would orbit the gravity source twice by courtesy of differential spin before the correct antidote came around again.

Continuous acceleration in a gravity field cannot be realized in unbroken orbits where a gravity shadow (or repulsion) is at regular intervals pitted against its own mirror image.

A closer distance achieved during the 'attraction' phase, turns a collective spinfield less boosted and attraction more massive. The implanted movement in the original direction would then be more than cancelled after 180 degrees with greater attraction but, depending on the angle, the accumulated still unopposed 'pull' may also bring it further away in an elliptical orbit. As both mattrons affect each other, the direction of the gravity 'attraction' misnomer becomes the perpetual result of changing vectors where the corrupted aim of distant falling masses causes them to miss each other by a wide margin.

If the independent universal movement by an approaching mattron brings it into a close encounter with an attractive mattron, much of the acceleration absorbed during the approach phase (that launched it mass into an elliptical orbit) cause deceleration during the recession phase though a succession of diametrical gravity implants.

One could envisage a distant mattron slowly entering the outer domains of a source of gravity and emission along a universal line tangential against the collective spinfield carried by emission. Such a path of approach might serve to curb its independent linear movement. Any motion in relation to a source of gravity or repulsion helps to shift focus and neutralize the field.

 For mattrons to enter into pure gravity bonds, any independent linear movement must be neutralized (and utilized). When differential orbits are induced at far distances, they become more perfect in the circular sense but if induced at closer quarters against fast linear movement with less orbital push, they turn more elliptical. In either case, once the orbit is closed, no further movement of approach or recess is called for.

After that, at any distance, the mass falls continuously around the source of gravity or repulsion at a specific speed that is realized by the strength relation of the interacting fields. And, if a mattrons collective spin ultimately changes, so too will the differential orbits of its faraway sidekick.

To create a gravity shadow, exchanged flux must come at two mattrons along lines that cut both. Massive boosted orbital flux passing repeatedly through a next mattron where a lion part is harvested cannot constitute a 'black spot' in the sky for other mattrons. Most of the drive that opens these orbits has also been linearly stopped midways between the two mattrons.

Two linearly head-on competing inter-mattronal orbits after activation of virgins on their first flight across a mattron grid of average 10^{776}ãr that yield 10^{388}¢ could occasionally be in tightly knit local group where flux may reach a 'wavelength' that reaches between such grid mattrons and therefore able to create a field unbalance between mattrons. In such rare cases, the flux can in principle create gravity shadows across the average local distance between a 'handful' of mattrons in a group but each would then find it impossible or 'forbidden' to be positioned on the same gravity orb as another mattron.

The main rule is that a small fraction of activated once-hit newborns (with intact linear drive) reaches distances where their orbits encompass several mattrons. This is the more usual root to a local gravity shadow and we shall later on look at how many such once-hit newborns reach orbits larger than the grid distance and become linearly effective as gravity shadows between other grid mattrons.

Maximum Mass

The mattrons compression by straypressure balances against its often less than constant expansion function. Broadly speaking; at some size and internal density, the mattrons stray-induced compression will come to equal its expansion excess and the mattron obtains spatial balance.

Despite its enormous population of primal atoms, the neutral mattron is doubtless subject to erratic expansions and contractions, given the uncertainties of primal fields.

A finite expansion excess spread by newborns to a finite number of resident atoms that occupy the mattron volume pushes its sphere outward while a fraction is center-oriented at the same time. If its internal emission field were the only effect on the scene, the neutral mattron would condense towards its center as it disintegrates outward into emptier voids.

With compression by strays and activated newborn orbits, its population becomes fairly well averaged upon the mattrons internal radius with greater density eventually gained around its central regions.

The neutral mattron has a definite cut-off level that its primal population cannot exceed. This cut-off level of primal growth comes when its atoms start to cluster in its central regions. This is a place where density is both set by incoming straypressure and by conflicting orbital pressure from internally emitted and activated newborns.

In a large 'dense' neutral mattron, the pressure creates a central mass that occupies a shrinking volume, which inevitably results in greater slow-down of parental mass and facilitates further physical blocking of primal mass in the mattrons (immense) central areas. No matter if the size of the mattrons sphere increases, the lines of primal distribution within the mattron dictate that a considerable portion of additional mass will collect around its 'center'.

Again, it is wise to remember that a mattron embryo is hardly growing more 'massive', - only less tenuous with mass embodying an atomic number of less than r√r (an average sized sun of that density holds the equivalent of 10^{21} nuclei within its volume; something like a milligram of mass).

In a compressed mattron center, parental spin slows and the escape rate of newborns out of central areas increases. Not only does the compressed center become more repulsive by emitting more straight line newborns into the outer layers but the newborns are less orbital and thus more repulsive.

A center will therefore return the pressure visited upon it by gradually becoming more transparent by the loss of parental movement and collective rotation. Up to a point, an aligned center redistributes newborns further out more quickly by straight-line emission and resists compression as long as its aligned mass allows a fair ratio of open paths for each straight-line newborn passing through it and continuing without collision.

As the density and size of the center increases, free-paths grow fewer and soon straight-line newborns emitted from a perfect center are increasingly captured within that central mass. The denser center starts to lose its ability to redistribute primal particles outward within the mattron. The pressure that bears upon the center with increased efficiency with its rising blocking ability can eventually bring central mass into double jeopardy.

In a relatively short time on the primal scale an overwrought center will increasingly capture its own straight-line newborns by activating virgin spin that is quickly spent among its great physical mass. With a 'sudden' inability of highly aligned central mass to emit newborns that can physically pass out of its central regions, the neutral mattron steps across the maximum mass threshold. The lost of straight-line emission with outward resistance within the wider mattron facilitates further collapse of its denser innards.

Paired with the perfect center's newly acquired ability (by sheer numbers) to trap passing strays at higher speeds while unable to redistribute mass outward, the influential central mass can no longer reach the immense middle population with effective outward pressure. At that point, emission out of the massive central area ceases, even while its outer population is driven inwards by the heavy straypressure that bears upon it.

Any part of a mattron, like a central core, must of course be viewed as an independent entity balancing between compression and expansion, adjusting its size accordingly but a center that loses its ability to redistribute mass outward can no longer resist the crunch.

The central density of an overweight mattron that this effectively loses its central emission field turns into something of a black hole. The small central core is so contracted (relatively) that its highly aligned particles endowed with diminished local movement may number more than the cores r^2.

Density may not initially deny a central core its independent expansion and balance (up to a point) but if the unopposed vast middle layers of the start to crowd upon a small dense central core, the passing flux from these massive areas sweeping through to be blocked, increase its compression.

The inability to disperse a massive neutral core causes the primal mass within its perfect center to start to cluster to such extent that free passages away from the individual primal atoms start to close off. Thus, ultimately, mass within the absolute core loses its ability to procreate and it primal particles are no longer able to split or emit newborns for the simple reason that there are no free paths for these newborns to leave by.

At that point there is no influence at hand to prevent rapid consolidation of the mattrons neutral central mass. In a relatively swift movement on the primal scale, the central core of a too massive mattron collapses to take up little more volume than the physical volume of its clustering mass.

We recall that each primal atom is an indivisible entity, a digit that keeps its shape even while passing through an adverse particle (merging volumes turned briefly into 'o' at the moment of passing), whereupon the particle (as the primary rule) kept its former image.

Any temporary cohabitation of adverse particles as they pass through each other does indeed cut down their mass presence. Although a marginal effect, the ability to split new particles in their own image is certainly hampered by the fact that the image is no longer there.

There is however nothing temporary about the cohabitation of a compact particle mass within a collapsed core of a neutral mattron that allows no detached coexistence of adverse particles. Upon the core collapse, adverse particles are pushed into each other's volume by allied collisions and here they simply merge to comprise a single unit. There is no release of anything when they annihilate; they simply disappear.

Other primal particles further out from the original central crunch may be pushed into the vacated space as the core of the neutral mattron crumbles into nothingness. Particles of both signs cease to exist as a double entity and return into the original form of 'o'. There is nothing mysterious about this death clot.

In the controlled annihilation of neutral mass at the perfect center of the standard mattron core, an excessively massive neutral mattron has an internal safety valve that automatically adjusts its mass to the greatest possible level of sustainable neutral mass it can maintain against internal and external strayfield pressures.

Looking to the neutral mattron as a whole, the rate of central annihilation equals its rate of procreation. Other losses and gains aside, the mattron has gained inherent ability to achieve a perfect balance of a constant massfield in a self-sustaining cycle of creation and annihilation.

The central collapse of an overweight mattron as described above would not under balanced circumstances indicate a collapse of the entire mattron of which the central core at each given is a tiny part indeed moment.

If the collapsing core contained a mattrons entire mass and if we threw in the mass of 10^{72} other mattrons for good measure, the volume taken up by all that physical mass would be 10^{984}ā, which equals the private space that every single primal atom has to itself within the standard mattron.

That a collapsing (not yet annihilated) core of rapidly dwindling size and parental spin cannot block as many strays keeps the central collapse more restricted than would otherwise be the case. While contracting inner layers gradually increase their pressure upon a vastly smaller unresisting central core, the core remains a tenuous place; even with a population density in excess of r^2.

While field disturbances may cause collapses that prevent a mattron from balancing in an orderly manner, increased pressure generally allows internal annihilation to rise in a steady process before the crumbling central mass is relegated out of existence with the center sporadically rendered transparent.

Just before collapse, a core can hold a sizable part of the mattrons mass as it rises swiftly to a zenith of opacity while straight emission resists the excessive crunch up to the point of core 'closure'. For a long time (on the primal scale), aligned increasingly straight-line emission streams out of the core to expand the middle layers.

In great upheavals, the outwards pushed middle bulk takes the form of an expanding corona that may turn into a vast external halo that is no longer part of the mattron, except for carrying its collective spin. And then, while this grand supernova takes shape, with the wave of a hand, the spell of the mattrons central field of outwards linear expansive emission is lifted from the overweight mattron.

Aside from loss of a collapsing core, a mattron may suffer a permanent loss of parts of its halo when residential mass proceeds beyond redemption but there are all kinds of reservations to a halo scenario, which we look at later. Halos, of a sort, are also a formation phenomenon and a necessary initial condition while mass is being bottled up inside mattrons.

It is important to note that though we conceive this kind of upheaval as fast and furious, this is an extremely slow and tedious evolution over vast eras of primal time where nothing much is happening.

100

The final closure of an enormous lot of central distances can be achieved in a 'relatively short' period by relatively 'few' particles at opposing angles. As the central hollow forms around the collapsed core, it may take long for the population of the outer tenuous middle layers to be chipped back into a void that is vacated and to fill it up again.

Normally, the population of a collapsing core would be small compared to the mattron that engulfs it or whatever vast spaces vacated by its material collapse. These empty transparent central spaces remain encased in the intact still populous inner and middle regions of the mattron.

In the relationship between size and density, the number of open internal stray paths is an important factor in regulating expansion and compression. For every particle annihilated at a standard mattron center, an entire 10^{632}âr average stray path turns into a free-path where no collisions occurs, which further cuts 'attraction' towards a now more tenuous center.

The number of annihilated atoms indicates a drop in central compression, but the resultant drop in expansion is greater. All particles annihilated at a mattron center (or put out of commission by clustering) were earlier endowed with entirely outward emission that transferred expansion coordinates to all the remaining outer mass of the mattron. Although the compression of the mattron has dropped, its expansion has dropped more.

After a major central collapse, there is a lack of uniform nearness within a vacated center. Initially this would cause inward circling particles from mass that borders a central hollow to be trapped within its lining, thus unable to exert inwards pressure from primal atoms in the surrounding material walls that are considerably aligned. Their spin is a smaller factor but makes the inner lining of the wall 'expand' more rapidly out from the central hollow. Though by no means empty, a vacated central hollow needs time to regain uniform nearness to enhance internal entrapment and eventually refill it.

The middle mattron faces the onslaught of activated emission from the disintegrating halo that helps the lighter mattron to balance with. Having described the collapse of an overweight mattron as a supernova explosion, I hasten to add that this would be a rare scenario and that a drop in incoming straypressure would be the most common cause for spatial unbalances that can elicit collective emissions of primal halos. The erratic annihilation during steadier core collapses affects the main mattron more marginally.

In a mild drop in grid straypressure, an expanded partly retrievable halo may keep a mattron to quarters by secondary emissions and inter-mattronal boost; the bigger the drop, the more decisive the expansion and the more massive the corona that parts into an independent halo.

The neutral mattrons reach active maximum possible mass population relatively fast and a drop in external straypressure lessens their ever ongoing central annihilation (or clustering) while a rise speeds it up; allowing them to maintain maximum mass in a more or less smoothly ongoing process.

Despite waning average intensity in a local strayfield (with atoms locked up in denser subclouds), stray speeds in voids deliver great pressures and large scale penetration where mattron sizes are set by exchanges in the local grid but all these relations are interactive variables. At some point, a fluctuating balance is set in primal masses and sizes of the local mattrons that deliver and harvest the pressures let loose by the escaping newborns from similar mattrons across average distances.

Mattron Disc in Two-dimensions

Trying to understand the mattronal balance, I found it helpful to slice the mattron volume into two-dimensional discs, each the thickness of a primal particle. One disc would conceptually exist in two dimensions only where all relative movement within the disc, linear as well as orbital, would be two dimensional.

Such a disc illustration cannot give a correct picture of the three dimensional mattron but helps to isolate some strands of the mattrons three-dimensional interaction field, showing the internal balances work in a less complex setting.

However, when a newborn leaves its plane of spin in a three-dimensional space, its boost becomes on average the square root of the distance traveled while the orbital boost experienced by an activated newborn in a single plane of spin increases by the distance traveled. I must therefore compensate by boosting the orbital speed on the two-dimensional disc by the square root of the distance only.

One can in theory separate the mattron volume into 10^{632} average two-dimensional discs, each the thickness of 1ãr. Away from the perfect center of each such disc (at a distance of its 10^{632}ãr radius), sits the disc 'surface' that is depicted by its 10^{632}ãr circumference.

A three-dimensional mattron with uneven mass distribution and a single central point for all its 10^{632} identical 'discs' show the flaws of a two-dimensional neutral mattron model.

In a two-dimensional coordinate system, a parent atom that causes a split of two newborns in opposite directions 'sees' them move away on the single plane that is the thickness of the atom. The individual birthspin of the two newborns are in that plane of spin and there can never be any movement out of or into that single two-dimensional plane except out through its circumference.

I populate each two-dimensional disc with an atomic number in relation to its spatial points. On each disc out of 10^{632} within a standard mattron there sit therefore 10^{280} primal atoms in each sign. For clarity, I use mass in one sign and ignore mass in the other. The model ignores the effects of emission interplay that ordinarily takes place between the two opposite masses. As no two mattron discs have the same center, there are, needless to say lot of effects that must be adjusted to get a more correct picture.

As primal mass on such a disc can only move in two dimensions, a model that does not address interaction between masses of different discs, renders a crude concept of mass behavior on each disc and by inference within the mattron as a whole. On each disc, unsolved parental spinfields cause the primal atoms to move in closed local circles within the disc plane.

On each disc $10^{280}m \times 10^{632}\tilde{a}t \times 10^{-1192}\acute{\kappa} = 10^{-280}$ splits occur in the $10^{632}\tilde{a}t$ period a newborn takes to leave a disc and 10^{-280} newborns exert 10^{-280}pu motion that ultimately becomes expansion during that period.

The average rate for one split on every one of 10^{632} two-dimensional discs within a standard mattron is once in $10^{912}\tilde{a}t$. To say that this newborn is on average split at each disc center is flawed as the disc contains 10^{280} atoms that are all just as likely to cause a split. Even when averaged upon the disc radius with more atoms the closer you get to its center, the overwhelming majority is situated far from that center.

Using macro-mattron facts with curvature of newborn paths set by the local circling of parents at $10^{-244}¢$, each virgin atom aims with 1pu push from its parental coordinate center in a universal line but the achieved boost has to be tempered (macro reality being three-dimensional). The exit orbits are on average $10^{316}\tilde{a}r$ wider but they allow only an average individual 10^{158} fold boost (to $10^{-86}¢$). The newborn escape rate out of such a disc that contains 10^{280} particles is $10^{280}m \times 10^{-244}¢ \times 10^{632}m \times 10^{632}\tilde{a}t = 10^{1300}\tilde{a}$. Versus the $10^{1300}\tilde{a}$ disc area, this still allows 10^{-36} of all the internally split newborns to escape.

Newborns are split in positions further from the disc center at intervals longer than the average $10^{912}\tilde{a}t$ estimated for central splits. In time, newborns will be split in the most peripheral outer positions housing fewest atoms per volume. For atoms on the discs circumference, it takes 10^{280} times longer for a virgin split to occur or within an average period of $10^{1192}\tilde{a}t$, - but splits will occur in these rarified places.

The overwhelming majority of all newborns are of course split well outside the perfect disc center. After their activation, each can carry smaller parts of disc mass along in their linear direction out of the disc. The direction can of course be initially 'inwards' (until the moved mass reaches the same depth on the other side of the disc center and becomes 'outwards' from there).

Activated newborns in outward orbits of smaller than disc radius carry smaller parts of disc mass outwards. Their coordinate systems move on narrower fronts and in more specific general directions (instead of doing so in all directions incumbent upon the disc circumference). The shallower the split of an activated newborn, the lesser mass it carries outwards; its specific direction is more focused and the speed of the smaller mass its sets moving is correspondingly greater.

The empty coordinate center realized in a physical collision is moving away from the collision point and this movement is of the same geared-down speed at which the orbits of the collision particles (that the new coordinate system 'governs') open up.

Thus the primal atoms that orbit vacated coordinate centers move away from them at the same linear speed at which the centers were set moving. The result is that their orbits are tangential to the original collision point in an almost constant relation as these orbits widen away from a coordinate center moving linearly further out.

The result is that orbiting mass, moving outward within the disc and eventually out of it, never reaches deeper (in back reach orbit) into the disc than to the birthplace of the newborn that set this mass on the move. The position of an emitted newborn decides how much mass it can carry and at what speed and affected mass cannot reach back deeper into the disc than to the birthplace of the newborn that originally instigated the movement (unless affected by strays). The point inside the disc circumference in that linear direction serves as a center point that sets the greatest sweep into the disc and the quantity of primal mass it can possibly affect and carry along.

From a center position, a newborn would move all disc mass in both the orbital and linear direction of its moving co-ordinate system. The subsequent collision orbits widen to disc radius size and broaden its aim, allowing it to set most of the disc mass on the move and rotating in that broad but unique linear direction. In the absence of incoming strays that oppose this motion, any unopposed central newborn can carry all or most of its disc mass in its sign outward from the disc center at 10^{-280}¢ or whatever higher speed that is relevant to a shallower position.

Of the mattrons 10^{632} newborns in 10^{912}ãt disc time, only 10^{352} are split on the disc edge (on 10^{352} out of its 10^{632} discs). On an outward course, these newborns at birthspeed encounter no or few internal atoms on their disc. An inward direction would carry all the disc mass in that direction at 10^{-280}¢.

One can assume that half of all peripheral disc newborns head inward and carry the main mass at 10^{-280}¢ out of the disc area in 10^{912}ãt before another virgin is split on the disc. Though not expansive, half that internal emission sets the whole mattron moving if this is not opposed within 10^{912}ãt.

One can assume that roughly half of all disc newborns, split at varying depths, head outwards. For example, of 10^{632} newborns in disc time, some 10^{572} are split within 10^{-60} of their disc radius. If outwards bound, they will encounter roughly 10^{-60} of the disc mass (on 10^{572} discs) and carry the 10^{220} resident atoms encountered along out of each disc at 10^{-220}¢. The mass on these discs will 'expand' or unravel 10^{60} times faster; namely in 10^{852}ãt.

In another example, in 10^{912}ãt disc time, 10^{-600} newborn split within 10^{-32} of their disc radius. On an outward course they encounter 10^{-32} of the disc mass or 10^{248} atoms they carry along at 10^{-248}¢ to expand out of the disc in 10^{880}ãt. That this is a marginal effect becomes clear when we consider that this takes place on 10^{-32} of the mattron discs (in10^{912}ãt) and affects 10^{-32} of their mass and only 10^{-74} of the mattrons mass expands therefore at this speed.

The main mattron mass is therefore mainly affected by the emission half that heads inwards in 10^{912}ãt disc time and can carry larger chunks of its mass in one linear direction at 10^{-280}¢ or somewhat less, exiting the disc before another virgin is split somewhere on the disc after 10^{912}ãt on average. This significant 'expansion' works to unravel the main mattron but only full out during 10^{912}ãt.

With variable parts of disc mass dispersed at different speeds out from its central areas, expansion is rather a question of what you are studying, the parts or the parcel. For an illustration of the mattrons collective expansion to work, it must deal with both the unraveling majority and fractions thereof. If the mattron is to survive in the long run, shallow mass that is pushed faster outwards must be held back by incoming stoppable stray pressure as well as all mass within disc time.

Weighing against this mass expansion on each two-dimensional disc is the random outside pressure received from outside its 'surface' circumference. Strays move onto the disc plane along relatively 'straight' two-dimensional lines through its outer edge. On entering the disc, strays, that are stoppable at 10^{36}¢ cross the disc along a single path out of all the other possible paths set by the disc number.

Each incoming flux atom passes from 'surface to surface' along one path. If occupied; the stray collides with a resident atom; the chance of encountering mass being greater, the closer the aim to the disc center. As the two collision atoms move on at half stray speed, they orbit locally, making their rounds in widening (square root) boosted parental orbits. When they affect more atoms, all are set moving in the stray's general direction while the linear speed slows faster than the orbital speed (that meets less cohesive opposition).

In many ways, stray effects are similar to expansion from the discs internal emission but with distinct differences. All strays are initially incoming and are therefore generally inclined to affect greater parts of the disc mass than are internal emissions, thus moving greater swathes of disc mass in their linear direction and at greater speeds.

An important distinction is that strays in large inter-mattronal orbits have more than one shot at affecting the mass, returning repeatedly in the form of ever more numerous fluxes divided upon more directions and becoming ever slower until able to be harvested by the mattrons main more central mass.

Expansion is neutralized by inducing motion in inter-mattronal orbits into numerous residential atoms as straypressure weighs against the expansion of large chunks of disc mass to whatever extent it is in linear opposition to a central point. If not in linear opposition, it works as dispersal.

Any linear particle can move disc mass inward towards the disc center along a straight path of fractional accuracy. The disc circumference is its two dimensional surface and a stray from whatever universal direction can take any of 10^{632} paths through the standard disc (10^{1264} paths through the standard mattron).

The chance for a stray to pass through the perfect disc center is 10^{-632} and on doing so, it has touched 10^{632}ã spatial points. Passing tangentially by the disc center, its actual path can in theory shrink to 1ã on simply touching the disc circumference. This renders the average path as the square root of these two extremes, namely 10^{316}ãr.

As an activated stray passes repeatedly through a disc in almost straight lines (depicted by its immense inter-mattronal orbits) and with disc mass at its densest near the center but diluted outward in line with the radius, the likelihood for the stray to encounter resident atoms is small. This likelihood is affected by both aim and speed. The speed sets the time the stray spends on the disc and therefore how long the internal parental movement can build up the spatial presence of local mass that facilitates collisions.

The sparse residential disc mass encountered in the outer regions is carried along shorter paths at greater speed and is more likely to be lost out of the disc to become cannon fodder for inter-mattronal flux. As the central regions are better equipped to stop fast inter-mattronal flux, the disc, for that reason loses more mass from its outer regions during increased stray pressure. A stray at speeds that are barely stoppable on more central runs are not stoppable on tangential runs closer to the disc periphery.

A mattron would therefore not as much shrink under pressure as instead be re-arranged across the inter-mattronal void with outer mass eventually deposited closer to the center, thus indirectly accomplishing compression (or rather increased density). To directly compress mass towards a disc center is more difficult than moving it closer by inter-mattronal proxy.

The main part of the spatial balance of mass on a two-dimensional disc, and thus the mattron, is arrived at when the discs single activated newborn in disc time yields 1pu expansion. This is in all directions incumbent upon its circumference. Held back in the same time by 10^{-36} strays passing onto each disc at 10^{36}¢ that collectively wield 1pu net compression equal to the average 1pu expansion in disc time; makes for its standard overall balance.

Another factor could be the disc transparency-ratio, i.e. the ratio of empty straight paths versus occupied paths. In principle, two resident atoms hit by one incoming stray (along one line) harvest compression, i.e. cast an effective internal shadow between the two residents and create an external gravity shadow along that line (if the stray is eventually stopped).

One residential atom hit by two strays in opposite directions along the same line receives no communication of internally linear net movement. The two opposite strays cast an external gravity shadow but cannot cast an internal gravity shadow as their opposite linear effect has been neutralized.

For a 10^{632}ãr wide disc with 10^{280} residents, there are 10^{352} empty lines for each line blocked and the standard transparency ratio is therefore 10^{352}. To downsize compression, a disc must harvest strays in sufficient numbers but over long periods each disc atom is nowhere near receiving two hits along the same line. That requires a multitude of strays to hit each of 10^{280} atoms from all opposite directions during disc time (instead of 10^{-316} strays for each).

The disc balance is affected by orbital forces in various ways. The single activated newborn that heads across the disc can push 10^{280} atoms (or a large fraction thereof) linearly along at 10^{-280}¢ out to the disc boundary in 10^{912}ãt. At the same time, its activated birthspin is communicated to the same 10^{280} strong mass and primally boosted 10^{316}-fold in orbital speed (macro mattron rules) along curved paths that span the disc radius.

The rotational power of a single newborn is so robust that, the remaining orbital speed brings the entire disc mass around a disc center for a complete orbit at 10^{36}¢ in 10^{596}ãt. All 10^{632} discs rotate in different directions, altered by new inputs in disc time. Were it not for constrictions of this model, discs mass would (in reality) interact extensively.

This orbital push carries the 10^{280} disc mass in large orbits at 10^{36}¢ at the disc circumference in one and the same orbital direction. A stray and its shower of internal flux, this side of a disc center, are therefore easily carried around to the other side although half the perceived disc spin stems from a stray's orbital approach. After 10^{316} strays interact with all 10^{280} atom disc mass and slow from 10^{36}¢ to 10^{-280}¢

The balancing off collisions occurs throughout the disc and all central mass is rarely carried out of the disc border. The 10^{280} mass that expansion works to push outward is individually circling at 10^{-244}¢ on the disc in closed local 10^{316}ãr wide circles. As only two spin directions are available on each two-dimensional disc, parental spin is equally in opposite local directions.

Using a classical three-dimensional method to look at the gravity shadow by 10^{912} mass, a mattron harvests 10^{704}pu straypressure in 10^{912}ãt and each atom casts a 10^{-208}pu gravity shadow upon every other atom, diluted by the inverse of the mattron radius squared. The resulting 10^{-1472}pu average single primal shadow cast by 10^{912} atoms upon all others with a backhand harvest by 10^{912} atoms, constitutes a cohesive 10^{352}pu collective gravity tug that holds a mattron together but is not entirely center-wise compression.

In classical theory, this standard 10^{352}pu internal stray shadow works against the collective expansion of 10^{352} newborns split within the mattron during 10^{632}ãt standard radius time. Again, not center wise compression in the classical meaning of the word, especially as mass is often rearranged across the inter-mattronal grid and deposited deeper within the mattron.

Compression, Annihilation, Balance

When a mattron of maximum 10^{632}ãr size in a mutually balanced grid faces increased straypressure in both signs; its only recourse is to shrink.

In a standard grid, activated newborn reach across the 10^{776}ãr average distance after being boosted to 10^{388}¢ and consummation of coronal flux is set by the exponential time the last and slowest orbit before blockage takes.

In a 10^{776}ãr grid during 10^{740}ãt consummation period, 10^{424} of the 10^{460} virgins escape the standard mattron to drive flux to 10^{388}¢. It takes 10^{776} flux atoms ripped out at each time for the resulting flux to slow to 10^{36}¢. This is equal to the number of available empty orbits at that distance.

Under greater flux pressure, there are not enough orbits to allow an empty orbit for each flux atom. However, (I assume) that flux atoms pass sideways through great many 10^{776} point long orbits in 'non-opposing' directions and are therefore not substantially weakened in orbital speed as the 'pressure' is divided upon more flux directions.

A mattron that receives local flux from multiple mattrons and directions may not avoid local field pressure by shrinking. In a small close cluster of mattrons within a larger void into which exchanged virgin pressure can be lost out of a constellation while group mattrons harvest the local flux pressure. Bus as local strays return repeatedly in inter-mattronal orbits, their orbital coordinate centers remain at linear standstill midways between mattrons until all are blocked.

There are many conflicting and interactive effects at play here. Generally, at some point flux is slow enough as it encounters mattron mass numerous and wide enough to break its orbital stride. Strays are blocked on paths across immense internal distances that become warped enough for linear flux to collide with large enough portions of mass to slow the internal orbits sufficiently for capture.

The most important factor in trapping traversing strays is the curvature out of internal collision points against slowing parental spin with weaker boost across shorter distances within a smaller mattron. With shrinkage, the shorter straighten r√r constricts its ability to bend stray paths and the stray push gradually shifts to become less compressive and more dispersal prone.

Smaller mattrons are less capable of blocking the inter-mattronal flux that increasingly pushes out the other side and works as expansion instead while potential internal 'attraction' (as a theoretical average) between primal atoms in a highly compressed mattron is greater than in a standard mattron. The shorter internal distances cast darker shadows that cement the internal grip through rearrangement of mass within a smaller sphere.

Mattron shrinkage in a mattron grid does not lead to a smaller flux harvest on account of more flux missing the mattrons. As exchanged orbiting flux is overwhelmingly stopped linearly midways between mattrons, it is eventually harvested by local mattrons at whatever sizes, transparency or masses with all exchanges coming home to roost in a mattron grid.

As internal parental spin is slowed in increased pressure across greater inter-mattronal distances with strays turning more dispersive, relatively more freed virgins escape a mattron and more primal mass is ripped out as flux to be either re-deposited closer to some other mattron center or at a denser level in the same mattron.

Mattrons are generally not compressed in the squeezing manner to which we are accustomed but by inter-mattronal proxy as individual coronal atoms are carried out as flux into large orbits to be exchanged in the mattron grid and deposited closer to some other mattron center and at greater density.

As a mattrons outer corona is most easily moved, it shrinks relatively faster under pressure; the central mass grows more slowly as it gradually houses more clustering mass within a potential cohabitation volume where the two signs move particularly closely among each other.

As the space afforded each primal particle within the mattron can hold all the primal mass of 10^{72} entire mattrons in a compact death clot, I assume that actual annihilation brought on by inescapable cohabitation is gradual. While primal mass can move relatively unhindered within a central volume, the clustering and mass annihilation is prevented. However, lesser parts of central mass is by allied cluttering put out of active commission; it remains within a death clot but cannot procreate or split free particles (to the extent it can, it facilitates smoother annihilation).

Death clots behave (largely) as a single primal particle, inasmuch as their 'entire mass' is moved in a primal collisions with the same easy as a single particle (according to 'distances closed', as treated in the Primal Code).

The inevitable clustering in death clots at the center of a neutral mattron becomes 'solid mass' that in a neutral death clot is no mass at all. As this clot is set by pressure, I call it the mattrons 'annihilation rate' instead.

Let us say that the sphere of a neutral mattron is compressed from its 10^{632}ãr maximum radius and ask how its central mass is affected in way of being put temporarily out of commission through clustering or by permanent mass annihilation.

Whatever I arbitrarily decide in this illustration will land me in a mess. I use the radius of a neutral mattron to set a cohabitation volume of central mass; the inverse radius change sets the size of a neutral death clot that consistently annihilates neutrally blended mass.

Using the 10^{632}ãr radius as a yardstick; when a neutral standard mattron shrinks 10^9 fold under pressure to 10^{623}ãr, its central mass cohabitation increases 10^9 fold and its neutral death clot claims all but 10^{-9} of its mass. In spite of a constant splitting rate, this faster annihilation rate leads to 10^9 fold standard mass in a neutrally compressed mattron.

All but 10^{-9} of 10^{912} neutrally blended atoms cease to exist, residual mass is downsized to 10^{903} actively emitting primal atoms. Noticeably, after the mass depletion, each of a compressed mattrons two-dimensional discs (now 10^{623}ãr wide) carries 10^{280} primal atoms and is populated by exactly the same number of primal atoms as it was before the shrinkage.

Increased central annihilation affects several mattron fields; among them internal mass production, parental and collective spin, internal opacity and virgin escape rate; none of which makes this illustration lucid. True to form, I launch this illustration without serious calculations, to be revised as the dust clears and I use harvested pressure in a standard grid during 10^{912}ãt as a benchmark measure for mattron spheres.

Internal resistance affects compression through transparency and I use a simple formula to illustrate mattron shrinkage in accord to any increase in the benchmark grid pressure. The pressure increase forms a smaller sphere but its shrinkage is decreased the square root of the radius of that sphere.

A 10^{24} fold (10^{412}¢) grid boost would compress a neutral mattron sphere 10^{24} fold from the standard 10^{1264}ã benchmark sphere, except that now the 10^6 square root of its 10^{12} times smaller radius facilitates resistance and so the neutral sphere is eventually compressed by 10^{18} instead to 10^{1246}ã. The new 10^{623}ãr radius allow the neutral mattron to keep merely 10^{-9} of standard mass or 10^{903} atoms.

There are various consequences in this wide grid. The parental spin slows $3,16\times10^{13}$ times to $3,16\times10^{-258}$¢ (from standard 10^{-244}¢) and internal virgins escape at a rate of $3,16\times10^{-23}$ instead of 10^{-36}. Now some $3,16\times10^{4}$ fold more virgins escape per ãnna in spite of 10^{9} times smaller mass, which jacks up the level of inter-mattronal flux.

Great distance increase secondary emission from flux constantly on the wing and this eventually affects compression. The trek across 10^{824}ãr takes $3,16\times10^{801}$ãt and each of $3,16\times10^{310}$ virgins released by a mattron in that time is slowed by ripping out $3,16\times10^{389}$ atoms and secondary emission by the $3,16\times10^{879}$ flux field between the 10^{623}ãr mattrons is only ten times away from eclipsing their allied emission. Increase the distance and the collapse goes too far and disperses the neutral mattrons.

As there is a point at which the mattron can no longer serve to adequately slow down and trap strays, I am tempted to set another conceptual limit of a minimum radius as I did for the 10^{632}ãr maximum mattron and let the neutral mattrons teeter on the brink of dispersal in 10^{824}ãr wide grids.

Having thus broadly evened up the internal odds and set up the sizes and behavior of my standard mattron model at varying distances from other grid mattrons in the grid, I have a model that, while imperfect and inaccurate, allows at least some measure of conceptual clarity.

I want to touch briefly on the speed at which mattronal compression or expansion takes place; the time it takes their mass to shrink under pressure. Central density brought on by inter-mattronal exchange by flux re-deposited at lower levels in other mattrons would be achieved (individually) faster.

A single newborn split in each sign in 10^{912}ãt pushes standard mass on each disc 'outward' at 10^{-280}¢. At that speed, it takes little more than 10^{912}ãt against deficient compression to push all mass out across the standard discs 10^{632}ãt radius periphery. This is also the time it takes to split the required number of newborns to perform the task on all mattron discs. The internal expansion period for a standard mattron is therefore 10^{912}ãt. If compression takes place from larger to smaller radius, the mass is moved 'inward' across the larger radius, requiring the same time.

The annihilation process within neutrally compressed mattrons is effective in spite of a relatively small affected mass gathered to the center at each time as mass keeps pushing inwards during the 10^{912}ãt compression period. Its mass can be doubled in 10^{1192}ãt by the creation constant 'ᴋ' (10^{-1192}) but the 10^{280} successive compression periods of 10^{912}ãt can in the meanwhile move large parts of mass into the center to be annihilated.

Collective rotation is based on mass and radius and a small circumference turns the other cheek faster; a compressed 10^{623}ãr neutral mattron does it in 10^{903}ãt (instead of 10^{912}ãt for a standard mattron). The collective spin is important in distributing stray compression as flux comes at mattrons in large orbits along opposite directions with coordinate centers between two mattrons, hitting them mostly from the side (perpendicular to the axis line between two communicating mattrons). Over time, a mattrons differential rotation spreads the flux harvest more evenly across its surface.

One full rotation of a 10^{623}ãr mattron allows straypressure to resist the ongoing expansive disruption of all surface areas within 10^{903}ãt (whereas a standard mattron requires 10^{912}ãt). As this is the time it takes a newborn to disperse disc mass out of a small lighter mattron, this is perfectly sufficient rotation to stave off its dispersal. If this surface shift does not happen in time, the mattron may unravels in unopposed directions while being overly compressed from the more active sideways flux directions.

If field pressure drops again, a smaller lighter mattron can expand as mass in both signs grows back to 10^{912} maximum while the annihilation process runs interference. The greatest mattrons splits most virgins of all mattrons but it also exchanges the smallest fraction of its internal procreation.

As emitted newborns aim in all directions and not only toward the nearest mattron, the exchange from a mattron can involve many mattrons in a local community. It is only for the convenience of a handy approximation that I let my illustration involve the direct exchange between two mattrons.

In my mattron model, 2×10^{912}m$\times10^{632}$ãt$\times10^{-1192}$ḱ $= 2\times10^{352}$ splits take place in 10^{632}ãt 'radius time'. This requires its 10^{1264}ã-sized neutral mattron sphere to gather 2×10^{704}pu in random straypressure in each sign for spatial balance to be gained while 2×10^{316} newborns in each sign issue out of it. The spaces around a mattron are not empty but rather teaming with 'material' activity. It is above all the presence of fast moving activated flux out between the mattrons in the local strayfield that can secure the activation of a virgin newborn that exits a mattron.

The 10^{460} virgin newborns that occupy the inter-mattronal field traverse 10^{776}ãr with a combined presence of 10^{1236}ãr. A single virgin that runs this flux gauntlet covers 10^{776}ã and the combined presence of both is 10^{2012}ãr in the 10^{2328}ãr inter mattronal volume. This indicates that 10^{-316} of all head-on virgins self-activate under way by colliding. This may not seem much but the activated flux from only 10^{144} newborns boosted to 10^{388}¢ per ãnna sweep more space than all untouched virgins, thus jacking up the activation rate.

The relative coordinate systems of two particles orbiting in perfect circles around two coordinate centers where the two head-on peripheries touch can yield everything from relative standstill to piston like movement in a two point coordinate system. Generally, the two orbits form into one convoluted orbit relative to the Universe around a common collision coordinate center but the much more simple linear drive is effectively slowed and eventually stopped in orbits that lead through both mattrons.

As the spin planes of two activated newborns or their flux mid between two mattrons are rarely the same, their orbits grow more locally complex with each new added collision (the universal relation being less affected). As each is countered in collisions by opposing flux at various speeds, its progress along the one universal line (the width of the mattron sphere) is slowed and stopped while hardly a dent is put in its orbital movement in any of the 10^{776} convoluted directions available to each orbiting flux particle. Out of 10^{776} available orbital orientations, only 10^{460} orbits are taken and annulment in contrary orbital directions is rare. The resulting flux keeps the 10^{388}¢ orbital drive of its virgin input, albeit downsized by ripped out flux.

Without linear slowdown, an atom sweeps 10^{388} points on the last sphere of primal thickness added to its journey on reaching the next mattron. The other mattrons area covers 10^{-288} of that sphere. Of each new orbit made, only 10^{100} points are within the mattron as the sweep progresses repeatedly through its radius. Doing so 10^{632} times on average equals 10^{100} full sweeps through the mattron. With activated flux spending too little time within a mattron to allow its mass to put residential atoms in its path, the flux rarely hits anything and passes further, carrying its meager shower along to ad to orbiting exterior flux.

In reality, the opposing linear push as the virgins activated between the neighboring mattrons had coordinate centers that aimed linearly against the oncoming newborns at birthspeed; a conflict in two broad opposite directions that affects linear progress of coordinate centers far more than the orbital phases of their eventual flux. The 'single' linear phase between the mattron will quickly slow while the orbital element still delivers the lion part of its original punch.

Again, a linear virgin aims 'perfectly' between two mattrons along a line that on average passes through their centers. If a virgin (or its flux) collides with another virgin head-on along the same mattron wide line, their relative linear motion stops in its tracks while the orbital phases of the activated virgins are not so easily mastered.

Flux orbits that no longer widen away from the originating mattron can make repeated orbital passes through both mattrons, ripping out more flux until even the orbital movement slows enough to be captured in the direct exchange. Flux from set moving by newborns heading out in many different directions is eventually harvested by several neighborhood mattrons and almost all exchanged virgins are locally harvested along with their flux.

The extremely rare activated flux in orbits greater than the average distance between grid mattrons can cause a gravity shadow between other mattron pairs when blocked. A vastly smaller fraction of once-hit activated pairs that can still retain much of their linear and orbital drive will make it beyond the next mattron to cast direct linear flux shadows between some other mattrons and we shall come to that later.

The direct repulsion upon a nearby mattron by the virgins that survive across the inter-mattronal void is limited by the inverse of the distance squared in mattron radii. Being non-boosted in parental orbit across this distance, the rest misses the next grid mattron.

If such virgins are activated between other mattrons further away, they do not add to the inter-mattronal gravity by their primal input in that local area but will add their parental punch and eventual collective inputs.

Duality of Mass and Movement

Hitherto when I have referred to a mattrons mass, I have been referring to the number of its primal particle population. As you have already seen, this in misleading and I would like to comment formally on this phenomenon.

From outside the mattron, its 10^{912} particles, or whatever, are somewhat difficult to count and can best be measured from outside by their external collective gravity shadow. That gravity shadow depicts the number of strays blocked by the mattron and would ordinarily depend on the paths blocked by physical particles.

Given the effects of all the other fields, such measurements would not be an easy task. Clearly, the less material element of the relative local movement of internal atoms seriously affects the outcome and indicates a presence of mass where there is none.

A residential atom within a mattron can move within its local spaces in a closed orbit at great speed without collisions. The primal atom, as we have seen, can occupy a great number of spatial points in one unit of primal time compared to a stationary atom.

A perfectly straight virgin strayparticle that whisks through a mattron at birthspeed occupies on average a number of spatial points (ãr) equal to the mattron radius. The speed at which it travels sets its time within the mattron. Aside from this, its collision potential depends on the number of spatial points blocked by the mattron population during its passage.

If a strayparticle moves faster than birthspeed it spends less time within the mattron although it sweeps as many spatial points. Its eventual collision and capture depends on meeting enough residents to block that linear run and if it moves fast, the time does not allow such mattron residents that are on the move enough time to move onto its relatively straight path.

Increased stray speed decreases its collision potential relative to the local spaces it passes through. The speed gives the stray greater mass presence on its greater orbital inter-mattronal sphere but not within local spaces. Instead the speed here impairs the ability of slower moving local atoms to move onto its path or crash into the locally fast non-boosted stray.

A fast stray has great potential in dislodging primal mass when a collision occurs but it has less potential to realize that effect.

The resident atoms that circle in closed primal orbits within any primal cloud at great speeds, i.e. the more spatial points each resident occupies during the primal ānna, the greater is the spatial presence of the collective mass of the cloud that consequently gains in blocking ability.

Primal movement therefore casts a definite shadow.

As a mattron ages, its mass becomes more aligned and loses some of this individual movement. As the mattrons original opacity dwindles, we would say it cools down or loses energy. In fact it loses part of its mass presence but not its mass and its gravity shadow suffers as consequence.

In its earliest anarchic days, the mattrons primal mass was active in fast and larger local orbits that were gradually slowed in collisions to a level that I earlier set at an internal average of 10^{-244}¢. The new slower collective spin of the mattron, distilled out of unpaired directions, is no substitute for the loss of this vast average parental movement.

Not only does a highly aligned mattron block fewer strays but it also 'emits' more of its newborns out of the mattron along straighter paths and therefore dilutes or obliterates its already weaker gravity shadow.

If you send a large shower of virgin newborns through a totally aligned mattron to be trapped, its residential mass would of course become highly excited anew. The induced internal movement in local orbits far exceeds any linear phase induced and though directionally incoherent, it would be seen to 'energize' the mattron that consequently casts a darker shadow.

When viewing spatial presence of any atoms against the spatial presence of other atoms, one cannot simply use the distance each travels (the spatial points they touch). Relative mass presence must take speed into account. The faster an atom covers a distance, the lesser the ability of other mass that is moving within that volume to boost its own spatial presence by movement. When I speak of spatial presence, I am referring to such relative space-time presence that unites both frames of reference into a single frame.

Bear in mind that great spatial presence is not the same as physical mass that alone has the actual ability to block strays and to split newborns.

The spatial presence of 10^{912} resident atoms in each sign moving at 10^{-244}¢ during the 10^{632}ãt traversing period by atoms at birthspeed is equal to 10^{1300} resident atoms at relative standstill in terms of opacity.

However, entrapped atoms share their coordinates through recreation with only 10^{912} physical particles within a mattron and a strays linear dislocation ability is unaffected by the increased spatial presence that facilitates its eventual capture.

Such is the duality of mass and movement.

The Expanding Mattron

Compressed mattrons can expand against dropping straypressure without initiating a halo until spatial balance is reached at larger radius that is smaller than 10^{632}ãr 'maximum size' but in my mattron model, primal mass that is pushed beyond its 10^{632}ãr radius must be viewed as a primal halo.

In my setup, the balance of a decompressed mattron in a stray deficiency would call for a cut in its expansion function. If the mattron loses mass, it loses part of its expansion excess and a balance is reached when potential expansion (driven by now lost mass) matches the lost compression.

Again, it is good to keep in mind that I put this maximum size in place (in a balanced mattron grid) instead of an immense span of a gradually thinning corona. Point objects are conceptually more difficult and this restructuring makes the mechanisms plainer while the truth falls between the chairs.

A threshold at 10^{632}ã is an approximation that gives a pivotal (conceptual) alternative instead of insensible mathematics. A widening maximum mattron expands unevenly; faster precursor halos deliver greater grid pressure than the ever fewer virgins that escape its main mass.

I consider any mass pushed (beyond 10^{632}ãr) into an outwards thinning mattron corona, not as part of the mattron but rather as a separate halo or parts thereof. This threshold dictum of convenience means that all resident particles pushed beyond 10^{632}ãr must be viewed as free atoms and a mattron mass cannot balance at a larger sphere except as temporary bridging action.

To find balance against dropping straypressure, expanding mattrons must either get rid of a relevant part of their internal expansion excess or acquire substitute compression. They cannot push mass beyond 10^{632}ãr without forming a halo, and they cannot emit a halo without losing mass (according to this rule; if not quite in reality).

In a grid, instead of easing its short term imbalance, an expanding mattron amplifies the imbalance as fewer virgins escape and mutual exchanges that deliver the flux field pressure dry up. Whether isolated or in a grid, the initial mattron response is similar; they must start to expand. The response to the loss of majority field pressure for isolated mattrons or mattrons in tight grids is also very different from a fractional drop in exchanged grid pressure.

In a fractional pressure drop; mattron expansion restricts direct exchange and thereby acerbates the initial drop in grid pressure on two fronts; by the virgin number, the lesser activation rate and ability to block flux at greater speeds. In other words, depending on the drop, it is not the initial shortfall that sets the eventual halo mass required but the accelerated response to an increasing drought in pressure that the expansion causes in the grid.

It is likely that greater fractional drops will at some point cancel enough pressure in the grid to fashion a less variable halo expansion cycle.

In the initial stages, halo formation is less of a collective phenomenon and more of a disc phenomenon. As a mattron responds to falling straypressure, its activated disc virgins start to push primal mass linearly in some direction out of individual disc planes and out of the mattron surface.

At first this is on relatively few discs and the exiting flux restricted to a few discs is not all that cohesive. The first batch is shallow surface emission; fast and meager precursor to more massive halos from deeper within. Generally speaking, these precursor batches are composed of 'surface' atoms with little from the middle layers and least mass from the mattrons main inner mass.

Activated internal newborns from shallow layers move at differing speeds in a newborns common linear and orbital direction on its two dimensional disc. The linear exit speed of a halo is mainly set by the depth of emission that drives the disc mass outward. Though highly aligned with mattronal coordinates, the precursor flux swiftly reaches great distances and orbital speeds and can no longer be considered an integral part of the mattron.

The hallmark of mattron expansion is that its unraveling surface does not experience enough incoming strays to be collectively held back but the initial expansion is surface mass exiting from two-dimensional discs in increasing numbers at decreasing speeds.

For every level of straypressure drop, these precursor halos follow each other in a seamless succession into more substantial coronal halos during greater pressure drops until eventually, the whole mattron mass is driven outwards from a disc center. In spite of facing total loss of its original field pressure, a mattron never quite turns into a point object field or fades evenly away from its center and out to no mass at all.

In a majority pressure drop, the delivery of boosted newborn pressure comes at the expanding mattron in seamlessly interlinked stages; at first from the most tenuous fastest precursor born of shallow surface emission and then denser coronal halos of somewhat lesser spatial presence. The secondary emission of this combination is well fitted to pick up the tab and bridge the pressure drop by highly boosted activated flux.

If the precursor pressure cannot hold back or recompress the main central density that casts darkest shadows, this culminates in a large corona driven by slower central expansion. Eventually, inwards aiming secondary emission chips out mass from the expanding inner corona, using it as cannon fodder before slowdown and harvest by the soon re-compressed central mass while only a small fraction of the decimated main corona and precursor halos is ultimately lost to the mattron.

Precursor Emission

We have now progressed from the pristine principles of the Primal Code to a complicated set of circumstances where various phases of mattron unbalance have enough embroidery crammed into its constantly interacting cycles to satisfy the most scholastic. Soldiering on in the mud, I emphasize that the mattrons balancing act, as drawn up, is a conceptual exercise and my advice is to avoid all mathematical temptations.

Let me first use a small permanent drop in straypressure to illustrate the system and thereafter a more substantial pressure drop. In principle, the magnitude of each drop determines how much primal mass a mattron must get rid of (mass-wise) to regain spatial balance.

There are so many interacting factors here that it is hard to isolate them lucidly enough for a reader to grasp, and while this is not a pretty picture, what it lacks in elegance, it provides as a roadmap to more elusive truths.

Although it is helpful to use the two-dimensional disc model, it is wise to keep its shortcomings in mind. A newborns linear direction can be confined to a single plane while its rotation cannot; to make the model work, I use the square root of the primal boost across a disc radius. I could also do this by screwing the disc plane into a twisted cone away from the mattrons center but that cannot provide a lucid illustration of 10^{280} atoms sitting on each of 10^{632} two-dimensional discs and all splitting secondary virgins.

The primal time unit 'ãnna' (ãt) is an accurate instrument to describe the advent, apex and ebb of the fleeting existence of such emission, whether as free precursors or more massive and stable precursor halos. Relatively extensive units for measuring time, such as the 10^{912}ãt disc time (with newborn pair split on each standard disc) are just useful markers.

For a small drop in straypressure

To demonstrate the principles, I begin by using a 10^{-100} drop in the straypressure. This may seem little; for a standard mattron, the pressure loss per ãnna in each sign is smaller than one atom at birthspeed. Even so, it equals 10^{604}pu during 10^{632}ãt radius time and accumulates to 10^{884}pu in each sign during 10^{912}ãt 'disc time', equaling 10^{456} freed and activated newborns, boosted to 10^{388}¢ in a standard mattron grid.

With a 10^{-352} compression ratio, the mattron must get rid of the number of atoms that in 10^{912}ãt can split 10^{884}pu÷10^{-352}=10^{532} newborns causing the mattrons excessive expansion. The number of primal resident that can split this many newborn pairs would then be its 'excess mass'.

It takes 10^{-100} of the mattron mass (10^{812} atoms)×(10^{912}ãt)×(10^{-1192}k) to split 10^{532} newborn pairs in each sign during disc time. In theory, the mattron must therefore get rid of 10^{812} atoms in each sign, which adds up to 10^{180} atoms on average to be shed from each of its 10^{632} discs.

The mattrons first response is to expand until its sphere harvests required pressure but at maximum size (according to my arbitrary rules), it cannot do so without hemorrhaging mass across 10^{632}ãr. This takes on the character of a fast expanding precursor of boosted flux on a few of its 10^{632} discs.

Some of the mattrons primal atoms are more easily carried out from its two-dimensional discs than others, and on average, precursor mass is carried along by whatever expansion incumbent upon the drop in straypressure. The virgin that started to drive disc flux along is soon just another atom among the flux it set moving and in no way discernible from that field of linearly slow but highly boosted flux.

Disc or precursor flux constantly orbits its driving coordinate center (as set by the single disc newborn) as that coordinate center moves out on the disc plane and out of the mattron. All that specific disc flux orbits a common coordinate center within the expanding precursor from which it repeatedly passes back into the mattron down to the depth where the original driving newborn was split on its disc plane.

Precursor particles are as any other atoms, except that the coordinate centers governing them are circling within and eventually outside the periphery of their disc planes. The primal wards (actual particles) of each coordinate center are soon making shallow cuts into those disc planes in orbits as they move outwards and emerge ultimately clear of the main mattron.

If drops in straypressure are fractional, the average speed of precursor expansion driven by a newborn within its disc is the product of the unopposed 10^{-280}¢ (central) expansion speed and the actual drop in straypressure. For example, during a small 10^{-100} drop, the unchecked outward speed drops from 10^{-280}¢ (total central push) to 10^{-380}¢ upon the entire disc mass that is being moved outwards in such a scenario.

All internal push aims eventually in a single linear direction shared by the encountered disc mass. If driven by one newborn from a disc center against a small drop, the entire disc mass can (in theory) become an exterior disc plane precursor with a slow 10^{-380}¢ linear movement outwards, clearing the disc border in 10^{1012}ãt. However, in that extensive period, 10^{100} new splits will have taken place on each disc and many of the new inputs are in linear conflict with the first direction (even if 10^{-380}¢ linear speed were applied, the mattron cannot mount any effective response to the pressure drop during its shorter 10^{912}ãt disc time).

But averages are not all there is. From each new pair split further from the disc center, on average one virgin of a pair aims 'outward' and carries fewer disc atoms (positioned outside of it) in its sign along at relatively greater speeds. Due to this minority effect, a considerable number of activated newborns escape the mattron in relatively short periods.

To probe the inner workings of the precursors, the best point to start is at the beginning and look at the smallest and fastest vanguard of any such precursors as that relatively small package of mass is sent packing in a sequence that familiarizes us with the elements at play.

There is for example a 10^{-270} chance that a newborn is born so close to the disc border that the outgoing one encounters only 10^{-270} of residential atoms on its disc on spiraling outward after being activated, potentially to reach 10^{10} disc atoms to carry into space at 10^{-10}¢. Earlier this expansion was matched by compression able to move disc atoms inward at 10^{-10}¢. After dropping to 10^{-100}, the outward speed of these few and shallow residents is 10^{-110}¢.

This 10^{-270} chance of collisions will see the same thing realized on every disc once in 10^{1182}ãt (10^{270} disc times) while in disc time such splits occur on average on 10^{362} two-dimensional discs (out of 10^{632}). Each split is an adverse pair where one compresses towards the disc center or drives the greater mass outwards in the other direction at a slower 10^{-380}¢ speed. Some 10^{372} atoms endowed with these shallow characteristics on 10^{362} affected discs are being pushed faster 'outwards' and the flux they carry out of the mattron builds a precursor halo and a single split on each disc means that this orbital push goes unchallenged.

Even at a relatively slow 10^{-110}¢ during a small 10^{-100} drop, it takes time for flux to clear the disc border and from there to leave the mattron as its fastest precursor. The shallow faster flux needs only to cover 10^{-270} of a standard radius or 10^{362}âr to exit the mattron and can leave it behind after 10^{472}ât; a relatively short period.

While this shallow emission is of little consequence to the gargantuan mattron, we can look at the 10^{10} disc atoms. Although reality may not be that particular, the two-dimensional model does not allow this flux to interact with atoms on other discs. The linear and orbital speed of the driving newborn is therefore impeded only by the flux it sets moving on the source disc.

After crossing the disc border on leaving the mattron, I allow that more of the disc flux can be rearranged onto other planes of the two-dimensional model by colliding exterior flux. This means that after exiting, the orbital flux can cut into the mattron repeatedly at high enough speeds to carry residential atoms into the open from many other discs (I do not dwell on to what extent).

As each disc flux coordinates center is moving outwards, the depth to which flux can reach turns gradually shallower and it cannot be geared down in sufficiently speed to be blocked by the mattron and so escapes its surface. Though the rip out from other discs is pretty certain as it is shifted across many discs by outside flux, it is unlikely that it affects all discs or even down to their full emission depth.

In a precursor, only the activated newborn that drives the disc mass delivers a growing boost across the distance it can reach. The add-on flux it carries along is simply cannon fodder in much the same way as flux being ripped out of mattrons in their ordinary mutual exchanges.

Once clear of the mattron surface, the outward speed of the exiting disc flux that is part of the precursor can be accelerated away from the mattron on the wings of virgin newborns that escape the mattron without collisions but collide instead into the relatively slower decamping disc flux and are thereby activated. Such escaping virgins do marginally accelerate the slower coordinate centers away from the mattron but due to the vast flux orbits and conflicting aim, this push is heavily downsized.

The mattron also rotates collectively on average once in half disc time and the relation of the new individual coordinates to the common collective coordinates shifts the movement of individual atoms from their exterior and inherent planes to make them effective upon other discs. Individual coordinates that are out of sync with a collective phase causes flux to rip primal mass out of some of 10^{632} available discs but I am not estimating the extent of how many or to what depth.

Secondary emission by exterior precursor mass is not equal in number to virgins that issue out of a mattron until the precursor halo contains 10^{-36} of its mass. By that point the precursor mass splits more virgins than escape even a smaller less opaque mattron (given time). While numerous deeply or more centrally placed atoms are already being carried outward on other discs, these have yet to reach the original mattron surface while the faster more meager precursors may be far from realizing the required level of secondary splits in what can be termed a more stable halo.

As the distance of each coordinate center that governs a given number of flux increases away from the mattron at a given linear speed, each individual flux atom is moving away from that driving coordinate center at the same speed. In other words, the flux continues to pass tangentially in orbit by the mattron all the time and to reach into it to the dept where the newborn that set it moving was born (albeit gradually shallower).

The period it takes a precursor particle to cover the medium distance back to a mattron is therefore irrelevant. The widening orbit of the atom passes repeatedly close by the mattron in widening orbit. In that sense, it does not take a precursor atom much time to backtrack across the distance back to the mattron. And any new secondary input that widens the orbit of the precursor atom will bring it on collision course with the main mattron itself.

Extremely few secondary newborns escape as virgins from a mattron out of a vast tenuous precursor corona or halo; which is an even greater point object field of extreme activated presence.

The flux from a more stable halo can ordinarily be re-captured by a standard central mattron if its orbital speed has on average been geared down to 10^{36}¢. Shallow flux cuts meet lesser resistance in the mattrons outer layers and required slowdown here needs be far greater. Each redirected halo atom rips out a great number of atoms from many discs, adding flux to the halo and raising the number of atoms that a driving newborn sets moving, thus slowing the linear speed of a 'decamping' halo.

Inasmuch as there is little resistance to orbital speed, the orbital drive of an activated newborn differs from its linear drive. Its spin coordinates are geared down in relation to the number of atoms it sets moving and the neutralization of opposite spin is marginal for decamping flux. The actual directions taken up by the flux in a standard grid are a tiny part of (up to) 10^{1552} planes available. Increasing distances supply more numerous discs and afford the flux an even better leeway than that to avoid collisions.

If birthspin equals birthspeed, the gear down for capture by the mattron can be approximated. With a coordinate center at the 10^{632}ãr border in the above example, as the flux starts to 'peek' out of the mattron in orbit (within 10^{-270} of its radius) the potential orbital sweep of each driving newborn is 10^{181}¢ about its coordinate center (10^{362}ãr distance).

As each newborn drives 10^{10} disc atoms, the 10^{181}¢ collective orbital speed of the disc flux has been geared down to 10^{171}¢ as the coordinate centers exit the mattron surface. With a shallow cut into a rarified outer plane of 10^{-270} disc radius (and mass); each flux atom must be slowed 10^{270}-fold from the original speed of flux atoms on a central trajectory (stoppable at 10^{36}¢) to a stoppable speed of 10^{-234}¢. From its 10^{181}¢ driving speed, a flux atom must then be geared down 10^{415} fold which depicts the level of flux to be ripped out of the mattron within 10^{-270} of its radius to do so.

The potential of a shallow virgin to kick atoms encountered on its disc out of the mattron to 'build' a precursors at an initial 10^{-110}¢ linear speed is excellent. However, to slow 10^{362} driving disc time newborns 10^{415} fold in orbit would require 10^{777} shallow residential atoms driven into flux. It also requires a driving boost at 10^{171}¢ to slow its 10^{-100}¢ linear progress 10^{415}-fold to 10^{-515}¢ with all that flux taking 10^{1147}ãt to reach a standard radius out of the mattron surface.

The problem is that all the 10^{632} discs (not just 10^{362} discs with shallow emission in disc time) hold merely 10^{642} resident atoms down to that shallow depth. On top of that, the flux can neither reach all those discs nor down to the full shallow depth. As the mattron adds 10^{632} newborns by internal procreation, stripping the entire surface down to its initial orbiting depth and shuttling 10^{642} atoms in disc time; causing it to lose weight by intermittent small precursors against the small drop in field pressure, is not possible.

Halo acceleration by virgins issuing out of the mattron, should not be viewed with everyday glasses. Each halo atom is now part of a private coordinate system with a central point situated one mattron radius out of the surface, moving linearly outwards at 10^{-380}¢ (a system set by the spin of the driving newborn). Again, the picture is of a flux atom orbiting its central coordinate point; its orbital path toward the mattron, sweeping by its surface, and then away in its orbit out to a mattron diameter away.

In fractional drops in pressure with sizeable exchange of grid flux, this is not realized. Due to its great spatial presence, precursor flux is quickly chipped from the inter-mattronal field by the ordinary flux exchange (albeit deficient) and the precursor flux provides the missing pressure.

If a precursor atom is hit by an emitted linear virgin, it is not the coordinate center that is hit but the atom that is moving within it. A virgin hit fashions a convoluted orbit by activated birthspin, circling the 'collision point'. From the precursor atoms vantage this is in a widening or tightening orbit at half the input speed related to its coordinate center. If the orbit is opening up at half the linear birthspeed while the pair orbits a precursor coordinate point at half the greater orbital flux speed, means that the pair still sweeps close by the mattron surface.

Depending on linear aim (away from or towards the mattron and whether engaged at the closest or furthest orbital leg of precursor flux), an average of a large number of virgin collisions at various points affects its orbital size; able to move precursor flux away from its mattron and out of contact or towards a mattron causing internal collisions and eventual re-harvest.

In the end, the 10^{362} driving newborns emitted into a precursor in disc time cannot match the 10^{-100} loss of pressure, even when driven to a 10^{388}¢ boost across a standard mattron grid. In spite of the small drop, the frail precursor falls 10^{134} times short of the 10^{884}pu needed to restore balance to the mattron.

At this point, secondary virgin emission within a first tenuously stretched precursor distributes its mass and widens it linearly away from various coordinate centers in all directions and facilitates possible re-harvest of its flux.

This however is the smallest and fastest precursor. Let's look at a larger deeper precursor and continue to use the 10^{-100} drop in pressure that accumulates to 10^{884}pu in disc time. From the time that deficit is felt, emissions have been advancing from greater depths, driving more numerous mass outward at slower speeds and taking longer to surface. More massive precursor surface step by step in succession behind the first precursor in disc time and these culls of hemorrhaging disc flux exit a mattron that is still unable to make an immediate dent in either the deficient pressure or its excess mass.

Given these circumstances, the most massive precursor emission possible in disc time is driven by one newborn on 10^{582} discs (10^{-50} of 10^{632}), each pushing 10^{230} atoms on their disc outwards. With the compression that is ordinarily able to move disc atoms inward at 10^{-230}¢ having dropped by only 10^{-100}, the speed fraction outwards for this deeper flux is 10^{-330}¢.

There is a 10^{-50} chance for a pair of newborns to be split at this depth. Each can encounter 10^{230} of the discs residential atoms in its sign to carry into open space across 10^{632}ãr in 10^{912}ãt disc time at 10^{-330}¢. This affects 10^{-50} of the mattrons mass (still a relatively thin outer layer) and the distance the disc flux must cross is 10^{582}ãr and more massive culls of flux cannot be pushed out into the open during disc time.

Of the mattrons 10^{632} discs, splits at this depth occur on 10^{582} discs in 10^{912}ãt and new coordinate centers with 10^{291}¢ orbital boost collectively drive 10^{812} atoms out of their discs at 10^{-330}¢ across 10^{582}ãr in disc time.

As the coordinate centers of this deeper disc flux exit, the flux orbits 10^{582}ãr in and out of the mattron with each driving newborn slowed by 10^{230} disc flux and orbital speed brought down from 10^{291}¢ to 10^{61}¢. At this depth, the 10^{36}¢ central blocking ability adjusted by a 10^{-50} radius sets stoppable speed at 10^{-14}¢. The orbital exit speed must be slowed 10^{75} fold from 10^{61}¢. Even so, chipping mass out of other discs until stoppable at 10^{-14}¢ won't slow it enough. It requires more than the mattrons entire mass to slow the 10^{862} flux sufficiently.

As its 10^{862} flux is 10^{50}-fold the 10^{812} 'excess' mass against a 10^{-100} drop in compression means that the mattron has shed its excess mass in a single precursor pulse in disc time while only adding 10^{632} atoms in the same time through procreation.

The balance to match a 10^{-100} fractional drop (10^{884}pu) compression wise is remedied by 10^{496} newborns in disc time driven across a standard grid at 10^{388}¢. This is served up on 10^{496} discs from a depth of 10^{-136} radius, driving 10^{144} atoms outwards on each. Tallying 10^{640} atoms, the mattrons mass loss falls far short of the 10^{812} atoms in each sign that it must get rid of against the small drop.

As 10^{496} driving virgins are collectively less than ordinary flux exchanges from nearby mattrons, the precursor flux is here unable to push back the ordinary collective flux. Although precursor flux can balance the books compression-wise against any fractional pressure drop, it is so quickly chipped away that it cannot form a stable physical halo.

The Precursor Halo

Aside from apparent stability in mattron grids by way of ephemeral precursor flux, there are certainly circumstances where the physical aspect of halos is paramount. Such would be the case in voids where compression disappears and no (effective) exchange is possible between mattrons.

It is easy to understand why halo compression is so powerful when pitted against the relatively stable net expansion function of a neutral mattron, as the directional compression driven by flux generated by newborns in an unraveling halo differs radically from surface flux. The linear compression upon a mattron rises from next to nothing (when a newborn is part of its surface) up to the orbital punch its flux can pack in vaster faster orbits from outside. Whereas a newborn's orbit within a corona is locally contained with its linear effect weighing almost equally in and out upon its mattron, highly boosted halo flux translates into direct compression of the center as it cuts it repeatedly against similar collisions from opposite directions.

There will always be mattrons that lose out on a share from the field while communicating with nearby mattrons. They may periodically orbit out of full exchange with nearby mattrons in uneven grids and harvest flux from fewer of virgins in tighter mattron groups. There are also tight grids where boosts cannot keep mattrons balanced due to insufficient grid compression even if they retain a steady inter-mattronal exchange.

A majority drop in straypressure can leave a mattron with varied harvests and in a void it would be all its harvested pressure. In a tight 10^{766}ãr grid, it would receive 10^{-5} of standard pressure; with a loss of 10^{704}pu pressure and left with a 10^{699}pu pressure in each sign during 10^{632}ãt radius time, its loss accumulates to 10^{984}pu during 10^{912}ãt 'disc time'.

The problem with this grid is the hundred thousand times lesser mutual boost of ordinary exchanged flux. Weaker precursor halos have their inter-mattronal flux chipped away by the ordinary flux, which prevents smaller precursor halos from getting physically off the ground. However, splits from deeper within the mattron are already driving all mass outward at full effect on every disc and without much resistance.

For a halo to mount a successful response to a massive shortfall calls for the delivery of 10^{601} driving newborns during 10^{912}ãt, activated and boosted to 10^{383}¢ across this tighter mattron grid. The disc time limits the effect; the more mass a newborn must drive, the slower the pace. The collective disc flux must exit before another pair is split on the same disc as new splits send virgins in conflicting directions and neutralize its 'expansion'.

In succession, precursors of deeper increasing mass follow the first precursors, long before neutralization steps in. For example, newborn pairs have a 10^{-31} chance to be split within 10^{-31} of mattron radius. Each will bring 10^{249} disc atoms outwards at 10^{-249}¢ to exit the mattron surface in 10^{881}ãt after the split. Within that shorter period, no neutralization is experienced for similar splits on the 10^{601} affected discs during the 10^{912}ãt disc time.

These 10^{601} driving newborn in disc time will eventually chip out 10^{249} disc atoms on their own disc planes. This would allow for a substantial physical halo of (in excess of) 10^{850} atoms to reach the 10^{766}ãr grid distance in more than a 10^{1015}ãt expansion period. This hefty precursor halo beats back the weaker virgin emission until an opposite grid halo meets it midways in linear conflict that alters the orbital orientation of their flux.

It would seem that the 10^{601} driving virgins in each sign (during 10^{912}ãt) are boosted to 10^{383}¢ across the grid to deliver the lost 10^{984}pu pressure. The same level of secondary emission issues from the robust and now physical precursor halo. The problem with this smooth outcome is that massive parts of the mattron have continued to expand within the grid during the 10^{1015}ãt formation period of this precursor.

As pointed out; my 10^{632}ãr mattron model derives from a collapse of a submist that was originally more of a 'point object'. The tradeoff is a simpler mental picture but it is right to question if a mattron can expand on a broad front without emitting (losing) the far reaching precursor halos that steam ahead of the general expansion. There is an intermittency here that prevents a re-expanding mattron from becoming a point object field that distributes gradually thinning disc mass smoothly away from a center out to no mass at all without any definable central mattron mass.

I therefore illustrate the precursor halo system by an isolated mattron in a total void where it goes without exchanged pressure and central expansion subsequently drives its mass outwards at 10^{-280}¢, eventually out across its 10^{632}ãr on all of its discs already in 10^{912}ãt disc time.

Without resistance at its constant initial 10^{-280}¢ speed, the entire mattron expands in size to 10^{735}ãr during 10^{1015}ãt. The expansion turns differential as the mattron increases in size with gradually lesser mass carried from less stocked disc planes while the expansion speed increases.

For example, at size 10^{735}ãr, the 10^{735} discs now hold 10^{177} atoms each and all mass can expand at 10^{-177}¢ (instead of the initial 10^{-280}¢). Using a square root of the product as average expansion, a faster average speed at 10^{-228}¢ could expand the mattron into a vast 10^{787}ãr globular subcloud in 10^{1015}ãt that thereby exceeds the tighter 10^{766}ãr grid 10^{21} fold. The inference is that halo formation comes about in a complicated interplay of massive expanding main mass coronas and flightier precursors.

For example, without compression in the isolation of a void, a corona at slowest 10^{-280}¢ speed expands to 10^{680}ãr in 10^{960}ãt. At size 10^{668}ãr, its 10^{668} discs hold 10^{244} atoms each and expand at 10^{-244}¢ (instead of 10^{-280}¢). Using the square root to glean an average expansion at 10^{-262}¢, the main mattron would already in 10^{960}ãt turn into a vast 10^{698}ãr globular subcloud.

Now take a conceptual snapshot. The former mattron is an expanded subcloud of 10^{698}ãr radius that splits 10^{418} virgins in its 10^{698}ãt disc time. $3,16 \times 10^{174}$ fold boost in 10^{349}ãr parental orbit sets central parental spin at $(10^{418}) \times (3,16 \times 10^{174})$ or $3,16 \times 10^{592}$ fold the 10^{-912}¢ collective central spin. Spin at the centers of parental coordinate system is on average $3,16 \times 10^{-320}$¢ and reaches 10^{-145}¢ in parental orbit of 10^{349}ãr; giving the subcloud a combined 10^{1465}ã presence during 10^{698}ãt, the time it takes a virgin to leave it (parentally boosted to $3,16 \times 10^{29}$¢ and gaining a significant $3,16 \times 10^{727}$ã presence on its own). The combined $3,16 \times 10^{2192}$ã presence in that 10^{2094}ã subcloud volume, allows only $3,16 \times 10^{-99}$ of internal virgins to escape it untouched (while linear strays are blocked at up to $3,16 \times 10^{69}$¢).

During the 10^{960}ãt formation period of this halo, faster lighter precursors also stream ahead of the 10^{698}ãr corona. In 10^{960}ãt, the slowest and heaviest of the 10^{680} splits realized in that time take place on 10^{-18} of the subclouds discs or 10^{680} discs, sending 10^{220} atoms on each outward at 10^{-220}¢ and 10^{150}¢ in orbit. In 10^{960}ãt, this forms the most massive 10^{900} precursor halo possible that reaches 10^{740}ãr from the main mattron. As precursor mass soon is out of sync with coronal rotation, driving virgins are ultimately unable to push precursor mass linearly away in any universal direction.

This 10^{740}ãr wide and 10^{900} strong physical precursor serves as a robust example but there are lighter faster and heavier precursor halos. It splits 10^{406} secondary virgins in the coronas 10^{698}ãt radius time ($3,16 \times 10^{86}$ times the virgins that escape the corona). Activated in widening orbits by precursor flux further away, this secondary newborn halo flux comes at the coronal mattron at 10^{370}¢ and delivers 10^{776}pu pressure in radius time.

My choice in a total void is to use a halo which basis is precursors where a corona is driven at 10^{-280}¢ to 10^{680}ãr in 10^{960}ãt with the 10^{680} discs of 10^{232} atoms expanding at 10^{-232}¢. The square root yields an average at 10^{-256}¢ that turns into a bigger more meager 10^{704}ãr wide globular subcloud in that time that splits 10^{424} virgins in 10^{704}ãt. The 10^{176} fold boost in 10^{352}ãr parental orbit sets $(10^{424}) \times 10^{176}$ central parental spin (10^{600} fold the 10^{-912}¢ central spin) that reaches 10^{-136}¢ in 10^{352}ãr wide parental orbits with a combined 10^{1480}ã presence in 10^{704}ãt. The escaping virgin (parentally boosted to 10^{40}¢) gains a 10^{744}ã presence and combined 10^{2224}ã presence in a 10^{2112}ã subcloud volume which lets 10^{-112} of internal virgins escape while linear strays are blocked at up to 10^{72}¢.

During the 10^{960}ãt formation of this foundation precursor, faster lighter precursors stream ahead of the 10^{704}ãr corona and the slowest heaviest of the 10^{680} splits in that time take place on 10^{-24} of its discs, sending 10^{208} atoms on each away at 10^{-208}¢ and 10^{168}¢ in orbit. This 10^{752}ãr wide and lighter (10^{888}) precursor splits 10^{400} secondary virgins in 10^{704}ãt. This secondary halo flux comes at the central mattron at 10^{376}¢ and delivers required 10^{704}pu pressure in the central mattrons 10^{632}ãt radius time.

I am not addressing possible variable compression of an expanded mattron against a faster growing central shadow of its main mass. In spite of size and opacity the cohesive body of an extended corona serves at first as cannon fodder to slow the boosted flux that stabilizes its central regions.

Unable to withstand fast secondary flux against then increasing central density and its darker shadows, coronal mass that is ripped out by flux is gathered faster by the center and this is at some point likely to create a rift in the mass distribution between the precursor as the central regions that collapse at variable speeds.

The expansion started by feeding 'slow' massive flux into a new corona and was in seamless succession followed by a rising crescendo of fewer faster flux atoms spawned on less massive discs; pushing the rarified envelope outwards. As rising precursor compression matches the coronal expansion (and in turn required compression upon the central mattron), virgin feed into new precursor halos drops off in perfect reversal of the aforesaid.

In this fashion, the emitted particle wave brings an effective stop to itself.

The less pressure that a mattron is paid by its surroundings, the greater is its unbalance up to lost majority pressure. The faster its central expansion is, the larger its corona up to a width of a 10^{752}ãr precursor halo to complete that maximum re-balancing act. Due to secondary input of virgins that widen flux orbits, some mass is ultimately lost into the wider void (matching procreation and lesser central annihilation).

The two 'entities' of the unbalanced mattron are both essential to regaining balance in this two-stage act. Although precursor halos constitute a collectively emitted globular particle wave of a highly composite character in mass and speed, the halo system is otherwise simple; to match the coronas and the central mattrons internal expansion, the blocked flux must deliver the correct compression without excess pressure.

The general mattron balance is set from within and without. Center and surface parameters lock neutral grid mattron into a constricted range of mass and size where central annihilation kicks in during compression and dwindling annihilation kicks in during deficits with precursor halos.

Precursors around isolated mattrons can easily activate and entrap any boosted virgins crossing their path and thereby cast virgin gravity shadows, but they are too tenuous and light to cast linear gravity shadows by highly boosted activated flux. The particle showers that such linear flux generates will simply pass on, which does not represent a shadow. It only signifies that more and slower atoms partake in the linear delivery of flux coordinates that could have caused a backhand shadow if blocked.

In principle, a diminishing precursor halo of dwindling presence drives flux that takes longer to dislodge field particles and this would lead to oscillations of different cycles depending on distances. Mattrons however are too small and too close in grids to be governed by such physical precursor cycles.

Decompressed mattrons in tighter than 10^{776}ãr grids (where boost supplies insufficient compression) cannot for example elicit physical precursors that have a structural frequency (even with majority pressure drops). The driven precursor flux constantly adjusts the pressure exchange across a short grid distance without the need or ability to form close physical intermittent halos. Instead the inter-mattron wide halos affords standard mattrons in tighter grids a much steadier balance if they emit more than the mattron itself.

I am tempted to put the illustrative combo of a 10^{704}ãr wide short-lived intermittent corona and 10^{752}ãr wide precursor halo as an average response by totally isolated mattrons; both formed in 10^{960}ãt but with a less rigid life span. The corona is rapidly compressed to standard size but the precursor takes longer to disperse and intermittency is limited to a brief disappearance before the next precursor. One can speculate that precursor dispersal takes between 10^{996}ãt and

10^{1008}ãt, in any case vastly longer than its formation.

A stabilized precursor halo surrounding a balanced mattron is a subcloud structure of its own and follows the laws that guide the internal movement of primal mass that slows from its original 10^{168}¢ orbital drive. The parental orbital speed of this stabilized precursor is ultimately governed by the splits in radius time (10^{448} virgins) times the parental orbit (square root of 10^{752}ãr precursor orbit) boosting its 10^{-888}¢ central spin (set by 10^{888} primal atoms). The parental spin within a precursor halo is therefore 10^{-64}¢.

The external emission from the precursor halo is tiny and somewhat odd; because the central virgins reach only 10^{-26} fraction of 1¢, 10^{-72} of all virgins emitted by the standard central mattron escapes out through the precursor or 10^{364} escaping virgins in radius time. The precursor similarly stops locally non-boosted traversing atoms at average speed 10^{72}¢. The secondary faster moving internal precursor emission reaches 10^{124}¢ exit sweep and merely 10^{-196} of these virgins escape into the strayfield (10^{252} virgins in radius time).

The precursor stops expanding by way of coronal input but may its edges do so to some degree from its secondary emission while the precursor holds a central mattron at a 'stable' 10^{632}ãr radius, often by depositing mass closer to the center rather than by classical compression. The precursor halo that grows to full strength and ebbs out is most extensive and has most presence at the end of a cycle. As lose mass plays a neutral part, this is true of any large field of lose primal mass; greater mass has great secondary emission, which drives the mass dispersal more effectively.

With all atoms being of the same size, all driving newborns in a grid have a $\sqrt{}$distance boost that sweeps d$\sqrt{}$d of the halo volume and their d$\sqrt{}$d sweep sets the ability of spawned flux to hit a single halo atom; this ability diminishes in larger volumes and increases in smaller.

Knocking about for a quick illustration at the end of an emission cycle where the unraveling of a precursor halo (field emission generally) picks up the pace, I opt to describe parental motion in the field by the distance 'd' where flux with its d$\sqrt{}$d sweep sets the frequency of a reoccurring field deficiency by inverse of $\sqrt{}$d.

One complete halo cycle must supply a massive enough field magnitude (primal mass and or distance) to stabilize the submist over time and since $\sqrt{}$d of the distance sets the distance boost of activated secondary emission, the field must compensate by the $\sqrt{}$; all in all a rough estimate to be sure.

One would see this clearly in tighter grids of direct exchanges (except that no close physical haloes of structural frequency are exchanged there). This free lost mass in a tight 10^{766}ãr grid (10^{-10} of standard grid) would need to split enough virgins or generate enough pressure to supply the required grid pressure;

although the intermittent exchange cycle has no structural halos.

Non-structural precursors in this tighter grid would have 10^5 times the driving newborns per disc pushing that much more mass across the field. As this precursor flux is depleted 10^5 times more effectively within the smaller volume, the structural frequency (if they had one) would be 10^5 times higher.

Far apart in a 'void' where neutral mattrons are without support from grid mattrons (and cannot stop linear flux) and can therefore have no meaningful exchange. These mattrons must instead maintain standard size by letting loose intermittent precursors halos of 10^{888} atoms that stabilize for a while in a vast precursor cloud of 10^{752}ãr radius.

There are certainly many effects I have missed or misunderstood that affect the advent, apex and ebb of a potential halos as various exchanges climax and fade, all of which must be looked at carefully in detail.

The Mattrons Field of Charge

Any collection of primal atoms with an excess number in either particle type is 'charged'.

By removing + and − atoms in pairs from any primal mass collection or submist, you may in the end be left with some primal excess of either type. This singular excess would then be the 'charge'.

Up till now I have treated mattron composition as being neutral. I have made it a perfectly equal blend of allied and adverse primal atoms. This is because the creation of highly charged structures is almost illogical when considering the number of primal atoms in these vast mattron submists in which every particle is indisputably created in the company of an adverse twin and the whole submist therefore in perfectly equal numbers.

However, once a sustainable charge is acquired, it cannot be done away with through any readily available process within the submist itself.

The internal creation of a neutral mattron will successively add new mass that is by nature uncharged. This would serve to make any initial diminutive charge relatively smaller, were it not for the annihilation at the center of a neutral mattron that causes the original charge relation to stay the same.

There is the complication that adverse atoms have no effective collisions while allied atoms 'collide'. Each type can therefore in collisions form loose allied scattering within a local area, which does not prevent atoms of opposite charge to collide within that same local area and turn it more neutral again.

Playing the odds, we may well see a great number of primal scatterings where atoms of the same charge chance to cohabit within a mattron. But though these areas are by definition charged; ephemeral scatterings are short-lived by nature. They only block allied atoms to the extent they can entrap and hold on to allied atoms at all. And there's the rub.

It takes a good part of an entire mattron (albeit of shifting masses) to gear down primal mass to catch; this is why a mattron is the smallest submist. No smaller mass gathering (than a heavily compressed, perhaps decimated mattron) has the required opacity to effectively block fast moving atoms.

Charged scatterings within mattrons are therefore not extensive enough to block strays or newborns or bring them to a complete stop. The ephemeral gatherings that by chance collect locally are unable to increase their charge within that local area and will ultimately be scattered.

A mattron can become subtly charged in one sign by chance trappings of more atoms in one sign. Such a charge would be distributed all through the mattron with insignificant effects that are lost in the mattrons gargantuan primal population. And yet, there is a way; if a slow shower of heavily charged passing virgins is narrowly directional enough, the neutral mattron has a rare chance of becoming singularly charged by outside means.

A planet moves under the influence of its fields with the 'soft grace' that derives from the enormous number of primal atoms working upon its resident atoms at every time. The further you descend down the ladder of material structures, the fewer atoms will affect any particular smaller mass gathering at a particular time. All movements become jerkier and functional lapses between contrary collision vectors become more notable as smaller sub-clouds such as mattrons are burdened by greater uncertainty in their collective fields.

An unpaired strayparticle caught by a balanced neutral mattron will have its linear movement transmitted to all its allied mass through the internal collision shower it creates. This is worth repeating; its particular coordinate motion is initially exclusively transmitted through collisions to allied mass.

The only effect that can transfer this movement to the adverse mass are newborns (allied to the stray); born of parents that have already gained part of this motion. When such a parent splits newborns, the implanted direction coordinates become part of the adverse mass within the mattron.

The propagation of the linear movement of let's say a heavily charged stray shower to a mattrons adverse mass must wait for splits of adverse newborns by parents that are allied to the singular stray shower and already brought on the move. The less time that has elapsed from the initial collisions, the fewer adverse atoms possess the linear coordinates of motion that is being step by step geared down by allied mass. When hit by a freak shower of singular virgins, the allied part of the mattron mass will be the first to start moving in its general direction, albeit quite unevenly to begin with.

Although the myriad collisions that achieve this general movement occur all through the neutral mattron, the most effective stray shower is composed of virgins that are very effective in shepherding along local mass in their sign when activated within a mattron. If parentally boosted across vast distances, these penetrate throughout the mattron from where they transfer and gear down a universally directional movement upon allied residential mass.

Allied mass in a mattron surface that faces a singular stray shower is carried inwards by the slowest strays. The more penetrating virgins (at say 10^{36}-fold birthspeed in parental orbits) carry mass out of the surface on the other side in accord with mattron opacity, speed and number of strays. If the stray shower charge is massive enough and fittingly slow and directional, the allied part of the mattron mass is eventually moved along while the adverse mass cohabiting within the same volume is initially hardly affected at all.

The erratic nature of primal fields is burdened with great uncertainty. The mattrons neutral mass, though gargantuan, is too small for its own emission field to be anything but erratic. While the local strayfield may in time provide perfectly even pressure from all directions; it can rarely provide a matching opposite push for an odd influence within a corresponding period.

Until an opposing force is provided against a directional singular shower of virgin strays or until enough newborn pairs are created within the neutral mattron to transmit the imposed motion to the opposite half of its mass, the one sided linear movement of the allied residents continues.

The mattrons own newborns transmit the new coordinates of the absorbed motion to both types of mass. The newborns are only 10^{-280} of the mass per ãnna and the communicators are in both signs. This continues to give mass that is allied to the imposed movement a tangible edge and newborns may be unable to transmit a faster movement to the other mass sign. This becomes more acute; the (relatively) shorter time frame is set by a massive directional singular virgin stray shower. The more substantial and linearly directional it is, the less the strayfield is to supply opposite linear effect in time.

If one part of allied mattron mass is in this fashion brought on the move in some universal direction, the adverse mass simply sits behind. Atoms in its sign are 'unaware' of what has hit its 'other half' as no strays and too few newborns have made their call upon that mass sign.

Two opposite mass bulges have now formed on either side of the mattron in the general line of the stray shower. The bulge facing the singular shower is built of mass that is adverse to the passing singular stray shower while the other bulge on the mattrons far side is allied to that stray shower.

The process is not really a matter of two bulges; a substantial part of mass allied to the singular virgin shower is linearly moved beyond the coordinate center that was earlier common to the mattrons entire mass. Mathematically speaking, the overlaying globular submists of both signs have been moved out of what earlier was a common center; now situated mid between two new centers, and (to that small extent) the two orbit their old coordinate center.

A mattrons collective spin turns it against the linear stray wind and works to distribute the charged virgin push more within and about its surface. The shorter the time in which it the shower operates, the less leeway it allows the slow collective spin of the mattron to turn the other cheek. If the stray wind is weak, this pushes the orb back towards the common center and creates a gentle polarization rather than linear dislocation of allied mass.

If the stray wind is powerful and the linear separation of the twin masses goes far enough, and the shift of one mass out of the space occupied by both is substantial enough within a short span of time, a new potent effect comes to unfold on the world scene.

The Electromagnetic Split

What happens is illuminating. The primal mass within each charged and bared sphere orb continues to emit newborns of both signs but now the sign that is adverse to each singular bulge will escape it more freely.

To begin with, the adverse newborns that escape the charged orbs that signify the shift of the two rotating masses tend to be activated by fewer remaining atoms in their sign until these are swept along out of the bulges (much as 'ethnic cleansing' in halos). Charged primal emission from the two orbs increases the directional shift within the main bulk of the enormous neutral mattron that sits between the two bulges.

Being directional in opposite directions, any increased shift in mass makes the mattron ever more vulnerable to further separation of its two signs. The bulges are independent of the mattrons inherent rotation as the allied bulge mass that is returned into the combined mattron is replace by other mass in that collective orbit.

If the singular directional virgin shower is persistent enough or acts swiftly enough to separate sizable parts of the mattron mass; thus baring two large mismatching singular orbs on either side of the neutral mass (shifting the average centers of all the cohabiting mass), an accelerating separation of its singular components can facilitate a total split of the neutral mattron.

Although charged bulges are not completely bared, they are too small to contain allied internal emission as allied newborns split within each bulge to cross-collide with allied majority mass. Being slower off the mark than adverse emission, allied emission is contained within the bulge for longer while adverse activated newborns escape the charged orb more swiftly.

A directional kick in the linear sense is delivered by birthspeed effect away from a charged orb when blocked within the main neutral mass between the orbs. While the universally linear 'away' phase remains intact, any parental and primal spin coordinates communicated by the same adverse newborns that escape a bulge in contradictory orbits are extensively annulled. As the neutral mattron had full consummated spin. the In a fully

The charged linear pressure upon adverse mass away from the opposite bulges pushes the two mass signs within the main mattron in two linearly opposite directions and the overall shift amplifies the effect of the initial and soon neutralized shift created by the singular stray shower.

If a persistent separation push is wielded within a relevant time-span in a linear enough mode, the resulting bipolar disparity acts to consummate a split of the entire neutral mattron into singular adverse twin mattrons. Each twin contains 10^{912} atoms; half the formerly combined neutral mass and the force ultimately unleashed by the split makes the original stray-shower, or consistent stray wind that triggered the upset, pale in comparison.

A newborn that is split within either singular mattron half is activated by allied singular mass and its subsequent boost ensures its addition to the singular mass in the allied mattron half. An adverse newborn emitted away from its parent, whether virgin or activated and in parental orbit, passes right through the entire adverse population without entertaining effective collisions to stop that linear progress along its way.

If the adverse virgin encounters a remaining minority mass in its sign and is activated, it carries these atoms rapidly along out of that mattron part, flushing it clean of that type of singular mass in primal ethnic cleansing.

When a split is consummated, the mattrons bipolar twins are side by side in close proximity and in the process of clearing the separation. All adverse virgins pass out through each singular twin mass at birthspeed and if they aim in the direction of the other twin, less than a tenth of them miss it. The rest collides effectively with the allied mass of the other twin (10^{-36} of them pass through it without capture). Here they deposit the spatial coordinates of birthspeed, birthspin, parental spin in a garbled singular input that will in time become the new collective spin of that twin mattron.

The collective linear splitting speed of the now two adverse mattrons, their relative movement away from each other, is set in the very last period during which a split is fully consummated and the speed set by the totally escaping adverse emission that issues out of and away from each opposite twin.

A linear harvest of escaping adverse singular virgins at birthspeed pushes the twins apart and sets the mutual splitting speed. The collective mattron spinfield that is identical and inherent to both twin makes the push effective in all directions while each twin covers one rotation in the time of the split. A blocked virgins inherent to the mattron spin aims in all universal direction as the twins are separated; away from the other twin. The original neutral mattron can be part of a joint inherent spinfield of larger mattron cloud and to the extent that these coordinates are also inherent to its mass, the virgins are also push-effective in all universal directions incumbent on that orbit.

Once blocked within the recipient twin, the linear repulsion phase of these virgins (here in an allied sign) is active along a line that cuts the center of both the singular twin mattrons and they do not suffer the drag of half an unaffected mass, being pushed faster by half than a neutral mattron.

Within a recipient twin mattron, the birthspin of harvested virgin emission is activated into primal orbits. In the time the split is being consummated, newborn with opposite spin supply the other singular twin with new unmatched inputs for its future collective spin.

The earlier collective spin of the neutral mattron around a common coordinate center is still an inherent part of each mattron twin as they move out of that center but nonetheless a minuscule movement compared to the new (opposite) collective spins independently acquired.

While the adverse twins were united, their masses had matching opposite spin in a single coordinate center of the neutral mattron; in other words, the mattron twins were from the beginning two independent units that cohabited in a single coordinate system. Governed by collective coordinates, the twins move outwards from the center of an empty common coordinate system; not that dissimilar from the split of a spinning primal particle.

Look at the split from the perspective of the twin mattrons (like the rest of the world did not exist); the virgins driving the split are endowed with the collective spin coordinates of the original neutral mattron at the same time as they are infusing new unpaired primal spin coordinates into the other twin where these now clash with the old coordinate system.

If the twins could be magically moved apart (without emission virgins) they would follow their inherent and identical collective coordinates; they are not going to start to start orbiting each other. It would take no input for either to orbit ever faster in relation to the Universe according to the inherent field of spin from which center they have been (albeit magically) dislodged.

There is however no magic involved; each twin is driven away from its earlier common coordinate center by mutually exchanged linear virgins at birthspeed; the repulsion that dislodges the mattron twin masses and sends them packing. When harvested, the new emission that drives the split alters the old common spin coordinates (already in 10^{948}ãt).

The relation of each mattron twin to the old common coordinate system is disturbed on two counts; the linear push away from center by newborns at 1¢ and by the orbital push of the same unpaired activated newborns at 10^{316}¢ (across 10^{632}ãr). The latter orbital inputs are in opposite directions and do not much alter the original status quo of collective spin coordinates.

The overwhelming push that splits a neutral mattron is delivered in one universal direction at the time when the twins are joined at the hip. No comparable repulsion is available at further distances and the linear motion implanted at close range carries the day.

The primal shower or stray wind that caused the split was seen to advance on the original mattron along an orbital path defined by say half the neutral mattrons 10^{-912}¢ collective center spin (it takes 10^{912}ãt to rotate once). By reasons of relativity, a shower pushes away from the mattron center in that orbital mode in accordance with the involved masses. The inherent 10^{-280}¢ surface spin speed is however almost imperceptible against the linear speed of the incoming strays that instigated the split.

Nevertheless, whatever tiny part that the stray shower ultimately has in the neutral mattrons final split, the rejected mattron twins move in perfect accordance with that (tiny) orbital motion. The orbital push of the stray wind upon the mattrons gargantuan population is an insignificant drag on its inherent coordinate system within which neither twin is at home anymore and that marginally affects their relative future motion. As the cohabiting masses are pushed out of their earlier common center situated between the two, both masses are not 'orbiting' their earlier common center.

The ultimate splitting speed of a neutral mattron is determined by primal emission from the mass delivering the push and the 'inert' opposite mass set on the move when receiving the push. The separating singular masses are therefore under all circumstances perfectly equal.

The size of a split neutral mattron does not affect the splitting speed any more than mass does. Although the gathering ability of larger spheres may block more emitted virgins during longer split consummation, the universal linear focus is pushing larger and more diffuse mattrons in more numerous contrary directions and linear speed is not affected.

For example, because of linear conflict, precursor splits do not experience acceleration of their initial 10^{-280}¢ speed away from the adverse twin. The speed gained by the mutual repulsion is constant without regard to size or bulk of the neutral mass; the generation of new virgin mass is always governed by the creation constant 'ƙ' (by 10^{-1192}). The bulk that emits the repulsion during a split is always equal to the bulk that blocks it so that the mass number cannot affect the splitting speed either.

This mattron split can be termed 'electromagnetic' and the two singular mattron twins of opposite mass are its smallest conceivable package. The resulting opposite linear motion of twins is always the same and represents 'the speed of light'. Of course, as the Universe is always accountable for half of the relative speed (even in two opposite directions at the same time), the speed of light is always constant from any Universal point of view.

The speed of light is better qualified as a constant than its instigator the birthspeed that is also a similar 'average constant' (quite a contradiction in terms). Still, the vast number of atoms that accelerate the twins primal mass to the collective speed of light makes for a vastly more stable 'constant'.

This may be the first direct connection of the Primal Code to the 'real world'; the basic neutral mattron model and 'ƙ' (10^{-1192}āt creation constant) sets the speed of light as 10^{-280}¢.

The 'split' of a neutral mattron takes 'time' and the propelling force by the mutually harvested virgins is uneven during the split as the twin masses are propelled at increasing speed across the original neutral mattron radius. In the early stages of a split, gravity shadows of neutral mass between the two singular bulges may also marginally slow the propelling force. After the twins separate, they do not block virgins in the same sign and therefore cast no effective backhand to gravity shadows upon each other.

The propelling force is weakest at the beginning when the two halves are initially pushed apart by small orbs with the main neutral mass situated between them. There is a gradual buildup that culminates when the halves are completely separated, side by side. After a lengthy separation process the actual splitting speed is set by the close unadulterated and mutual virgin emission from each fully separated twin.

The period of 'side by side' twin masses, whatever the buildup period, is when the respective repulsive virgin majority harvest is soaked up; during which the linear separation speed reaches 10^{-280}¢.

During standard 10^{912}āt, the acquired 10^{-280}¢ speed transports each twin away from the other twin across a distance of 10^{632}ār standard size, adding another diameter between them. This causes more adverse virgins to miss the other twin and the next 10^{912}ãt period supplies less virgin push. There is negligible acceleration of the initial speed (I let it cover the less than tenth of virgins that miss standard twins side by side).

The singular twins reach a relatively leisurely speed of 10^{-280}¢ in a single Universal direction. I am, as you have gathered, not partial to mathematics as a method of understanding and sweeping descriptions in a field where all is precisely interconnected merit confidence in philosophical outlines only. I don't know where this logic is going but as 'electromagnetic' twins are built of 'material charges with heavily charged adverse emission', it feels odd to consider them as 'photons' that are supposed to be without charge.

We cannot view electromagnetic twins as primal particles that move away from a spinning parent in a universal direction where their inherent orbital coordinates cause the virgins to follow their original common inherent spin. For departing atoms, their equal non-boosted linear speed and orbital spin brings the atoms to cover outside fields with average boost of the square root of the distance measured in primal radii. This is also true of electromagnetic twins except that their inherent inputs of linear and spin speed are seriously unequal out of the starting gate.

With this scenario, mattrons of totally singular charge take their place as a new phenomenon on the Universal stage and their presence will be felt for all times to come. The splitting of a neutral standard mattron in a mattron grid is nonetheless utterly improbably because the general lack of disparate stray winds in such grids cannot set it off.

I assume that almost all electromagnetic splits of neutral mattrons take place outside mattron grids in extensive voids and only if an when they face disparate stray winds during otherwise insufficient pressure in one sign with intermittently expanded neutral precursor halos engulfing them.

Even after an electromagnetic split, singular twins must eventually keep singular coronas and singular precursor halos to support their balance in the void. The split and departure phases of electromagnetic twins present a complicated and fluid exchange field that alters in various ways before the twins are out of mutual conflict.

A precursor halo in a void is made up virgins of both signs and harvests disparate boosted virgin stray winds on behalf of the central mattron and in various opposite directions (a halo cannot permanently hold onto activated flux). When blocked by a neutral mattron, the linear aim of such virgins that constitute a wind is in the same universal direction away from a distant emission source (these are usually in both signs but one can ignore a sign that is pushed exponentially slower).

Pressure wise, the harvest of passing virgins from the void is negligible. Their primal spin is activated by the fast precursor flux and the orbital flux shower eventually deposits its foreign coordinates into central mass to blend with massive secondary emission of precursor virgins that generates the flux that keeps central mattron to quarters. Local flux arrives from 'all' directions and imparts no linear push to the central 10^{632}år wide mattron.

A standard precursor is built of shallow mass in a neutral blend, driven in 10^{960}åt as flux at 10^{-208}¢ outwards to a width of 10^{752}år to balance with the great spatial presence of 10^{-64}¢ parental sweep. Before the central mattron can harness directional virgin winds or secondary halo virgins, the 10^{888} strong precursor (or less for a lighter mattron) activates and redirects them.

This would be the most common type of electromagnetic splits. The twins are initially two elements within overlapping relatively immense adverse precursors that may continue to affect the unfolding situation after the twins are fully separated. This most common split does not lead to a local static environment as the twins and the precursors continue to affect the local field exchanges for some time.

The Electromagnetic Threshold

Under any circumstances the split of a neutral mattron facing a disparate stray wind takes time and I am referring to distant constant emission which the recipient neutral mattron is not orbiting at high speed; it is more or less at standstill in relation to that disparate source.

The linear repulsion focus is important; I am assuming that if a mattron at standstill is being pushed away by a harvest of virgins with sideways boost that is greater than their linear speed, the linear push is reduced by square root of the distance; namely by the boost. If the mattron is also orbiting that same emission source the effective linear long range push is reduced to the square root again; namely to square root of the boost.

It is not merely for clarity of illustration that I place the shifted mass within a large mattron corona into two bared orbs but also out of necessity. for example, the harvest of a distant disparate virgin wind is always a tiny fraction of the secondary input by a precursor halo. A singular orb of each adverse globular twin at its center must come to mutually protrude before a split can take wings, which means that the magnitude of the mattrons linear harvest must reach a certain threshold level.

With no orbs being bared, the mass allied to the dominant stray wind that has been shifted marginally in that linear direction within the mattron bulk is carried with the collective rotation to the neutral mattrons other side. It is not that allied mass receives an antidote to its earlier linear push but rather that the immensely greater internal emission in both signs within a mattron can communicate the achieved movement to both signs.

With too small orbs laid bare, overlapping mattrons also start to gain each their own rotation and each orb is periodically carried into the main neutral mass where more is communicated to both signs. If the orbs are too light, they cannot mutually blow the twins asunder.

To generate an electromagnetic split within one rotation, the outside wind must linearly push one cohabiting mass sufficiently out of sync so that the bared adverse orbs generate enough mutual bipolar repulsion to accelerate the linear repulsion during one rotation.

If that level is reached; if the shift is great enough to form massive enough singular orbs in half the rotation period, it does not help to turn the other cheek. A fast enough linear motion cannot be turned into mild polarization by overlapping mattron rotation and internal splits cannot communicate the decisive linear motion to the other half of its adverse mass in that time.

The dynamics of a powerful disparate virgin wind that forms singular orbs dictate that incoming linear repulsion units harvested by a neutral mattron shifts allied mass at corresponding linear speed and the singular mass thus laid bare generates exactly the number of virgins that match the repulsive units deposited by the stray wind. The internal bipolar thrust is undiluted by directional diffusion and is also more independent of the remaining rotation of the neutral bulk.

The internal communication of linear movement in one of the two signs lags behind by two factors. An activated virgin and its flux must repeatedly engage with a great number of atoms among its allied population before its orbital drive is brought to the relative harmony of 10^{-244}¢ parental spin. First thereafter will the linear move be communicated to the adverse population in accordance with the 10^{-280} creation lag per ãnna.

The difference of a 10^{402} slowdown to 10^{316}ãr parental radii (from 10^{158}¢ boost in that orbit to 10^{-244}¢ parental movement) is 10^{50} times greater than the 10^{352} virgins that in a standard mattron relays enforced movement to the other sign in radius time.

Thus my offhand estimate of an 'electromagnetic threshold' is that when during 10^{912}ãt, a continuous incoming linear stray wind delivers $10^{352} \times 10^{50}$ or 10^{402} singular virgins in the same linear direction into a neutral precursor in a void, its central mattron mass becomes hardwired into building a bipolar split of its own.

I shall use this lag in the adverse sign combined to the mattrons internal slowdown as an illustration of the critical linear virgin harvest that must be eclipsed to set off an electromagnetic split. After a critical threshold harvest of 10^{402} linear virgins is reached by the incoming stray wind in 10^{912}ãt, the already bipolar mattron has crossed the electromagnetic threshold and has no independent way of thwarting its inevitable separation.

A linear 10^{402}pu virgin wind harvested by a precursor within 10^{912}ãt in a void represents the threshold for a bipolar electromagnetic split.

The dynamics are that the small shift of allied mass into orbs achieved by a virgin stray wind within 10^{912}ãt releases a perfectly matching bipolar push; in the next similar period, the additional push from the orb gives twice that of the former short period, then four times, etc. etc.

The rising internal repulsion between the charged areas becomes ever more singular and ever more effective in kicking its mirror image mass away until it ultimately triggers an electromagnetic split of the two adverse halves. The final repulsion turns devastatingly effective when the twin areas are only joined at the hip and then driven by the full force of the other twin's linear up close adverse virgin emission.

An electromagnetic split fashions an intricate and fluid field where blunt illustrative tools are woefully inadequate and the split is important enough for some tortuous conceptual points on the interactions that play out and must be taken into account when the field is treated mathematically.

The Electromagnetic Cycle

It is the central mattron with its massive and stable mass that is split; not a fleeting corona during an expansion or a longer lasting 10^{752}ãr precursor. There is also no acceleration of that splitting speed across larger inter-twin distances in longer periods with more adverse emission gathered.

If 10^{402}pu linear virgins in 10^{912}ãt set 'all' mass in one sign moving at 10^{-510}¢, it uncovers a 10^{402}ãr wide orb on each side of a 10^{632}ãr wide central mattron, laying bare 10^{682} atoms of its standard mass in each singular orb. This orb mass will in turn generate an initial bipolar push of 10^{402} virgins in 10^{912}ãt; rising to 10^{632} virgins during the final split.

The average push during the entire separation period is by 10^{517} virgins and the out of sync gestation that leads up to an electromagnetic split takes much longer than an ultimate split of side by side twins.

Since the linear push by singular bipolar virgins is on average 10^{517}pu and shifts two 10^{912} masses in opposite directions at 10^{-395}¢, the push across a 10^{632}ãr takes 10^{1027}ãt (spanning many intermittent precursor halos) while the subsequent electromagnetic split will be executed in 10^{912}ãt thereafter.

During a split, a neutral mattron meets its compression needs by neutral precursors. After the electromagnetic split, intermittent blended precursors of various sizes overlap at first, affecting the twins until they reach the edge of the overlapping 10^{752}ãr wide precursors at light speed after 10^{1032}ãt.

The aim from a parent particle remains (more or less) uncut as long as it does not leave its parental orbit on activation. The 10^{-244}¢ parental sweep in 10^{316}ãr orbits within a central mattron allows virgins a boost to 10^{-26}¢ out to a 10^{752}ãr precursor. Activated by allied precursor mass (in orbits at 10^{-64}¢) or by its secondary virgins, the flux has lost most linear aim of the driving virgins when harvested after myriad collisions.

A precursor that cannot stop fast flux on its own is also likely accelerated away by activated emission from the central twin and secondary emission by faraway precursor mass is less likely to reach back to a central twin. Closer halo mass takes its place and initial adverse emission from a twin alleviates some pressure loss by secondary emission but there is much precise work to be done on mattron stability during an electromagnetic cycle.

While the 10^{-280}¢ speed separates the singular twins with their full active mass of 10^{912} atoms further in the void, each twin is soon forced to form a singular precursor to provide for its own stability.

The 10^{1027}ãt out of sync gestation that leads up to an electromagnetic split affects the forming globular orbs according to sign, pushing one faster away than the other. During most of 10^{1027}ãt, this linear push is communicated in part to the other mass. Each orb is part of a main bared singular mass that has been shifted out of its original center with opposite spin although they at this time mostly occupy the same space. The spin causes the orbs bared at each time to periodically enter the neutral majority mass in between the two.

The linear direction of a disparate stray wind will aim the linear direction of an electromagnetic split by aligning the orbs on the standard mattron (and its precursors). Although this is downsized by communication between the signs within the main mass, it is nonetheless effective during the 10^{1027}ãt gestation to aim the orbs towards and away from the source of the disparate emission. The ultimate electromagnetic split in opposite directions during 10^{912}ãt will then aim along this single Universal line.

The Singular Mattron

The singular mattron starts out as a photon propelled away by an identical opposite twin at the speed of light has several of its fields compromised. The singular mass of standard 10^{912} atoms can only harness half the strayfield pressure of a neutral mattron as the other part is blocked by adverse mass and then communicated to the rest of its mass by adverse emission.

The fields of collective spin have effectively doubled in speed in a recovered mass. Collective spin in a neutral mattron comes about through the effect of the unpaired direction of one atom working upon all the mass (both signs) where a single unpaired direction at 1¢ was eventually shared by 2×10^{912} atoms (of both signs) and geared down to 5×10^{-913}¢ at the center of a single unified coordinate system.

Now each unpaired spin direction needs be shared by only half the earlier 10^{912} mass, which is all there is. The collective spin of a singular mattron allows a single activated newborn of the allied sign to rotate its entire mass at 10^{-912}¢ (at the mattron center). To glean the parental spin from this faster 10^{-912}¢ collective central spin in a singular mattron, I increase the average central parental spin in each individual coordinate system by the product of two values.

First the number of newborns split in 10^{632}ãt radius time that, now unpaired, constitutes the average addition of 10^{352} newborns in the allied sign in that time. Second the speed of ongoing boost in average local orbit. I set the local orbit entertained by a parent as 10^{316}ãr with actual orbital movement having been boosted on average 10^{158} times.

The individual parental coordinate systems increases the central input of collective spin by the product of the two values, making parental spin $(10^{352}) \times 10^{158}$¢ $= 10^{510}$ times greater than the collective spin.

While a central atom in a singular mattron spins at 10^{-912}¢, the atom at the center of the average parental coordinate system orbits 10^{510} times faster in a central coordinate system at 10^{-402}¢. Using an average 10^{316}ãr orbital radius that is boosted 10^{158} times from that coordinate center to 10^{-244}¢; the parental movement is the same as within a neutral mattron.

Contrary to the twice faster collective spin, the parental spin is the same as in a neutral mattron, albeit here all in the same direction. The chance of escape for allied newborns is therefore also 10^{-36} and singular mass blocks allied strays at the same average 10^{36}¢ speed, given same size and mass.

The internal emission within a neutral mattron is twice the emission within a singular mattron but the now halved emission works to expand the allied singular mass that is only half the number. Because of the opposite internal utilization of emission in both signs, the level of internal pressure is relatively stronger in neutral mattrons, which affects their mass distribution and density.

It is important not to simply view a neutral mattron as two cohabiting singular mattrons; each element of charge is affected by adverse emission of another intertwined but adverse singular element.

Although four times as much internal emission is utilized in a neutral mattron than in a singular one, the balance is complicated by the different physical presence in each sign that makes for a different blocking quality for its two components. Albeit marginally smaller, it suffices for my purpose to say that a singular mattron keeps the maximum 10^{632}ār size broadly intact inside its precursor halos.

As central pressure and density is mostly governed by internal emissions, the central density in a singular mattron is probably lesser and the singular mass more evenly distributed than within neutral mattrons. Although the chance of escape for an allied newborn remains 10^{-36}, internal production of newborns in the allied sign is effectively halved and escaping emission in its singular sign therefore also half compared to that of a neutral mattron.

Aside from these marginal changes, the most distinguishing feature of a singular mattrons is that all its emission in the adverse sign escapes into the local strayfield. This dual aspect of a highly charged emission field that also drives its external collective spinfield is significant.

It is easy to appreciate the disparity in escaping emission in the two signs compared to the originally neutral mattron. After the split, the two singular twins have 10^{36}-fold standard emission (less if they suffer mass annihilation in any compressed neutral phase between the bulges). Early on, after a split, escaping emission blends to neutral in a local field but as the distances between the twins grow, the singular emission of each twin 'charges' local fields it passes though.

The larger strayfield is fairly stable. Its enormous source of straying atoms (from neutral and singular mattrons) turns more neutral the bigger it is and where directional strays arrive orbitally boosted from faraway to pass linearly through local spaces. The penetration of orbital emission in a mattron grid dwindles rapidly with distance.

Relatively speaking, large primal collectives are situated 'far apart'. The closer an emission generator (such as a singular mattron); the more it 'charges' a local strayfield and the vicinity of a singular mattron is always significantly charged with pressure in its adverse sign.

Their gravity field has this dual aspect; residents in one sign block only allied strays and cast a gravity shadow in that sign but none in the other. The combination of no gravity shadow and totally escaping emission in the same sign affects nearby mattrons by repulsion with a collective spinfield.

To stop the relative movement a singular mattron twin moving along at the speed of light requires a similar level of opposition that sent it packing. Large mattron grids are far more likely to stop them than individual mattrons, even two adverse singular mattrons scoring a direct hit on each other.

A singular mattron has no ongoing central annihilation and when it grows too massive, it may turn unbalanced. It can in principle emit excess mass in intermittent burst of slow moving allied halos, although some of this mass must be re-harvested.

We can in our mind-eye see wavelike singular halos enter the mattrons perpetual adverse emission field, intermittently altering its composition of charge. A massive slow moving allied halo superimposed over a fast but perpetual adverse emission field where, during the collective emissions of an overweight singular mattron, its emission field alternates between the two charges.

Although halos take extensive periods to form and fade, they may eclipse the constant adverse emission field if much of their mass is lost during their existence. No annihilation awaits the collapsing center of a singular mattron. Otherwise the collapse cycle is the same as that of a neutral mattron.

After the increasing initial stream of straight-line newborns that precede the termination of outward emissions from a central core, if halos cannot get rid of the mass (without generating excess pressure) the singular core collapses.

It is less likely that the unbalance lends wings to an explosive expansion after a central collapse that would also blow the singular mass into a vast halo where the singular mattron regains balance after losing a great deal of its mass into the strayfield.

With allied atoms compressed into a ball of little more than their combined volumes, a collapsing central region is vacated and turns more transparent. One likely scenario of mass loss is therefore that the 'tiny' but massive black hole at its center can more easily be kicked clean out of the mattron during periods of great instability; as the core responds like a single primal atom.

Such a loss would essentially result in the same level of mass loss that is achieved by a mattrons constantly escaping adverse emission. Although any allied mass loss is of a temporary nature due to the continuous creation, a singular mattron can in theory get rid of the equivalent of million mattron masses in one go through losing its central death clot.

Polarization

The ordinary split did not happen in the earliest widest mattron grids and, if it did, polarization of nearby mattron by their adverse emission would be locally isolated and very short lived.

A polarized neutral mattron has one sign compressed closer to its center. That inner more compressed sign is engulfed in an outer corona of the other sign. Polarization of neutral mattrons by a fluke stray wind could also have taken place early on in a few local spots but cannot be widely communicated on the Universal stage.

The problem with stable polarization at an early stage, by whichever way it arrives, is that conditions are not ripe at great grid distances. They generate a boost that has neutral mattrons teetering on the brink of dispersal and can easily render them into primal clouds of vast inter-mattronal size.

Sustainable mattrons cannot form until enough primal mass is situated at distances that let them to manage the tear and stresses of inter-mattronal flux effectively enough to form a stable submist.

In a stable mattron grid at smaller distances, the initial dynamics of a freak and heavily charged stray showers (from many directions) dictate that the more such straypressure is initially delivered upon a neutral mattron; the greater the initial width and mass of the 'charge' layer of coronal mass that spreads over its surface.

Whenever an extensive charged corona forms on a mattron surface, this 'coronal' mass can continue to capture its own allied secondary newborns while adverse newborns increasingly escape the outer layer and the mattron or are blocked deeper within it, moving adverse mass inwards. This speeds up the ethnic cleansing of the charged corona but is not boosted sufficiently to compress the interior mass (as would a distant exterior halo).

It is important to note that the annihilation volume of the two signs does not overlap in the same way as in a neutrally compressed mattron where the central mass density is identical in both signs. The mass in both signs is still being annihilated in accordance with the shrinkage of the coronal radius.

For central mass annihilation, it takes two to tango. The 'denser' charged central mass in one sign experiences greater cluttering but with less mass present in the other sign, the additional annihilation is less robust.

In a heavily compressed charged core (or singular mattron), the atoms in the two signs within a collapsed center do not clutter to the same degree. It is only at the perfect center that the more compressed core atoms can clutter to such an extent that there are no free paths away from the parents and they cannot procreate. Fewer of the compacted coronal atoms are housed permanently within the death clot and cannot therefore annihilate although many of them pass through and also cohabit within the core that is not singular, albeit heavily charged.

The remaining core atoms are actually equal in number to coronal atoms but the majority of the core mass is housed in a central death clot where it cannot procreate and is effectively put out of commission.

If a fluke charged stray shower in some corner of the Universal grid of mattrons is powerful and slow enough to cause a broadly charged corona in one local mattron, that single corona emits enough adverse virgins to inflict a similar disparate compression upon great many of its neighboring neutral mattrons.

While inter-mattronal distances allow, these mattrons perpetuate the same shift in polarity outwards within that tight mattron grid.

Polarization changes the outwards demeanor of a neutral mattron radically and in turn of great many of the surrounding grid mattrons as they respond to the first mattrons charged coronal emission. Disparate compression that arrives locally from many directions does not need extensive periods to work and compared to electromagnetic splits, such improbably shifts are in one sign. As charged straypressure pushes allied mass deeper, the differential collective mattron spin smoothes the disparity to an average depth across the mattron surface.

Let me use an excessive initial compression to illustrate; say that a neutral 10^{632}ār standard mattron within a 10^{776}ār standard inter-mattronal grid has one mass sign compressed by a charged stray wind to say a hundredth part of its former sphere. With primal mass averaged upon its exponential radius, a core sign compressed to ten times smaller 10^{631}ār radius while the corona keeps its original 10^{632}ār radius; the left behind ten times wider corona still contains the lion's share of the mattrons primal mass in that sign.

I name the allied emission of singular coronal mass subservient as fewer of its newborns escape the corona. I call the freely escaping adverse emission in the coronal sign dominant.

While the minority coronal mass can activate its virgins, the weak coronal boost harvested within the double system does not add much compression upon the charged inner core. Even if all 10^{352} radius time newborns were boosted to 10^{316}¢ within the corona (which is not the case), they only yield 10^{668}pu in collective pressure, which pales against the charged stray pressure that is herding one sign inwards within the mattron.

As the outer corona turns ever more singular, the number of minority virgins in parental orbits that find no collisions grows and this insignificant pressure drops further. With all minority mass eventually flushed out of the corona, adverse virgins are no longer activated within it. If the core in their sign blocks a hundredth part of their horizon, with 10^{-86}¢ parental sweep, the adverse virgins have one chance in a hundred to collide into the central core while most miss.

With adverse emission increasingly escaping a singular corona; away from a disparate mattron, virgins activated on an inter-mattronal trek are boosted to 10^{388}¢ between mattrons in a standard 10^{776}ār grid where the adverse flux is easily harvested and causes similar polarization. Totally escaping adverse coronal emission increases the pressure upon a shrinking core exponentially and ultimately delivers 10^{36} times more pressure between grid mattrons and heavily compresses cores. A disparity can spread like wildfire through vast local grids of neutral mattrons until it comes upon greater inter-mattronal distances or voids that block the spread.

My preliminary estimate is that no stable polarization of mattrons can take place until mattrons have gathered into grids of less than 10^{805}ãr average inter-mattronal distances. Compared to the greatest 10^{824}ãr grid distances that can support neutral mattrons, disparate grids are relatively tight.

There are two phases to minority atoms in a charged corona, the ratio of which affects the character of disparity. Self-activation of minority newborns is possible as long as the corona has too little minority mass to block them but enough to activate them. The minority is activated within the corona and sets the more meager coronal mass on the move in orbits that are gathered in the mattrons lower stratosphere where mass in that sign is more plentiful.

As the corona turns more singular, adverse newborns split by majority mass deliver less activated and more merely parentally boosted pressure upon the central allied sphere, until most miss it.

Although a singular corona contains the overwhelming majority of active atoms, the highly compressed 'central' sphere still holds a small fraction and the core is relatively opaque in both signs. As the disparity deepens, the core contains ever fewer active atoms in both signs but the rest of the coronal sign cannot be flushed out of a joint space which is situated at the center of both partly cohabiting masses. A this small subcloud at the perfect center of a vast singular adverse corona turns heavily charged, its separation cannot turn completely singular like that of the corona.

The allied emission escaping the core does not interact with the engulfing coronal mass but is boosted across the width of the corona to exit along with the greater number of adverse virgins in their sign split by coronal mass. The core emits adverse virgin in its minority sign that interact within the corona and are mostly activated and blocked and may augment its expansion edge.

The core has slower collective spin in its majority sign but completes one collective differential rotation earlier than the corona. Due to the constricted communication between the halves, this is likely to be in a direction that is opposite to the rotation of the corona.

I do not intend to pursue these inner balances in great detail; they are in the realm of mathematics. It suffices to say that although the escape rate of fewer newborns split within the inner core grows and their focused outwards push from center may be effective in expanding a vast allied corona, they do not pack a wallop. I assume that the hidden central core is not an influential component in a heavily polarized neutral mattron.

The separation of two signs into a singular corona and a small charged center affects a neutral mattrons linear long-distance gravity shadow. The corona is opaque to strays in one sign but constitutes no hindrance for strays in the other sign that pass right through it. Slow long distance virgins in equal numbers from background strayfields cause a polarized mattron to cast two gravity shadows. A large opaque corona casts a darker larger linear gravity shadow than a smaller central sphere; it is affected by both greater size and the better blocking ability that goes with bigger subcloud.

With the same allied emission, disparate standard mattrons can keep full coronal mass and size and harvest 10^{316} allied newborns in 10^{632}ãt radius time in a standard grid where the next mattron corona across the average 10^{776}ãt distance covers 10^{-288} of the sky. The greatest character change is that in polarized mattrons, adverse newborns split in the corona pour out of it in numbers closely matching those of a singular mattron.

A heavily compressed singular death clot, the former main mass generator is now a poor retinue where every outwards path is blocked by innumerable atoms in the same sign. It is best described as a singularity that has largely the character of a single atom.

An insignificant part of the 10^{2328}ã space allotted to a single primal atom within a standard mattron can probably stove away all primal mass of the entire Universe in a death clot.

The Primal Code shows that relative dislocation of colliding atoms is tied to distances closed, not the number of pushing atoms. A single atom can as easily push a 'solid' death-clot with all distances closed around as it would a single residential atom. It does not matter if it contains all the other atoms of the Universe.

The relative movement of this atomic ball is not a property of its mass and it has no inertia. Although a death-clot in a standard mattron receives flux bombardment from all directions, and massively so, it can zigzag in a mild highly erratic fashion within the mattron center due to primal uncertainty.

The death-clot is not a significant influence; it throws a more distinct shadow than a single atom because its sphere blocks more strays but cannot return the compliment by emitting newborns. Its ability to block strays is a marginal part of the mattron shadow; set by the extent of its tiny collective sphere within which all that 'compacted' allied mass is hidden.

There are therefore no paths of death-clot width out of the larger volume of the main core that are free of residential atoms; making it less likely that a death-clot can be kicked clean out of the mattron, which otherwise is quite feasible. But a straight line stray that crosses the mattron at high speed can easily move a death-clot along and often push it apart to leave large chunks behind as the rest floats away into the wider core to be collided asunder or pushed back into the death-clot. Allied primal atoms are in no way 'attached' and collisions at opportune angles can divide a death-clot into parts that are in turn ever more easily collided asunder if the compression is released.

The internal central collapse in heavily polarized or singular mattrons does not provide permanent solutions to excess weight. A death clot that ends up in more open or less pressurized areas is dispersed with relative ease. Until then, it remains a massive lump from which nothing escapes.

A local neutral strayfield may be erratic for a variety of reasons; collective emissions of neutral or singular mattrons can cause the pressure to arrive in pulses instead of a steady flow. There may be fluctuations in spatial boost due to the rotation angles of singular planes of spin; mass passing through such planes experiences vacillating pressure.

Generally, the smaller a local field, the more susceptible it is to erratic behavior and shifting levels of charge. In any large chunk of the strayfield, space includes an immense assortment of mostly neutral mattrons with a tiny portion of singular ones. Taken together, their emission represents a perfectly neutral strayfield.

As a singular corona cannot provide enough internal boost to compress its own core, *neutral mattrons cannot keep their polarization in a void* when their cores expand again. In a void, both signs expand towards greater coronal size where extensive precursor halos yield enough secondary emission and boost to keep them individually balanced without exchanges.

Although not annihilated within the core, much of the cluttered central mass is put out of commission by disparate central density. As before, in a field of polarized mattrons, there is direct annihilation of both signs in step with a shrinking coronal radius. On top of that, the charged core suffers decommission of mass in step with the square root difference between the coronal and core radii. Decreasing compression adjusts core sizes upwards by the same square root difference and releases primal mass from a death clot at the same rate.

There are other changes than relative spheres in store for effective cores in polarized mattrons. Parental spin is dictated by central collective rotation of commissioned mass, boosted by the square root of radius and complicated by the newborn input in radius time. More virgin coordinates that affect parental spin will arrive as adverse coronal newborns than newborns split within the core itself.

In grids of large inter-mattronal distances, less effective core blockage leads to longer flux consummation and more inter-mattronal core flux in flight between two disparate mattrons. As the level of free flux rises with the distance, the secondary emission by free flux on the wing increases. At far distances, this field emission exceeds allied coronal pressure exchanged by the mattrons. When that happens, the corona starts to collapse as the core disintegrates with the disparate mattrons dispersed into the inter-mattronal void. Disparate mattrons cannot be sustained as structural entities in grids of more than 10^{805}ār.

The center of the universal grid that first condenses to less than 10^{805}ār in inter-mattronal distances has implications for the Universal mist as a whole. From there on down, any experienced disparity is ever more stable.

Disparate Rules

The rough method I use for disparate spheres forged by harvested pressure is to use the standard mattrons 10^{1264}ā coronal sphere as a benchmark to set all other disparate sphere sizes.

The main compressive effect that compresses spheres is inter-mattronal boost that is set by the grid. Using just the boost is of course incorrect as it brings different mass annihilation with mass put out of central commission. To have a broad rule to set *coronal* mattron spheres without prior knowledge of their ultimate masses, I use a shortcut that gives a rough outcome that can be tightened up with the facts in hand.

For mattron *cores*, the same rules apply by using the different grid boosts in concert with the additional emission input that is also likely to compress the cores by adverse emission steaming unimpeded out of coronas.

Standard neutral and singular coronas exchange 10^{316} allied newborns in radius time, boosted to 10^{388}¢ across a standard 10^{776}ār grid distance which delivers a mutual 10^{704}pu collective pressure that sets a 10^{1264}ā standard coronal sphere that is my benchmark (1).

For coronas, any change in grid boost brings a different and preliminary pressure sphere that due to the ongoing decimation of active mass is only academic. To account for the mass change, the actual sphere is smaller by the square root of the radius change in the *pressure* sphere compared to the benchmark sphere (1).

An example; polarized mattrons in a wider than standard 10^{784}ār grid have coronas that are compressed by 10^4 times greater boost. The new *pressure* sphere would then be 10^4 times smaller or 10^{1260}ā but this is downsized by $\sqrt{\sqrt{}}$ of 10^4 due to the diminished emission exchange. The new coronal sphere is therefore ten times greater 10^{1261}ā with a $3,16\times10^{630}$ār coronal radius.

We can look at this actual corona that would have $3,16\times10^{910}$ atoms where $3,2\times10^{33}$ fold adverse emission of $1,78\times10^{315}$ emitted newborns is boosted to 10^{392}¢ across the grid and delivers $1,78\times10^{707}$pu which is $1,78\times10^3$ times greater than the benchmark pressure. This actual mass emission number would fashion a

$5{,}62\times10^{1260}$ã sphere with 1,33 times smaller $2{,}37\times10^{630}$ãr coronal radius; close enough for illustrative purposes.

Disparate cores across the same 10^{784}ãr grid are affected by the same 10^4 fold boost times the adverse coronal exchange from $3{,}16\times10^{910}$ atoms which is $3{,}16\times10^{34}$ fold standard allied emission. This delivers $3{,}16\times10^{38}$ fold extra compression and a new pressure sphere would be $3{,}16\times10^{38}$ times smaller than the 10^{1264}ã benchmark sphere or $3{,}16\times10^{1225}$ã that must be downsized by $4{,}22\times10^9$ ($\sqrt{}\,$ of $3{,}16\times10^{38}$) due to annihilated mass. This yields an actual sphere of $1{,}33\times10^{1235}$ã with $3{,}65\times10^{617}$ãr radius and $1{,}07\times10^{904}$ emitting atoms. The rest of the core mass (the difference compared to corona mass) has been put out of commission in its central death clot.

In a standard 10^{776}ãr grid of stable disparity; standard mattrons of 10^{632}ãr radii have 10^{912} coronal atoms which 10^{36} fold adverse emission compresses cores 10^{27} fold to $3{,}16\times10^{618}$ãr with $1{,}78\times10^{905}$ actively atoms where all but $1{,}78\times10^{-7}$ of their atoms are put out of commission in a central death clot.

Disparate Fields

Rare as they are, singular mattrons with their mostly singular emission are a new decisive factor on the universal scene and a powerful addition to the local inter-mattronal straypressure. Created in equal numbers, they allow the Universal strayfield to remain neutral; even if singular mattrons of one sign would be destroyed, their primal masses remain.

Nevertheless, a singular mattron charges its immediate surroundings with freely escaping adverse emission and highly disparate straypressure. The reaction of the vastly more numerous neutral mattrons in the local strayfield is therefore of great importance.

A neutral mattron in the vicinity of a singular (or highly disparate) mattron is subjected to vastly more strays in one of the two signs bearing upon its surface as newborns stream unhindered out of nearby singular mattrons in the adverse sign (or out of a disparate corona). The corona is the place where the formation of a charged layer is most expeditious.

The collision of a primal atom takes it to wherever its coordinates and those of its collision partner dictate. In the smallest stable mattrons that assimilate strays, there are therefore no permanent internal forces to hold the two separated layers of internal mass charge apart. I leave the possibility of layers in larger than standard mattrons aside for now.

In the absence of incoming disparate pressure, the two signs are easily collided back together again by simply dispersing into each other. The charged corona and adversely charged interior blend again into neutrality unless the outside strayfield keeps up its disparate pressure.

A point of order here; a stray charge that leads to disparate distribution of mass within a neutral mattron, as illustrated, can just as easily come about through stray deficiency in one charge as an excess in the other charge.

Thus during a drop in one sign of incoming straypressure (make it in a heavy singular gravity shadow), less of the mattrons primal population allied to the shadow sign is affected. During deficiency, the excessive expansion function steps in to bring on a similar disparity (or to increase a disparity brought on by excessive pressure in the other sign).

A neutrally or disparately compressed mattron can allow its mass to move outward while it does not cross my arbitrary 10^{632}är threshold with the exceedingly tenuous outer layers leading the field. The more robust the stray charge drop, the more charged and broader the rarefied corona.

The disparate charges that arise in a neutral mattron are much the same by whichever route they arrive; singular stray deficiency or excessive singular pressure. In both cases, a disparately compressed mattron forms a charged corona in the deficient sign and a charged interior in the other.

Thus internally adjusted, as a unit, the mattron remains neutral. In broad terms, the mass charge of its disparate corona matches the opposite charge of its interior (coronal emissions and decommissioned mass aside).

When a mattron is disparately compressed by a mild stray charge in one sign or slightly expanded by a mild lack in the other, there would be little manifestation of a singular corona or charged center. This would instead lead to primal mass in each sign having slightly different average distance to a mattrons common center.

While I leave the question open of whether a second layer of charge can under moderate pressure remain intact, temporarily or constantly, in a huge heavy mattron; in a standard mattron, a second layer is indistinguishable from its central areas.

In a mattron grid, the upshot of a corona of great initial width and charge is that in the mutual exchange, ever more adverse coronal newborns are driven deeper into the interior where that mass sign builds a second layer of opposite charge below the corona.

A singular mattron can cause an upheaval as it approaches from a void into a grid of neutrally compressed mattrons; gently beginning to turn the grid mattrons disparate from afar. Once formed, a charged corona within a mattrons grid can swiftly turn a large number of neutral grid mattrons deeply disparate in the same sign.

Aside from polarity, a mattron can easily gain smaller and fluctuating mass charges. Simply by emitting a singular coronal halo, a mattron can easily turn somewhat charged and the charge balance is skewed in various other and often temporary ways; for example by emitting freely escaping adverse coronal virgins while holding on to allied newborns.

Mass-wise (counting atoms), I still consider the average disparate mattron to be on average 'more or less' neutral.

Most grid mattrons form by shrinking to a balance where they hold on to maximum 10^{632}ãr size and 10^{912} standard masses and many eventually turn disparate one way or the other. This can in turn cause strayfields around large grids to be composed of heavily charged stray winds.

It may seem easier to split a polarized mattron than a neutrally composed mattron, given that the two spheres like sails of different sizes will be pushed differently by equal linear stray winds. However, although the spheres are formed by a stray wind and the linear thrust would then be the same for both, the empty void affords little change of disparity but calls instead for the mattrons to expand and form precursors of both signs.

Flux Consummation

I have already commented on the time it takes for exchanged grid emission to be geared down sufficiently enough in speed after activation to be blocked and harvested in an average grid. In this new and heavily charged world, a neutral field can rapidly change character and averages lose their luster.

There are so many ways to look at flux consummation that it makes your head spin. As all rests on an entirely logical foundation, I have no doubt that mathematics can put these interactions precisely. It is not really the volume of a field with infinite number of points that counts but the defined number of primal particles. As conceptual logic must nonetheless pave the way, I merely try to point out what components these equations should include.

Externally, the overwhelming emission by a polarized mattron in a grid is delivered by the same singular corona in both signs; (in this respect) one can largely forget the hidden adverse central core.

An important part of the balance is the flux consummation period. I have used consummation periods of inter-mattronal flux based on emitted virgins that after activation are carried into large orbits across grid distances that gives their flux great spatial presence. The actual presence between the grid mattrons after activation is the same for a once-hit newborn as for the vast number of slower moving flux that its multiplying collisions spawn.

In the 10^{776}ãt a newborn traverses the average grid distance, a standard mattron emits 10^{460} subservient newborns. Consummation of non-linear allied flux is 10^{740}ãt when blocked at 10^{36}¢ by 10^{632}ãr sized mattrons of 10^{912} primal atoms after a last slowest 10^{776}ãr wide orbit. In the case of singular or disparate mattrons, the 10^{496} dominant newborns during 10^{776}ãt, consummation for standard core flux blocked at $1,78 \times 10^{29}$¢ is set during a last 10^{776}ãr wide orbit during a longer $5,62 \times 10^{746}$ãt.

As the parental boost of virgins grows to 10^{-14}¢ (a fraction of birthspeed) across 10^{776}ãr inter-mattronal grid distance, boosted spatial presence does not increase interaction. Two emission fronts of exchanging mattrons meet head-on midways in a grid with a tiny fraction activated by their own kind. If this self-activated virgin fraction has greater collective spatial presence than the collective virgin majority in the same time; activation rate is jacked up.

All flux in both signs is harvested in a relative flash (10^{740}ãt-$5,62 \times 10^{746}$ãt). This would hamper activation in a grid if flux levels had not begun to climb earlier through an exponential string of consummation periods for already activated flux in that field.

I opt for an illustration of gradual buildup and let each virgin put all its potential presence into the pot. The combined presence during 10^{776}ãt then becomes the product of the virgin number emitted throughout that time and the individual presence achieved by each driving newborn; its active life on the wing; for example through 10^{740}ãt multiplied by its 10^{388}¢ boost.

With that way of measuring; the collection of 10^{460} subservient activated driving newborns where each sweeps 10^{1128}ã in 10^{776}ãt (boosted to 10^{388}¢ during 10^{740}ãt) reaches a collective 10^{1588}ã presence. A virgin in that field has a presence of 10^{776}ã. Combined with the flux sweep, 10^{2364}ã swept out of the 10^{2328}ã volume activate all but 10^{-36} subservient virgins across the field.

10^{740}ãt subservient consummation is 10^{108}-fold standard radius time in which the mattron harvests 10^{-108} of 2×10^{812}pu of the orbital flux pressure. In 10^{632}ãt, the harvest share is 2×10^{704}pu, which is the pressure I set earlier as required for standard balance in standard radius time.

The input of 10^{496} dominant newborns in 10^{776}ãt that in $5,62 \times 10^{746}$ãt each covers $5,62 \times 10^{1134}$ã or $5,62 \times 10^{1630}$ã collectively between disparate mattrons, reaches a combined $5,62 \times 10^{2406}$ã presence in the 10^{2328}ã inter-mattronal volume and activates all but $1,78 \times 10^{-79}$ of dominant virgins.

If the collective presence of all the virgins during their passage is greater than the spatial presence of the fraction of virgins that is activated for any reason, I assume that the majority of the virgins will be activated.

After majority activation is reached, parental spins of kicked out mattron mass add its coordinates to stir the soup and shift the orbital drive among ever more flux particles. Although more cross-collided, this delivers orbital coordinates at full push potential. As before, in a disparate standard grid, I assume no orbital slowdown in delivered pressure whereupon the 10^{36} fold dominant flux delivers 10^{36} fold pressure. The subservient flux in a standard grid covers exactly the available number of independent orbits before harvest and is for that reason less cross-collided.

Subservient flux across the average inter-mattronal grid is slowed from 10^{388}¢ to 10^{36}¢ after a first orbit in 10^{388}ãt at unstoppable speed and spread into large orbits by 10^{352} flux collisions mostly within mattrons. The final shower of subservient flux leaves a mattron to circle in a last largest 10^{776}ãr inter-mattron orbit at 10^{36}¢ before capture the next time it crosses a 10^{632}ãr mattron; completing that last orbit in 10^{740}ãt.

An activated subservient newborn at 10^{388}¢ is part of a linearly stopped front of the same weight. It must pass through a mattron 10^{352} times before hitting any of its mass. When it hits a local atom circling at 10^{-244}¢ in a 10^{316}ãr-wide differential orbit, those new orbital coordinates broaden its single line of local orbital thrust to 10^{776}ãr while that local speed input grows 10^{300}-fold to 10^{-14}¢. During 10^{740}ãt consummation, the orbital line of a driving newborn is broadened to spread over a 10^{726}ãr thick skin of that globular orb between the mattrons that are situated near the middle of that 10^{776}ãr wide flux orb. As mattrons can at each time exchange with a handful of other mattrons, these flux orbs are likely to interact.

Activation of an exchanged virgin majority on an inter-mattronal trek comes early. During the first 10^{758}ãt period of a 10^{776}ãr journey across to the next mattron, a virgin covers 10^{758}ã points while 10^{442} other virgins enter the field, each sweeping 10^{1078} spatial points before flux is withdrawn from the field. In that time, 10^{1570} spatial points are swept of 10^{2328}ã exchange volume and 10^{-54} or 10^{2274}ã lie within the volume radius of the traversing virgin.

Already on this first leg of its journey, 10^{-18} across the average grid, the subservient virgins face a fifty-fifty chance of activation. Continuing from there and across 10^{776}ãr, being exposed to the same activation level, they are at journeys end again left with the average 10^{-36} escape rate.

It is helpful to replace the corona of a disparate mattron with an image of our Sun while placing one molecule (holding half the solar mass) at the solar center as the core; which may be the core size relative to a corona. Do not be misguided by talk of compression; even after the mattron volume shrinks to that order, a standard core remains inconceivably tenuous and its interior provides immense spaces for each primal atom to cavort in.

The dominant flux field has $1,78 \times 10^{43}$ times more presence per ãnna than the subservient flux field. The obstacle race through these two highly disparate flux fields allows very few virgins to make it untouched out of the average exchange volume but the dominant field is the greater obstacle. The 10^{-36} chances for a subservient virgin to cross the divide (after 10^{-36} escape the singular mattron) and $1,78 \times 10^{-79}$ for a dominant virgin mean that for each dominant virgin that slips untouched out of an exchange volume and into the wider grid, $5,62 \times 10^{43}$ subservient virgins slip out and virgin seeding is clearly in the subservient court.

It would be amiss not to mention that 10^{-36} allied virgin escape based on $10^{-244}¢$ internal motion in 10^{316}ãr-wide average differential orbits takes no account of internal movement driven by activated newborns until an average balance is struck. I assume that such independent local motion caused by each activated virgin is brought to rest well within the period of passage for harvestable inter-mattronal flux and that this unrequited local movement is smaller or equal to the broad average. This however is an approximation that must be looked at; if faulty, the consequences are universal.

All of this presupposes that orbiting atoms in vast and fast orbits between larger subclouds, such as hydrons, suns or galaxies do not kick the flux that build up between mattrons away at immense speeds in numbers that can skew the results of this illustration.

Mattron Stability in Open Spaces

By virtue of past material growth and Universal expansion, at some point the primal grid gains the overall density to collapse into mattrons. This does not happen all at one go. Simultaneously in a slower process, the original mist and mattron grids are collapsing into denser local submists (that have mattrons as their largest building blocks).

Mattrons forming in more tenuous grids are neutrally compressed and of lighter masses where no disparity is possible. There are in the condensing primal mist many mattron grids of diverse inter-mattronal distances but the mist will condense. My standard illustration concerns mattrons in present denser situations as part of larger macro subclouds that we have wrongly come to call 'elementary' or of larger suns and galaxies where the voids between the structures can are even more dispossessed of mattrons.

No stable neutral mattrons can form in grids at 10^{825}ãr distances or larger. This distance would represent the average formation threshold for mattrons in the primal mist. Below that, the field stabilizes and neutrally compressed mattrons that earlier teetered in the flux storm on the brink of dispersal become ever more at ease. The stability threshold has been passed and a full quantum split into stable mattrons can proceed in the primal field. As the grid distances continue to shorten, mattron stability improves.

Neutral mattrons in a grid of 10^{824}ãr average distance would balance at 10^{623}ãr radius size under subservient pressure in both signs. The extensive annihilation of neutral central mass leaves 10^{903} atoms with parental sweep slowing to $3,16 \times 10^{-258}$¢ in $3,16 \times 10^{311}$ãr local orbits. Virgins are then boosted $1,78 \times 10^{256}$ fold across 10^{824}ãr to reach a sweep that is 17,8 times slower than linear birthspeed and becomes non-boosted across that field.

In this widest grid with matching emission, the mattrons shrink to 10^{623}ãr with the consummation period rising to $3,16 \times 10^{801}$ãt while each harvests mutual flux driven by 10^{490} virgins. Stopping flux at $3,16 \times 10^{22}$¢ after slowing it $3,16 \times 10^{389}$ fold from 10^{412}¢ requires $3,16 \times 10^{879}$ core atoms constantly on the wing. The secondary emission of the free flux is now only ten times less than the coronal emission. With neutral mattrons in a slightly wider 10^{825}ãr grid is a step too far as all of their mass is dispersed by the secondary flux emission into a mass that covers each entire inter-mattronal span.

Disparate mattrons are unstable in wide grids before ultimately reaching the disparity threshold as neutrally compressed mattrons continue to draw closer. I propose that on reaching 10^{805}ãr inter-mattronal distances whatever disparity that is inflicted upon a mattron can be maintained by it. Should the one mattron for any reason turns disparate, the same disparity will then spread like wildfire through the local mattron grid.

Close Encounters in a Void

Imagine two mattrons traveling through a void that is unfathomably large on the mattron scale. We can begin with two neutral mattrons approaching from afar in an inter-galactic void and later look at the less straightforward meeting of one neutral and one singular mattron. All types of lone mattrons in a void are engulfed in precursor halos, neutral or singular.

Say that the two neutral mattrons of 10^{912} atoms traveling in opposite directions are moving at half the speed of light. In the widest 10^{824}ãr grid of mattrons, the orbital sweep of their emitted virgins does not reach close to the linear speed due to slow parental speed in highly compressed mattrons. In mattron grids, head-on flux fronts push from all directions and this is a key factor in breaking their linear stride; a key prerequisite for a complete harvest of all the mattron flux in grids.

Virgin emission out of isolated precursors across excessive distances is not as easily tamed (especially during unequal emission). In this illustration, two standard central mattrons with 10^{752}ãr wide precursors pass by each other in an empty void and have a rare 'close' encounter 10^{824}ãr apart.

The precursor halos take 10^{960}ãt to form and though their dispersal takes longer, they have become intermittent with myriad halos during 10^{1104}ãt; the time it takes for the two mattrons pass each other at light speed, which renders the periods they go without halos less significant. Allied emission out of precursors is relatively tiny but even here the neutral mattrons affect each other across large distances by such emission.

The escape rate out of precursors falls into two categories; virgins split in a central mattron and virgins split in a precursor. Because of slower parental spin, 10^{-72} of virgins emitted by the standard central mattron escape out of a precursor that stops traversing atoms at 10^{72}¢. This releases 10^{364} central virgins in radius time with 10^{-26}¢ exit sweep. Virgins split in the precursor gain 10^{124}¢ exit sweep and 10^{-196} escape into the strayfield or 10^{252} in radius time (10^{112} times fewer than 10^{24} time more numerous central virgins).

For the two approaching neutral mattrons, escaping allied virgin emission from both central mattrons starts its trek out from widening 10^{316}ār orbits with 10^{-244}¢ parental sweep. When a central virgin reaches the 10^{824}ār midway point, boosted 10^{254} fold to 10^{10}¢, it has already made 10^{10} orbits around its ancestral mattron while a precursor virgin has made 10^{160} orbits.

Of the 10^{582} released central virgins, 10^{-134} hit the other precursor during a 10^{1104}āt flyby and this repulsion can repel the other mattron at 10^{-330}¢ which does not significantly affect its linear course at light speed.

When the two virgin fronts meet and activation starts, it is small. Jacked up precursor virgin activation may and may not be on the cards as the flux presence of all the self-activated virgins equals the virgin presence. But, even if activation of precursor virgins is jacked up to a majority, the presence is not massive enough to seriously affect central virgin emission (10^{-46}).

The flux drive of 10^{448} precursor newborns with 10^{412}¢ orbital boost during 10^{752}āt consummation rips out 10^{788} atoms of precursor mass before this small 10^{-100} fraction of its mass is re-harvested equally by both of the two precursors at 10^{72}¢, somewhat compressing them.

As the halos diminish with increased external emission, their eventual dispersal and harvest by the central mattron is inevitable and the central mattron provides even greater emission input and is soon turning heavily compressed and losing mass in the mutual exchange.

Such exchanges in a grid support heavily compressed neutral mattrons but here more of the exchanged virgins are lost from exchanges that would be somewhat lighter. Without grid support, when a haloed neutral mattron has a close encounter with another such mattron across a 10^{824}ār void, the effects are too meager to disperse them or alter their movement.

But this is a fluid situation; as the exchanged neutral flux fields heavily compress both mattrons, the parental spin slows (to roughly 10^{-252}¢), which means that the virgin boost across the void reaches roughly 10^{-2}¢ and does not eclipse the linear birthspeed. The path of an escaping virgin is now a 'straight' line away from a compressed mattron. A smaller presence without grid support from all directions simply turns off activation in the field.

For a while the two neutral mattrons will expand anew with or without precursor halos until another activation burst is possible when the virgin presence has reached a certain critical point. That two oscillating neutral mattrons will pass by each other, more or less concludes my arguments on close encounters in a void of two neutral (and two allied singular) mattrons.

The exterior flux fields that form between distant standard mattrons are fuelled by virgins escaping the central mattrons or their precursors. Both types of virgins travel more slowly linearly across the void than they do in orbit around a standard home mattron and its precursor; 10^{160}¢ in orbit for a precursor split virgin and 10^{10}¢ for a centrally split virgin.

Activated flux generated by precursor virgins is traveling at high speed around the home mattron and precursor. Flux generated by the 10^{12} times more numerous central virgins would quickly lose its orbital ability around the ancestral mattron and be carried by precursor flux in various opposite directions.

Outside a totally isolated precursor mattron, this exposes any such flux to a universally linear push that during each whole orbit is contradictory and annuls the 'widening' pace. For this reason, the flight of a widening flux front around an isolated precursor (if at all activated without another oncoming emission front) is slowed and brought to standstill.

When two mattrons meet in a void, their mutually activated orbiting flux fronts engage with each other mid between the mattrons. This focuses flux into an interaction area situated between them and lets new virgins therefore gain a constant linear aim. The widening of the inter-mattronal interaction front is now instead held back by a head-on flux front that prevents it from widening (if the newborn input is equal by both).

In a flux field between neutral mattrons at 10^{824}ãr apart, the individual 10^{412}¢ orbital drive is divided upon all ripped out atoms. After chipping out say 10^{100} atoms of 10^{340} required for harvest, the orbital speed of precursor flux atoms in orbits around its mattron has slowed (in theory to 10^{96}¢).

When these flux atoms orbit into the interaction area, they are too slow to move across it (a fair part of 10^{824}ãr) and out from between the mattrons before consummation in 10^{752}ãt and are harvested long before it can embark on another round. The flux interaction area becomes focused between the mattrons and closer to the mattron with less powerful emission.

If one of the two stray mattrons is singular and the other one neutral, it alters the exchange. By again making 10^{824}ãr the starting point (ignoring the massive prior exchanges), the input between the two different mattrons as they pass by each other during 10^{1104}ãt, one singular and another neutral, is far more robust than between neutral mattrons.

The governing factor here is the totally escaping and highly orbital adverse emission that streams unhindered out of the singular standard mattron; an effect that makes its mark across far greater distances.

Surviving adverse virgins work to reflect the approaching neutral mattron. Virgins split within a singular precursor gain great boost by parental (or unresolved orbital) motion and since all such adverse virgins in principle hit a neutral precursor (at first with this 10^{824}ãr initial setup), they could deliver powerful collective repulsion. During the 10^{1104}ãt flyby, some 10^{800} adverse precursor virgins are fielded, which in principle can send a neutral mattron packing at 10^{-112}¢ (versus 10^{-280}¢ linear advance) and an approaching mattron could in theory be reflected far earlier.

But long before this 'setup', massive precursor virgin presence activates the virgin field between the faraway mattrons on meeting a lesser virgin field in that sign. The interaction starts to jack up the activation of precursor and ultimately even centrally emitted virgins with 10^{10}¢ sweep, whereupon the flux field tones down effective repulsion (only 10^{-46} of adverse central virgins are untouched and only 10^{-134} of these central survivors hit a precursor).

Of 10^{824} central virgins fielded during a 10^{1104}ãt flyby, merely 10^{-180} would hit another precursor with effective repulsion by 10^{644} surviving virgins at birthspeed, which could in theory cause electromagnetic splits and alter the neutral mattrons course. This will not come to pass because the precursors cannot remain in place during the flyby.

In the wink of an eye during 10^{752}ãt flux consummation, 10^{472} newborns escape the singular mattron. Activated and boosted to 10^{412}¢, the flux storm chips 10^{812} allied atoms in the adverse sign out of the neutral precursor or 10^{-76} of its 10^{888} mass before being re-harvested by the neutral precursor.

Due to my scant thoughts on halos (and Universal flux that could remove high presence particles from the local scene), I cannot be categorical; but it seems that after 10^{752}ãt, the chip out effect does not lead to dispersal of the adverse singular sign from the neutral precursor and the adverse emission cannot deplete it by chipping out atoms (lost mass takes 10^{1192}ãt to grow back, which is longer than the 10^{1104}ãt flyby).

What turns the table is the disparate inwards pressure by re-harvested orbital flux that swiftly shifts one sign deeper inwards within the neutral precursor while atoms of the other sign stay in place; turning the outer precursor singular in the sign of the singular mattron.

With excessive one sign compression, the neutral standard mattron is first to lose its precursor in one sign and the standard center becomes polarized. While its adverse emission does not affect the singular mattron at all, its allied input has become equal to that of the singular mattron, whereupon the singular precursor becomes superfluous.

Grid mattrons at this distance do not require precursor support and turn heavily compressed and, if polarized, are dispersed. With isolated mattrons, the initial interaction area between the two mattrons is moved at almost birthspeed towards the neutral mattron by the virgin input from the singular mattron until it is gradually slowed by chipping atoms out of the neutral precursor (albeit with less orbital drive if carried closer).

There are important stages (using 10^{824}är distance as a starting point) as the exchanges adjust but the overpowering linear stride of adverse flux in the face of weaker oncoming subservient push loses ground. When the two adverse virgin fronts match, the interaction area regains the middle ground between the mattrons.

With neutral mattron polarized without grid support, things get dicey and there appear to be few holding points of initial stability with so many urgent reactions waiting in the wings. This would show how 'light' is affected in empty space and mine is a rough assumption for an interplay that deserves to be thoroughly thought through and calculated.

Although a central mattron has 10^{24} times more mass to give than has its precursor, it blocks flux at 10^{36} times' slower speed and if the flux storm persists as the precursor is dispersed, the central mattron mass follows. In the void however, the flux storm drops off when the mattrons lose their precursors and are compressed. The secondary emission of ripped out flux in the adverse sign gives no extra effect as the singular and polarized mattrons already emit all adverse newborns and this input cannot rise.

As both mattrons become compressed, the smaller parental movement will no longer allow their orbital boost of escaping virgins to eclipse linear speed and their path becomes a 'straight line' where few are activated and few hit the other small mattron. There is an oscillating balance in there somewhere after the flux storm ceases and the mattrons expand; parental spin increases until activation sets in again and is subsequently turned off again.

The void is swept clean by the flux in both signs even adverse virgins with 10^{10}¢ sweep streaming outwards from standard mattrons remain untouched and few (10^{-374}) hit another distant standard mattron and the overwhelming majority would be lost further afield in a virgin state. They are no longer part of the mutual exchange. But there could possibly be exceptions to an initial clean sweep, especially between adverse singular mattrons that effectively exchange only adverse emission (while allied emission has no mutual effect).

A 10^{36} times weaker allied emission that issues out of a standard singular mattron is most concentrated in the local areas surrounding the mattron and is here countered by the greater but much diluted faraway input of virgins in that sign in the opposite direction.

At some distance away from the singular mattron, the number of outgoing allied virgins equal the number of incoming adverse virgins from the singular mattron (in its originally more massive adverse sign) and matches its linear push. When virgins are activated in this volume around the singular mattron the flux interaction area becomes at local standstill.

This interaction area works much as a precursor without being one. As the activated number of adverse virgins from the singular mattron equals the number of activated virgins from the opposite mattron in that sign, the flux is no longer accelerated linearly away but quickly harvested by the singular nearby mattron and with a lesser local primal boost. Flux in this sign does not stand in the way of allied emission in the other sign (it is oblivious to it).

My assumption is that during such mutually uneven exchanges in a void, an activation area or volume forms around the mattron with the weaker emission. The flux winds in one sign, driven by greater numbers, beat back and carry along the oncoming flux and clear the field of flux. This causes the virgins that follow on the heels of the more massive front to increasingly miss the shrinking activation area to continue deeper into the void and be lost from the mutual exchange.

If this were a mattron grid of mismatched mattrons, such activation areas could not shield mattrons from excessive pressure; all the virgins exchanged between grid pairs end up within one activation area or another. Being paid in kind, they deliver the excessive compression with a slight delay.

There are two orbital movements to keep an eye on; parental orbits in all directions around a singular mattron at 10^{10}¢ and activated orbits around coordinate centers out of each collision point between the singular mattrons. With the slow parental drive, this is not an effectively orbiting flux field.

The activation area of the singular adverse virgins is swiftly moved closer towards the opposite mattron and flux is harvested before embarking upon another round. The fractional compression of the globular activation area by virgins from opposite directions with a 10^{10}¢ sweep is already accounted for in a matching number of outgoing virgins with expansion push upon the interaction area. Parental boost is not equal to a greater number of incoming virgins.

The activation field might reach 10^{801}ãr out from the adverse mattron and be hit by 10^{-36} of adverse virgins emitted by the another opposite standard singular mattron (boosted to 10^{10}¢ across 10^{824}ãr), forming a Van Allen belt of a kind. The locally incoming virgin wind matches the virgins emitted by the singular standard mattron. And lo and behold; between the mismatched mattrons at this distance in the void, instability takes the form of stability.

The local primal flux is boosted according to the width of the activation area and delivers $3,16 \times 10^{400}$¢ corresponding local boost to the mattron in that sign. 10^{316} virgins from both mattrons (in 10^{632}ãr) deliver $3,16 \times 10^{716}$pu in their signs (harvested in 10^{765}ãr) but even this $3,16 \times 10^{12}$ fold standard benchmark pressure would heavily compress the singular mattrons. The outcome stands my first impression of the massive mid distance interaction flux on its head. Instead of mattrons being compressed by the entire virgin input, the activation area in the allied sign allows 'only' 10^{-36} to take part and with a lesser boost.

This mattron compression does not take much time and with central mass put out of commission, the virgin input drops from both sides. With lesser parental spin, more of the fewer virgins miss the local interaction area, and for both reasons, the pressure subsides until the mattrons expand anew.

It is possible that interaction areas apply to other exchanges, which is a fascinating subject with alternative activation areas and sign reversals but nonetheless outside the scope of this treatment.

I hereby leave close encounters in a void, assuming that this controlled instability, especially when one mattron is singular, allows the mattrons to get out of each other's hair during passage without permanent damage, although depending on distances they may retain a somewhat decimated mass.

Flux Gravity Input

This brings us to the potential gravity shadow in a mattron grid. Gravity inducing flux is whatever part of flux or parental virgins that reaches further than the grid distance with an uncompromised orbital drive. If this is all mathematically crystal clear, it is a conceptual nightmare.

The activated subservient flux presence is 10^{144} times greater than what presence the residential mass of the next corona puts in the way of a stray within its tighter space in 10^{776}ãt. The mattrons 10^{-244}¢ average parental movement provides its mass with 10^{532} times its physical 10^{912}ã presence or 10^{1444}ã but this is constricted to 10^{-72} part (10^{1372}ã) of the 10^{2328}ã exchange volume of mattrons. The flux is active within 10^{2328}ã volume that is 10^{432}-fold the 10^{1896}ã coronal volume. It therefore takes 10^{288} coronas to match the collision chances given by inter-mattronal flux. Although the globular flux orbs can to some extent interact, I do not consider this gravity inducing.

The overwhelming flux majority is linearly stopped by opposing flux and eventually captured by the next mattron. The flux atoms are adjusted in orbital directions so that a single composite orbit carries multiple direction inputs, including those from mass ripped out of mattrons with a slower push delivered by each single flux atom in a composite of orbital directions.

An easiest way to describe a potential gravity shadow is the number of once-hit activated newborns passing through the inter-mattronal flux after reaching wider than 10^{776}ãr paths; once hit activated newborns that escape the exchange volume without more collisions. Many driving newborns do achieve this in little more than 10^{776}ãt and continue at half birth speed.

Here 10^{460} subservient newborns during 10^{776}ãt have a spatial presence of 10^{1588}ã (10^{460}n$\times 10^{388}$¢$\times 10^{740}$ãt) while a single activated newborn in widening orbit covers 10^{1164}ã out to 10^{776}ãr in that time. Together, sweeping 10^{2752}ã of the 10^{2328}ã exchange volume, 10^{-424} reach further than 10^{776}ãr or 10^{36} once-hit subservient atoms in 10^{776}ãt on orbits that can pass through numerous other mattrons and give rise to orbital gravity shadows between them.

The massive adverse 10^{496} dominant input during 10^{776}ãt have a spatial presence of $5{,}62\times 10^{1630}$ã (10^{496}n$\times 10^{388}$¢$\times(5{,}62\times 10^{746}$ãt)) as the newborn in widening orbit covers 10^{1164}ã. Sweeping $5{,}62\times 10^{2794}$ã of 10^{2328} ã exchange volume, $1{,}78\times 10^{-467}$ of them reach further than 10^{776}ãr and this counts $1{,}78\times 10^{29}$ once-hit dominant newborns that give rise to orbital gravity shadows between numerous local mattrons.

A shortcut is to divide the 10^{388}ã activated spatial presence into the virgin survival rate (subservient 10^{-36}, dominant $1{,}78\times 10^{-79}$); this leaves 10^{-424} subservient and $1{,}78\times 10^{-467}$ dominant local flux gravity input.

Average Local Repulsion

Surviving virgins are all repulsive and easily captured if they hit nearby grid mattrons but their number dwindles as they initially spread out by the inverse square of the distance in mattron radii. Boosted in parental orbit across a void, they need large distances (rare for mattron grids) before their parentally boosted side sweep eclipses their birth speed. When that happens, virgin repulsion is effective in all directions incumbent on its inherent orbit.

In the average grid, 10^{-36} of all emitted subservient virgins keep a repulsive character after having crossed the standard divide with their potential push diluted by the inverse of 10^{144} mattron radii squared. The mutual repulsion by 10^{460} virgins during 10^{776}ãt between two standard mattrons is further cut (on top of the 10^{-36} left after activation) by 10^{-288} and delivers 10^{-324} of the original input 10^{460}pu input. The actual repulsion effect by the exchange of linear virgins between two disparate standard mattrons is then 10^{136}pu.

If the Universal focus is not shifted by collective spinfields or other relative motion, this weak repulsion accumulates at the receiving end. Even ignoring the gravity shadow, it requires an extensive 10^{1272}ãt period to accelerate the next standard coronal mass to 10^{-280}¢ (the speed of light).

In a grid where similar pressures arrive from all directions, where gravity shadows are present and mattrons are likely to orbit and therefore shift their linear position, such linear repulsion would rarely be realized.

On the dominant side by the original 10^{496} newborn in 10^{776}ãt, $1,78 \times 10^{-79}$ of virgins keep their repulsive backhand with a mutual repulsion further cut across the grid by the inverse of 10^{315} upon $3,16 \times 10^{618}$ãr wide cores with the push diluted to deliver $1,78 \times 10^{-394}$ of the original 10^{496}pu input and actually hit the next mattron core. The dominant virgins exchanged between two disparate standard mattrons would repel cores by $1,78 \times 10^{102}$pu.

The subservient virgins that mutually repel coronas deliver $5,62 \times 10^{33}$ fold linear repulsion above dominant repulsion of cores. Due to lack of coronal penetration at 1¢, virgins 'flatten' a coronal surface marginally after activation within in the time it takes a mattron corona to rotate its far-side back to the front. The next period allows repulsion to flatten another part of the corona as the far side re-adjusts to internal expansion.

Activated repulsion is less relevant in stable grids between balanced mattrons as the exchanges have lost linear focus with little of the original linear push left to work as repulsion upon another mattron. Although all the activated flux easily 'finds' the next mattron, it has already been met head on by equal opposite flux and its repulsive ability goes largely for naught.

Average local Gravity Shadow

A once-hit activated newborn activated in a collision center in its home domain that reaches orbital size in excess of the average grid distance has the potential to cast orbital gravity shadows between mattrons on somewhat larger orbs. By driving orbital flux in large fast orbits beyond the average grid distance, it fuels backhand gravity shadows between other local mattrons. Even so, one would ordinarily associate gravity shadows between mattrons with long-distance virgins from the deep background strayfield.

10^{36} subservient activated newborns or 10^{-424} of the input reaches further than the standard 10^{776}ãr distance (let's say twice that distance) and creates gravity shadows between grid mattrons, a shadow that fades at the inverse of the distance in respective mattron radii squared.

If the 10^{36}-strong gravity input of once-hit driving subservient newborns is boosted to say $1,41 \times 10^{388}$¢ and first slowed in collisions with coronal mass on that orb and its flux harvested at 10^{36}¢ to cast $1,41 \times 10^{424}$pu gravity shadow by each 10^{632}ãr corona in all orbital directions on that larger orb. This input affects all local coronas that are situated on that same orb.

The $1,41\times10^{424}$pu local gravity push is diluted between grid mattrons by the inverse of 10^{288} across 10^{776}år on that orb; yielding a mutually effective backhand gravity push upon the next corona that in 10^{776}åt accumulates to a backhand push of $1,41\times10^{136}$pu collective 'attraction'.

An observer 'standing on' a mattrons coronal surface, measuring the weak (absence of) gravity inducing flux blocked by the 'underlying' mattron would find its blocking effect at roughly the same 10^{136}pu in 10^{776}åt. Of course, in that same time, 10^{460} subservient virgins 'shoot' out of the coronal surface (where the observer is standing) with 10^{460}pu linear repulsion; rendering the 10^{136}pu gravity pull a weak (10^{-324}) effect, albeit still there.

A $1,78\times10^{-467}$ strong dominant gravity input of $1,78\times10^{29}$ atoms are also boosted to $1,41\times10^{388}$¢. Slowed by core mass and blocked at $1,78\times10^{29}$¢ they cast an effective $2,51\times10^{417}$pu gravity shadow by $3,16\times10^{618}$år cores in all orbital directions, affecting other cores on that orb. This $5,62\times10^{6}$ times weaker dominant gravity input into nearby fields is diluted by the inverse of 10^{315} across 10^{776}år and yields a less mutually effective backhand gravity push of $2,51\times10^{102}$pu in 10^{776}åt between core spheres on that orb.

The core backhand attraction and core repulsion is dwarfed by the coronal shadow and coronal repulsion and both are in turn dwarfed by the local field pressures. In a mattron grid it is likely that positional changes are initially brought on by more than one mattron. The movement of a disturbed mattron is affected by several vectors of attraction and repulsion and its enforced motion is tangential to several mattrons (albeit aiming closer to one than others).

Let us snap our fingers to make the 10^{136}pu subservient repulsion between coronas disappear (along with other opposite pull or motion that neutralizes by changing focus). With that, the subservient backhand would inexorably push the mattrons closer than the average equilibrium grid distance.

What changes when the two grid mattrons move closer than the average distance is interesting; as gravity shadows grow stronger the local repulsion increases even faster. Fewer virgins are activated across smaller distances while more hit the closer mattron with repulsion rising on both counts.

The fact that more once-hit flux or virgins would now escape the local inter-mattronal volume to serve as gravity input, it would not affect the two interacting mattrons; only other mattrons on a larger orb.

On the other hand, if the two mattrons are instead moved apart, the gravity shadows gain relative influence (even as they diminish). Ever fewer virgins will now make it far enough across the grid to exert repulsion and fewer of them will also hit the mattrons.

Due to mutual responses to altered distances and direct adjustments in grid mattron exchanges; there is a direct incentive for mattrons to keep to their respective grid distances.

For either imbalance, the winner ultimately takes all unless the mattrons can break the new dominance of repulsion or gravity by changing the focus and periodically pit the fields against their own effect.

Without repulsion and collective spin, the gravitas in subservient shadows between coronas of 10^{-640}pu per ãnna accumulates to 10^{524}pu in the 10^{1164}ãt period it takes them to fall across the exponential 10^{776}ãr grid distance; at first slowest at 10^{-388}¢ (which consequently sets the time).

It would take cores that are reliant on dominant push far longer to join up but the cores are not left behind as they are easily kept to quarters by the internal pressure balances of a disparate mattron that in this context is an overwhelming 'force'.

Most grid mattrons are likely moving in orbits of varying sizes on paths of least resistance at orbital speed described by some average (like the parental movement within mattrons). But unlike primal atoms within a mattrons, the fastest linear speeds achieved by mattrons in grid orbits are only marginally able to increase their harvest of passing primal flux.

The Local Spinfields

There is much to say about local strayfields and they are the hardest thing to get your hands around with all the external spinfields in the grid vying for influence in a garbled collective field that impairs speeds and directions.

As to linear internal movement, I allowed the average mattron to expand and stabilize within 10^{912}ãt and picked ($M^2 \times \sqrt{r} \times k = 10^{948}$ãt) for collective spin consummation. Collective spin consummation for all material systems built of mattrons is shown by the product of mattron consummation and the mattron count of the larger system; typically off the top of my head.

For a 10^{632}ãr mattron, a new unmatched spin orientation can become a collective inherent 'property' common to all its 10^{912} residents already after 10^{948}ãt. Once established, the inherent 5×10^{-281}¢ collective rotation of the surface of a 10^{632}ãr mattron is in this case boosted 10^{144}-fold in accordance with the average widening of the orbit from 10^{632}ãr to 10^{776}ãr.

Instead of making the inter-mattronal trip itself, each residential particle within the mattron sends a newborn emissary that (in addition to its own new and totally independent coordinates) carries all collective coordinates of parent and mattron. Aside from its individual input, each newborn that exits a mattron would on reaching the next mattron orbit on its two-dimensional disc plane at 5×10^{-137}¢.

The differential plane of the newborn orbit is itself rotating around the emitting mattron so that each resulting flux particle will at some time (given the chance) cross every point on that distance sphere. If unchallenged, the communication of its collective spinfield onto another (perfect) mattron, will in time deliver the relative coordinates of the 5×10^{-137}¢ speed in a differential orbit about the mattron of its birth.

The operative word here is time. For example; when all 10^{912} residential atoms of a mattron have communicated their average position to another mattrons 10^{912} residential atoms in the relevant sign during an extensive period of say 10^{1192}ãt (when all the particles have put in their vote), the emitting mattrons spinfield may have changed its course 10^{244} times.

This indicates that the collective spinfield can deliver no coherent message except for each relatively short 10^{948}ãt period for the consummation of its new collective spin; if this new spin is different from its former spin.

During one consummation period its escaping subservient emission of 10^{632} driving newborns delivers a consequent sideways push of 5×10^{495}pu and moves the other neutral mattron or disparate corona in its differential orbit to a speed of $2,5 \times 10^{-417}$¢.

The speed brings a corona across a distance of $2,5 \times 10^{531}$ãr during that spin consummation with an shift orbital shift across 10^{-101} coronal radius. This renders the mattrons collective spinfield almost inconsequential in a mattron grid. This insignificant collective spinfield message is carried in great numbers by virgin and activated flux alike in mattron exchanges. Subservient coronal emission imposes change upon outside mass up to 10^{36} times slower than the dominant collective spinfield (in both cases the same coronal spinfield).

As new spin in unpaired directions constantly works to change the collective spinfield, even the exterior subservient message lags behind and is heavily garbled before maximum speed in outside orbit is reached. Bear in mind that even a stable collective strayfield that delivers a full potential acceleration over long period can never exceed its own inherent speed (no matter how much emission is accumulated in the other mass over time).

Between interacting mattrons where one mattron perhaps approaches in a tangential fashion due to some initial motion, the corrupting influence of a collective spinfield does not have sufficiently long periods at its disposal to shift the focus of gravity or repulsion effectively (unless both are moving with embedded inherent coordinates of some larger spin system).

The linear corruption of the relative approach path of another mattron is therefore mostly too small to neutralize either an experienced pull or push.

Setting the Strayfield

We have now before us a relatively even Universal mist where the primal field has undergone its momentous quantum transformation that came with the formation of mattrons. This local transformation was eventually to be seen throughout the condensing center of the Universal mist. Where earlier in the creation process we would see evenly populated patches of primal matter we now see 'extensive' empty spaces dotted with mattrons.

The remaining primal clouds that lace the bowels of the immense neutral grid will continue to grow in primal mass while more unsubstantial clouds are depleted through dispersal by fast moving strays as collisions ultimately gather these primal remains into some cloud of mattrons or another. Primal mass is only annihilated at the centers of neutral mattrons, not on the wing.

The weakening general strayfield allows mattrons to sustain larger masses without risk of central collapse until they grow to maximum size against the reigning grid pressure and most mattrons in the present time are therefore not in danger of collapse when exposed to increased pressure.

Although the universal pressure has eased with the bottling up primal matter into mattrons, making for a weaker strayfield, a lot of the more flimsy rapidly breeding primal subclouds that lace this field have yet to consolidate. At the outer Universal border on the outskirts of this great mist of neutral mattrons there remain extensive remains of the original primal mist that is too sparse to condense into quantum waves that are massive (and close) enough to build sustainable mattrons.

The primal halo that engulfs the Universe sends incessantly orbiting newborns our way in a neutral mix at immensely boosted speeds but this orbital pressure is still unlikely to measure up to massive local pressures. A good deal of the primal population of the central Universal mist is by now bottled up in mattrons large enough to withstand scattering, situated in consolidating and often unstructured grids of various shapes and sizes.

Some early mattron grids may be polarized by fluke in either sign. As times pass and the first electromagnetic split takes place outside of submits with adverse twins head in opposite directions at light speed. Passing tangentially by large or small mattron subclouds, a single twin can start a chain reaction in these clouds and spread polarity in its image across bridgeable distances within each such grid of neutral mattron.

With disparity, inter-mattronal strayfields are driven by dominant flux generated by adverse coronal emission from the overwhelming number of neutral disparate mattrons. This jacks up the inter-mattronal pressure and locks the inner cores in place, sustaining the polarization of each grid.

Larger fields have far too many pockets of voids for exclusive polarity. The more immense macro subclouds like galaxies or suns are less likely to have mattron populations divided into large areas of opposite polarity.

Concerned with the individual interaction of mattrons, I am not dwelling on balancing acts that demand extensive time periods and there are various long-term aspects that I skip. For example, a neutral mattron in a disparate grid would be exposed to charged pressure for a long time and its primal mass composition can undergo long-term consequences of internal charge.

Disparate mattrons are grid creatures with direct exchanges to provide balance and do not ordinarily require intermittent halos. Neutral mattrons build up mass in the subservient sign by emitting more dominant atoms. Although most of these are simply exchanged, the long term effects of tiny escape fractions can lead to disparate masses but I do not for now address eventual potential asymmetry.

Even singular mattrons supply a local grid with adverse emission but as they cannot build singular mass ad infinitum, they are at intervals forced to emit excess mass in halos or to face collapse. This adds the other charge to the local grid to be soaked up by neutral mattrons and thus ultimately annihilated at their centers.

If singular mattrons cannot re-polarize a surrounding gathering of already disparate neutral mattrons in the opposite sign; that local strayfield disparity remains 'indefinitely' intact as the original disparity would hold until enough singular opposite mattrons muster collective emission to eclipse that of the neutral disparate mattrons. As this is possible first when singular mattrons are as many as neutral mattrons, it takes an eternity in anybody's book.

Without reversible polarization, Universal disparity will sub-divide through different processes into myriad smaller areas of opposite charge.

Already disparate opposite neutral mattrons (with the outwards character of a singular mattron) that are compressed but not wholly re-polarized by a singular mattron undergo short term annihilation at their centers where an existing polarity is forced into cohabitation (and would certainly do so if polarity can be toggled).

Although a singular mattron can clearly play havoc in a local field of neutral mattrons; how and whether its adverse emission can reverse an opposite polarity in a grid that is already disparate (and in sheer numbers has a gargantuan edge in disparate mattrons) is clearly central to Universal evolution.

To look at polarization and re-polarization of neutral mattron fields and to study flux seepage between adversely charged areas in local grids, I find it helpful to use the concept I call a 'singular domain'.

Singular Domains

On assigning an emission 'charge' to a disparate mattron, I named it after the sign of its coronal mass (and its subservient emission sign); this is a matter of choice. I now need to understand how the singular mattron casts its local spell by dominant emission and compression upon other domains. We must look at several grid versions to get a better sense these situations.

Let me first illustrate by following what happens when a singular mattron (make it a negative twin photon) approaches a gargantuan globular cloud of say 10^{228} neutral mattrons with stable $3,16 \times 10^{627}$ãr radii in a wider grid of 10^{800}ãr inter-mattronal distances where their $3,16 \times 10^{907}$ atoms in both the cohabiting signs has never been polarized.

It should not play a role whether the singular negative mattron has less than a standard 10^{912} singular mass (it might be split far off and grown to full 10^{912} singular mass or been split locally with $3,16 \times 10^{907}$ atoms). The negative photon approaches a mattron submist grid across the void and, unsupported by the grid, surrounds itself with a precursor halo.

The main question is whether or how this negative singular mattron affects the disparity within the 10^{228} mattron grid and much of the forthcoming universal evolution rides on getting that answer right.

Polarizing a Neutral domain

Each of the 10^{228} slightly neutrally compressed mattrons in the globular gargantuan grid contains $3,16 \times 10^{907}$ primal atoms of both signs due to the greater grid pressure across 10^{800}ãr that annihilates its mass at neutral centers. The inter-mattronal grid distance is well below the wider 10^{805}ãr disparity threshold.

The single negative mattron on its linear course through the void. Let's give it $3,16 \times 10^{907}$ atoms as it 'slowly' approaches the surface of the vast neutral grid domain at 10^{-280}¢ light speed; assuming that it is locally split. If it has been procreating from faraway, it could have reached standard 10^{912} mass but that really does not affect the outcome. It starts to influence the neutral grid first when close enough for its massive positive adverse virgin emission to spice mutual exchanges of subservient coronal inter-mattronal flux within the mattron grid where these are activated.

It's already orbital adverse virgin emission affects the 'surface mattrons' in the approaching globular grid of mattrons from which little primal emission escapes.

Grid mattrons serve each other with pressure that compress grid spheres of both signs and they have no precursors. When extra positive virgins that are emitted by the approaching singular negative mattron and increasingly added into the flux field from the outside void, these virgins are activated and harvested as extra compression between the local grid mattrons.

The approaching singular mattron is as yet separate from the grid without mutual exchanges with grid mattrons. For example; at 10^{813}år away from the mattron cloud (10^{13} fold average inter-mattronal grid distance); its spreading orb of adverse positive virgins can affect 10^{26} surface mattrons (of the grids 10^{228} neutral mattrons).

Neutral mattron coronas of this size emit $5,62\times10^{-30}$ of their subservient neutral virgin emission while the negative photon emits all of it or $1,78\times10^{29}$ times more. The extra positive input into the inter-mattronal flux exchange is therefore $1,78\times10^3$ times the collective allied emission between the 10^{26} neutral mattrons within its reach.

Generally speaking, the neutrally compressed grid mattrons have been and will gradually experience increasing positive compression while the negative compression within the grid has started to drop from the affected mattrons.

With coronal surfaces turning ever more highly polarized, the 10^{26} grid mattrons respond with parallel changes that strengthen the ongoing effect; with widening cores, their outer spheres contain ever more negative atoms and consequently less singular volume of positive atoms which in turn allows less negative emission to be mutually exchanged.

The mattron grid that lies beyond the 'small' affected surface area, a vast sea of stable neutral mattrons, gets no whiff of the change that the negative singular mattron heralds for the local grids mattron coronas. The few virgins that escape out of the submist grid may engage or activate positive virgins from the approaching negative mattron but cannot alter the outcome.

What happens is paramount; from the start of top layer disparity, the local surface mattrons become ever more heavily negatively polarized by the approaching negative photon; first the top layer in the mattron grid where singular coronas grow in depth as cores continue to collapse under a rising pressure. Flux exchanges from these top layer with mattrons in the second grid layer forward the disparity sideways and deeper into the neutrally compressed grid and they achieve this faster by mutual self-polarizing than does the approaching negative singular mattron.

A fully polarized mattron within a 10^{800}år grid keeps $3,16\times10^{627}$år coronal radius, same as the neutral mattrons did in both signs. One mass sign would initially be compressed to $4,87\times10^{615}$år. This becomes the core radius where all but $3,92\times10^{901}$ coronal atoms (of the original $3,16\times10^{907}$ mass) are put out of active commission.

Although these few initial $8,06\times10^5$ times lighter primal mattrons cannot match the subservient emission by the rest of the neutral mattrons, their local adverse emission pulls a $1,78\times10^{29}$ times greater punch and there is really no competition. The grid wave of negative coronas spreads out from the first effective compression orb; transporting the wave of mattron disparity further into the vast mattron grid.

Driven by the unbalance, the non-material wave is a domain pulse of negative coronas, transported into the neutral grid in all directions in and around the submit from the direction of the initial unbalance; upsetting ever more grid layers where the pressure in one sign is radically jacked up. Within a neutral grid, there is no stopping this wave.

Once this globular wave front passes any point in the mattron grid, the left behind polarized mattrons have no reason to revert back to their original neutrality; they are perfectly balanced in a mutually supported disparity. The easy polarization of the neutral grid by a domain pulse continues to spread in all directions; it reaches the other side of the vast submit and its center in roughly the same time (or it comes to a stop if it reaches a divide across which mattrons cannot effectively exchange emission).

I earlier set 10^{912}ãt as the time needed to polarize one standard neutral mattron but a domain pulse works concurrently on a globular front that eventually reaches the entire grid. I therefore use the product of the square root of the mattron number (10^{114}) and standard 10^{912}ãt polarization as the time required to polarize all of the 10^{228} mattrons; namely 10^{1026}ãt.

In spite of the now two disparate spheres in all the 10^{228} neutral mattrons, it is coronal compression and boost that dictates central annihilation which barely affects the disparate corona. It is the mass of the compressed core that is put out of commission while the singular coronal mass remains the same in that sign as long as the corona maintains its size.

It takes the approaching negative mattron perhaps 10^{1093}ãt to reach across the last 10^{813}ãr leg of its approach and by the time it arrives into the grid surface, all 10^{228} grid mattrons have already been negatively polarized for a very long time on the primal scale (after 10^{1026}ãt).

As the approaching negative singular mattron enters the grid surface and settles between surface mattrons at the average grid distance, it is outwardly indistinguishable from the mattrons its meets (aside from the fact that it has no core). Its arrival does not in any way disturb the present collective peace in the gargantuan and now entirely negative domain.

Mattron Re-polarization

We have seen how easily a 10^{800}ãr wide neutral mattron grid is polarized for the first time in either sign and how the approaching photon allied to that polarity can freely join a grid of its own image without much ado.

Now assume instead that the singular negative photon is approaching a submist grid of 10^{228} mattrons that has already been positively polarized (by an earlier positive singular mattron). The approaching negative singular photon is now opposite to the polarity of that massive grid.

The impending entrance of the negative singular mattron into that opposite grid will now run into a serious obstacle in re-polarizing adverse mattrons on the grid surface, if it reaches that far, which depends on circumstances and is not given. This is a fluid situation where the coordinate fields must be more closely studied while I merely provide several conceptual snapshots as the situation unfolds to illustrate the logical threads governing the outcome.

Every mattron in the grid communicates primarily in mutual exchanges with up to 8 nearest coordination mattrons around it. The flux exchange between these mattrons comes about by virgin emissions at birthspeed away from a parent within the mattron. Each newborn atom on its linear and non-orbital path has its birth spin eventually activated and an ever increasing swarm of flux continues thereafter at rapidly dwindling linear speeds away from a parent. The orbital phase in a greater array of possible orbital paths is less geared down and allows more boosted speed in widening orbit away from each new a virgin coordinate center in incessant collisions.

Not to complicate things, I use an approaching negative photon that has been split from a similar local grid and its primal mass has had no time to grow from $3,16 \times 10^{907}$ atoms. If in spite of the grid emission it can penetrate into the grid, it starts exchanging activated flux with 8 nearby coordinate mattrons of $3,16 \times 10^{627}$ãr coronal sizes and the same $3,16 \times 10^{907}$ atoms.

(If the approaching negative photon was split far off and had time to grow to standard 10^{912} atoms, things play out differently in the short term but the final outcome is similar but exchanges are more robust and more like those in maximum 10^{805}ãr wide grids where disparate mattrons teeter on the edge of core and coronal dispersal at the sustainable disparity threshold).

A negative photon of $3,16 \times 10^{907}$ atoms that penetrates into a 10^{800}ãr wide grid of positively polarized mattrons has already lost its precursor. It affects nearby coronas with adverse emission but is heavily compressed in return and has most of its singular mass put out of commission as it faces the full blast of negative flux brought to stop between 8 coordination mattrons. It collapses to $4,87 \times 10^{615}$ãr like mutually compressed negative cores that hold $3,92 \times 10^{901}$ active atoms.

The negative grid exchanges after a negative photon enters the grid are clear; it cannot regain control of its centrally decommissioned primal mass and remains in a weakened state until it can diminish and or toggle the positive polarity of its 8 coordination mattrons. This may not seem difficult when comparing the escaping allied positive emission from a positive corona, to the initially $2,21\times10^{23}$ fold positive emission of the singular mattron (this adverse positive emission is $1,24\times10^{-6}$ of a positive coronas adverse negative emission). Nevertheless, there are extreme hurdles that the negative photon must overcome to toggle the adverse grid disparity.

The flux exchanges here between coronas are roughly 10^{770}ãt after orbital flux is brought to full stop mid between them; all flux particles continue to orbit through both mattrons around an average center mid between them.

In 10^{770}ãt, after ripping 10^{370} flux atoms out of each corona, the $2,21\times10^{23}$ times greater flux from a negative photon can put that number of flux atoms on the move at birthspeed. That it takes 10^{1147}ãt at 10^{-347}¢ to carry the flux out from between the 10^{800}ãr exchange area while harvest takes 10^{770}ãt, is not the full picture. If the orbit from mid between a neutral and a singular mattron is merely widened more than the $3,16\times10^{627}$ãr sizes of neutral coronas (in $3,16\times10^{627}$ãt), that linear outwards going flux can no longer pass through the neutral mattron as it travels further afield in widening orbits.

To this extent, perhaps all but a $2,21\times10^{23}$ part of the flux wind from the singular photon does not have its back broken by the positive wind offered by allied coronal emission. Its orbits widen too fast away from a singular photon to be stopped midways between the mattrons and the overwhelming majority of activated flux is shifted fast enough to avoid such local capture. This is my estimation without calculation; a tipping point that matters when only a tiny fraction of the singular emission has time to orbit through its 8 coordination mattrons.

By avoiding local harvest, the singular negative photon cannot significantly compress their coronas. If merely $2,21\times10^{23}$ hit, they keep their coronas and are not forced to form precursors to keep coronal balance and harvest more flux. Their coronas continue to keep the singular photon overly compressed and unable to release more of its centrally decommissioned mass.

As the activated flux from the singular photons travels through the local grid, ever more positive coronas resist its linear outwards drive with allied emission. This is a part wise zero sum game where you add a second, a third and myriad new tiers of coordinate mattrons up until that spherical volume contains of $2,21\times10^{23}$ mattrons and the linear flux drive from the singular photon is matched in an opposite direction. The vast gyrating flux orbits within this sphere deliver their payload of boosted coordinates into all of the $2,21\times10^{23}$ neutral mattrons within this globular 'entourage'.

The exchange, on its merits alone, reaches across an average $6,05 \times 10^7$ times larger distance or $6,05 \times 10^{807}$ãr where flux is $2,46 \times 10^3$ times more boosted. On average, this renders the coronas of all the $2,21 \times 10^{23}$ neutral mattrons more compressed to 18,69 times smaller $1,69 \times 10^{626}$ãr radius. A smaller sphere lowers a neutral mattrons resistance to singular photon flux and this is reinforced by lighter $1,69 \times 10^{906}$ atom masses of the $2,21 \times 10^{23}$ neutral mattrons.

With more neutral mattrons involved with higher boost and slightly lesser mass, the mattrons still keep their massive edge of coronal emission (above allied emission) largely intact, which keeps the singular photon in check. A slightly smaller corona has a fractional effect on the number of virgins that slip by a slightly larger core and become escaping adverse emission. There are all kinds of ripple effects but none, that I can see, which can broadly affect or revert this situation.

If more flux is harvested by a first coordinate tier, causing compressed first and subsequent tiers, it seems insufficient in toggling their old polarity. Even with a majority blocked, the singular photons adverse emission ($1,24 \times 10^{-6}$ of neutral allied coronal input) divided upon 8 mattrons is $1,55 \times 10^{-7}$ times weaker than the pressure upon a core and a coronal radius is still more than hundred fold the size of the negative core. In such a scenario, there is no increase in flux boost in widening orbits. In the other direction from a third tier, allied and adverse coronal emission upon a second tier is little changed except that a smaller coronal emission from the first tier eases flux pressure upon the second tier by a considerable fraction and causes the more massive outer tier coronas to expand slightly instead.

The drastic consequence of a $3,16 \times 10^{627}$ãr corona shrinking until it meets a $4,87 \times 10^{615}$ãr wide core turns the mattron neutrally compressed for an long period of primal time with its mass is annihilated from former $3,16 \times 10^{907}$ to $4,87 \times 10^{895}$ atoms. If the former core continues to expand, the mass loss can first be corrected by procreation in vastly longer periods than any of the local transformations (it takes 10^{1192}ãt for its small mass to merely double). Due to the smaller masses, the toggled polarity would then run into heavy flack of more massive coronal resistance from further off mattrons.

It is difficult to isolate each effect in brief snapshots as all is interactive but my rough take is that if a photon of adverse polarity reaches into an already polarized grid, it does not even come close in toggling the polarity of nearby coordination mattrons and much less transplant re-polarization through an already polarized grid.

The likely outcome seems closer to the first scenario of a flux driven wave that compresses a large number of neutral mattrons slightly before the flux pulse of greater coronal pressure peters out into the vast original grid. Using disparity (like $2,21 \times 10^{23}$) to assess the initial flux share of all the affected mattrons is rough and to be adjusted up by $2,46×10^3$ fold boost.

The gargantuan positive domain of 10^{228} grid mattrons is hardly affected at all by this tiny local spot of compressed positive coronas (not even a negative domain) that has the singular negative photon at its heart. For illustration, the surface spot of a slightly compressed *'entourage'* might cover a surface area of say 10^8 tiers across and affect some 10^{24} mattrons out of 10^{228}.

If the negative singular mattrons departs from the submist grid or is for any reason destroyed, its new compressed domain quickly erodes without its support and the engulfing more massive grid mattrons swiftly reinstate the correct sizes of the disparate neutral mattrons.

Thus, while one singular mattron with adverse emission can easily claim 10^{228} neutrally compressed mattrons as its disparate domain; one singular mattrons entering a vast domain of opposite polarity is likely to compress a pittance of neutral mattrons and remain highly compressed itself.

Postscript to the Mattron

With this, I conclude the discussion on mattrons and their environs, emphasizing as always that this is a carefree illustration of what inevitably flows from the Primal Code.

The uncertainty of mattron fields is ill suited to describe interactions like gravity 'attraction', virgin repulsion and spin forged by mattrons pairs. I hope such so called 'properties' of mattrons come into better focus as we move up to the submist ladder to larger systems.

Because individual mattron interactions are more difficult in a mattron grid where mattron responses are more often than not of a collective nature and not those of individual mattrons, I am not going through the exercise of setting up individual relations of different mattrons at this time.

I reiterate that nothing of all the aforesaid should be of my own choice. The Primal Code does not lend itself to question like; 'what if two mattrons repel each other'. The mattrons do what they must by the laws set by the Primal Code or they do not and my present quest is to have no say in the matter. Mattron behavior, as you have noted, is a great deal more complicated than that of primal atoms. Having made the wrong call more often than I care to remember, I can but faintly hope that I have not unwittingly departed from the strict logical path of the Primal Code.

Although I do not know (as I write this) to what kind of Universe this chain of logic leads, my illustration seems to be fashioning a world from the bottom up that appears to have similar material 'properties' as our real world.

I grant you that my artless method has most likely failed to estimate or spot many wrinkles in this nap. This does not overly trouble me if others eventually see fit to set me straight. I may not be flying as blind as I would with mathematics but I have gone astray often enough to realize that there must be flaws to be corrected.

The time has come to move another step up the system ladder to the level of larger material structures in the Universal primal mist; a submist that has mattrons as its main building block (but not its sole building block).

Although the pitfalls are many in slavishly following a logical chain, these larger globular structures have the benefit of handy reference to quantifiable reality. I turn forthwith to more familiar entities of 'elementary particles' that I treat under the group name 'Hydrons'.

Part 4: The Hydron

The Hydron

One step above the mattron in the chain of collective systems is the hydron, a globular submist to the form but composed of mattrons as its main building blocks; also an inevitable consequence of the Primal Code.

Relations are more intricate on this higher level where three interacting 'structures' vie for influence. I am assuming that hydrons are our everyday subatomic entities but at this point there are no guarantee that they are not parts thereof or a another stepping stone between mattrons and 'elementary' particles. Should I be mistaken that Protons, Neutrons and Electrons are all hydrons; the error should become clear along the way.

Contrary to popular belief, a hydron is too complicated to describe by a single formula to fit all circumstances. This so-called particle comes in many guises; forged by various fields and circumstances, initially carved out of vast grids of neutrally compressed mattrons.

I view a hydron much like a mattron; a vast submist composed of tiny building blocks. The rules that govern consolidation of mattrons into vast gaseous structures carved out of a homogenous mist are much the same as the rules for primal atoms gathering into mattrons; only more gradual and structured and perhaps more dependent on mass availability.

That mattrons are composite building blocks should have a natural feel; we regard subatomic particles as the building blocks of stars - and stars as the building blocks of galaxies.

Although all mattron 'fields' apply to hydrons when adjusted to account for altered relationships between emissions and gravity shadows, the rules that govern the mutual intercourse between these different proactive building blocks do not deliver the static responses of primal atoms that cannot alter their size, their rate of 'emission' or their level of 'charge'.

Hydrons forged by the same laws as mattrons are shaped by proxy and governed by the more indirect representation of primal fields. In this, the hydron is like a senator through which actual conflicts on the subordinate scene are relentlessly channeled. Like a senator, a hydron must be tuned to the will of the interacting primal collective or be destroyed in the process.

The first hydron embryos would start to form near the universal center of the homogenous mattron mist. That nursery is likely found at a center of a future solar embryo near the somewhat denser center of a galactic kernel.

While aspiring formations of other 'particles' like suns and galaxies or even greater structures are already discernible as diffuse contours, an aspiring hydron embryo is tiny compared to such immense mattron grid; comments on the hydrons gargantuan size and mass refer to the mattron scale only.

The most average center within say a 'galactic mist' of mattrons has the greatest average presence of mattrons. Early, I thought that a mists center would supply the heaviest local field pressure within a mattron grid. I soon found that it supplies the least. The primal emission near galactic centers that escapes into the empty spaces around neutrally compressed mattrons will gain the least boost due to the smaller inter-mattronal distances.

Due to greater proximity, the universal center is where darkest long-range linear backhand shadows are cast. Although the backhand push is weak, it causes the distances between mattrons or more flimsy clouds of mattrons to gradually decrease; causing greater mist density, allowing mattrons to draw closer, up to a point. This backhand, by the way, is not what we call gravity.

All the fields dealt with earlier in the collapsing primal mist are at hand. To avoid repetition in studying the formation and evolution of hydron embryos within a vast uniform mattron grid, I refer you to the relevant ideas on the gathering of primal particles into a mattron embryo.

In the Universal mist of mattrons; as myriad mist grids of mattrons gather ever closer, the densest point of a vast volume populated by stable mattrons eventually starts to collapse faster than the rest; a variation of the earlier quantum split. Such immense mists are hydron nurseries where the point of a first hydron collapse sets off an exceedingly slow quantum split that leaves a vast number of hydrons. Due to the eventual lack of mass, this collection may often turn out too meager to form an aspiring sun, let alone a galaxy.

As with mattrons creation, the initial point condensation or 'collapse' creates an empty hollow in a uniform mattron mist, such as it is, and creates an irreparable rift in the 'smooth' fabric of that material space.

On this new level in the system chain, we witness for a second time the breaking of the symmetry of material distribution as our Universe (or rather parts thereof) passes into a state where uniform distances are again lost and more massive 'particles' come into being. Enormous mattron populations are hidden away deep within these gathering structures.

During this early hydron creation, I see no electromagnetic splits of neutral mattrons into singular adverse twins as the disparate stray winds that are required are not present. Hydrons cannot even at that stage turn disparate by fluke which can first happen when grid distances have shrunk to below 10^{881}år (from 10^{900}år) and a fluke mattron disparity cannot spread through that or other hydron grids. If a disparate neutral mattron leaves such a overly large hydron to head across a void, it would simply expand and lose its disparity.

Rough Orientation

My ideal Universal (unit^2)^2 chain of average systems

Submist type	Ideal number of building blocks	Primal mass
Mattron	10^{912}	10^{912}
Hydron	10^{228}	10^{1140}
Sun	10^{57}	10^{1197}
Galaxy	10^{14}	10^{1211}
Galaxy Clusters	10^{3}	10^{1214}

For an ideal chain where mass units are based on Our Suns hydron count, I multiply the 'maximum' radius of a lower submist structure by the average number of internal building blocks in successive higher links to set the 'maximum' radius in the higher submist.

As to the possible span of radii within each tier, I use the square root of the lower systems radius range for the radius span of the higher system and allow submist sizes to vary within that interval. At the top of the structural ladder above star level, mass availability starts to affect the spans and any groupings may require a wider stability range.

Submist	Radius range	Span	Average radius
Mattron	10^{612}ãr - 10^{632}ãr	10^{20}	10^{632}ãr
Hydron	10^{850}ãr - 10^{860}ãr	10^{10}	10^{852}ãr
Sun	10^{912}ãr - 10^{917}ãr	10^{5}	10^{914}ãr
Galaxy	10^{923}ãr - 10^{926}ãr	10^{3}	10^{925} ãr

Reality places us in a Milky Way of fewer suns than anticipated by my ideal chain (not necessarily fewer hydrons); a shortfall that may affect mass numbers and sizes of 'lower' links and annihilation within neutral mattrons.

I also intended to make my material chain relative to our world by fixing it to a precisely measured hydron size. Regrettably I find that the sizes of all the 'fundamental particles' still remain a mystery. The Internet tells me that size is not a useful concept but volunteers that a Proton may have a radius of 10^{-25} meters set by the stretch of its 'quarks'. A reluctant footnote point out that this mathematical result fails outrageously when used in any other respect. It advises me to stop thinking in outdated terms and use 'interaction areas' instead; size being such a bore.

Having geared the structural chain to 10^{632}ãr maximum mattron radius and 10^{860}ãr 'maximum' hydron radius, I shall therefore put my faith in an interaction area where my size fails. Were I to embrace the uncertain interpretations of lepton or quark radii (smaller than 10^{-22}km), I must adjust my mattron radius and measures of time and distance and a host of other fundamental factors. And of course, if anything in the material chain is altered, all the interrelated values are affected and the whole set piece of illustrations must be adjusted to every last detail based on a dodgy mathematical result. Without the luxury of time and given my oversized basket of untreated factors, I don't have the stomach for it.

As measured star sizes of in our Universal backyard are also somewhat uncertain, I elect Our Sun as the average sun and with that the ideal chain becomes set in stone. From there, I have the relativity of every step from primal atoms up through galaxy groups in hand.

A more relevant objection to my ideal chain is the creation constant 'ќ'. As a 'time' based constant, my measure of primal time can be linked to the size of an 'average' sun to allow me to measure a kilometer in primal distance units (ãr) and the duration of an Earthly second in primal time units (ãt). It is troubling that using the ideal chain with Our Sun as an average sun, the Universe will multiply its mass hundred fold every Earthly second. Even with the extremely effective safety valve of mass annihilation at the center of neutral mattrons, many will find this situation uncomfortably racy.

With various aspects waiting to throw a monkey wrench into my heavenly plan (such as singular mattrons); I opt to continue this evolutionary tale, warts and all and refrain from tweaking my creation constant to allow the Universe a more comfortable billion years or so to double its mass. Looking at galaxies, we see average suns, collapsed suns, dispersed suns, planets that almost made suns and a variety of exotic spin-offs. Reality on the hydron scene allows a similarly varied scope but I must pick some measured radius to represent hydron size.

The obvious choice to fit the chain is that a hydron built of 10^{228} times mattron mass should have a radius 10^{228} fold the 10^{632}ãr mattron radius; namely 10^{860}ãr wide radius. This will however soon be seen as at odds with the stability of a standard mattron as 10^{228} evenly spread mattron within the 10^{860}ãr wide hydron are compressed by boost across larger inter-mattronal distances and can neither keep their 10^{632}ãr radii nor 10^{912} mass.

This rock bottom stability appears within **10^{852}ãr** sized hydrons and they shall appropriately serve as 'standard hydrons' of 'standard mattrons'. As the 10^{860}ãr wide hydron is within the hydron stability range of the ideal chain, I shall call a hydron in that expanded state for a 'maximum' hydron.

Measures of Time and Distance

Endorsing Our Sun as the average sun in this perfectly linked universal chain, its 7×10^{914}ãr radius stands in direct relation to its actually measured 7×10^5 km radius and a simple division yields a primal radius of 10^{-909}km.

One kilometer equals 10^{909}ãr.

The constant speed of electromagnetic splits of neutral mattrons into two adverse singular mattrons remains independent of the chain. The Primal Code puts the relatively constant Universal speed of a twin at 10^{-280}¢ that is measured at 3×10^5 km per second. This sets the average primal birthspeed at 3×10^{285} km/sec and links the intimately related measuring units of one kilometer and one second directly to the primal particle.

If one kilometer is 10^{909}ãr and light speed covers 3×10^5 kilometers per second, the primal path traveled by a photon in that second is 3×10^{914}ãr. While a virgin at average birthspeed takes 3×10^{914}ãt to travel that same distance, a slow moving photon at 10^{-280}¢ requires 3×10^{1194}ãt to do so.

One Earthly second equals 3×10^{1194}ãt.

With 10^{-1192} creation constant (k), a mattron corona of 10^{912} primal atoms splits 3×10^{914} newborns in each sign during one 3×10^{1194}ãt long second.

Were it not for the security valve of central annihilation in mattrons, their neutral mass would multiply 100-fold every second (as would all Universal mass). The constant annihilation dampens this creative burst and balances the books. Singular mattrons suppress excess mass into central death clots or release it in primal halos to be harvested by neutral mattrons.

The period of 10^{912}ãt in which an electromagnetic split is consummatedcounts 3×10^{-282} Earthly second. The time a newly split twin takes to travel the width of an 10^{852}ãr hydron radius at 10^{-280}¢ is 10^{1132}ãt and this is a more tangible 3×10^{-62} of a second. For an electromagnetic photon to traverse the 10^{-18}km (10^{891}ãr) often mentioned in textbooks as the size of a hydrogen 'atom' would take 10^{1171}ãt or 3×10^{-23} of an Earthly second.

My 10^{909}ãr measure of a kilometer would give the average textbook reach of Proton 'quarks' at 10^{-28}km a 10^{881}ãr wide radius. This is way out of line although this happens to be the size where 10^{228} mattrons condensing in a 10^{903}ãr hydron embryos can support internal disparity of coronas and cores; an odd coincidence as hydrons change character at this size. All parameters of what today's science calls 'elementary particles' are far below these sizes except during embryonic formation and the term 'interaction area' may be aptly named.

The Collapse into Hydrons

Things were different at the beginning. A hydron embryo condensing into a standard hydron contains enormous multitudes of tenuous mattron clouds that gradual consolidate with in a volume shrinking from a vast embryonic size. To a lesser extent this would also be true at the time of more diffuse submist structures destined to become suns and galaxies.

All links in the chain of universal substructures are in the formative stages in a gradual overlapping progress where a lower system affects condensation and structure of the larger system. In relation to primal population, external emission of a higher system drops but it may still move the building blocks inward at an exceedingly slow pace. The formation of hydrons is always dependent on the larger environment but smaller systems are always first to gather and react and they will collectively govern the larger systems.

From the bottom of the chain, the primal atom governs everything.

The large scale Universal primal mist is not like evenly distributed perfect gas. The original strands of primal particles crisscrossing empty Universal spaces at birthspeed have through eons of time spawned more primal mass along and around their routes that eventually forms an irregular patchwork or consolidating strings of primal clouds across the Universe.

Local quantum splits into mattrons within these less cohesive corporeal mists may leave great enough numbers of mattrons to form a vast number of hydrons but grids big enough to form galactic clouds may be rarer. But local primal clouds continue to beget more primal mass and, exceedingly slowly, such quantum splits would consolidate into a galactic mist of hydrons.

I find that the uniform mattron polarity I envisaged throughout a galaxy runs into a snag of distances and numbers. Neutrally compressed mattrons cannot be polarized until they collapse into hydrons and I must reject my earlier idea of a uniform fluke disparity spreading on a galactic scale.

Even with optimistic mass numbers, our galaxy cannot support disparity before average inter-mattronal distances have shrunk to 10^{805}är. Uniform disparity within a sun becomes possible when each forming hydron embryo of 10^{228} mattrons or more has shrunk to 10^{881}är (excessively large) radius where disparity can be supported in an internal grid of 10^{805}är distances.

Long range emissions drive the slower compression of larger submists in vast primal orbits; gravity drive of a Universal submist, part and parcel, is aided by neutral emission and flux from leftover primal mass surrounding the Universe, galaxies and suns; mass that is too sparse to form mattrons.

Looking out into the Universe, we are unable to detect primal mists but detect denser lumpy patches or clouds where mattrons have already formed hydron embryos. The scarcity of primal mass would prevent quantum splits on a greater scale and, more often than not, the formation of suns may wait for gathering hydron subclouds to consolidate enough to collapse into suns and to repel leftover mist. In such quantum splits, a solar nursery of forming proplyds grids in the Orion Nebula might serve as an illustration.

In the quantum split into hydrons, a lack of mass is less of a problem. A first local collapse at the center of a mattron mist into a hydron embryo creates a 'vast' exterior void carved out of that mattron mist and this crucial ingredient affects both the immense mist of mattrons engulfing the emptier hollow and the hydron embryo at the center of it.

The rift causes the grid to form a denser 'skin' of mattrons on the inner lining of the outside hollow where mattrons in local orbits periodically enter into a more substantial 'wall' and cause it to condense linearly outwards from the hollow instead of inwards; increasing the width of the hollow.

As seen in the formation process of mattrons, the orbital nature of flux did gather more mass into the primal wall that lined the hollow by sweeping exterior primal mass away from the void to be deposited into that denser gaseous shell rather than to be pushed further out into the void.

When a hydron collapses, the reigning orbital stray winds outside an inner hollow lining move orbiting mattrons into more stable exchanges within that mattron 'wall'; impeding linear movement into the emptier void as mattrons in this new 'density wave' grow more numerous on the inner lining.

Mattrons situated in the original mist on the other side of the new density wave (now outside the embryonic hollow), experience similar thrust towards the more opaque outer lining of that density wave (on the other side of it). And outside the first spherical wall, on its further side, orbital condensation gathers more mattrons to that lining and a new void is gradually carved into the mattron mist.

The subsequent gathering into another tenuous inner lining of a second 'wall' outside that second void creates a third void by depletion and so on as a quantum split into denser orbs of mattrons moves outwards in a spherical fashion (until mattrons give way to primal population).

At that point and in perfect accordance with the rules that govern primal atoms, the density wave stops propagating (it never really moved linearly), whereupon the quantum split of that particular mattron wall comes to an end. The spherical walls like uncountable onion skins (surrounding the first hydron embryo) has a sufficient mattron number in each say 10^{903}ãr wide wall patch to sustain a new hydron that are significantly closer within the wall than they were in the galactic mist before the quantum split.

A somewhat wider, say 10^{904}ãr, initial void between the density waves is vast on a hydron scale, especially as primal virgins emitted by compressed mattrons out of such a density wall turn orbital at around 10^{828}ãr and circle right back into the wall grid. Instead of linearly dispersing the immense quantum walls, this emission speeds up the collapse into spherical hydron embryos that soon dot the fading spherical mattron walls. Eventually, no walls of mattrons are to be seen, only an enormous fairly evenly structured grid of hydrons. In whatever direction you look from each hydron embryo, you see another hydron of 10^{228} mattrons or more forming roughly 10^{904}ãr away.

The inter-hydronal is within density walls is smaller than the quantum wall distance. As hydrons communicate with the closest hydrons, it is likely that opposite pulls and pushes facilitates pairing of hydrons. Although the hydron majority is unlikely to form pairs, a sizeable portion will form closer individual bonds and even this hydron mist has a long way to condense.

If a mattron mist is massive enough, a quantum split creates a vast mist of hydrons at a decreasing 'wavelength' out to its edges. On the periphery of the last density wave, all that remains is an amorphous mist of haloed mattrons and further away, the tenuous mist of primal particles extends across to larger neighboring submists, serving as cannon fodder for exchanged flux.

Reality is not always this clean; quantum splits can start early with less mattrons or simultaneously in multiple denser local lumps, rippling from many collapses out to the border of smaller subclouds in a garbled situation affected by insufficient mattron number or grids that are potholed by voids.

As more twists muddy the waters, one potential nut to crack is a present value of long-distance gravity shadows (background emission from the vast primal halo surrounding our Universe).

If some hydrons should rotate in a single plane, not with differential spin, the multitude of mattrons in quantum waves that are gathering into walls of increasing wavelength may take on a very rudimentary form of spirals.

With 10^{904}ãr as the inter-hydronal formation distance of hydron embryos in mattron clouds, it is interesting to note that modern cosmology sets inter-galactic mass density as roughly one nucleus (hydron) per cubic meter volume (radius 10^{906}ãr); hundred times the formation threshold for hydrons.

I have not given much thought to how hydrons unravel but a population of one hydron of 10^{228} neutral mattrons spread into a cubic meter volume must be supported by unraveling into massive precursor halos where the mattrons could easily be ripped apart, possibly along with the hydrons.

Throughout, one thing remains the same. Outside and within each hydron; with primal mass bottled up inside functioning mattrons, primal atoms run the show. Only a primal particle is capable of the inherent disturbance of the random split of a newborn pair. In this capacity, the primal atom is the only engine that keeps the Universe going.

I assume that hydron embryos are built of neutrally compressed mattrons and no singular mattrons play any role in this early evolution. Somewhere in the Universe a few hydrons inside a dense enough mattron grid may turn disparate by fluke as an extremely rare exception. Electromagnetic splits of neutral mattrons are impossible outside vast hydron embryos but smaller sizes soon become more common with rare hydrons polarized by fluke.

It is not the purpose of this treatment to ascertain the actual values of material structures or to combine them in precise equations; it simply to fathom how the principles of The Primal Code govern eventual equations.

To focus on whatever influence a single mattron has within or outside a hydron, I return to the familiar fields of collective influences that also hold court in the world of hydrons, albeit with subtle changes where the primal effects are tuned and sifted through vast structures in interactions between larger more complicated building blocks.

Having accounted for collective mattrons fields earlier, I feel that extensive comments on identical subjects are entirely too repetitious. A direct run of the hydronal fields is ill suited for clarity and I have chosen to focus on the topics that govern hydron structures in a succession that is most favorable to a general understanding of the spell cast by the Primal Code.

The Mass Neutral Hydron

The hydron is a submist built of mattrons. Its structure has a radically different relationship to its building blocks than mattrons have to primal particles. When turning a blind eye to the primal particle and viewing the mattron as a building block, never forget the underlying heritage.

Stepping up from the mattron level to the hydron level, one notes the vast improvement in field continuity where invisible effects from the lower level are 'constantly' at work between the building blocks of the 'higher' system.

The hydrons collective bulk emits and blocks primal atoms 'continuously', which is not to say that mattrons as building blocks are devoid of the erratic nature that plagues lower level continuity. It is just less conspicuous as the hydronal fields act with less noticeable absence of effect and are less prone to the erratic uncertainty that characterizes mattrons. But by no stretch of the imagination can we view this as an unruffled environment.

Each fully formed hydron contains roughly 10^{228} mattrons each built of 10^{912} atoms (in a standard grid). The spatial presence (local movement) of its atoms and flux exchange are part and parcel of its properties. The blocking power of primal movement is real but should not be equated with mass; it cannot split newborns, which remains the property of physical particles.

A primal parent emits a newborn that can kick a nearby atom away at half birthspeed. It can be said (with reservations) that a hit particle casts a total shadow. This massive repulsion and its parallel massive shadow occur in an entirely erratic fashion - sometimes or never.

The primal field is thus extremely effective in communicating its coordinate systems to other primal particles when - once in a blue moon - the field has something to communicate with.

The step-up from the mattron level to the hydron level brings a great drop in the relative effect of the fields of the building blocks; similar to the great drops in field-strength between two primal particles versus the exchange between two mattrons. What the hydron field lacks in brute relative strength is offset by the incredible multitude of primal messages that make its weaker fields immensely more stable and consistent.

I do not know how much loose primal mass engulfs the structured center of the Universe but it must be a source of penetrating primal emission that cuts through its central bulks. With no way of measuring the background pressure, I wave these estimates until a better understanding of the hydron structure is at hand. It is likely to be less problematic to approximate the average primal emissions from local mass that reaches nearby fields where it can serve as inter-hydronal local gravity input.

I cannot yet tell whether the background stray pressure eclipses linear local strays that penetrate the distance of one meter to represent 'G'. But it should be possible to glean the emission levels that penetrate further than one meter and give rise to gravity shadows between hydrons with one meter linear accuracy. For now, I simply assume that 'G' is driven by local galactic fields and not by primal mist engulfing our material Universe that is immensely bigger than commonly imagined (a virgin passes out of the observable Universe in 10^{-260} of a second and has been expanding its size ever since its first primeval atoms were split).

A hydron has no easy way to multiply its building blocks in the way that a mattron does (where primal mass breed up to a limit set by either its central collapse or by the disintegration of its surface). Continued hydron growth by multiplying mattrons is hard in a grid where primal surplus swiftly becomes cannon fodder for exchanged flux but Mattrons stand a better chance of forming in opaque primal clouds in external spaces with primal mass out of the firing line.

Hydrons must deal with a variable blend of internal and external emission inputs from singular mattrons, disparate mattrons, collective emissions of unstable or overweight mattrons and secondary emission from loose primal mass. It is a situation far removed from the static emission and blocking ability of a primal particle.

Bear in mind that it takes less than a second for a singular corona to double its allied mass (and emit similar adverse mass). Facing insufficient pressure, even an overweight singular mattron must get rid of excess primal mass upon loosing balance (or gather a death clot at its center) and it must hemorrhage all or part of that excess into the bowels of a hydron at regular intervals (or eventually have a too large death clots kicked out).

The primal annihilation within compressed mattrons is affected by altered hydron size but in the wrong direction so to speak; i.e. contrary to mattrons. During hydron expansion, inter-mattronal distances grow fastest for surface mattrons and greater boosts in a hydron corona compress the mattrons there and increase the ongoing annihilation.

Even in our times, singular mattrons remain an inconceivable minority. Their peripheral mass production (subsequently annihilated within neutral mattrons) is largely a zero-sum game and it may be common for hydrons to keep singular mattrons of both signs.

Virgin emission that leaves the hydron is re-harvested in exchanges within other hydrons and singular mattrons or constantly disparate mattrons are therefore unlikely to grow uneven masses in the short term. I ignore the long term effects mostly due to the scarcity of singular mattrons; a significant disparity in mass is unlikely to be notable for a long time to come.

Primal mass is randomly emitted for various reasons throughout a hydron interior. The 10^{228} mattrons grid that is a hydron adds mass equal to 10^{230} mattrons every second. This mass is rarely lost out of hydrons but is, as a rule, annihilated at neutral mattron centers. Heavily compressed hydrons do not build central death clots (out of mattrons); less boosted internal pressure cannot push mattrons into close enough proximity to force their masses to occupy the same space for extended periods of time.

Present day hydrons contain so few singular mattrons (relatively speaking) that if separated into singular areas, they are too scarce to play the coronal or core role of singular mass within disparate mattrons. In the most charged hydron conceivable, the insignificant measure of singular mattrons could be theorized but never measured as weight.

Asymmetry

There is a point to be made about the possibility of large scale asymmetry in mass charge; I am yet to see how one polarity can take hold throughout great swathes of our galaxy, Universe or parts thereof. An asymmetry would leave stubborn traces when exchanged emission can locally distort mass balance in the two signs (unless the universal field is of a single disparity).

Just as with small scale asymmetry in disparate mattrons, large scale asymmetry in vast fields of permanently disparate mattrons allows virgins in one sign an easier escape. Adjacent galactic grids with mattrons of adverse polarity would with the passage of time soak up disproportionate numbers and become more massive in the hidden core sign. This sign would escape more easily through the adjacent galactic grid where it is the coronal sign.

A perfectly equal emission in both signs from a hydron requires perfect mass symmetry; if the mattron population is disparate, its arrangement within the hydron is vital; neutral external emission presupposes symmetry.

If the hydrons mattron grid is polarized in both signs, the adverse domain distribution need not be small scale symmetric or uniformly scattered within a hydron. Adverse hydron domains can also be layered on a large scale or even have two-polar mirrored positions equally close to the surface. Mostly, in polarized hydrons, a relatively few singular mattrons in either sign do not play a significant emission role.

I don't yet know if the two types of mattron mass within in our galaxy are of equal numbers but if it has exchanged with another galaxy through eons of time, I must assume that mismatch could have reached proportions that can affect disparity. There are however other things that can affect this such as when mattron disparity is toggled or it forms a precursor halo and a more massive sign becomes an emitting corona instead.

The Hydrons Collective Spinfield

Return for a moment to the wholly fictional model of a perfect and neutral mattron that rotates in a single plane of spin (the opposite orbital directions of the signs having been totally communicated). In such a perfectly fictitious mattron each primal particle is endowed with a symmetric orbital path while all primal spin has already been annulled in internal collisions.

Within a 'perfect' (and imaginary) hydron, things are different. The building blocks are perfect mattrons that are endowed with differential spins but that rarely collide. Their orbits are aligned or annulled by the exchange of mutual primal emission rather than in direct collisions of mattrons.

Mattrons in a perfect hydron model cannot be the non-spinning building blocks like atoms in a perfect mattron model. If a mattron stops spinning, its mass is blown apart by its own straight-line emission. That perfect mattron model must be upgraded, as the perfect hydron is built of perfect mattrons that are differentially spinning; only not locally orbiting within it.

The hydrons gargantuan and 'constantly' emitting population guarantees a fairly smooth distribution of internal virgin emission. The coordinates carried away from spinning mattrons by each newborn are gradually relayed to the rest of the hydron mass where multitudes of eccentricities test their influence in the primal collisions amongst mass of both signs.

The collective spin of an entire Universe results from <u>one</u> primogenitor atom that is the common ancestor to all of its primal atoms. Such common unmatched heritage is hardest to shake (or discover); everything off that pedigree has exactly the same relation to those original coordinates.

In the smaller systems, inherent spin is an echo of latecomer primogenitor particles; common ancestors to great many of primal atoms of that particular structure. New collective spin coordinates in opposite directions are regularly communicated and relatively swiftly consummated in smaller systems like mattrons and hydrons but ever more rarely in the largest submits.

As primal input within and size grows, must directional pushes are already broadly at hand and few overall changes are possible, not least as the drive of a single newborn will be heavily diluted as its flux leaves that path. One would therefore assume that the original inherent spin of a submist is very hard to shake, if not impossible.

The hydrons inherent collective spin is firmly rooted in the coordinates of all unpaired orbital directions brought to bear upon primal residents of its mattrons. Eventually, the process of primal collisions enters all spin denominations (relative to the hydron center) into the collective field, along with a few odd or unmatched ones. In time, whatever orbital coordinates remain unchallenged become the hydrons collective spin of the 'moment'.

The realignment within a hydron is in one sense different from that within a mattron; it comes about without collisions of building blocks that are being 'continuously' affected by their surrounding fields.

The linear movement of a straying mattron is soon thwarted or brought to standstill in the hydrons vast internal grid of repulsion and attraction where constantly shifting spinfields work steadily to corrupt any straight path that is taken by a mattron across a hydron interior.

In a 'perfect' hydron, all the local orbits of mattrons of have been annulled and they form a statuary grid. Any odd influence that could not be annulled within the hydron (as no antidote was found), eventually becomes part of the inherent coordinates of every primal atom in every mattron.

Local internal orbits are entertained by all mattrons but these fall within the collective shadow of hydron mass and cannot cause linear opposition in regard to that collective mass (the relatively tiny orbit of the local coordinate system is itself moving in orbit around the hydron).

All primal splits deliver zero net effect and (relatively) few virgins escape the largest systems and inherent unique spin is rarely realized collectively as an antidote is already aspiring within that mass, if not yet communicated to the whole system. But when primal mass in immense submists gains early common coordinates, it is hard to beat from within.

Needless to say, new spin inputs that are waiting in the wings are ever less likely to be unique on vaster scales. Even so, the common denominator of the past always counts as a submists collective spin of the present.

I arbitrarily used $(M^2 \times \sqrt{r} \times \acute{\kappa})$ to illustrate the consummation a new collective spin for mattrons, which allowed a standard mattron to take on a new inherent spin after a relatively short period of 10^{948}ãt. I used the polarized mattron grid number to increase the communication period of a new unified collective spin; adjusted by the mass and distance relation. The maximum hydrons 10^{228} strong mattron grid then requires 10^{1176}ãt to take on a new collective spin if an odd new and unopposed spin input was available.

This could in time bring surface orbital speeds of all maximum systems up to half the speed of light. Given the slow pace of light, the realization of fast collective spin is not a problem but observation confirms that we live near a sun that spins at a slower speed within a galaxy that is short of mass compared to my ideal chain.

While average values are made to be broken, the time to fully establish a new altered inherent spin in larger submists take long to be realized. Suns would take 10^{1233}ãt (10^{31} Earth years) alter their spin.

In a perfect hydron, after the dissemination of odd influences, the mattron majority is spatially at rest. To an outside observer, this is observed as a slow coordinated rotation of the hydron body while the Universe (that has undergone no such alignment) has gained a new coordinate relation to the hydron. An observer within a perfect hydron would note absence of linear motion of its spinning mattrons while an observer within the hydron would note that the entire Universe has taken to orbiting the hydron.

Although the differential spin projected upon a hydron may be less than perfect, stable hydrons are unlikely to spin on single axes. I assume that each surface mattron for the most part makes a two-faceted circumference before its original spatial relation to the rest of the mattrons comes up again. The perfect hydron can be said to spin 'twice' before its mattron bulk arrives at the spatial coordinates from which it started.

The inherent hydronal spinfield is the concerted orbital movement of all its mattrons about a common center without external prompting; slowest at the hydron center where a central mattron traverses its 'orbit' by rotating in the same time a peripheral mattron completes a relatively immense orbit.

Intensive halo emissions within compressed hydrons as mattrons expand in darker shadows while lesser boosts deliver straighter penetrating emission that communicates an inherent spinfield more effectively outward. This may aid in faster alignment of central spin under pressure.

The communication of inherent spin to mass outside the hydron depends on its external emission field. Even a newborn that escapes a perfect hydron along a straight-line parental axes is also following a differential orbit about a parental mattron, which soon turns into an open differential orbit that at first leads through but ultimately around its ancestral hydron.

Outside a hydron, a parental path can form any number of scatterbrained orbits but when it passes through its parent's orbital plane, it speeds up in the parent's orbital direction in accordance with inherent coordinates. By being temporarily absent from or present at any relevant plane of single spin on its linear journey outward, a newborn receives lesser or greater boost and spatial presence.

This relationship affirms the average $\sqrt{r}$ rule of increase in spatial presence for emitted newborns. Away from its surface, the hydrons spinfield is diluted by more numerous sphere points but the escaping virgins and eventual flux forms a continuous field without blind spots (except for inevitable primal uncertainty).

The hydron is an incredibly empty place; within its 10^{2556}ã standard volume, each mattron is surrounded by 10^{432} times its own volume of 'empty' space. Even if all 10^{228} mattrons were situated on a hydron surface (leaving the gargantuan rest of its volume empty), its surface would still be so empty that on a string of 10^{220} mattron volumes around its circumference only 10^{-8} of its 10^{228} mattrons need share a whole surface orbit with another mattron.

A hydrons external emission communicates the common direction of its inherent spinfield and tends to nudge foreign mass outside the hydron in a consistent unopposed, often differential, direction. Its distribution dwindles by the inverse of $r\sqrt{r}$ for a sphere and by $\sqrt{r}$ for each additional circumference on that sphere. The gravity shadow dwindles by the inverse of the distance squared or by distance (r) for each additional circumference; its backhand attraction diluted by r^2 while repulsion by emission is diluted by $r\sqrt{r}$.

Within a mattron, enforced orbiting of atoms in a central shadow is not possible since primal particles are carried to wherever a single primal collision takes them.

A mattron as a unit can experience enforced movement against backhand virgin shadows cast by hydrons (mattrons rarely block fast linear flux). In larger submits such as suns, enforced orbital motion of hydrons is induced by both virgins and fast flux when expansion and gravity do not match and leave a residue of linear push towards or away from a system center.

For example, during any given period, a surface hydron in the shadow of a solar mass falls in one universal direction (towards the suns center) but is soon caused to fall in another universal direction towards the same center as it is pushed by penetrating flux shadows in forever changing directions.

When such a collective orbit is closed under the spell of a gravity field, we view the motion as part of system spin but this is not accurate as enforced spin is not differential by nature; it is not even spin by nature.

However, any change in system size as it shrinks or expands will affect this 'apparent' spin as the enforced rotation bridges the gap and prevents further collapse or expansion by periodically pitching the accumulated linear gravity push against itself from an opposite direction. The catch is that the inherent spin slows with shrinking system radii while enforced spin speeds up. For example during contraction, a slowing inherent rotation serves as a drag on increasing enforced speed.

Not to put too fine a point on it, a building block in a larger system accrues gravity backhand over time but not all mass lends an effective hand to its enforced spin. The internal backhand to a gravity shadow is cast according to $M \div r^2$ that builds a cohesive pull. However, only a square root of the building block number tugs perfectly center-wise without linear conflict within the system. Potentially, enforced spin builds therefore to maximum speed in accordance with $\sqrt{m} \div r$ (losing all circular constants).

However, if we could isolate a single orbit 'r', we see how a backhand push would drive a surface hydron slowly at start, fastest midways until all drive in the original linear direction is cancelled after the completed orbit and with this specific $\sqrt{r}$ restriction of the accumulated drive towards a system center, achievable point speed of hydrons in enforced orbital free-fall is downsized by $\sqrt{r}$. This would happen with a planet falling around a solar center or a moon around a planet. The Earth for example has an enforced orbit around Our Sun during a relative free-fall rotation period of $\sqrt{m} \div r\sqrt{r}$ which yields a $\sqrt{m} \div \sqrt{r}$ speed at any given point in that relative 'r' orbit, circular or elliptical.

Enforced mattron orbits are part of the hydron rotation but signal no orbital movement by escaping emission. What is signaled is each mattrons actual linear universal direction of the moment. When this is implanted into another local hydron through emission, it signals no coherent orbital message as all the forwarded directional pushes are eventually in internal conflict and any long term accumulation is neutralized.

In principle, building blocks within larger systems, such as hydrons within suns, need not fall in the same direction around a system center. Were it not for the systems inherent spin that 'steers' the sheer internal grid bulk in one given direction and orchestrates the enforced free fall, they could do so. If the enforced spin grows strong enough to eclipse inherent spin as a system shrinks, the spin is therefore likely to turn less differential.

Many effects disturb enforced spin around a large submist center, such as uneven mass distribution, shadows and emission exchanges with nearby systems. But the main factor remains that old or new inherent spin depends solely on the odd influence of unmatched spin input. If that is not present, no inherent spin will be experienced.

Broadly speaking, a single central newborn endowed with inherent spin can rotate one standard mattrons mass once in 10^{912}ãt. At the surface of a standard hydron, the coordinates of a single central newborn can drive the hydrons entire mass of standard mattrons at 10^{-288}¢ around its center in 10^{1192}ãt in 10^{852}ãr wide (differential) orbits. This fundamental principle of the standard submist chain is broken in 10^{860}ãr wide maximum hydrons where the compressed mattrons are

31,6 times lighter than standard. A single central newborns inherent spin can rotate its $3,16 \times 10^{1138}$ primal mass in orbit at the surface where it generates a faster $3,16 \times 10^{-279}$¢ speed around its center; rotating in $3,16 \times 10^{1138}$ãt.

My standard hydron is out of step with the ideal chain but its 10^8 times smaller 10^{860}ãr radius does support standard mattrons while its inherent spin suffers a 10^8 shortfall to 10^{-288}¢. Its enforced spin speeds up to whatever fractional extent is possible within a hydron.

Hydron Transparency

Despite the apparent transparency of mattrons, parental motion renders them opaque to passing strays at standard 10^{36}¢. I have assumed modern hydrons to be built entirely of either neutral or disparate mattrons that come packed in two sizes (coronas and cores) while primal atoms come in one.

A hydrons primal transparency ratio that treats all primal mass as a single cloud takes little account of mattron disparity or primal parental movement and being the same in both signs, is of little practical use. A hydron may at first glance appear transparent to linear strays but shows otherwise when measured by two different methods that are better tuned to disparity.

1. The hydrons *structural transparency* ratio measures the chance for a stray to interact directly with its building block, the mattron.
2. The hydrons *flux transparency* ratio measures the chance for a stray to interact with its inter-mattronal flux.

The 10^{228} mattrons with maximum 10^{632}ãr coronal radii 'block' collectively 10^{1492} out of the hydrons 10^{1704} radial lines (at 10^{36}¢), which leaves the standard hydron wide open to passing strays. On every (mattron-wide) path that blocks a stray by a corona; numerous other such paths are empty. Of 10^{440} mattron-sized standard paths available through a hydron, there are 10^{228} coronas in the way. This leaves the equivalent of 10^{212} mattron-sized (coronal) paths open for each path blocked by a standard mattron corona.

The structural transparency ratio (mattron coronas) of a standard hydron is 10^{212} but this presupposes that the hydron has a single polarity and the outcome would be modified to the extent a hydron carries a different blend of domains (each stray interacts only with its own kind).

At the same time, there is the equivalent of 10^{467} unique mattronal paths through a standard hydron in the dominant sign (measured by $3,16 \times 10^{618}$ãr wide mattron core radii). With 10^{228} disparate cores for 10^{467} paths, for each path blocked in the dominant sign there are 10^{239} open core paths. The standard hydron structural (core) transparency ratio is therefore 10^{239}.

An incoming stray that traverses a hydron and hits a mattron can move that mattron linearly within the hydron while that push remains unopposed by strays from the opposite direction. Mattron motion is a prerequisite for hydron expansion or compression but takes a long time to build and its aim is rarely towards a hydron center.

Given the hydrons structural transparency, the overwhelming majority of strays that hit an internal mattron has not passed through another mattron and when another stray hits the same mattron with matching linear push from an opposite direction, the earlier imbedded motion is neutralized.

In that case a mattron shadow is thrown in both directions to become part of the hydrons collective external gravity shadow but the movement and push upon the affected mattron has been cancelled within the hydron (as far as the two involved strays are concerned) and the mattron stays put.

A mattron is in time affected by inter-mattronal shadows; the backhand to obstructed stray push; to push a mattron towards another mattron requires a stray on a path that leads through both mattrons. For this reason, the ability of strays to move mattrons in internal shadows is trimmed by the structural transparency ratio of a standard hydron to a tiny 10^{-212} fraction of whatever stray pressure its internal mattron coronas harvests (a 10^{-239} fraction for the cores). Nevertheless the structural transparency is highly relevant to hydronal balance.

This is not a static environment; the second method in measuring hydron transparency is the blocking effect of the primal inter-mattronal exchanges between mattrons that are constantly on the wing as activated flux. Much like parental motion within mattrons, the flux presence impedes strays that traverse the hydron; only vastly more effectively than the slow parental motion ever did in mattrons. Were it not for flux, the structural transparency would let most strays pass unimpeded through a hydron.

Being restrained by averages, the exchange of newborns within a standard 10^{852}är hydron is between neutral mattrons across 10^{776}är where activated newborns gain an average 10^{388}ä presence per änna during consummation. The blocking power of a hydron is then set by how many inter-mattronal domains a single flux atom passes though across a hydron and the presence of flux in its sign encountered in each during the time of passage.

A hydron can expand, contract or have asymmetric mass distribution, all of which changes these averages. I look at flux transparency as if a hydron had a uniform mattron polarity. In reality, if its radius expands to 10^{881}är, the internal mattrons lose disparity. If recompressed, a new disparity can be well garbled if there are many singular mattrons present in both signs.

We can look briefly at the flux transparency pitted against a non-boosted linear stray that passes at birthspeed through different hydrons built of 10^{228} mattrons; first the standard 10^{852}är hydron. Note that these models need not be actual hydron sizes and properties but serve as a benchmark for the real thing.

The Standard hydron

In this 10^{852}är wide standard hydron of 10^{1140} atoms that rotates at 10^{-288}¢ in 10^{1140}ät, the subservient emission by its mattrons is mutually harvested by coronas of 10^{632}är radius in one unified domain. 10^{652} subservient virgins are emitted by its coronas ($10^{228} \times 10^{424}$) during a 10^{740}ät flux consummation period. This constant subservient flux level boosted 10^{388} fold moves among its mattrons as collective sweep covers 10^{1040}ã per änna within the hydron.

The 10^{424} subservient virgins in individual exchanges during 10^{740}ät that are on the wing simultaneously see 10^{1588}ã swept in 10^{776}ät on an average virgin run across the 10^{776}är inter-mattronal distances. 10^{-36} of subservient virgins cross intact across 10^{776}är into nearby mattron domains. At 1¢ a stray passes 10^{852} points through the hydron in 10^{852}ät while 10^{764} emitted newborns sweep 10^{1892}ã with a combined 10^{2744}ã spatial presence within the hydrons 10^{2556}ã volume. This marks all but 10^{-188} of such subservient strays collision. A stray at 1¢ has merely a 10^{-188} chance to avoid hitting flux and the speed it needs to avoid the first collision is therefore 10^{188}¢ or more.

Another way of calculating this is that if a flux atom crosses the 10^{852}är double distance, it traverses 10^{76}är inter-mattronal domains that exchange positive subservient flux. Per änna, these domains have a flux presence at 10^{388}¢ during 10^{740}ät consummation from 10^{-316} activated virgins in 10^{76} inter-mattronal domains or 10^{888}ã collectively. A flux atoms 10^{388}¢ period of passage raises this collective inter-mattronal flux presence by 10^{664}ät to 10^{1552}ã and a single flux atoms 10^{852}ã point path combines with that as a 10^{2404}ã collective presence; the same point number as the collective volume of the 10^{76} inter-mattronal domains it crossed. And again, the Neutron engages passing flux at or below speeds of 10^{188}¢ in the positive sign.

The 10^{496} dominant virgins in individual coronal exchanges during 10^{776}ät sweep 10^{1642}ã collectively on the average inter-mattronal run and during a shorter $5{,}62 \times 10^{746}$ flux consummation, $1{,}78 \times 10^{-79}$ of all dominant virgins make it untouched across the exchange distance.

During 10^{852}ät some 10^{800} dominant virgins ($10^{228} \times 10^{572}$) are emitted within the hydron. With 10^{388}¢ boost during a shorter $5{,}62 \times 10^{746}$ dominant flux consummation, the 10^{800} collective dominant flux sweeps $5{,}62 \times 10^{1934}$ã within a hydron while a stray at 1¢ passes 10^{852} points through it. The $5{,}62 \times 10^{2786}$ã combined presence within a hydrons 10^{2556}ã volume marks all but $1{,}78 \times 10^{-231}$ dominant strays at 1¢ for collision. Dominant strays have $1{,}78 \times 10^{-231}$ chance of avoiding flux and the speed they need to keep in order to pass through without collision is $5,62\times10^{230}$¢ or more.

To set flux opacity of standard hydrons as 10^{188}¢ in the subservient sign and $5,62\times10^{230}$¢ in the dominant sign, collisions at these speeds by linear strays must lead to capture. The average parental sweep to 10^{24}¢ within a hydron increases presence 10^{24} fold; once-hit pairs meet 10^{24} new collisions andexponential collisions. It takes the resulting flux longer to exit the hydron, which further spikes the spatial presence as linear flux traveling ever slower across a hydron, sets up more ever collisions. In the case of first collision, the odds are in both signs stacked against escape out of the hydron and I assume that a single collision beds for sufficient repeat collisions for an ultimate capture of all that flux.

Although a hydron is opaque to strays at lesser speeds in the respective signs and the resulting flux is redirected into mattrons, the internal shadows from long distance primal exchanges (between hydrons) are no match for the vastly more intense and local (inter-mattronal) exchanges.

The direct blockage by 10^{228} mattron spheres is irrelevant compared to the long-distance (excess of one meter) gravity shadow cast by *hydron flux* and any calculation of 'G' must use flux opacity to measure hydron gravity. That means that an opaque hydron sphere dictates its external gravity shadow; not the collective spheres of its mattrons. The hydrons ability to block dominant strays is $5,62\times10^{42}$ times greater; its flux traps strays at $5,62\times10^{230}$¢ while subservient flux traps $1,78\times10^{-43}$ of strays at that speed. As to virgin gravity shadows; hydrons can easily block 'all' linear virgins; even at greatest possible 'parental' boosts.

A stray that is carried out of its push-effective line as it traverses a hydron and is deposited into a mattron is unlikely to be able to push any mattrons together as in theory they bypass standard disparate mattrons in 10^{212} to 10^{239} of all cases (structural transparency ratio). And if their coordinates do not aim through two mattrons to begin with, they cannot create a <u>full</u> linear shadow between them when redirected by inter-mattronal flux although some of that linear pressure is effective on a larger than mattron scale.

The hydrons structural transparency renders internal shadows ineffective upon its mattrons and this gives rise to a difference between external gravity fields and the internal gravity compression distilled from straypressure. If a stray majority derives from a hydrons relatively puny exchange of primal emission, its gravity pull on the macro scale pales against the raging storm of similar exchanges within a hydron. Fast orbital strays from the local field pass repeatedly through the hydron (initially at over 10^{450}¢) until they are geared down enough by flux to be harvested by some of its mattrons. Away from a standard hydron, blocked strays give rise to local gravity shadows where the backhand pressure equals the product of all blocked strays and their average speed.

I set **10^{188}¢ in the subservient sign** (which is both signs if a hydron is not polarized) as stoppable stray-speeds for a standard hydron model populated by standard mattrons while it is $5,62×10^{230}$¢ **in the dominant sign**.

The Maximum Hydron

I am now deep in the maze of rough assumptions in an expansive submist of 10^{228} mattrons; a stable functioning 10^{860}ãr hydron that rotates at 10^{-280}¢ in 10^{1140}ãt. I am concerned about broad assumptions on core responses that earlier seemed of less consequence but here suddenly turn decisive.

Hydron embryos collapse out of mattron grids all passed the 'maximum' 10^{860}ãr stage (hundred million fold the 'standard' radius) as they continue to shrink into standard hydrons. Here the average mattron grid distance is 10^8 times larger than 'standard' and newborn exchanges gain 10^4 fold standard orbital boost (10^{392}¢) across a 10^{784}ãr average inter-mattronal divide.

Its dynamic building blocks respond in average logical unison to exterior effects where the initial response to 10^4 fold inter-mattronal pressure is to compress coronal radii of disparate mattrons to $3,16×10^{630}$ãr where mass is annihilated to $3,16×10^{910}$ and parental spin slows to $5,62×10^{-247}$¢. With disparity, the cores are compressed to $3,66×10^{617}$ãr where mass is put out of commission to $1,07×10^{904}$ atoms and parental spin slows to $6,56×10^{-260}$¢.

The coronal blocking speed by $3,16×10^{630}$ãr mattrons drops 178 fold to $5,63×10^{33}$¢ and subservient flux boosted to 10^{392}¢ allows consummation across 10^{784}ãr in $1,78×10^{750}$ãt. The 10^{228} mattrons with $3,16×10^{910}$ strong corona split $3,16×10^{806}$ virgins in each sign during 10^{860}ãr and $1,78×10^{-34}$ of the subservient virgins escape out of coronas, namely $5,62×10^{772}$ virgins.

During 10^{860}ãt, $5,62×10^{772}$ subservient virgins are let out of coronas within a hydron and with 10^{392}¢ boost during $1,78×10^{750}$ãt consummation, the flux sweeps 10^{1915}ã collectively while a stray at 1¢ sweeps 10^{860}ã. The combined 10^{2775}ã presence in a hydrons 10^{2580}ã volume marks all but 10^{-195} of these strays for collision with 10^{-195} chance of avoiding flux and in order to avoid a first collision the subservient speed must exceed 10^{195}¢.

During 10^{860}ãt, $3,16×10^{806}$ dominant virgins are emitted within a hydron and with 10^{392}¢ boost during $5,23×10^{756}$ãt consummation, that collective flux sweeps $1,65×10^{1955}$ã while a stray at 1¢ sweeps 10^{860}ã with a combined $1,65×10^{2815}$ã presence in a hydrons 10^{2580}ã volume marks all but $6,06×10^{-236}$ dominant strays at 1¢ for collision with $6,06×10^{-236}$ chance of avoiding flux and stray speed must eclipse $1,65×10^{235}$¢ to avoid a first collision.

The flux ripped out of each mattron core and held constantly on the wing at all times has secondary emission well within what core mass can give up and cannot therefore compress coronas while on the wing. As it is internal flux that traps most linear strays and not mattrons, the hydron casts different long-distance gravity shadows in the two signs and this long distance harvest is further affected by their size.

This shows up important differences between maximum and standard hydrons. It does not matter if the mattron cores are vastly smaller and have less active mass than coronas when affected by linear strays; it is the push upon the internal flux in either sign that sets all mattrons and therefore the hydron as a whole on the move.

Long-distance shadows may seem different from local highly orbital and linearly thwarted flux exchanges between hydrons that are harvested at any hydron size after passing through them repeatedly until sufficiently slowed for harvest. This however is true of all orbital universal flux.

For a long distance gravity shadow, I consider the harvested long distance stray push to be equal to its stoppable speed; the higher the stopped speed, the higher the pressure and its backhand shadow. It may be better to use another sphere method and call it a 'dynamic sphere' if the blocked flux from a constant input in the two signs is not shared equally by all the hydrons in the field. It poses a problem to use a sphere for flux shadow backhand when a hydrons high speed blocking ability governs a cast shadow and backhand; not its sphere area. By hard-wiring the 10^{1704}ã blocking area of a standard hydron to its 10^{188}¢ blocking speed, I could for example translate shadow areas of other hydrons into 'dynamic spheres' of relative size.

Whether or not a linear long distance gravity shadow is of equal strength in the two signs, an individual hydrons dynamic sphere can share differently of the finite flux input. That hydrons of each sign are of a similar size and flux is divided upon the entire hydron population sets an average gravity shadow but not the specific one.

The long distance dominant flux shadow from a 10^{860}ãr wide maximum hydron is harvested $2{,}94 \times 10^3$ times more effectively at $1{,}65 \times 10^{235}$¢ than the standard $5{,}62 \times 10^{230}$¢ while its 10^{16} times larger sphere, does not gather $2{,}94 \times 10^{19}$ times the linear gravity shadow gathered by a standard hydron. Its $2{,}94 \times 10^3$ fold dynamic sphere casts a $2{,}94 \times 10^3$ times darker gravity shadow, which would mean that it has a dynamic sphere that covers $2{,}94 \times 10^{1707}$ã (versus the standard 10^{1704}ã dynamic sphere).

On the receiving end of this gravity shadow, the $31{,}6$ times lighter maximum hydron is pushed $9{,}29 \times 10^4$ times more effectively by its long distance gravity backhand than is a standard hydron.

I leave **10^{195}¢** in the subservient sign as stoppable stray-speeds for the maximum 10^{860}ãr hydron model (that is internally unstructured and with a mattron population of uniform polarity) while it is **1,65×10²³⁵¢** in the dominant core sign. But with these scribbled estimations unchecked for correctness, it would be a mistake to take my word for it. Note that as the Universe is likely to contain hydrons polarized in both sign, few flux subservient atoms would be left to be blocked in the subservient sign.

I do not wish to complicate things in general by counting extra presence within the hydron gained by the relatively slow linear movement of mattrons in the time it takes a stray to traverse a hydron. Given inter-mattronal flux opacity, a mattrons blunted motion plays a minor role in stray capture.

Flux boosted across a galactic radius can reach 10^{463}¢ (it can reach 10^{926}¢ if left in its original plane but additional paths continuously spreads the coordinate thrust in all directions underway and the effect slows towards the average). Even using the 10^{463}¢ average boost and the extremely unlikely scenario that all the flux of an activated newborn pushes one mattron in its idiosyncratic orbit at 10^{-449}¢ shows the induced movement to be insignificant and the mattron encounters many strays from broadly opposing directions.

The pace at which a blocked stray can move mattrons is extremely slow. The 10^{36}¢ highest stoppable speed can move a mattron at 10^{-876}¢ and that backhand takes 10^{1508}ãt to move it across one radius (practically an eternity and too marginal to affect the rate of entrapment). The speed mattrons gain in local orbits (and that neutralizes mutual attraction and repulsion) is therefore not fast enough to significantly increase blockage of passing strays.

The Minimum Hydron

This smallest stable 10^{850}ãr hydron of standard mass rotates at 10^{-290}¢ in 10^{1140}ãt with 100 times shorter than standard radius and is at the end of the stability tether. I assume that hydrons can shrink towards this end of the hydron size span in the tightest of all grids where pairs of hydrons or groups of hydrons gather under heavy local mutually exchanged virgin pressure.

Using 10^{632}ãr average coronal radii for mattrons might be understood as some coronas are smaller (which is all right) while other are larger (which is not). In this small hydron, mattrons must gain balance from secondary flux emission (which is also all right). I am more at ease with a hydron where mattrons can uphold their maximum 10^{632}ãr coronal balance but the world is not construed to make me feel at ease.

If this small hydron contains 10^{228} standard mattrons, the average 10^{774}ãr inter mattronal distances have shrunk hundred fold and yield insufficient boost and the inter-mattronal flux cannot hold back the standard coronas. As no physical haloes of structural frequency are exchanged in a grid the precursors lose free mass out into the grid, enough to split 10 times more than the standard mattron allows to escape and this carries between the mattrons as a substitute for 10 times smaller inter-mattronal boost. This causes a temporarily compressed grid hydron to lose its precursor, either individually or in more collective bursts, only to be forced to repeat that performance.

Here the activated subservient emission by free mattron mass is mutually harvested by standard 10^{632}ãr coronas and 10^{651} such virgins ($10^{228} \times 10^{423}$) are emitted in the shorter 10^{738}ãt flux consummation. The subservient flux is constantly on the wing and boosted 10^{387} fold amongst its mattrons. The sweep of each activated newborn covers 10^{1038}ã per ãnna within the hydron.

At 1¢ a stray passes 10^{850} points through the hydron in 10^{850}ãt while 10^{763} emitted newborns sweep 10^{1888}ã with a combined 10^{2738}ã spatial presence within the small hydrons 10^{2550}ã volume; marking all but 10^{-188} subservient atoms at birthspeed for collision. A stray at 1¢ has 10^{-188} chance to avoid hitting flux and the speed it needs to avoid the first collision is 10^{188}¢ or more; this is the same blocking ability as a standard hydron.

The 10^{494} dominant virgins in individual coronal exchanges during 10^{774}ãt sweep $5{,}62 \times 10^{1625}$ã collectively on the average inter-mattronal run and during a $5{,}62 \times 10^{744}$ flux consummation, $1{,}78 \times 10^{-78}$ of all dominant virgins make it untouched across the shorter inter-mattronal exchange distance.

During 10^{850}ãt some 10^{798} dominant virgins ($10^{228} \times 10^{570}$) are emitted within the hydron. With 10^{387}¢ boost during a shorter $5{,}62 \times 10^{744}$ dominant flux consummation, the 10^{798} collective dominant flux sweeps $5{,}62 \times 10^{1929}$ã within a hydron. A stray at 1¢ passes 10^{850} points through it and the combined presence of $5{,}62 \times 10^{2779}$ã within the hydrons 10^{2550}ã volume marks all but $1{,}78 \times 10^{-230}$ dominant strays at 1¢ for collision. Dominant strays have $1{,}78 \times 10^{-230}$ chance of avoiding flux and the speed they need to keep in order to pass through without collision is $5{,}62 \times 10^{229}$¢ or more.

The flux blocking power of a minimum 10^{850}ãr hydron as 10^{188}¢ in the subservient sign and $5{,}62 \times 10^{229}$¢ in the dominant sign. All hydrons start as embryos at vast sizes up to 10^{900}ãr and collapse towards standard size in a grid. Looking at the blocking speed by internal flux of three benchmark hydron models within a stable hydron span, the blocking speed of their opacity is as follows;

	Subservient sign	Dominant sign
Maximum size (10^{860}ãr)	10^{195}¢	$1{,}65 \times 10^{235}$¢
Standard size (10^{852}ãr)	10^{188}¢	$5{,}62 \times 10^{230}$¢
Minimum size (10^{850}ãr)	10^{188}¢	$5{,}62 \times 10^{229}$¢

The Hydron Aura

To provide a conceptual crutch, a case can be made that the empty space outside all spherical collective primal bodies is composed of spatial layers or belts of varying thickness and with different characteristics. Although this is also true of mattrons (in theory at least), their 'belts' are for obvious reasons burdened with far greater instability and unwieldy variance.

For a hydron, the closest such belt would begin at the periphery outside its outermost top layer mattrons and stretch away from that theoretical surface from the hydron to one additional hydron radius further out.

For lack of a better word, I call this first distance belt for the hydrons 'aura' and view this 'empty' padding of an extra hydron radius (whatever the hydron size) as an integral part of the hydron rather than an independent spatial belt outside it.

In my structural models I use this conceptual brake-off point called the 'surface' in lieu of a point object. The presence of my building blocks does not graduate from a central density into vacuum but drops off abruptly at some distance at which point space adjacent to that conceptual surface becomes empty of structured building blocks.

When a maximum 10^{860}ãr wide hydron shrinks ten billion times to reach standard 10^{852}ãr width, its 'aura' shrinks with it.

Although mostly lacking in structural entities like mattrons, the aura is never completely 'empty' and to study this is useful in understanding what happens within the aura where several influences vie for supremacy. This is a place where a hydron wields its most massive virgin shadow and where its locally boosted outgoing emission can affect wayward mattrons. Studying the aura helps to lay bare the major field lines of how mattrons that wander into the empty aural spaces are affected by the various stray winds.

Great spatial presence affords activated local flux reaching into the aura in orbits out of the hydron surface little staying power. The output of activated emission by 'stable' hydrons gains a boost of from 10^{426}¢ to 10^{430}¢ within the aura, causing it to interact in the aura. Being swiftly swept into the hydron or inter-hydronal orbits, flux peeking out of a hydron has little staying power and leaves scant activated presence in either sign within the aura.

The spatial presence of long distance inter-hydronal flux emitted by other hydrons and activated under way, is too small within the aura to extensively interact with the hydrons newly emitted virgins passing out through it. This dilution trivializes the role of flux as an activation agent for newly escaped virgins passing out through the hydron aura.

External Hydron Emission

Hydrons exchange primal pressure that escapes hydrons directly or issues from mattrons that escape the hydrons. Hydrons can expand and release halos that are relatively slow moving; built of mattrons that in turn emit a secondary primal emission that increases pressure in the hydron grid and resist hydron expansion. The hydron system depends heavily on the levels of pressure that are exchanged across the average grid distances.

As to directly escaping hydron emission, virgin newborns at birthspeed in parental orbits have the best chance to escape from deep within hydrons due to a relatively slower (parental) boost.

Activated primally boosted newborns have a tiny chance of reaching and ultimately escaping a hydron 'surface'. Linearly outwards aiming newborns activated within a hydron with a presence of 10^{426}-fold physical sweep have trouble escaping out of and away from the hydron (even from its top surface layer) as outwards driven flux circles innumerable times in and out of each mattron about coordinate centers that remain within the hydron.

I want to underline that though flux and newborns that exit a hydron gain great spatial presence, it does not increase the inherent 'away' message that is delivered by the driving newborn. Unlike boosted sweep, linear repulsion by a virgin or its activated flux does not increase although the sweep assists in the carrier's ability to communicate the original finite 'away' message. The end effect of the Omni-directions repulsion delivered by one virgin or all the myriad bits of its activated flux are, with all counted, only the original single digit linear push from the newborns coordinates.

That message is more effectively distributed outside a hydron, as activated flux is better equipped to locally distribute the inherent one digit linear repulsion than is a straight traversing stray from far off at great speed with no local boost. The problem with the linear efficiency of collision prone flux is that it is quickly implanted with other contradicting linear directions that tend to neutralize its linear push.

Every virgin that ultimately takes its leave of a standard hydron makes on average 10^{24} rounds about that ancestral hydron (and its parent within it). How many outside orbits a virgin covers before it reaches another hydron in the grid depends on the inter-hydronal distance; for example 10^{42} orbits across 10^{888}år. It also depends on hydron size that sets mattron sizes and parental movement that gives virgin boost its starting point input.

While the linear aim of activated flux is widely neutralized in its swarm by a head-on direction, a virgin that hits the next hydron can push it away with a full birthspeed kick in whatever direction incumbent on its parental orbit. Its untainted movement is always 'away' from its parent.

Initially, the activated flux is similarly focused linearly in all but no specific linear direction around and away from the emitting hydron. But after myriad collisions in mainly two head-on linear directions between two hydrons; the directional push has been heavily decimated.

The tiny part of the activated flux that travels further afield than the average inter-hydronal distance continues to be boosted orbitally by the square root of the distance traveled (now measured in hydron radii). But the overwhelming flux majority is already linearly slow enough to be harvested by the nearest hydron with heavily downsized linear push while its orbital kick remains largely intact.

Escaping primal emission and collective gravity shadows are projected and harvested by outside mass in accordance with the property of the interacting material system. There is a variance in how a primal particle, a mattron or hydron (all with different relationship between mass, spheres, volumes and stoppable speeds) react to primal emission in virgin state or activated state.

Mattrons within a hydron surface are subjected to very different types of pressure than mattrons just outside or clear of a hydron. Pressure in the hydrons intense internal strayfield is shared in local exchanges between its mattrons while mattrons outside its 'surface' drop out of contact with this ordinary flux and form supportive primal halos. Extremely close to the top layer surface the mattrons will drop more gradually out of flux contact but we will come to that later.

In massive oscillating upheavals where hydrons can float large numbers of mattrons into the strayfield on a cohesive front that allows emitted mattrons to stay sufficiently close to mutually exchange inter-mattronal pressure in a lattice grid to harvest mutual flux without ripping each other apart or form precursor halos.

Hydron opacity derives from the internal movement of activated flux between its mattrons whereas the passive blocking power of 10^{228} mattrons is marginal, keeping collective areas that cover 10^{-212} of the hydron area.

Even in the rare case of a linear stray scoring a direct hit on a mattron; it is blocked at no more than 10^{36}¢ whereas the collective inter-mattronal flux fills the hydron and covers the entire area of its target disc; stopping linear strays at standard $5,62 \times 10^{230}$¢ in the dominant sign and 10^{188}¢ in the subservient sign.

A standard hydrons freely roaming internal emission during 10^{852}ãt radius time (escaping its 10^{228} mattron coronas) in a permanently and uniformly polarized hydron counts 10^{764} allied newborns and 10^{800} adverse newborns.

Activation curtails the chance of flux to escape out of a hydron; exposing linearly slowing flux to ever more local more flux due to 10^{402}-fold difference between 10^{24}¢ parental boost and 10^{426}¢ primal boost in the time it takes to cross the hydron. This being said, it is a misinterpretation to call external escape of activated flux negligible, given the enormous numbers of newborns but they face more daunting hurdles than do virgins. Of all dominant or subservient virgins emitted within a hydron a 'notable' fraction in both signs escapes the hydron.

Of 10^{764} subservient newborns emitted by its 10^{228} mattrons during 10^{852}ãt from deep within a standard hydron of uniform polarity, 10^{-614} of the activated newborns escape it without a second collision and 10^{150} once-hit pairs exit the its surface in large fast orbits at 10^{426}¢ while moving away at half birthspeed. These 10^{150} activated newborns provide a meager spatial presence outside the hydron surface where they collectively sweep 10^{576}ã spatial points per ãnna.

As to the dominant sign; of 10^{800} newborns emitted and activated deep within the hydron, $1{,}78 \times 10^{-657}$ can exit the hydron in 10^{852}ãt and merely $1{,}78 \times 10^{143}$ newborns reach outside the surface with a 10^{426}¢ sweep that collectively covers $1{,}78 \times 10^{569}$ã per ãnna.

There is an exception of flux in the lowest aura that we come to later but the presence in both signs of escaping activated flux from deep within the hydron is too small to interact extensively with outside flux or activate the virgins that exit the hydron surface at the same time. Escaping flux or flux that 'peeks' out of a hydron in large (internal) orbits may also occasionally end up as easy cannon fodder for enormously fast inter-hydronal flux.

Virgin Emission by Standard hydrons

Look at the spatial obstacles facing newborns in a virgin state that escape from deep within a standard 10^{852}ãr hydron of permanent positive polarity. Mattrons are too few, too small and far between to achieve significant effect in direct blockage with their bodies as even with 10^{24}¢ side sweep, 10^{-188} of passing virgins hit a corona while 10^{-224} can hit a mattron core.

In the 'subservient' sign, the 10^{740}ãt consummation of inter-mattronal flux within a standard hydron allows flux from each virgin to sweep 10^{1128}ã while in the 'dominant' sign; a $5{,}62 \times 10^{746}$ãt consummation allows flux to sweep $5{,}62 \times 10^{1134}$ã. In standard 10^{852}ãt radius time, collective internal emission out of mattrons is 10^{800} dominant and 10^{764} subservient virgins. Dominant flux sweeps $5{,}62 \times 10^{1934}$ã points in that time while the subservient input sweeps 10^{1892} points; a $5{,}62 \times 10^{42}$ fold difference.

We recall that the individual spin at the center of the standard parental coordinate system is 10^{-402}¢. Across the 10^{852}ãr hydron, a virgin is boosted 10^{426} times to an average orbital speed of 10^{24}¢ and sweeps on average 10^{876} points on exiting the hydron, while moving outwards at 1¢.

10^{764} allied coronal virgins are emitted by the 10^{228} mattrons in 10^{852}ãt, all making a similar run. The flux from each virgin of the activated majority sweeps 10^{1128}ã (or 10^{1892}ã collectively) and each virgin has swept on average 10^{876} points on exiting the hydron in orbits larger than its radius to leave its 10^{2556}ã standard volume. The average outwards escape rate at 1¢ from deep within a standard hydron of stable disparity is 10^{-212} and **10^{552} subservient virgins** exit intact in 10^{852}ãr-wide orbits (that on average have the hydron center as their coordinate center) and 10^{24}¢ orbital speed.

The exit path of a dominant virgin faces a combined $5{,}62 \times 10^{2810}$ã presence within the hydrons 10^{2556}ã standard volume and $1{,}78 \times 10^{-255}$ of dominant virgins escape the hydron without colliding with flux. Of 10^{800} virgins split within a standard hydron of stable disparity, **$1{,}78 \times 10^{545}$ dominant virgins** escape the hydron surface to join the $5{,}62 \times 10^{6}$ times larger subservient crop in 10^{852}ãt.

Most of the 10^{552} virgins that exit the surface of the singular hydron in an 10^{852}ãt orbit around its center at 10^{24}¢ will continue outwards until they hit a faraway hydron although many can also re-enter the ancestral hydron. The collective repulsion pressure of all outgoing virgins, from deep within, deliver the average birthspeed push held by each newborn away from the hydron.

Most escaping virgins are eventually activated by flux between hydrons and captured in the local hydron grid. A few virgins penetrate further afield, a minority embarking on a long trek out among suns and ultimately galaxies but their chance of hitting another hydron underway is slim and the number continues to dwindle in collisions with the own activated kith.

Emission by Minimum Hydrons

The smallest 10^{850}ãr hydrons may be stable in close pairs. In the 'subservient' sign, the 10^{738} ãt consumption of inter-mattronal flux within a standard hydron allows 'dominant' flux from one virgin to sweep 10^{1125} ã in $5{,}62 \times 10^{744}$ãr consumption where flux from each split sweeps $5{,}62 \times 10^{1131}$ã. In minimum 10^{850}ãt radius time, the collective internal dominant emission is 10^{798} out of standard mattrons. I internal subservient emission substitutes for 10 times lesser boost with tenfold splits or 10^{763} that drive primal surface mass to escape and act between mattrons (no physical haloes of structural frequency are exchanged).

With 10^{-402}¢ individual spin at the center of a standard parental coordinate system, virgin boost across 10^{850}ãr is 10^{425}-fold at orbital speed 10^{23}¢ that sweeps 10^{873} points on exiting the hydron at linear 1¢.

10^{763} subservient coronal virgins escape 10^{228} mattrons in 10^{850}ãt and flux from each virgin of the activated majority sweeps 10^{1125}ã across the smaller inter-mattronal distance (or 10^{1887}ã collectively) and each virgin has swept on average 10^{873} points on exiting the hydron in orbits larger than its radius to leave its

10^{2550}ã minimum volume. The average escape rate from within a minimum hydron of stable disparity is 10^{-210} and 10^{553} subservient virgins exit its 10^{850}ãr wide sphere in radius time with 10^{23}¢ orbital speed. Though tenfold the emission of a standard hydron, it is thousand fold per ãnna and a ten times slower 10^{23}¢ side sweep reaches 10^{24}¢ first 10^{852}ãr away.

The exit path of a dominant virgin runs the gamut of $5,62\times10^{2802}$ã total flux presence within the hydrons 10^{2550}ã minimum volume and $1,78\times10^{-253}$ of dominant virgins escape the hydron without colliding with flux. Of 10^{798} dominant virgins split within a minimum hydron, $1,78\times10^{545}$ dominant virgins escape it to join the $5,62\times10^{7}$ times larger subservient emission. This equals the dominant emission of a standard hydron in its radius time and is therefore hundred fold per ãnna with a ten times slower 10^{23}¢ side sweep that also matches the standard boost first when 10^{852}ãr away.

Virgin Emission by Other Hydrons

10^{854}ãr hydrons virgin emission

I want to look briefly at a few hydron sizes before standard size of full internal mattron mass; starting with radius 10^{854}ãr. In the 'subservient' sign, $3,63\times10^{742}$ãt consummation of inter-mattronal flux within this hydron lets the flux of each virgin with 10^{389}¢ initial boost sweep $3,63\times10^{1131}$ã while $1,74\times10^{749}$ãt dominant consummation lets its sweep $1,74\times10^{1138}$ã. In 10^{854}ãt radius time, the 10^{228} mattrons with $4,22\times10^{911}$ atoms collectively emit $4,22\times10^{801}$ dominant and $1,53\times10^{766}$ subservient virgins into its bowels. The dominant flux sweeps $7,34\times10^{1939}$ã in radius time while the subservient input sweeps $5,55\times10^{1897}$ points.

Across the 10^{854}ãr hydron, a virgin is boosted to an average orbital speed of $3,41\times10^{24}$¢ and the virgin sweeps on average $3,41\times10^{878}$ points on exiting the 10^{854}ãr wide hydron while moving outwards at 1¢ in opening orbit. With $5,55\times10^{1897}$ subservient flux presence by 10^{228} mattrons in 10^{854}ãt, the $3,41\times10^{878}$ã presence of each virgin making a similar run combines as $1,89\times10^{2776}$ã before it leaves the 10^{2562}ã hydron volume. The average escape rate from this polarized hydron is $5,29\times10^{-215}$ and $8,09\times10^{551}$ subservient virgins exit intact in 10^{854}ãr-wide orbits at $3,41\times10^{24}$¢ orbital speed.

The $7,34\times10^{1939}$ã dominant flux presence combines with the $3,41\times10^{878}$ã virgin presence as $2,5\times10^{2818}$ã and the average dominant escape rate is here 4×10^{-257} of virgins. Of the $4,22\times10^{801}$ virgins split in the hydrons radius time, $1,69\times10^{545}$ dominant virgins escape the hydron to join the $4,79\times10^{6}$ times larger subservient crop in 10^{854}ãt.

10^{855}ãr hydrons virgin emission

In the subservient sign, the $7,03\times10^{743}$ãt consummation of inter-mattronal flux allows a virgins flux with $3,16 \times 10^{389}$¢ boost sweep $2,22 \times 10^{1133}$ã while a $3,09 \times 10^{750}$ãt dominant consummation lets its flux sweep $9,76 \times 10^{1139}$ã. The 10^{228} mattrons here hold $2,74 \times 10^{911}$ coronal atoms and emit collectively $2,74 \times 10^{802}$ dominant virgins and $1,93 \times 10^{767}$ subservient virgins into its bowels in 10^{855}ãt radius time, during which time the dominant flux sweeps $2,67 \times 10^{1942}$ã and the subservient input $4,28 \times 10^{1900}$ points.

Across the 10^{855}ãr hydron, a virgin is boosted to an average orbital speed of $6,26 \times 10^{24}$¢ and sweeps on average $6,26 \times 10^{879}$ points on exiting the 10^{855}ãr wide hydron, while moving outwards at 1¢ in opening orbit.

The collective $4,28 \times 10^{1900}$ subservient flux presence and the $6,26 \times 10^{879}$ã presence of each virgin making a similar run combines as $2,68 \times 10^{2780}$ã on leaving the 10^{2565}ã hydron volume. The average escape rate from within this polarized hydron is $3,73 \times 10^{-216}$ and $7,2 \times 10^{551}$ subservient virgins exit intact in 10^{855}ãr-wide orbits with $6,26 \times 10^{24}$¢ orbital speed.

The dominant $2,67 \times 10^{1942}$ã flux presence and the $6,26 \times 10^{879}$ã presence of a single virgin combines as $1,67 \times 10^{2822}$ã and the average dominant escape rate is $5,99 \times 10^{-258}$. Out of the $2,74 \times 10^{802}$ virgins that are split in radius time, $1,64 \times 10^{545}$ dominant virgins escape the hydron to join the $4,39 \times 10^6$ times larger subservient crop in 10^{855}ãt.

10^{856}ãr hydrons virgin emission

The $1,34 \times 10^{745}$ãt subservient consummation and 10^{390}¢ boost of activated virgins lets its inter-mattronal flux sweep $1,44 \times 10^{1135}$ã while $5,44 \times 10^{751}$ãt dominant consummation covers $5,44 \times 10^{1141}$ã in 10^{856}ãt radius time. Each of 10^{228} mattrons holds $1,78 \times 10^{911}$ atoms and collectively they emit $1,78 \times 10^{803}$ dominant and $2,05 \times 10^{768}$ subservient virgins. The dominant flux sweeps $9,68 \times 10^{1944}$ã in that time while subservient flux sweeps $2,95 \times 10^{1903}$ points.

Across the 10^{856}ãr hydron, a virgin is boosted to an average orbital speed of $1,15 \times 10^{25}$¢ and sweeps on average $1,15 \times 10^{881}$ã points on exiting the 10^{856}ãr wide hydron, while moving outwards at 1¢ in opening orbit.

The $2,95 \times 10^{1903}$ subservient flux presence combines with the $1,15 \times 10^{881}$ã presence of each virgin as $3,39 \times 10^{2784}$ã before leaving the 10^{2568}ã volume and the average escape rate is $2,95 \times 10^{-217}$ with $6,05 \times 10^{551}$ subservient virgins exiting intact in 10^{856}ãr wide orbits with $1,15 \times 10^{25}$¢ side sweep.

The dominant $9,68 \times 10^{1944}$ã flux presence combines with the $1,15 \times 10^{881}$ã presence of each virgin as $1,11 \times 10^{2826}$ã before it leaves the 10^{2568}ã hydron volume. The average dominant escape rate is $9,01 \times 10^{-259}$ and of $1,78 \times 10^{803}$ radius time virgins, $1,6 \times 10^{545}$ dominant virgins escape the hydron surface to join a $3,78 \times 10^6$ times larger subservient crop in 10^{856}ãt.

Emission by Maximum Hydrons

The maximum hydron stops strays at 10^{195}¢ in the subservient sign and $1{,}65{\times}10^{235}$¢ in the dominant sign. Making a run for it are virgins in both signs emitted from 31,6 times lighter coronas with slower $5{,}62{\times}10^{-247}$¢ parental spin in $1{,}78{\times}10^{315}$ãr sized parental orbits. Across the 10^{860}ãr hydron, the parental movement is boosted $2{,}37{\times}10^{272}$ times to $1{,}33{\times}10^{26}$¢ as its exits its surface to take leave of the hydron, having long before taken to orbiting its center.

In the subservient sign with 10^{195}¢ stoppable speed, 10^{-195} of all strays at birthspeed are activated but with $1{,}33{\times}10^{26}$¢ side sweep, all but $7{,}52{\times}10^{-220}$ subservient virgins escape a maximum hydron of steady mattron disparity untouched. With $5{,}62{\times}10^{772}$ subservient virgins emitted in steady disparity by the 10^{228} mattron coronas, $4{,}23{\times}10^{553}$ subservient virgins escape in 10^{860}ãt. To compare, this radius time emission is 42,3 times more than the 10^{552} subservient virgins that escape a standard hydron in its shorter 10^{852}ãt radius time but also $2.36{\times}10^{-7}$ times less per ãnna.

In the dominant sign, $1{,}65{\times}10^{235}$¢ stoppable speed activates $6{,}06{\times}10^{-236}$ of all strays at birthspeed and with $1{,}33{\times}10^{26}$¢ side sweep, all but $4{,}56{\times}10^{-262}$ of dominant virgins escape a maximum hydron of steady mattron disparity.

With $3{,}16{\times}10^{806}$ dominant virgins split by 10^{228} lighter mattron coronas in a maximum hydron, $1{,}44{\times}10^{545}$ dominant virgins escape it in 10^{860}ãt. This is similar to the $1{,}78{\times}10^{545}$ dominant virgins than issue out of a standard hydron in its radius time but in real time, a maximum hydron has $8{,}09{\times}10^{-9}$ dominant emission per ãnna.

Surface Virgin Emission

This leaves one other place worth pursuing from where virgins stand a plausibly change of escaping a hydron; its outermost surface, Unfortunately this is the place where averages are least applicable and virgin escape is hardest to estimate; while much rides on it. It is wise to recall that the freely escaping adverse emission from a single disparate mattron can be 10^{20} fold the deep subservient emission of an entire standard hydron!

When a hydron does not receive the correct incoming pressure for balance, its surface mattrons will be first to be affected; carried outwards into its aura on reigning stray winds or, depending on the specific blend of boosted virgins and activated flux, carried inwards away from the aura.

Virgin emission by stable grid hydrons, and eventual release of mattrons with secondary emission if a hydron hemorrhages mattrons as it fluctuates in size under field influences play a role in a complicated balance. I shall run you through what I perceive to be the main points of emission directly from a hydron surface; set by the number of mattrons in its top layer; so situated that no other mattrons are found outside them.

The standard hydrons 10^{852}år center-surface radius spans the width of 10^{220} mattron coronas. By averaging 10^{228} mattrons upon 10^{220} layers of 10^{632}år thickness, 10^8 mattrons would be so positioned on the hydrons outermost skin that no other mattrons are situated outside them. However, the arrangement of 10^8 mattrons on a mattron thick surface layer; sitting marginally raised from the 'flat' surface by 10^{-4} of the curvature of a hydron circumference is problematic to say the least.

Although 10^{776}år may be the inter-mattronal average within a hydron, no sensible average can put 10^8 mattrons 10^{776}år apart on a surface where they are 10^{848}år apart sideways. After the majority flux in inter-mattron orbits is trapped by the closest coordination mattrons and decreasingly beyond that, virgins emission slipping through the flux works in all directions.

That all mattrons within a perfect standard hydron are 10^{776}år apart does not mean 'on average' or that mattrons at the hydron center are closer while surface mattrons are further apart (although this can be the case). *It literally means that all mattrons within a perfect standard hydron are 10^{776}år apart.*

Such is the nature of averages; they do not add up when it comes to the specifics. Divide mattrons upon units of hydron radius, a tenth of its mass is housed within 10^{851}år (a thousandth part of its volume) while the 10^{227} inner mattrons are on average 2×10^{775}år apart, which puts another damper on any detailed surface logic.

When a hydron surface is compressed by virgins that penetrate through one or a few surface layers, a large number of top layer mattrons are moved inward faster than the rest of the bulk, thus evening up average distances and shortening them throughout the outer layers which may even cause the mattrons within the top layer skin to change places (boil).

Ordinarily, a standard globular hydron has a 'diamond smooth' surface where its lattice grid of standard mattrons with heavy local flux and virgin exchange meets the aura where mattrons are largely absent. The aura can be seen (as a conceptual picture only) as an extremely thin 'atmosphere' out across an additional hydron radius. The incoming virgin winds are the same throughout the aura while the relatively few emitted or peeking virgins with some meager flux is mainly found at the bottom of the aura.

Illustrating by averages, I approximate by dividing the perfect standard hydron into 10^{76} layers; each with 10^{776}år inter-mattronal span. The 10^{152} top layer mattrons with 10^{632}år coronas on a 10^{1704}å wide hydron surface are within 10^{776}år sideways and inwards; a repulsion lattice that allows insignificant local movement or mattron orbits in the mid top layer.

The top layer mattrons are in direct communication both downwards and sideways across 10^{776}ãr. The subservient virgin survival out of a standard mattron domain is 10^{-36} or 10^{-72} counting those activated within a standard mattron with $1,78 \times 10^{-79}$ survival rate for dominant virgins.

The inter-mattron grid distances are too short to bend a virgins path as it heads linearly away from a parent carrying the slow inherent coordinates of a parental orbit that inexorably alters the originally 'straight' aim. At 10^{804}ãr from a parent with $\sqrt{(10^{804}$ãr$\div 10^{316}$ãr$)}$ boost, the standard parental sweep at 10^{-244}¢ in 10^{316}ãr wide orbit realizes a first full circle at birthspeed around its parent and the top layer mattron within which that parent resides.

Virgins that re-enter a standard hydron in orbits after 'peeking' out of its surface at 1¢ to 10^{24}¢ are extensively captured on re-entry. Hydrons block flux at 10^{188}¢ (at $5,62 \times 10^{230}$¢ in its dominant sign if polarized) and few of the peeking virgins avoid activation; not least for virgins from below its top layer which mandates ever more orbital sweeps in and out of the hydron.

The parental orbits that govern orbital paths of virgins are seldom perfectly differential; virgins that leave a top layer along a hydrons radial line are rare birds. At 10^{804}ãr, the outgoing virgin completes one orbit about its parent in 10^{804}ãt. When a peeking orbital journey brings a virgin across all the 10^{852}ãr outer aura, it would thereafter orbit the hydron at 10^{24}¢ on a continued trek that has moved free of the hydron. This escape is similar to deep emission virgins from central areas except that central virgins never re-enter a hydron after once leaving it; they take the annihilation up front, so to speak.

All virgin paths remain 'straight' on an inter-mattron run where mattrons exchange surviving virgins in one direction with a companion at 10^{776}ãr and other directions with all coordination mattrons within 10^{804}ãr; keeping their linear virgin push. For the vast hydron, central coordinates from a parent at a standard center eventually 'causes' such a virgin to orbit the hydron center at 10^{24}¢ on leaving its surface; having by then already at increasing speed made 10^{24} whole orbits among its mattron population and activated flux.

Every virgin leaving a top layer grid mattron to peek out of the hydron surface must pass through the activated flux within its domain. Between mattrons, flux in a polarized hydron will have touched and activated all but 10^{-72} of newly split subservient virgins and $1,78 \times 10^{-79}$ of dominant virgins.

Surviving outwards going virgins that leave the hydrons top layer surface with a linear push at birthspeed from a top layer parent widen the distance in accord with a parental coordinate center out to 10^{804}ãr where a virgin orbits back into the surface at the same speed which is opening the orbit. Up until that moment, its linear coordinate repulsion exceeds its orbital boost.

Top layer virgins that peek out of a surface as external emission will like virgins from the hydron mid layers orbit 10^{24} times through hydron body before eventual escape free of the hydron and risk potential activation by marginally less flux. As 10^{228} central mattrons collectively deliver 10^{76} times more virgins than 10^{152} surface mattrons, the surface input of virgins plays an insignificant part in the hydrons collective external virgin emission.

Before one is more familiar with the evolution of hydrons and the effects of garbled domain charges, it may be premature to dwell on specifics. Neutrally charged hydrons with singular mattrons of both signs could be polarized in opposite domains of small scale symmetry throughout a hydron or in areas of equal sizes in two places of mirrored distribution. Virgins of opposite sign from deep within such a hydron would then traverse many adverse domains and their chance to exit as external emission would drops ever closer to the small rate of external dominant emission.

Local small scale symmetry of polarized areas leads to greater asymmetry in large-scale polarity. Less small scale symmetry indicates greater symmetry in large scale disposition of polarized mattron, perhaps gathered in central and surface areas or opposite poles.

The reach out of a hydron along its radial line is 10^{804}ãr from its top layer parent; here the virgins turns orbital and returns back into the top layer to be overwhelmingly blocked within the 10^{776}ãr top layer; a few of these virgins reach their potential depth of 10^{28} top layers. Within a top layer, linear push away from a parent is not out from a hydron surface but rather away from a parent within a hydron; a push that is mutually neutralized. Inwards push upon surface mattrons at 1¢ in orbit on re-entry is countered by the related outwards virgins at 1¢ as they exit the top layer but peeking top layer virgins are fewer due to substantial flux within 10^{804}ãr but the flux will deliver the coordinates. The intact peeking virgin remain 'repulsive' out to 10^{804}ãr.

This surface tension on a hydron skin diminishes rapidly with depth but does not disappear; its smallest most penetrating effect is from the last and largest potential 10^{852}ãr virgin orbit. This orbital surface compression has evened surface grid distances by resisting surface expansion. A few virgins penetrate 10^{28} top layers to yield weak deeper compression down to 10^{-48} of the hydron radius. To put the thickness of this hydron skin in perspective; 10^{-48} of our solar radius is the thickness of a single Proton.

Were we to move a neutral mattron out of the hydron top layer surface to within 10^{804}ãr, it would experiences an abrupt drop in the compressive flux field and form a precursor. However, due to substantial variable activated flux within

10^{804}ãr, fed by 10^{-76} virgin input from (up to) 10^{56} top layer mattrons, this is not the full 10^{752}ãr wide precursor one expects to find beyond the 10^{804}ãr distance.

A word of warning on using averages as a blunt instrument; hitting anyone over the head with a blunt instrument usually leaves a mess and is almost invariably wrong. However, seeing how effective the hydron is in bottling up virgins, its surface is not a significant source of external emission.

Between hydrons, any outwards orbiting virgins activated in the void will drive flux in a differential orbits around a hydron after activation and these orbits favor harvest by the nearest hydron companion (if flux is not chipped away). Coordinate centers of flux that are exchanged with a close hydron are situated mid between hydrons that tend to communicate in pairs or groups.

To recap the external main body virgin emission by various hydrons;

Ext. emission average	Subservient	Dominant
Standard in 10^{852}ãt	10^{552}	$1{,}78 \times 10^{545}$
Maximum in 10^{860}ãt	$4{,}23 \times 10^{553}$	$1{,}44 \times 10^{545}$
Minimum in 10^{850}ãt	10^{553}	$1{,}78 \times 10^{545}$

Hydron Expansion Drive

Before I venture a guess on how incoming field pressures, activated or virgin, compress a hydron when harvested, I need some conceptual guidelines on how its own internal effects work to expand a hydron.

When I estimated <u>mattron</u> expansion, I let all its internal emission count as expansion within a 'disc time' during which all the motion was expansive with no internally spawned force countering it; one newborn could kick all the mass along out of the mattron and was in that respect all expansive.

The two-dimensional mattron model is not applicable to hydrons; it takes no account of that hydrons consist of and are building blocks that must be viewed as a single particle.

A standard hydron has 10^{220} two-dimensional discs (of 10^{632}ãr thickness) that each could hold 10^8 mattrons delivering disc expansion by emissions. As primal particles within mattrons had long periods of no internal kicks in opposite directions, each push could work for extended periods upon outer disc mass. In a hydron during a period needed to marginally move a mattron by large number of harvested primal atoms, a similar numbers of primal kicks are delivered in from an opposite direction.

Another problem with the two-dimensional model is the three-dimensional boost across an average grid distances. I circumscribed this in the mattron model by using disc planes adjusted for orbital boost. In hydrons, the ever more complicated boosts lead to different flux opacity in the two signs while exchanged emission affects coronas and cores differently.

Unlike a mattron that experiences a drop in pressure and alleviates this by 'emitting' precursor halos from some of its disc planes, an expanding hydron cannot solve pressure drops in the same manner. When a hydron expands, its internal mattrons experience rising instead of dropping pressure. Taking all this into account, a two-dimensional model seems more problematic than a more natural three-dimensional hydron model.

One question to be asked is how newborn virgin push is delivered upon mattrons at widely differing depths within a standard hydron and from there eventually implanted as an average residue of hydronal expansion.

With inter-mattronal exchanges between mattrons of similar mass and size, the flux majority has its linear push neutralized, rendering it unable to deliver an effective universally linear push. Only the direct linear push by a virgin away from its mattron retains its potential to deliver a linear push.

If singular mattrons are part of hydronal mass, there is in principle a choice of more than one type of expansion generator, especially if singular mattrons gather at a hydron center but the overwhelming prevalence of neutral disparate mattrons leads to a great dominance of subservient virgins that would drive the hydron expansion.

The outward push harvested by any particular mattron is determined by its internal position. A mattron at a hydron surface receives linear emission from within the hydron that has 'unopposed' outward motion in relation to its body; as 'every' kick comes at the surface mattron aiming 'outwards', the motion induced into that mattron is not opposed within the hydron. This is nevertheless always a universally directional push, which is apparent when hydron rotation brings a linearly repelled mattron to its other side.

The hydrons collective expansion must therefore be set by overall average outward movement exerted upon all its mattrons. A mattron at the hydrons center receives 'all' linear kicks towards its center with unopposed inward motion but every kick it receives is eventually matched by an opposite kick that cancels any movement of that mattron out of its central position.

The collective average expansion has the same relation to all the internal emission as the hydrons 'average radius' has to its sphere. The square root of the mattron population could in principle supply the hydrons collective expansions drive from 10^{114} of the 10^{228} mattron. However, with the linear neutralization of activated flux leaving most mattron exchanges linearly dead in the water, this is not possible and 10^{114} disparate mattrons would deliver a marginal linear expansion by their exchanges of activated flux.

Perhaps the best indicator of standard hydron expansion are the **10^{552} subservient** virgins and **$1,78 \times 10^{545}$ dominant** virgins that escape from deep within it in 10^{852}ãt radius time with 10^{24}¢ orbital sweep.

The escaping 10^{552} subservient radius time virgin emission seems the best candidate for a reasonably generous and unopposed linear expansion drive, It is not really diminished by the 10^{24}¢ orbital sweep as they peek out of the surface and enter it again with a deep aim at 10^{24}¢. The orbital thrust evens up outer grid distances but the 1¢ outwards thrust of every virgin still remains. The standard 10^{-300}pu per ãnna by expansive subservient virgins is shared by all 10^{228} mattrons or 10^{1140} primal atoms. Accumulated during a rotation in 10^{1140}ãt (half of that to be correct), this can drive a hydron apart by 10^{840} virgins that can in theory with other things aside, widen its entire mattron lattice at 10^{-300}¢ across 10^{552}ãr which would then equal its standard radius.

However, much like backhand to a gravity shadow cast by a hydron upon another submist, orbital speed becomes static when a submist at every point in orbit faces an earlier linear opposite effect. The rotation caps the hydron expansion from its average center when an identical effect is harvested in the opposite universal direction. Instead of continuing to collect more expansion virgin over longer periods that push its mass outwards at accelerating pace, the expansion remains a steady 10^{-300}¢ per ãnna in all outward directions.

If this expansion speed could continue to expand the mass outwards, its size would reach 3×10^{870}ãr in one second and form a vast embryo the size of the original embryos as they started to be formed in a galactic grid. For various reasons, the 10^{-300}¢ speed drops as the hydron size increases as the number of escaping virgins with greater boost drop per ãnna. A grave loss of compression can easily blow a hydron asunder to exponentially greater radius when it is not held to quarters by both virgins and flux.

The expansion excess is a variable part of a hydrons mass field set by the level of escaping subservient emission. Its balance against compression is best shown by the speed at which it can drive a hydrons variable primal mass outwards.

The expansion drive by escaping virgins takes into account the increasing volume with more compressed mattrons and annihilated mass with dropping emission; although this drive is pushing a lighter mass in longer radius and rotation time. Hydrons are generally far less volatile than erratic primal fields and adjust more smoothly to pressure changes than do mattrons; the shifts in hydron sizes are more consequent and the forces at work are smaller in relation to mass to be moved.

I set the hydrons expansion function by the level of subservient emission out of and across the entire hydron; measured by the speed of outwards driven mass during one rotation when incoming compression falls short of expansion. The standard 10^{852}ãr benchmark hydron of 10^{1140} primal mass emits 10^{840} virgins during one rotation and generates an expansion speed at **10^{-300}¢** per ãnna.

In 10^{860}ãr hydrons, a single central newborns inherent spin rotates the $3,16 \times 10^{1138}$ primal mass at faster $3,16 \times 10^{-279}$¢ speed in orbit around center during $3,16 \times 10^{1138}$ãt and the subservient escaping $4,23 \times 10^{-307}$ emission per ãnna allows expansion in that period by $1,34 \times 10^{832}$ virgins and this generates a drive at **$4,23 \times 10^{-307}$¢** speed per ãnna; an expansion speed which is $2,36 \times 10^{8}$ times slower than that of a standard hydron.

During the extensive eras of collapsing oversized non-polarized embryos of vast spheres that allow enhanced virgin harvest, they must gain sufficient compression to collapse under virgin compression and find eventual balance at standard size; or else the embryos remain less structured submits.

The Leverage Point

Early on, I had a problem with my sense of scale in long distance pressure from activated flux and virgins. Looking at interacting hydrons as two circles on paper, it is reasonable to assume that orbital flux gains little elbowroom to cut linearly deep into either circle.

On paper, mid between two identical hydrons, flux at linear standstill and endowed with the coordinates of ripped out mass seems to be spreading its shallow orbits across a differentially rotating hydron. Here on a small piece of paper, it is hard to see how orbital flux can break out of this tight circling mode extensively enough to penetrate deep into the other circle.

One can 'see' that a thin globular flux orb of atoms that is perpendicular to the inter-hydron line and moves in widening orbits the atoms hits the other circle sideways (from opposing direction). Individual atoms that move a bit further eat sideways into that circles bodies imagined polar axis in 'thin' orbital slices. The pressure problem is that sideways cast shadows between surface mattrons do not deliver a push between mattrons towards the circle center except by the linear drive; a far weaker holistic compression.

This is a fallacy of scale. Earlier, I used a Sun and galaxy to illustrate the 10^{13} fold difference in size between mattron core and corona. The distance between two standard hydrons in a sun is 10^{43}-fold their radii (empty of all but primal exchanges). This can be illustrated by an 10^{891}ãr wide hydron embryo at the center of the observable Universe with the only next hydron found in orbit at a distance greater than the observable Universe.

Local flux at high speed rips mass into vast orbits in repeated passes and the spin coordinates widen the spread far in excess of a hydron thick orbs (the slowing orbits take longer but so too does the linear widening of each orbit) in the end allowing all to pass through entire width of the hydron (if not chipped away by flux in wider orbits).

This brings me to the compression value of locally emitted orbital virgins that come at a hydron from a source in 'close' proximity to the hydron; for example a singular mattron or another hydron. Virgins that collide into a standard hydron in tight orbits from its aura are boosted to 10^{24}¢ and cut into its main body in 'thin' orbital slices.

The greater 10^{24}¢ thrust hits the hydron sideways in tangential opposite orbital directions that are not inwards within the top layers. The virgin is activated by flux between surface mattrons but the boosted 10^{24}¢ orbital speed does not push any mattrons inwards. The 1¢ linear speed widening a virgin orbit delivers the compressive effect with an aim that is mostly not towards center and is heavily downsized as center-wise compression.

At some distance from a hydron, depending on parental sweep and hydron size, a majority of parentally boosted virgins from an emission source starts to miss the hydron. It is first from outside this *'leverage point'* that harvested virgins can aim through the entire hydron and its center to deliver somewhat some more effective linear compression.

Emission sources at the outer edges of a hydrons aura deliver orbital push that is largely useless as compression while boosted virgins from a further away sources are increasingly compressive upon its outer layers.

From a singular mattron at a standard hydrons aural edges, its adverse virgins are extensively captured by its main body but the escape rate grows with distance. A virgins orbit widens at constant rate; call it the linear step. For an orbital virgin, a relatively close hydron is not a cardboard cutout that falls under the linear inverse-square law. The hydron is a globular structure that each orbiting virgin may have more than one go at hitting as it returns to the same point in orbit and to the hydron position after one 'linear step'. Its chance of hitting a hydron grows by the number of active orbits the virgin makes before it passes linearly beyond the width of the globular hydron.

From a source in the outer aura of a standard hydron, an emitted virgin makes 10^{24} orbits before passing by the main body hydron. Whereas a single orbit ensures that 'all' such virgins hit a standard hydron (it blocks out most of the sky) on shallow tangential paths through its outer layers; their orbital coordinate paths reach potentially at best 10^{-24} of its width when stopped in the top surface layer.

Further from the hydron, the virgin orbits affect more of the hydron body with compression; the collision frequency is set by the number of orbits made by a virgin before it passes further than the hydron. The loss of virgins that slip by the distant hydron (using the inverse-square law) is mitigated by the number of orbits that each passing virgin makes.

For example at $10^{868}\alpha r$, a standard $10^{852}\alpha r$ wide hydron blocks 10^{-32} of the cardboard sky by the inverse-square law and a virgin reaches 10^{32}¢ boost. Making each $10^{868}\alpha r$ wide orbit at 10^{32}¢ takes $10^{836}\alpha t$. As it passes across a $10^{852}\alpha r$ wide hydron orb in $10^{852}\alpha t$, it makes 10^{16} full orbits before moving beyond the standard hydron and is thereafter unable to hit it. The 10^{-32} rate of collisions by the inverse-square law is mitigated by 10^{16} orbits and 10^{-16} of the emitted virgins hit the hydron at this distance.

From $3,96 \times 10^{861}\alpha r$, a standard $10^{852}\alpha r$ wide hydron blocks $6,38 \times 10^{-20}$ of the cardboard sky and a virgin reaches a $6,29 \times 10^{28}$¢ boost. While it passes the hydron orb by at 1¢, this boost allows it $1,59 \times 10^{19}$ full orbits before it has moved beyond the standard hydron in $10^{852}\alpha t$. Instead of $6,38 \times 10^{-20}$ collision frequency by the inverse-square law, the $1,59 \times 10^{19}$ orbits from the $3,96 \times 10^{861}\alpha r$ cause half of all virgins with standard parental spin to hit a standard hydron and half miss it; this is the 'leverage point.

Up until the standard $3,96 \times 10^{861}\alpha r$ leverage point for an outgoing emission source, standard virgins hit a standard hydron surface ever less tangentially with rising center-wise compression that is aided by their $6,29 \times 10^{28}$¢ boost.

Parental spin that is greater or smaller than standard requires different leverage. For example, a virgin split in a $10^{752}\alpha r$ wide precursor of an exterior photon has 10^{-64}¢ orbital motion in its $10^{376}\alpha r$ wide parental precursor orbit boosted from there. Although most such virgins may fall offer for activation because of great local presence, they cause little compression as they sweep on shallow sideways paths into hydron surface, leaving a tiny part of their linear 1¢ push as hydron compression.

Note that a 'leverage point' is very different for standard emission going the other way from a standard hydron and towards the same singular mattron with its $10^{752}\alpha r$ wide precursor at the same distance. With $6,29 \times 10^{28}$¢ boost it sweeps an insignificant part of the $3,96 \times 10^{861}\alpha r$ orbital distance while it passes linearly across $10^{752}\alpha r$; beyond a precursor a wide orb in the sky. The $6,29 \times 10^{28}$¢ boost renders the capture of virgins emitted by the main body hydron marginally more effective than the $6,38 \times 10^{-220}$ collision rate set by the inverse-square law upon a photon precursor.

Hydron compression drive

A simple conceptual illustration that covers the various elements of hydron compression is not easy to get your head around when you leave most of the particulars wide open in a field that requires detailed mathematical study.

Look at the center-wise aim in a uniformly polarized mattron population; a disparate mattron sphere has 10^{13} fold radius difference between core and corona. To conceptualize this level of mismatch, you can picture Our Sun of 10^{914}ãr radius as being a mattron core surrounded by the primal mist of a 10^{927}ãr wide corona; a corona of inter-galactic proportions. The aim of each approaching virgin towards a mattron at a hydron center on its way through a standard hydron that is 10^{220} fold its radius is ridiculously diffuse.

Peeking Virgins

The fact remains that while mass is denser deeper within a submist, that is where orbiting virgins from outside have a better chance of being brought to a stop. This is not really a compressive drive; it evens up the grid distances near the surface while the virgins later outwards function of a swarm it sets moving remains during a second half of that internal path.

During the gradual formation of hydron embryos, recapture of orbitally boosted virgins emitted out of a 'surface' from more shallow splits (outside a hydron center) have a natural corralling function. Even if the virgins hit just before they exit they cannot widen the surface as the activated local flux will continue along the virgins path and deliver all the virgin coordinates back into the surface by way of its peeking orbit.

This corralling effect is greatest for the initial vast hydron embryos at sizes around 10^{900}ãr as they first start to collapse out of an evenly spread mattron mist. At that time the internal mattron grid has a highly uneven internal lattice with smaller distances between mattrons at the hydron center that increase out towards the surface.

As 'peeking' virgins return in boosted orbits they can (depending on the depth of a primal split) reach an orbital radius up to the hydron radius and orbital speeds almost that of free virgin emission side sweep. The number of all peeking virgins is far greater than freely escaping virgins; shallow primal splits cause a similar effect but do not reach far out of a surface and their boost and compression drive is weaker and its coordinate aim has a shallow and less penetrating reach.

The peeking virgins orbit a parent outside the hydron center and gain an 'inwards' aim when captured on their way back into the surface (outwards if captured before that). Their orbital sweep pulls a far greater punch that the linear birthspeed push away from their parents but this falls by the wayside as actual expansion.

While the natural corralling function may appear ineffective against the shatterproof resistance of the surface mattron lattice, the peeking virgins coordinates are delivered into the top layer, the greatest orbits can carry a coordinate drive that reaches close to the hydron center. Without 'peeking' virgins, a hydron would become a point object and would have to depend on other virgin pressures to gain a corporeal shape by evening up the larger internal inter-mattronal distances closer to the surface.

Many effects play into the temporary compressive ability of peeking virgins resisted by a tightening internal lattice as hydron embryos form exceedingly slowly from a thinning surrounding mattron mist and creates a broadening void as its primal mass withdraws deeper. As the embryo continues to shrink, its internal mattrons are gradually less compressed and heavier and the boost of peeking virgins slows across a smaller radius.

This ongoing corralling phenomenon stops when a hydrons internal mattron grid has gained equal average distances from center to surface. In even grids, peeking virgins still compress the surface, just as a similar fraction expands it but the blocking by mattrons does not turn more opaque with depth.

I am assuming that this happens when its mattrons reach the rock bottom stability of a standard 10^{852}är hydron where mattrons in its 10^{776}är lattice where it can keep full 10^{912} primal mass. There is another type of balance; if during virgin pressure, a tighter surface grid comes to contain a denser mass than then rest of the hydron, the massive internal exchange of virgins begins to be blocked more effectively by a denser surface and this pushes outwards and resists changes to the internal grid on the account of peeking pressure.

Virgin pressure

The pressure input by foreign virgins, gauged by the product of number and incoming speed is marginally compressive towards the hydron center and this important effect can speed up the contraction of hydron embryos.

Incoming virgins usually arrive in vast orbits that appear locally linear and aim but rarely through a hydron center or penetrate a single lattice 'layer' on a hydron surface. Though all such incoming virgins are compressive to some degree, their center-wise residue is extremely small; each delivers their set of coordinates to mattrons in the top layer surface that will in time become part of the entire hydron mass; locally linear coordinates that are thus moved out of line after top layer activation do not alter the original linear virgin aim but flux carried off course adjusts their orientation to the original linear effect.

As with expansion, initial virgin compression is capped by hydron rotation that brings snippets of linear virgin coordinates to its other side where the original linear universal thrust is no longer compressive but rather expansive as it is met by the same level of compression from any direction.

My conceptual illustration is brutally simple; the harvest of virgin pressure upon a surface has the same base relation as a hydrons 'average radius' has to its sphere; the inwards input becomes the inverse of 10^{114} mattrons where the aim of incoming linear virgins ranges from tangential to center-wise and leaves a square root perfectly center-wise virgin thrust or 10^{-57} which when set as r^3 allows r^2 as the compression value; namely 10^{-38}.

The harvest of incoming virgins at various speeds and aims delivers 10^{-38} of the harvested pressure as holistic compression. Hydrons of all sizes must harvest virgin compression to match its expansion drive per ãnna after the pressure harvest has been downsized to 10^{-38}.

This center-wise incoming compression is independent of the temporary inwards pressure delivered into the surface by peeking virgins which is merely a gathering function whenever a surface grid starts to deviate from the standard lattice.

If a hydron were kept to quarters by virgin pressure alone; when it loses that virgin compression, it expands by the input of its expansion drive. Even when virgin pressure compresses a hydron below standard size, expansion will bring it back when the pressure subsides, other effects notwithstanding.

Flux

The field of activated flux permeates our Universe but its effect is first felt in the form of a powerful compression in the hydrons last formative standard stage where a powerful internal flux backhand steps in to grasp the hydrons mattron mass in an iron grip. For larger primeval hydrons, flux compression is of less importance. It is likely that the inner core in an $10^{860}\alpha r$ maximum hydron has reached the state of flux lockup; aided by virgin pressure that has fashioned its deeper interior into a uniform lattice while its surface may still be molded by corralling peeking virgins. Further virgin compression is needed to bring the hydron to $10^{852}\alpha r$ where flux shadow are dark enough to reign unchallenged.

A hydron is an empty place. A mattron wide primal path through a hydron is rarely occupied by two mattrons and, for that reason, when backhand flux push is experienced by a mattron that blocks a traversing stray at stoppable speed, this backhand is eventually countered by another stopped stray from the opposite direction that has not encountered any mattron either.

When two opposite flux pushes enter a hydron to hit the same mattron, it remains unmoved and this does not cast a shadow *within* the hydron but the hydron has blocked the neutralized flux and the backhand to both blocked flux atoms casts primal shadows along backhand lines *out of the hydron.*

Most orbital high speed flux sweeps through a hydron without hitting anything but if it is hit by a local inter-mattronal flux atom, a blocked flux atom remains true to its coordinates (even when channeled by internal flux into a mattron far off its linear path and with different orientation to it).

If the orbital path of a linear stray atom does not originally lead through two mattrons, it cannot cast a full collision shadow between them. Therefore, as potential compression, orbital flux pressure delivers a shadow backhand within a standard hydron that is first downsized up front by the hydrons transparency ratio. Few orbital paths of flux compression leads through two standard mattrons; 10^{-212} for 10^{228} coronas that measure 10^{632}år and 10^{-239} for the mattron cores that measure $3,16 \times 10^{618}$år.

In the embryonic hydron stages, the compression by flux in large and fast orbits is for obvious reasons weak due to the high transparency ratios of the large hydrons coupled to smaller compressed neutral mattrons; all of which leaves little effective center wise compression.

When blocked by a mattron, flux atoms in large fast locally linear orbits cast primal their shadows between the mattrons within a hydron and the backhand to such a penetrating shadow pushes mattrons closer on that common line. These lines rarely aim between center and surface mattrons and a tiny fraction of deposited mutual backhand shadows is center-wise; by the inverse square root of its mattron population (here 10^{-114}).

High speed flux or the lack thereof is discernible between hydrons if both block flux at that same speed; boosted backhand shadow flux will marginally affect another local hydron that cannot block the same flux. Activated flux, harvested at for example standard $5,62 \times 10^{230}$¢ dominant or 10^{188}¢ standard subservient distributes equal orbital backhand between all the mattrons and the whole effects is downsized by standard 10^{212} coronal and 10^{239} core transparency ratios (this part is included in the shadow dilution below).

The penetrating internal flux shadows between mattrons are least diluted between close mattrons (for example at standard 10^{776}år apart) while those between mattrons across the width of the hydron radius (standard 10^{852}år) are most diluted. As mattrons cast shadows upon all other mattrons in the lattice, the average standard distance is 10^{814}år.

Dominant shadows are created almost exclusively by cores that block the traversing flux. This finite flux harvest is not cast across 10^{814}år between

$3,16 \times 10^{618}$år standard cores and heavily diluted to 10^{-391}; it rapidly becomes communicated to the greater coronal mass which broadens the scope of its projected aim. I use the average shadow dilution between standard 10^{632}år mattron coronas across the 10^{814}år average distance as the inverse of 10^{182} squared whereupon the backhand compression for subservient shadows becomes 10^{-364}.

My method to account for the linear compression drive is to use the (standard) 10^{-364} coronal coordinate shadow of flux backhand. Dominant shadows cast across 10^{814}år between $3,16 \times 10^{618}$år standard cores are more heavily diluted to 10^{-391} in standard hydrons. Almost all the flux is brought to local harvest by core exchange fields across standard 10^{776}år. And on top of that, merely 10^{-114} of this penetrating backhand is center-wise.

There is an aspect of the core drive that increases its center wise push; the flux coordinates harvested by cores are communicated to coronas. Although this does not increase the shadow, it does increase the center wise push by the wider corona by broadening the scope of a finite push.

My method to account for this linear drive is to take the average of 10^{-364} coronal and 10^{-391} core shadows and use the $3,16 \times 10^{-378}$ as the standard communicated effect of flux backhand shadow coordinates. The finite drive is transferred to the $5,62 \times 10^6$ times more massive corona but does not increase the backhand push; the coronal width broadens the compressive aim and brings the center wise compression to $3,16 \times 10^{-492}$ of the flux input.

To sum up my shorthand illustration of collective flux compression within hydrons. Use the medium coronal-core width for the shadow distance and apply the inverse square law and the mattron populations square root. This renders the actual collective center-wise flux backhand; the average pull per ānna towards the hydron center that I shall use in my illustrations.

For maximum 10^{860}år wide hydrons, the penetrating internal flux shadows between mattrons are least diluted between close mattrons 10^{784}år apart in the maximum hydron while those between mattrons across the 10^{860}år hydron radius are most diluted. As mattrons cast shadows upon all other mattrons in the lattice, the average standard distance is 10^{822}år.

Here the average shadow dilution between $3,16 \times 10^{630}$år coronas across the 10^{822}år average distance is by inverse of $3,16 \times 10^{191}$ squared and backhand compression is 10^{-383}. Dominant shadows between $3,66 \times 10^{617}$år cores across 10^{822}år are diluted to $1,34 \times 10^{-409}$ and the average backhand is $1,16 \times 10^{-396}$. To top it off, the 10^{-114} center-wise aim brings the compression of maximum hydrons to $3,16 \times 10^{-510}$ of the flux input; a rather inefficient compression fraction against its $4,23 \times 10^{-307}$ expansion drive.

For Clarity; Gravity Flux Drive

We should not confuse the flux backhand pressure with the flux gravity backhand delivered by the same flux and cast by the same hydron.

All orbital flux is blocked within cores. Within a standard hydron, 10^{-239} of all flux boost is shaved off its potential flux backhand thrust at the outset and its push neutralized by opposite directions by each core; unable to cast any shadow out of a hydron. Even the remaining shadow is not a holistic hydron shadow but the shadows from 10^{228} cores that have been diluted by the inter-hydronal distances upon another faraway set of 10^{228} cores. The backhand shadows from such orbital flux components are a weak but still a marginal effect between relatively close hydrons.

Gravity on the other hand comes from the original linear drive of each virgin split in the inter-galactic void. This original inwards drive from outside the galaxy delivers effective snippets by vast flux storms of linear push from every side of the galaxy towards a galactic center although the majority has been blocked before reaching the center. The exponential side sweep of the original virgins will be extensively neutralized by the blending flux winds.

Like the virgin spin coordinates that are driving the flux storm, the linear coordinates are embedded as part of all flux. Linear compression from the same original virgins carries influence across the galaxy. The gravity push is delivered in coordinate snippets by an absurd number of flux atoms and first completely neutralized at its center.

You see this better in your mind's eye by eliminating all flux and virgin boost from the picture; what remains is a perfectly linear virgin emission from outside the galaxy. Any blockage of virgins in such an illustration cast their linear backhand shadow between hydrons that block the virgins.

In reality, every flux atom carries myriad tiny one way snippets of linear push from these inter-galactic coordinate centers around the galaxy and deposits these garbled coordinates into every mattron within every hydron along with all other coordinates embedded in the same flux atoms (mostly from ripped out galactic mass that slows such flux).

One can estimate the quantity of linear snippets by dividing the total flux pressure by the average orbital flux boost which shows the original virgin input from outside (or within) the galaxy. The total linear drive is delivered quite evenly between hydrons and this is what we call gravity, the linear backhand that is ever more effectively communicated by groups of hydrons or larger submits in accordance with the inverse-square law.

The non-cancelled linear coordinates deposited in equal proportions into every hydron are the sum of all the garbled snippets of linear coordinates pressing upon every hydron within a galaxy from outside it. The backhand is cast between hydrons according to how much of the passing flux is blocked. leaving the consequent linear backhand. This is not as simple as it seems as orbital flux does not come at a single hydron in any coherent way.

One can assume that a standard hydron stops a stray that traverses it at $5,62 \times 10^{230}$¢. The stray will be hit on its path and at this speed harvested in follow up collisions with the orbiting internal flux. So far, so good, but can this flux passing through a hydron, aiming at another nearby hydron, even in principle be blocked and create backhand by 'orbital flux attraction'.

If a passing flux atom hits at speed a tad over $5,62 \times 10^{230}$¢ which is too fast for a hydron to block through internal collisions, it generates a flux shower that leaves the hydron. This flux shower is no longer aiming at a next hydron along the original path of the incoming flux atom which has been divided on an immense number of highly convoluted orbits.

On exiting the presumed 'next to last' hydron, the linear pace is now the pace at which the new flux orbits open towards the next local hydron and this is the slowest movement among the exiting orbital flux. The flux shower is no longer slow enough for harvest by the next hydron along the original orbital flux path and cannot therefore create a backhand shadow.

Take for example the greatest flux presence by an internal flux atom that is part of standard hydron mass; say a recently activated virgin exchanged between mattrons with 10^{388}ã presence per ãnna. The original $5,62 \times 10^{230}$¢ speed of the flux atom that carries the internal flux atom from a 10^{776}ãr wide orbit and 10^{388}¢ orbital speed out into a wider orbit will become ever more dwarfed by that increasing speed gained in different orbital directions.

An exchange atom orbiting at 10^{388}¢ is hit by the original linear flux atom at $5,62 \times 10^{230}$¢. If the pair were to leave this last to next hydron, it would gain orbital speed 10^{429}¢ as it heads across to an extremely close hydron at say 10^{858}ãt. To slow a newly gained orbital speed to $5,62 \times 10^{230}$¢ to allow its harvest on passage through the next hydron on the original line, a flux pair leaving the next to last hydron must have undergone over $1,78 \times 10^{198}$ collisions with internal flux.

This also means that the original $5,62 \times 10^{230}$¢ speed that opens the first orbit will be slowed to $3,16 \times 10^{32}$¢ by the same $1,78 \times 10^{198}$ collisions. The flux shower at harvestable speed does gain a local presence of $1,78 \times 10^{1056}$α while orbiting across to the 10^{2574}α volume to the next hydron in $3,16 \times 10^{825}$αt. If all hits came from 10^{776}ãr internal orbits at 10^{388}¢, only a small fraction of flux orbiting across to another hydron at 10^{858}ãr (extremely close distance) avoids being chipped into larger orbits by faster galactic flux.

The 10^{388}¢ exchange speed is an extreme example and the $5,52 \times 10^{358}$ collective exchange flux atoms that are ripe for harvest by mattron cores at $1,79 \times 10^{29}$¢ have the same presence by numbers, if not individually. The blended input from all kinds of ripped out mass would allow somewhat greater linear speed of the exiting flux showers in opening orbits.

There is of course always some part of the flux that is not chipped out from between close hydrons and that in principle can cast tight orbital shadow from that hydron surface but such shadows do not lead through other local hydrons and fall by the wayside until they are chipped away.

A neighboring hydron will of course block flux that would otherwise be blocked by its neighbor. In that sense all hydrons cast a shadow by blocked flux upon its neighboring hydrons with a completely incoherent backhand in any linear direction except one that has become inherent to the flux.

It delivers an average hydron harvest of inherent snippets aiming into the galaxy from an immense assortment of weak linear virgin coordinates drives divided upon a flux swarm that neutralizes all the contrary virgin boost and it cannot neutralize a linear drive from one side of a galaxy, all in the same direction, until these winds meet at the galactic center.

This short clarification is needed to keep the effects separated; gravity is an exceedingly minor part of a hydrons backhand shadow and compression.

Another reason that prevents the field of activated flux that permeates our Universe in immense orbits from casting a weak flux backhand between two hydrons is that its own immense orbital flux presence comes into extensive conflict with local flux that it chips out from between two local hydrons into vast orbits. This chipping away is lesser at small distances as the local volumes set the portion of galactic flux presence at work against local presence of greater flux but chipping away local flux is easier at larger inter-hydronal distances. However, blocked orbital flux works fine as backhand between mattrons within hydrons.

I am assuming that virgin activation within 10^{895}år hydron grid distances is more prevalent by local than inter-galactic flux and that although locally emitted virgin majorities escape intact across 10^{895}år to enter a next hydron domain, this virgin seepage is less coherent between hydrons there while its greater presence activates its number into a minority among local virgins.

Hydron Balances

Hydrons do not have the straightforward mechanism to deal with weight problems or excessive pressure as do mattrons with the security valve of their central annihilation. But, as mattrons, hydrons have a density they cannot exceed and that is set by the concentration of neutral central mass.

An overly compressed hydron or one burdened with too much mass will eventually see its main mass forced to collapse towards center while its outer mass is slower of the mark. If cohesive virgin compression of a hydron brings its central mattrons too close for mattron exchanges to gain sufficient boost, the mattrons support their balance by unraveling into halos that they hemorrhage as primal mass.

If the pressure persists and central mattrons become neutrally expanded we see them ultimately unravel into a unified primal subcloud at the hydron center. This central subcloud of a multitude of mattrons has neutral primal mass that forms a central 'black hole' that is very effective in annihilating primal mass.

The central mattrons are unable to keep their mass together if primal mass fills the void and leaves no corporeal structures for the halos to return to. A black hole would ease the crunch and avert a collective crisis for the overly heavy hydron.

Conversely, in a roundabout way, if the outer layers of such an overly stressed overweight hydron pushes a mattron crest outwards, the opening lattice boosts increase in the outer layers with increased central annihilation within these compressed mattrons.

Within its stability range, a hydrons internal expansion excess pushing its mattron bulk outward and arrives automatically at a balance for its main mass against incoming virgin pressure. Only standard hydron size will eventually have its size is completely regulated by incoming penetrating flux after making use of virgin pressure to get a grid that tight.

A clarification is in order; there is no fixed inter-hydronal distance between paired hydrons or in groups, and therefore no fixed virgin boost between them. A 10^{914}år sun of 10^{57} hydrons of perfect hydron gas without closer ties provides a suggestion of 10^{895}år average distance (hydrons may originally form at vast 10^{903}år distances or more). Although 10^{894}år coincides with the familiar 10^{-15} kilometer idea of a hydrogen atom, it is premature to imagine that as the width of its 'interaction area'.

After solar formation, many hydrons are paired or in groups at closer than average inter-hydronal distances. Outside closer local bonds, the average lattice distance of 10^{895}år remains in place and I use it as the backdrop for my initial illustrations.

One can speculate that flux boost may be $3,16 \times 10^{447}$¢ and parental boost $3,16 \times 10^{45}$¢ within 10^{895}år interaction areas and lesser between nucleus and Electron and lesser still in a closely-knit nucleus. Air density of at Earth's surface is one kg per cubic meter where 10^{27} (none paired) hydrons would crowd within 10^{2715}ã, each with 10^{2688}ã private spatial points and exchange flux at 10^{448}¢ and virgins at 10^{46}¢ across 10^{896}ãr inter-hydronal distances but what that tells us is limited.

Using hydrons of different original sizes and masses at variable paired distances may be a more proper way to illustrate grids but I shall stick to the standard 10^{852}ãr hydron and set lattice distances accordingly. Evolved grids can rarely keep a perfect average distance between hydrons as various winds cause pairs or groups to form closer company over time. Hydrons in close constellations exchange virgins across the lesser distances but at the same time they seed the wider grid and nearby hydrons with surviving virgins.

A large hydron embryo can remain temporarily balanced by its corralling peeking virgins but these cannot bring it to standard 10^{852}ãt size even after a time-consuming contraction. Its collapse must be aided by direct additional virgin pressure, downsized to 10^{-38} or by internal flux shadows. If a standard 10^{852}ãr hydron loses all incoming virgin pressure, it remains balanced by inter galactic flux that keeps even its top layer mattrons from wandering off. It adjusts by gracefully widening its lattice without any great upheaval on that account.

A standard hydron surface blocks most arriving dominant virgins already in the top layer while subservient virgins at speeds above 10^{36}¢ penetrate marginally deeper on the immense hydron scale. To be held to the 'surface', each of 10^{152} top layer mattrons gathers 'compression' shared by 10^{228} grid mattrons when collectively downsized to 10^{-57}. Being 10^{-76} of the hydrons total mass, top layer mattrons collect compression in spades and channel it deeper against resistance from the next lattice grid layer.

Note that if virgins arrive into a hydron on tangential paths, these effects arrive tangentially from opposite directions within the top layer where that push is largely neutralized as compression. If boosted tangential movement is greater than linear movement, more of the unopposed linear movement counts as compression upon the hydron surface. If inwards heading virgins are completely linear locally in opposite directions relative to the hydron on hitting its surface, then all the virgin boost inputs hydron compression (before being downsized to 10^{-38} for center-wise compression).

There may be some need to regulate compression for standard hydrons where no boost neutralization occurs for virgins split outside the hydrons leverage point but total neutralization of tangential boost as compression input from within the leverage point. It may play a role for compression of vast hydron embryos that block the boosted virgins across short inter-hydron distances (relatively speaking) but from far within their vast leverage points. Though virgin boost can be great on distinctly tangential virgin paths, it can yield temporary compression.

The orbital coordinate paths are in and out of the hydron, even if blocked within the surface; they are like peeking virgins in this respect. The deeper an inwards coordinate path is projected before turning outwards again, the more temporary compression is delivered. The effect ultimately neutralizes but this takes time and, until then, there is some input into temporary compression from the boosted orbital speed.

It depends on the distance from a hydron aura to where the virgin is split and the relation of this to the hydrons leverage point. A virgin split in a standard aura with 10^{24}¢ boost has its compression totally neutered while a virgin split just outside its $3,96\times10^{861}$ãr leverage point keeps all compression input from its $6,31\times10^{28}$¢ boost.

If the need arises, one could use a conceptual method to set the distance from an aura to the hydrons potential leverage point (variable for hydrons) as the measure of potential temporary cut in virgin boost. Across this extra distance, a virgin boost as compression input upon a stable hydron drops from 100 percent in the aura until no cut is applied at the leverage point from where compression input is 100 percent.

For a standard 10^{852}ãr wide hydron, virgins split outside its $3,98\times10^{861}$ãr leverage point arrive on locally linear paths across the extra $3,98\times10^9$ fold distance. The inverse $2,51\times10^{-10}$ of that extra distance represents the range from no compression input from the aura to 100 percent compression input from the leverage point. As $2,51\times10^{-10}$ multiplied by $3,96\times10^9$ at the leverage point allows no cut in compression value, the virgins $6,31\times10^{28}$¢ orbital speed from the standard leverage point sets its full input into preliminary compression as the product of virgin number and $6,31\times10^{28}$¢ boost.

For a split originating outside the hydron in the 10^{852}ãr aura, compression input at 10^{24}¢ delivers no compression as $2,51\times10^{-10}$ multiplied by 1 allows only linear pressure input. The virgin number gives the linear pressure and the 10^{24}¢ virgin boost is totally neutered as a compression input. The effect is set without any boost by the number of incoming virgins at birthspeed.

At 5×10^{856}αr from the aura at a standard hydron, the span is 5×10^4 fold and the incoming $2,24\times10^{26}$¢ orbital virgin boost is cut in its compression effect input by $2,51\times10^{-10}$ times 5×10^4 which leaves a compression value of incoming virgins cut to $1,255\times10^{-5}$; a very rough conceptual illustration. And until I see the need to differentiate, I consider all incoming orbital virgin pressure that originates outside the leverage point of a stable hydron to have full compression value (before a cut to 10^{-38} center-wise compression) while only the linear drive applies for virgins split within the leverage point.

Inter-Hydron Communication

As already pointed out, flux is spread so fast and far afield that it becomes an insignificant factor in communication between hydrons that stay in touch by free virgin emission. A lot of these virgins are activated between the hydrons due to locally focused flux presence and are swept into larger orbits by penetrating flux from further afield. A hydron grid is fashioned by virgin exchanges and partly based on the local virgin majority surviving the trek.

It is nevertheless proper to comment on some aspects of flux exchanges to see how they differ from exchanges between mattrons where flux is a more important element in mattron balance while non-boosted virgin exchanges there are mostly insignificant.

An emitted virgin leaves its hydron at a given moment at birthspeed and orbits away. The next emitted virgin follows at birthspeed after some average time. Any two virgins emitted by a hydron are therefore on average nowhere near orbiting on the same sphere at the same time. As the virgin orbit shifts across more planes, it does not much facilitate collisions if virgins are not on the same sphere.

If virgins from an opposite emission front from another hydron are heading the other way and the two fronts start to mingle midways in half time where boosted spatial presence is greatest between hydrons, the joined presence yields in virgin collisions by some 'self-activated' fraction; albeit in this case a small fraction that cannot jack up the field activation.

Contrary to a mattron where the greatest initial available flux speed opens collision orbits to inter-mattronal distances and brings the opposite flux to a linear standstill mid between them, flux here in larger locally focused orbits is instrumental in moving local collision prone flux swiftly into larger than local grid sized orbits which affords flux little staying power midfield. This flux will nevertheless activate a sizable minority of locally emitted virgins.

The property I wanted to comment on is that of adverse exchange fields where one polarized hydrons subservient virgin emission is met head-on by another hydrons smaller dominant virgin emission in the same sign. Here the virgin winds and activated fractions do not match when they meet and while they intermingle mid between the hydrons. For mingling activation, I have chosen to use the average presence of head-on virgins (*the square root of the product of the conflicting front numbers*).

To whatever extent the virgin fronts are activated, a million times greater flux front would (in principle) beat back the weaker front to where it came from while keeping its 1¢ linear (virgin input) movement largely intact (except for the millionth part). The beaten back flux that has not been chipped into wider orbits is blocked by the adverse hydron in its dominant sign at high speed during short but intensive dominant consummation.

There is a balance in the input into a local activation pot; when a majority of virgins on the wing between two hydrons has been activated by the locally focused flux, the virgin majority can no longer increase the local flux presence by new activation; the rate of annihilation drops and more virgins slip out of that inter-hydronal domain.

The remaining effective virgin exchanges between hydrons are close to a majority but not quite. These virgins orbit across distances that generate greater speeds in parental orbits. As the virgins return (by proxy) at higher speeds, I let majority virgin crop harvest stand across its first domain.

Exchanged virgins cannot mutually balance two hydrons, even when side by side, given that virgin harvests must be downsized to 10^{-38} as holistic compression. If virgin boost across relatively small inter-hydronal distances is largely tangential, it does not count as compression; only if the inter-hydron distances are such that they have become locally linear at the receiving end. An exchanged virgin brings the same effect to a hydron across a void whether its sign is allied or adverse when coordinates are harvested within a top layer.

The size of a hydron sphere is mostly an extremely tiny spot in the sky across an inter-hydron distance. If virgin exchanges were linear, the number of such spots on the distance sphere would set the rate at which the emitted virgins would hit the other hydron (in accord with the inverse-squared law). The hydrons are however aided in capture of exchanged virgins by their 'depth' and by parental boost which plays an important role in virgin ability to hit that next hydron.

The aiding effect is determined by how many passes an orbiting virgin has makes in orbit by a far off hydron (or any other submist). Across $10^{894}\alpha r$, a virgin from one standard $10^{852}\alpha r$ hydron upon another is a case in point; a linear virgin would here have 10^{-84} chance to hit the next hydron. However, it is boosted from $10^{24}¢$ exit speed to $10^{45}¢$ speed in $10^{894}\alpha r$ wide parental orbit as it passes by the other $10^{852}\alpha r$ hydron.

Its $10^{894}\alpha r$ wide parental orbit is opening at $1¢$ and at $10^{45}¢$, each orbit takes $10^{849}\alpha t$ to complete. The virgin is therefore moving outwards by $10^{849}\alpha r$ in each orbit and it takes merely 10^3 orbits to move far enough outwards to move past the other $10^{852}\alpha r$ hydron (after which it cannot hit it). Its ability to hit the next hydron has nevertheless increased 10^3 times and increases the closer the two hydrons are and decreases the further apart they are.

Were we instead of a hydron look at this effect on a $10^{752}\alpha r$ wide mattron precursor at $10^{894}\alpha r$ where each parental orbit widens by $10^{849}\alpha r$, the virgin passes outwards more than that in each orbit and has only a single one shot at a small precursor. A perfectly linear non-boosted virgin would show the same result according to the inverse-squared law.

For the hydrons, an 10^{-84} escape rate, albeit cut 10^3 fold to 10^{-81} indicates that virgin seepage would complicate the picture if the majority had not been activated. With the majority activated, escape will not turn virgin exchanges into a group experience across multiple hydron domains where they would jack up local pressure upon grid hydrons by greater input at faster speeds, albeit with less orbital aid and more misses due to greater distances.

The seepage of orbital virgins in a hydron grid is limited in number by the times it takes virgins to gain enough presence to be activated by the galactic flux within a larger volume of grid hydrons. This gives emitted virgins another chance at hitting local hydrons after missing the nearest one. These somewhat more boosted orbital virgin pressure between many hydrons is exponentially greater than delivered between two close hydrons even if the hitting frequency is diluted by the inverse-square law without orbital gain. Being returned by proxy, I let the local effect of the virgin majority harvest stand whole with the hope that this assumption does not return to bite me in the tail.

Although these slightly more boosted virgins that reach across a greater patch of a hydron grid cast greater orbital backhand shadows between local hydrons, they will be shown to be weaker than the snippets of gravity that is also carried between hydrons by inter-galactic flux.

Effective mutual shadows between close grid hydrons are much weaker than mutual repulsion generated by the exchanged virgin harvests. What troubles me is that mutual exchange of virgins causes all hydrons to repel each other with an effect far greater than the attraction by virgin backhand. All hydrons find it hard to approach each other closer than the average grid distance. If the mutual repulsion were resisted only by weaker flux gravity backhand, even hydrons of opposite polarity apparently fly apart out to the average grid distance and are held in a lattice by reciprocated repulsion.

Allied standard hydrons harvest roughly 10^{42} times greater flux backhand in the same sign than an adverse hydron pair and are largely oblivious to each other's flux sign (except for a weak gravity component). While orbital virgins can push hydrons together by a shadow backhand blocked by their different sized spheres, what sticks out is that these weak mutual shadows between hydrons are extremely diluted by the inverse-square law across the great grid distances.

This inability to push hydrons closer (not least two adversely polarized hydrons) does not sound right if one looks askance at measured 'particle' behavior. If the construction of a logical universe built on the Primal Code were reverse engineering, this would be a good place to give up.

The Transparent Hydron Field

Play with the thought that in a standard sun of 10^{914}ãr radius built of 10^{57} standard hydrons in a grid of an average inter-hydronal distance of 10^{895}ãr (although 'most' would be arranged in closer constellations at up to 10^{894}ãr) where long distance flux presence across the 10^{895}ã between paired or group domains activates all but say 10^{-14} of emitted virgins on average. As a collective obstacle, the 10^{57} hydron bodies block but one in 10^{12} virgins during their 10^{55}¢ orbital exit from this sun; hydron bodies make hardly a dent in the number of surviving virgins that stream out of a sun.

10^{552} driving newborns in radius time sweep $3,16\times10^{447}$¢ per ãnna during 10^{707}αt consummation with $3,16\times10^{1706}$α spatial presence. In the 10^{62} times longer time it takes a virgin to exit the sun, local presence is $3,16\times10^{1768}$α in each closer domain and $3,16\times10^{1825}$α collectively in 10^{57} domains. A virgin at 10^{55}¢ gains 10^{969}ã presence and the combined $3,16\times10^{2794}$α presence within a 10^{2742}α solar volume allows $3,16\times10^{-53}$ of the virgins to escape.

From such tiny virgin escape rates; one might be tempted to conclude that for example a solar body is opaque to any incoming strays particles. This is far from the truth; suns and greater macro bodies are utterly transparent to strays.

Arriving from outside the sun, a virgin on a linear path that traverses the sun at 1¢ shows this picture by crossing the sun in 10^{914}ãt. In that time the $3,16\times10^{1825}$ã activated subservient presence combines with a smaller 10^{914}ã presence to sweep $3,16\times10^{2739}$ã of the suns 10^{2742}ã volume. This tells us that only one in every 316 traversing virgins at birthspeed is activated on a linear passage through the sun.

Then consider that most standard virgins that enter a sun are extensively boosted ($3,16\times10^{58}$¢ between suns) and crossing a sun in $3,16\times10^{855}$ã will cause no more than $3,16\times10^{-62}$ of the virgins to be blocked by that sun. As such macro bodies cast gravity shadows by flux backhand and weakly by virgins where a gravity shadow can be expected of 10^{-62} of passing virgins (on each passage). The interstellar flux at speeds from 10^{457}¢ and upwards therefore 'detects' the transparent solar body only on rare occasions.

As to gravity flux, each of the 10^{57} standard hydron bodies is more opaque than a sun as they stop strays at between 10^{188}¢ and $5,62\times10^{230}$¢. As one in 10^{67} linear primal paths through a sun lead through a hydron, the sun casts a powerful external flux gravity shadow set by the backhand to linear virgin push by flux that hits hydron bodies directly.

As a solar gravity shadow is cast directly by its hydrons and not by the solar area itself and the solar volume has little influence. A gravity shadow from a macro submist is therefore marginally affected by altered density. A sun can be compressed to million times smaller radius without it significantly affecting its collective gravity shadow. The effect upon internal hydron sizes due to lesser inter-hydronal is inconsequential with a 10^{61} fold margin.

The Event Horizon

We saw earlier than collective virgin emission heading out of the surface from deep within a hydrons main central mass, having already orbited out through its flux fields in orbits of hydron size radius, never reentered the hydron surface after exiting. If blocked by an exterior mattron in a hydron aura, these virgins signaled no coordinates aiming towards the hydron but became linearly repulsive in that specific linear direction from a parent; on average from the hydron center.

Peeking virgins have passed through similar flux fields but could exit into a hydron aura and reenter the surface up to 10^{24} times; a last few stragglers reaching almost to the top of the aura. If peeking virgins are blocked by an exterior aural mattron, they signal reentry coordinates back into the hydron surface that cannot be <u>totally</u> neutralized within an aural mattron harvesting such virgins in opposite orbits that are never quite head on except at 10^{852}ãr where even peeking parental coordinates gain full repulsion without any aim back towards the hydron. The linear repulsion is cancelled by the remnants of cancelled reentry speed applicable at each distance.

Peeking virgins deliver no repulsion within the hydron aura except within the 10^{804}ãr thick belt outside the top layer; that I call the 'event horizon'.

One aural aspect deserves therefore a closer look; the repulsion by orbital virgins peeking into a hydron aura after being split within the hydrons top layers. In my earlier illustration of standard surface emission, subservient virgins from a top layer domain counted 10^{-76} of 10^{-280} per ãnna virgins split by a standard top layer mattron. These 10^{-356} surface virgins (and a dropping number from layers below) peek out of the surface to 10^{804}ãr at the edge of the event horizon before returning into the surface. For angled emission, the average linear reach is closer to a top layer when its path turns orbital.

There exist therefore, outside the top layer, a linearly repulsive element in the form of virgins that are not yet orbital but linearly repulsive (albeit from various sideways angles). Even here the inherent orbital coordinates from parent particles within a top layer mattron in mirrored opposite directions are not totally neutralized. Orbital speed at less than 1¢ cannot cancel the linear outwards motion that as the fastest motion leads to linear repulsion. The implanted orbital coordinate difference is not lost; as part of a mattrons coordinate system when the potential orbital sweep exceeds 1¢ on reaching outside 10^{804}ãr, further linear repulsion is cancelled for peeking virgins while the initial and already implanted repulsion will remain in that direction.

Neutral surface mattrons in standard hydrons are held in the grip of inter galactic flux that is opposite to hydron polarity; harvested by cores with the drive transferred to mattron coronas. As this flux backhand is instrumental in holding neutral top layer mattrons up against the hydron surface and any mattrons within the event horizon are likely incoming high speed mattrons; not outgoing mattrons, singular or neutral. Were we for a moment to imagine that such an incoming precursored mattron is resisted and stopped within the event horizon, it would face top layer virgin emission within the event horizon ('evhoe' emission for short); a massive short range repulsion drive.

There is also another effect within the event horizon that can offset this repulsion effect fully or partly. This is the extra local field activation of the many virgins peeking out into this thin band and being mutually activated. The pressure is much weaker than further out in the aura but the increased flux pressure can reduce mattron precursor sizes within the event horizon and thereby abrogate the need for a full 10^{752}är wide precursors to keep its central mattron mass from dispersing. The precursor size sets the mattrons virgin harvest and repulsion speed.

During a drastic loss of incoming virgin pressure, the surface of a standard hydron can expand considerably without causing inter-galactic flux to loosen its holistic grip on hydron mass. General expansion to a fraction of a hydron radius does not seriously weaken flux backhand push upon mattrons across a hydron radius. A marginal expansion may be much wider than the event horizon with its 10^{-48} Proton radius and if the surface is brought far beyond that, the event horizon simply moves along in front of the top layer.

A <u>singular</u> mattrons adverse virgin emission within the event horizon is blocked in a top layer and shared by a vast number of neutral mattrons in the surface lattice below. There is no re-polarization if the surface is of another polarity but coronal compression may lead to slightly smaller coronas on this small patch.

A singular mattron held by virgin winds within the event horizon is barely affected by general hydron expansion or contraction. It can compress its top layer locally by adverse virgin emission but can hold back a small patch set by its minimal reach on the hydron scale; perhaps forming a small valley on the local surface. And with that minimal influence, I leave the event horizon and focus on the aura as an average holistic entity.

Disparate Virgin Winds

The paths of split electromagnetic twins in a standard aura do not aim in random universal directions. The robust bipolar push by the enforced linear wind forms the bulges as a gradual accumulation of the subservient virgin wind aims the allied twin away from the hydron which accelerates the twins into an electromagnetic split at light speed.

Revisit my simplest scenario of hydron embryos forming as 'tiny' lumps carved out of vast mattrons clouds in the galactic grid. The hydron grid is separating into spherical walls, divided by voids, but simultaneously drawing together on a vaster greater scale into suns and other structures that evolve much more slowly.

At the beginning of the quantum collapse that formed most hydrons, that shrinkage was initially driven from outside sources in the inter-hydronal void such as dispersing mattrons. The presence of unraveling mattrons gave rise to activated emission and allowed fewer virgins to cross the void. Local hydrons could not support polarity and with no wayward singular mattrons to sow such polarity, disparate virgin winds that could hit a hydron were weak at the time.

Later, as the void cleared, direct emission exchanges between oversized hydron embryos became the main source of virgin winds. Early hydrons are almost exclusively made up of neutrally compressed mattrons aside from the rare but important few where the mattron population was toggled by fluke into a uniform disparity of one sign or the other.

Virgin winds out of hydrons during this time were therefore not disparate upon eternal neutral mattrons. If neutral hydrons shrunk to standard size, the stray winds would intensify but remained equal; with the important exception that a few hydrons gained disparate internal domains by fluke.

Outside polarized hydrons, the disparity of opposing virgin stray winds striving for supremacy in empty auras grows as they shrink. Any neutrally haloed mattron in such an aura has one mass sign pushed harder away from the hydron by the share its precursor halo can harvest of the hydrons disparate emission. Pushing a neutral mattron towards a disparate hydron is a backhand push to the hydrons virgin shadow (here actually the mattron harvest of surviving virgins from other hydrons). The result comes down to that a hydrons untouched virgin emission remain intact out through the aura and further. Inter-hydronal activation under way secures dominance of repulsive head-on stray emission out of a hydron aura.

The collective linear inwards push within a hydrons rarified aura (mattron backhand against the hydrons virgin gravity shadow) is here reduced by the inverse of the hydron area and the area of the mattrons precursor halo. The aural boost aids entrapment by precursors marginally (the boost is across a hydron; not a precursor width) with virgin repulsion this close to the hydron carrying the day with a wide margin in both signs upon a neutral precursor that presents the same size for both stray winds.

There is little need to look at a next grid hydron; virgin winds in the two signs upon a neutral mattron in a hydron aura are not in opposite directions but consistently away from the hydron in the same linear direction; so *that if a linear wind in one sign is more powerful, this is relatively the same thing.*

Linear Virgin Repulsion

Keep in mind that before a virgin eventually escapes free of a hydron, it must cut through the main hydron in myriad hydron-sized orbits around a parental coordinate center that can be anywhere within the hydron or at its surface. Until a widening virgin orbit moves entirely free of the main body hydron, its orbital push upon a mattron in the hydron aura can push it in both outwards and inwards directions which then equal out.

The hydron repulsion upon an *'exterior'* mattron experiences no opposition to that directional push along a Universal line except when the orbits of that mattron (or any other subcloud) bring it around a hydron where harvested push is in an opposite direction that neutralizes the first input.

The problem is that mutual repulsion finds it hard to form an orbital bond on its own as it aims in one linear direction at each time instead aiming at one central point at all times as do linear backhand shadows. Repulsion of an object that orbits a repulsive point source aims ultimately in all universal directions. A subcloud may move independently in relation to the repulsive source which will but partly neutralize a push that can only be neutered in the appropriate orbital bond.

Relative orbital motion can rarely form independently; it is most effectively formed by different modes of 'attraction' aiming at one central point. A slow already set orbital motion set by weak attraction upon a distant submist would only fractionally neutralize its linear repulsion.

The orbital virgin sweep between hydrons of similar sizes can allow more of the emitted virgins to 'find' another hydron but this augmentation, as earlier explained does not apply to smaller submists like a precursored exterior mattron as it cannot return fast enough in orbit for another potential hit to increase the rate of the harvest of such boosted virgins.

Repulsion by a single virgin away from a parent upon an exterior submist is not linearly diluted. Multiple virgins that issue out of a large hydron sphere aim at the sphere of a receiving submist from various directions. This conflicting virgin aim from an emitting sphere plays a greater role at close distances for potential virgin repulsion than at further distances.

Let me illustrate aural repulsion by looking at polarized hydrons at varying sizes as they affect a neutral standard mattron, approaching in through their auras, shrouded in a neutral precursor. Shying away from accuracy in the ebb and flow of halos, I use the 10^{752}är wide standard precursors in a void for all isolated mattrons. Such a precursor has enormous spatial presence of fast moving primal mass that easily entraps virgin winds by great spatial presence and ultimately delivers all the harvested coordinates by their own local flux into the central mattron.

10^{876}ãr Embryo Snapshot

Take a snapshot when an oversized collapsing hydron embryo has shrunk to radius 10^{876}ãr at a size where 10^{800}ãr internal inter-mattronal distances can sustain mattron disparity. Assume that the hydron has a single polarity. In this embryo the mattrons are $3,16 \times 10^4$ times lighter than standard after being compressed and annihilated to $3,16 \times 10^{907}$ atoms.

During 10^{876}ãt some $3,16 \times 10^{819}$ subservient virgins ($10^{228} \times 10^{591}$) are split within this hydron and $1,78 \times 10^{790}$ are released out of mattrons. With 10^{400}¢ boost during a shorter $5,62 \times 10^{770}$ subservient flux consummation, the flux sweeps 10^{1961}ã collectively within a hydron while a virgin reaches $2,37 \times 10^{30}$¢ and sweeps $2,37 \times 10^{906}$ points within it. The combined $2,37 \times 10^{2867}$ã presence within the hydrons 10^{2628}ã volume allows $4,2 \times 10^{-240}$ free subservient virgins to escape the oversized hydron embryo.

If the embryo is a polarized domain, some $3,16 \times 10^{819}$ adverse dominant virgins ($10^{228} \times 10^{591}$) are split within this hydron; all released out of mattrons coronas during 10^{876}ãt. With 10^{400}¢ boost during $4,57 \times 10^{776}$ dominant flux consummation, the flux sweeps $1,44 \times 10^{1996}$ã collectively within a hydron while a virgin sweeps $2,37 \times 10^{906}$ points within it. The combined $3,41 \times 10^{2902}$ã presence within the hydrons 10^{2628}ã volume allows $2,93 \times 10^{-275}$ part of the dominant virgins to escape the embryo.

During 10^{912}ãt, this hydron embryo therefore emits a linear virgin wind of $7,48 \times 10^{550}$ subservient and $9,26 \times 10^{544}$ dominant virgins out of its surface and out through its aura (a difference in these virgin winds of $1,24 \times 10^{-6}$).

A wayward neutral de-compressed mattron in the hydron embryos rarified mid aura would here be balanced by its 10^{752}ãr wide precursor and receive (10^{752}ãr÷10^{876}ãr)² or 10^{-248} of the $7,48 \times 10^{586}$ subservient virgins emitted in 10^{912}ãt. The subservient emission is roughly million times greater than the dominant virgin emission

The subservient virgin wind delivers $7,48 \times 10^{338}$pu linear drive during 10^{912}ã to the neutral standard precursor in the mid aura; a relatively weak disparate stray wind of $7,48 \times 10^{-574}$ subservient virgins per ãnna outside the oversized 10^{876}ãr hydron embryo. Note that the peeking virgins are equally outwards and inwards so they do not much affect the outcome.

As to the *electromagnetic threshold* that requires a harvest of 10^{402}pu in linearly effective virgin push in 10^{912}ãt in order to split a neutral mattron, a $7,48 \times 10^{338}$pu linear repulsion within the aura is nowhere close to the electromagnetic threshold. If one uses an internal mattrons from this large hydron, it contains $3,16\times10^{907}$ atoms but I am using a standard neutral mattron that approaches the hydron from far off.

An approaching standard mattron that enters the aura engulfed in its precursor will during $3,65\times10^{1180}$ãt accumulate a harvest of $2,73\times10^{607}$ virgins with 1pu linear repulsion for its standard central mattron of 10^{912} atoms. The relatively weak subservient wind can drive this neutral surface mattrons out of its aura at $2,73\times10^{-305}$¢ in $3,66\times10^{1180}$ãt and first stop and then repel the standard mattron back out of its aura.

A singular mattron in a 10^{876}ãr embryos subservient sign that approaches from far off faces more diluted repulsion but nevertheless, instead of seeking additional mass by usurping straying mattrons from the outside strayfield, an oversized 10^{876}ãr hydron embryo whether polarized or not already spurns additional neutral mattrons that approach at slow speed and actively pushes any exterior mattrons away. Such rising repulsion is true of all hydrons that resist new mattrons unless the speed of approach outguns a hydrons virgin repulsion.

The $1,24\times10^{-6}$ times fewer emitted dominant virgins lower the acceptance speed for singular mattrons in the dominant sign as these are accepted in the aura at $8,06\times10^{5}$ times slower speeds (and slower by half for neutral mattrons). There is a given acceptance speed in either sign for mattrons that accumulate virgin resistance and are resisted by hydrons. If the mattron does not exceed that speed; its linear advance will be broken.

A subservient mattron approaching at far below penetration speed can be stopped in its tracks long before coming close to the oversized embryo and thereafter repelled at its original speed of approach at an angle of reflection that is perfectly equal to its angle of incidence.

If the aim of an approaching mattron is close and its linear speed sufficient to penetrate deep into the repulsion zone, it meets growing resistance that gradually slows its approach and may ultimately stop it (approach across a hydron aura gives greatest resistance). If successfully resisted, the mattron is brought to standstill somewhere within the repulsion belt where it cannot stay unless it reaches a correct annulment in orbit around the hydron to neutralize any excess of harvested repulsion.

Without the orbital annulment, a halted mattron is forced back out beyond the borders of the repulsion belt and out to the borders of the hydron grid where it feels neither repulsion nor attraction (or equally from all directions) and the even less significant traces of the hydrons collective spin are lost.

Once rejected back beyond a repulsion belt, mattron capture is unlikely as it demands cancellation of linear motion. A mattron that is linearly repelled out of a repulsion belt is set to gain speed on its way out through that belt and accelerate until it re-crosses the boundary of the reach of the hydrons repulsion. This newly acquired acceleration is equal to the speed it lost while penetrating this deep into the repulsion belt.

On exiting the repulsion belt, the repelled mattron is in possession of an outward motion that secures its permanent departure; its speed of reflection mirrors its incoming speed. The hydrons insignificant collective spinfield is not equipped to capture fast moving mattrons. The time it took the repulsion belt to break a mattrons advance is also the time repulsion takes to reject it.

The ability of a straying mattron to ram through a repulsion belt depends on initial velocity and mass and odd openings offered by the uncertainty of primal fields. If a mattron (most wayward mattrons have re-grown masses to standard numbers) coming at a hydron from far off rams through head-on repulsion with a broad aim towards a hydron from distance of 10^{894}ãr, needs only be marginally off to miss as it has but 10^{-36} chance of a direct score.

If the mattron misses, the winds in its sails on the far side turn repulsive and accelerate the mattron away on a slightly crooked path while it regains lost speed. Chances are that most mattrons that due to relative speed or primal field uncertainty make it through a repulsion belt ultimately miss the hydron to embark upon an elliptical orbit on the far side and be permanently expelled.

Given the right speed on an elliptical path around the hydron, a mattron might be captured in a fragile bond if its orbit can be perfectly closed but this would probably be rare. In elliptical orbits, bonds may continuously cancel repulsion across the further parts of the elliptical orbit while motion across the closer parts of the same orbit would instead cancel attraction.

10^{860}ãr Embryo Snapshot

Take a snapshot of an 10^{860}ãr wide 'maximum' hydron that challenges incoming stray winds by emitting $4,23\times10^{553}$ subservient and $1,44\times10^{545}$ dominant radius time virgins that gain $1,33\times10^{26}$¢ sweep out through its 10^{860}ãr aura. A wayward neutral mattron faces disparate aural stray winds as its precursor captures $(10^{752}$ãr$\div10^{860}$ãr$)^2$ or 10^{-216} of all virgins it emits into the maximum aura.

The neutral precursor harvests $4,23\times10^{389}$ subservient virgins in 10^{912}ãt $(4,23\times10^{-523}$ per ãnna) and a neutral wayward surface mattron (with lighter mass) receives $4,23\times10^{389}$pu in disparate linear push in the 10^{860}ãr aura. This is still well below the 10^{402}pu electromagnetic threshold and wayward neutral mattrons do not (even with the hydrons lighter $3,16\times10^{910}$ atom mattrons) become hardwired into electromagnetic splits. In the dominant sign; $1,66\times10^{381}$pu repulsion is further from the electromagnetic threshold.

In $1,43\times10^{1147}$ãt with $4,87\times10^{-523}$pu repulsion per ãnna, an approaching standard mattron of 10^{912} atoms that enters the aura gathers $6,96\times10^{624}$pu subservient virgin repulsion that could repel it at $6,96\times10^{-288}$¢. Crossing the 10^{860}ãr aura at speed $6,96\times10^{-288}$¢ requires $1,43\times10^{1147}$ãt and repulsion cannot stave off the approaching mattron with a perfect aim that joins the hydron at that speed (most approaches are tangential and easier to reflect).

$1,44\times10^{-531}$pu per ãnna dominant repulsion accumulates in $2,63\times10^{1151}$ãt to $3,79\times10^{620}$pu and can stop a standard mattron at below $3,79\times10^{-292}$¢ and then expel it from the hydron aura in a bit more than $2,63\times10^{1151}$ãt from the lower aura. Lower speeds would be returned before reaching this close.

The slow repulsion shows that the 10^{860}ãr hydron still accepts singular mattrons in both signs although it is more partial to dominant virgins. Once accepted, a dominant adverse mattron would become a surface phenomenon in far too tiny numbers to affect the hydrons collective external emission and they cannot toggle its main polarity and at best slightly compress coronas of neutral mattrons within a marginal entourage.

10^{855}ãr Embryo Snapshot

Take a snapshot of an 10^{855}ãr wide collapsing hydron. In the subservient sign, its $7,03\times10^{743}$ãt consummation of inter-mattronal flux allows flux boosted to $3,16\times10^{389}$¢ to sweep $2,22\times10^{1133}$ã. In the dominant sign in $3,09\times10^{750}$ãt consummation its flux sweep $9,76\times10^{1139}$ã. The 10^{228} mattrons here hold $2,74\times10^{911}$ coronal atoms collectively emit $2,74\times10^{802}$ dominant virgins and $1,93\times10^{767}$ subservient virgins into its bowels in 10^{855}ãt radius time, during which dominant flux sweeps $2,67\times10^{1942}$ã and subservient $4,28\times10^{1900}$ã.

Across the 10^{855}ãr hydron, a virgin is boosted to an average orbital speed of $6,26\times10^{24}$¢ and sweeps on average $6,26\times10^{879}$ points before exiting the 10^{855}ãr wide hydron in an orbit that opens at 1¢.

The collective $4,28\times10^{1900}$ subservient flux presence and the $6,26\times10^{879}$ã presence of each virgin combines as $2,68\times10^{2780}$ã on leaving the 10^{2565}ã hydron volume. The average escape rate is $3,73\times10^{-216}$, whereupon in 10^{855}ãr some $7,2\times10^{551}$ subservient virgins exit it intact.

The dominant $2,67\times10^{1942}$ã flux presence and $6,26\times10^{879}$ã virgin presence combines as $1,67\times10^{2822}$ã with $5,99\times10^{-258}$dominant escape rate. Out of the $2,74\times10^{802}$ dominant virgins released internally in 10^{855}ãr some $1,64\times10^{545}$ escape to join the $4,39\times10^{6}$ times larger subservient crop in 10^{855}ãt.

A wayward standard neutral mattron faces disparate aural stray winds that by way of its precursor captures $(10^{752}$ãr$\div10^{855}$ãr$)^2$ or 10^{-206} of all virgins it emits out through the aura.

The neutral (standard) precursor harvests $\mathbf{7,2×10^{402}}$ <u>subservient virgins</u> in 10^{912}ãt (or $7,2×10^{-510}$ per ãnna) and the approaching neutral mattron (or a wayward surface mattron of $2,74×10^{911}$ atoms) receives a disparate linear push well above the 10^{402}pu electromagnetic threshold and is hardwired into an electromagnetic split (in the dominant sign, the $\mathbf{1,64×10^{396}}$pu repulsion is still well below the electromagnetic threshold).

In $1,18×10^{1138}$ãt with $7,2×10^{-510}$pu repulsion per ãnna, an approaching singular subservient photon of 10^{912} atoms crosses the aura at $8,5×10^{-284}$¢ or above gathers $8,5×10^{628}$pu subservient virgin repulsion. The repulsion on crossing the 10^{855}ãr aura cannot stave off an the approaching photon with a perfect aim and it joins the hydron at speed that is 1180 times slower than the speed of light.

An approaching neutral mattron at the same speed takes $1,18×10^{1138}$ãt to cross the aura but is already in 10^{912}ãt hardwired into an electromagnetic split where a twin photon in the subservient sign is aimed away from the hydron while the twin dominant photon is accepted by the hydron.

An approaching singular dominant photon gathers $1,64×10^{-516}$pu per ãnna on crossing the aura and accumulates $4,05×10^{625}$pu in $2,47×10^{1141}$ãt. It can stop a standard mattron at below $4,05×10^{-287}$¢ and expel it from the hydron aura in a bit more than $2,47×10^{1141}$ãt from the lower aura. Lower speeds are reversed before reaching this close.

Although the 10^{855}ãr hydron is a game changer in its ability to set off electromagnetic splits, it still accepts singular mattrons at a fairly high speed in both signs (and splits neutral mattrons with perfect aim).

Once accepted, dominant adverse singular mattrons become a surface phenomenon in too tiny numbers to affect the hydrons collective external emission (and unable to toggle its main polarity more than marginally).

10^{852}ãr Standard Snapshot

Take a snapshot of the more repulsive but stable 10^{852}ãr wide 'standard' hydron. An aural precursor would harvest (10^{752}ãr÷10^{852}ãr)² or 10^{-200} of all virgins it emits and gathers 10^{412} virgins in 10^{912}ãt to a neutral precursor of any surface mattron that has wandered into the aura from its surface and is repelled outwards. Its standard mass gathers linear push that is 10^{10} times above the 10^{402}pu electromagnetic threshold (even its $5,62×10^{6}$ times smaller dominant repulsion exceeds the electromagnetic threshold).

During the 10^{1132}ãt it takes a subservient photon (or a neutral mattron) to approach at light speed across a standard aura, its precursor gathers 10^{-200} of the 10^{832} subservient virgins emitted by the standard hydron and harvests 10^{632} repulsive subservient virgins.

The 10^{632}pu linear repulsion does not allow an approaching subservient photon to penetrate its aura; its 10^{912} singular mass is stopped in its tracks within the aura and with a perfect aim repelled head-on at its original light speed. If the approach is tangential, the photon is reflected from further away at an angle of reflection that is equal to its angle of incidence.

The $1,78\times10^{545}$ dominant emission in radius time ($1,78\times10^{-307}$ per ãnna) is above the electromagnetic threshold but is still the weaker virgin wind. An approaching neutral mattron (or a singular dominant photon) that crosses a standard aura in $2,37\times10^{1135}$ãt gathers $1,78\times10^{-507}$pu per ãnna that with $4,22\times10^{628}$pu can stop a standard mattron at below $4,22\times10^{-284}$¢ and expel it from the mid hydron aura in $2,37\times10^{1135}$ãt. Lower speeds are reversed before reaching this close but the standard hydron still accepts dominant photons that hit it with a perfect aim.

 An approaching neutral mattron that wanders into its aura from anywhere at any speed is already in 10^{912}ãt hardwired into an electromagnetic split. Even now, the dominant twin photon is accepted and the subservient twin aimed away but for a neutral mattron approaching from outside with such a perfect aim would be a rare occasion.

At further distances from a standard hydron, mattrons are less aware of its disparate repulsive virgin winds and larger mattron clouds, either gathering without precursors in forming and less repulsive subclouds (or isolated in open space with precursors) can gather at the middle distance in the 10^{895}ãr inter-hydronal grid where they may ultimately form hydrons of their own.

Part 5: The Galactic conversion

<u>The Singular Population</u>

Setting up a lucid working model is not easy with so few verified facts. I have for example no reliable number of split photons or singular charges in our part of the Universe in our time; a minuscule number compared to the population of neutral mattrons.

To sketch the progress of hydronal field properties, I must make a stab at the current proportion of singular mattrons in our galactic field; a number larger than it was in early times and smaller than it will ever be again (assuming that photons are stable and not ultimately dispersed).

The Internet tells me that background radiation in the present Universe from any region of space has roughly 2×10^9 times more photons than there are baryons. I am uncertain as to the accuracy of this indicator derived from calculations based on various assumptions of a 'Big Bang' model but I shall run with it as it seems to be the common consensus.

So let me pick this elegant number out of my hat and match every hydron in the Universe with **2×10^9 photons** spread throughout inter-galactic space and among local hydrons, albeit not necessarily in equal proportions or matching numbers in each signs.

The bases for a photon count is a galactic mass estimate for a Milky Way of $8,38 \times 10^{68}$ baryons. Here the tiny dose of 2×10^9 singular mattrons in our Universal back garden for each baryon has roughly 10^{219} neutral mattrons for each pair of singular mattrons and does not perceptibly alter hydron mass. With the same ratio, Our Sun would hold 10^{66} singular mattrons among its 10^{285} neutral mattrons and if you could gather that singular population into a one hydron; it amounts to 10^{-162} of its mass.

With such a massive advantage in numbers, singular mattrons that are locked up in hydrons cannot compete emission-wise with neutral mattrons. Areas of local internal charge carry little direct influence; an example could be 2×10^9 singular mattrons within a hydrons outermost top layer that holds 10^{152} neutral mattrons on its own.

It is helpful to keep in mind that despite a gargantuan mattron population, a hydron is an empty place. Place all its mattrons on its surface sphere (of mattron thickness) and there would still be 10^{212} empty mattron size spatial slots around each mattron on that sphere with nothing inside the hydron.

Even so, in reality, a hydrons internal neutral but disparate mattron grid is almost impregnable for a traversing mattron. A singular mattron at light speed that scores a direct hit upon a vast 10^{864}ãr hydron embryo after an electromagnetic split would end up lodged within its top surface. When a singular mattron at light speed hits a hydron and the kick is not matched from the other side, it does indeed create a weak linear push and the hydron recoils as a unit in that universal direction at a very slow 10^{-508}¢.

Electromagnetic splits of neutral mattrons can in principle take place anywhere but rarely deep within a hydron where powerful uniform prevailing stray winds prevent gate crashing by adverse singular mattrons that could cause electromagnetic splits.

If a split for example occurs in a hydrons top layer, the singular twins are quickly trapped in the local grid where the combined emission in both signs is of adverse proportions. Any singular mattrons that approaches that area must advance on a repulsion that far exceeds the grid average.

Such occasional temporary specks of neutral charge deep within a hydron enhance the gridlock of repulsion and spinfields and provide further nails in the coffin of extensive buildups of neutral charge in local areas by cornering mattrons. For a traversing singular mattron to join a charged local scattering and thus consolidate the charge type is racked with problems. It must force repulsion and spinfields for a long time, which is next to impossible.

Given time, mattrons can be moved by consistent stray winds that for the hydron as a whole can be compressive or expansive. Charged penetrating strayfield pressure affect the positions of opposite areas of hydron charge and, unlike neutral mattrons where one component is moved (and the other by proxy), singular mattrons carry no extra baggage and this makes their displacement twice the faster.

A major difference between hydrons and mattrons is the building blocks. A primal atom can traverse the width of a mattron before it collides with another atom and is trapped. Prior to that collision, it is not affected by any outside influence. In a hydron, a mattron traverses a grid of attraction and repulsion, guided at every point by constant interaction with myriad primal fields that are communicated to it by an enormous number of primal atoms.

Mattrons that are gathered into closer proximity within a hydron must be moved across a multitude of contradictory barriers in their paths, governed by the net field surplus. A disparate mattron corona has the biggest sail but requires a great persistency from a charged field to shift its position within a hydron and to allow it to work its way around myriad other influences.

Nevertheless, singular mattrons are not only common on standard hydron surfaces but have from early on been gathered by the surfaces of the larger hydron embryos that accept all singular mattrons.

Due to lesser primal uncertainty in a hydron field, it is incalculably harder to cause an electromagnetic split of a massive hydron body. Linear speed of light in electromagnetic splits is far too slow to cause a chain reaction in a hydron that is rarely exposed to a charged external strayfield that can shift its primal mass persistently and linearly enough to give its gargantuan stable of neutral mattrons a jolt into an electromagnetic split.

From that point, it is impossible for the relatively weak external strayfield to separate all interlaced singular masses after a vast electromagnetic split in an exploding grid to mold all those the adverse components into two hydrons of adverse singular charges. I therefore posit that such structural electromagnetic splits remain a mattron franchise.

Photons

A hydron can cause an electromagnetic split of a mattron entering its aura. A neutral mattron can of course collide into the hydron aura and be split but the more common cause the collapse of a hydron embryo which is most rapid near its surface and can leave a straggle of neutral mattrons less protected in the aura. An electromagnetic split is followed by the emission of the subservient singular twin photon that departs at 'constant' light speed in standalone incidents.

If more than one neutral mattron is split at the same time, these are still standalone incidents; even if an outgoing collective pulse builds a widening globular form in many universal directions on an expanding sphere, this is not an actual wave but a front of independent photons moving outwards at the same speed.

A photon body at light speed does not represent its real power which does not preside in its bulk but rather its field effect. The collision pressure it delivers by hitting an object such as a hydron is small and derives from the transfer of photon momentum per unit time and area. The photon holds for example 10^{-228} of a hydrons mass and travels at a relatively slow 10^{-280} part of primal speed. The field effect around the singular photon (before it enters a collision) represents its real power; its 'constant' adverse emission and the consequent generated repulsion.

As an electromagnetic photon moves away, the singular mattrons form precursor halos in the weak outside strayfield while its internal creation of newly split atoms continues to pour out adverse emission that is the most important secondary component of electromagnetic emission; orbiting out with parental boost from both its precursor and central mattron.

Emitted primal virgins are an integral part of singular mattrons. They are in possession of the mattrons entire collective coordinate system. Each newly emitted atom is inherently oblivious to what we (as a universal observer) perceive as linear motion of the singular submist body that contains them. The primal residential coordinates of their mass cover all aspects of their photon origin; primal, mattronal, hydronal and so on.

For example, a singular electromagnetic photon exudes allied intermittent halos to keep spatial balance but the next primal halo has exactly the same center as the preceding halo. The lost parts of a precursors would leave the mattrons in spherical intermittent crests at identical intervals or wavelength all around the emitting body (unaffected by what we perceive as universal motion of that body). This independent perspective holds even if frequency and speed of primal allied precursor halos gradually subsides; the walls of lost halos, emitted in succession, do not overlap. Like a primal particle, a photon is a universe on to itself and at complete standstill.

The smallest 10^{912} atom photon parcel of a singular mattron can supply virgin emission in its adverse sign that is 10^{20} times greater than that of a standard hydron. The pressure field of a receding photon can easily affect the intermittency or lapse before a new electromagnetic split by its initiating hydron as well as from nearby hydrons.

When a field of collective spin is moved out of a common coordinate center by new coordinates when electromagnetic twins split from the hip and are moved out of a common center in opposite directions on an electromagnetic front, the further the crest travels in space, the greater the side-shift in an open differential orbit. This is in full accord with the original coordinates inherited from the primogenitor neutral mattron.

The inherent spin of a photons ancestral mattron is 10^{-912}¢ central orbital motion but its primal mass keeps various distances with side sweep speeds up to 10^{-280}¢ at surface. The neutral mattrons mass had a 10^{316}ãr median distance from where it collectively delivered 10^{-596}¢ median inherent spin (in opposite directions for the two cohabiting signs). All that pertains to free electromagnetic twins, unencumbered by other inherent spin coordinates; for example those gained as a fully integrated part of hydron coordinates of inherent rotation.

An electromagnetic photon heading out of a hydron aura has its orbital motion boosted by square root of the distance traveled as measured by the median 10^{316}ãr. However, to the extent it is fully integrated with the hydron coordinates of surface spin (like a Protons 10^{-288}¢ orbital motion about its center), the departing e-photons path will be seriously compromised.

Across the 10^{932}ãr width of an observable Universe (measured by 10^{316}ãr), the 10^{308} square root of 10^{616} fold distance boosts the photons 10^{-596}¢ side sweep to 10^{-288}¢. As the Universe is immeasurably wider, the inherent side sweep grows to equal and eclipse the 10^{-280}¢ speed of light which happens at 10^{948}ãr if a photon does not hit anything underway in differential orbit. If the width of 10^{16} observable Universes seems daunting, remember that the Universe must be incredible wider than the part we observe; the virgins that build it have been streaming outwards at 10^{280} fold light speed from the time of the initial primal split.

A roving electromagnetic photon is affected by any primal particles it encounters as it moves in relation to the Universe and the experienced 'drag' can slow a photons 10^{-280}¢ linear relative motion but whatever rare primal mass it encounters at stoppable speeds underway is likely to render this slowdown insignificant. Singular photons can of course answer in the same manner and accelerate nearby hydrons to close to light speed.

Generally, half of all new photons are captured by a hydron during the split. The escaping twins may find a final resting place in a mattron cloud or hit an oversized hydron embryo. The photons can be affected under way by directional repulsion and shadows and though close encounters are few to start with, rising inherent sweep causes more of the photons to be captured by hydrons underway.

The twin of a split neutral mattron repelled out through a hydron aura can be far less linear than the twins of free e-photons split in the open without additional structural coordinates. This is especially significant if the mattron is in possession of the hydrons collective inherent spin coordinates that will affect the path of the outgoing twin with relatively fast and gradually rising side sweep.

Even neutral mattrons will from time to time be propelled to light speed or above but they are neither photons nor carriers of local charge.

Polarizing the Hydron Field

Even before a singular mattron quite reaches into the embryo, the forming hydrons acceptance of it is immediate and it takes merely $10^{10^{26}}$ãt to polarize all of its 10^{228} mattrons. If the embryo is already polarized in an opposite sign, that original disparity is secure. An adverse singular mattron is unable to transport a domain pulse deeper into a hydron and can at best slightly compress the coronas of a small entourage of nearby mattrons.

My premise is that the mattron grid was quantum split into large hydron embryos long before electromagnetic splits got off the ground; that singular mattrons were not part of an environment as hydron embryos had barely started to gather into cold solar subclouds within diffuse galactic submists.

Individual hydron embryos that stabilize around 10^{900}ãr radius are built of highly compressed neutral mattrons that are too far apart to be polarized and the embryos in local grids communicate at the time with their closest and similarly oversized neighbors which outermost layers are all gradually condensing from the collaring effect of peeking virgins.

The condensation of the largest embryos is almost exclusively driven by peeking virgins that gather their mass towards greater inner density during extensive periods; the embryos do mutually exchange virgins but it is a weak effect from within leverage points that allow 10^{-38} of the linear component and not the boost to work as compression. Without inter-galactic photons, a few galactic virgins fall way short of speeding up this condensation process.

The more progressed embryos are likely found around the galactic center where mutual exchanges are greater and shrinking may proceed to pick up. At the center of a forming galaxy where some hydron embryos first shrink below the 10^{881}ãr threshold (where mattrons can support polarization), there are still no singular mattrons to set off hydron polarization. As the embryos shrink, their external emission increases per ãnna but diminished sizes causes more virgins to miss and they gain less orbital aid although increasing virgin seepage can significantly increase the exchange.

In the earliest most progressed part of the <u>premier</u> Universal area of hydron formation, at the center of the first galaxy, a special hydron claims the field. The 'Unetron' gradually takes form; a hydron built of 10^{228} neutral non-polarized mattrons. From its formation above 10^{900}ãr radius, the first Unetrons that after almost an eternity succeed in shrinking below 10^{881}ãr would still remain non-polarized as there exist no photons at that time to polarize them.

The Unetrons continue their slow condensation in a non-polarized state and a few reach standard 10^{852}ãr size with equal subservient emission in both signs so that even a standard Unetron is unable to split neutral mattrons in its aura.

Within a Unetron shrinking below 10^{881}ãr where mattron polarization is sustainable, the chance is small but real that some of 10^{296} galactic mattrons are eventually polarized by fluke in both signs. The gargantuan numbers at the center of an advanced premier galaxy make this rare occurrence fairly common. After such a mattron fluke, the entire Unetron turns mono-polarized in 10^{1026}ãr; so swiftly that there is next to no chance that another mattron within the hydron embryo becomes polarized by fluke during that period (in one sign or the other during 3.33×10^{-169} second).

Which sign is 'negative' and which sign is 'positive' is semantics but this premier polarity, born of a coincidence, fashions the Universal evolution and gives the premier polarization an upper hand in the race for Universal polarity. So let's make it positive.

The relatively few positively and negatively polarized Unetrons, during final collapse past 10^{855}ãr hydron size, generate massive photon emission when a few neutral mattrons stragglers are left behind in the auras and split during the rapid collapse. The electromagnetic photons shower local embryos in the premier galaxy with the first burst of photons. The first burst of positive photons will soon be followed by negative photons from other Unetrons.

When a neutral Unetron is hit by a singular photon, it cannot resist or repel the photon in either sign. If the Unetron is smaller than 10^{881}ãr, it immediately becomes mono-polarized. If it is larger, it can continue to gather photons of both signs which, on reaching 10^{881}ãr, would rapidly polarize various different parts of its mattron population in opposite signs.

In this first polarized collapse in the premier galaxy with electromagnetic splits, each positive subservient photon leaves on a straight path at 10^{-280}¢, aimed by subservient emission away from a polarized hydron while the adverse 'dominant' twin photon is aimed into the polarized surface against the hydrons weaker dominant emission.

On average 2×10^9 singular electromagnetic photons are split in the aura of a mono-polarized Unetron during its collapse to standard size by subservient virgin emission. The massive adverse virgin emission from 2×10^9 receding photons hit the left behind collapsing hydron in boosted orbits as pressure grows out to the leverage point and then starts to fade as the photons depart the local grid. Spreading out to further away, neutral oversized embryos accept all such linear photons at light speed.

Of perhaps 10^{68} hydrons in a premier galaxy of shrinking hydrons, the first Unetrons polarized by fluke emit positive photons and consequently, 2×10^9 (out of 10^{228}) embryos become positively mono-polarized. On their continued collapse, some will in time polarize that many oversized hydron embryos.

An important point is that the first positive photons take flight a tad earlier and inseminate a few hydrons before the first Unetron that became negatively polarized by fluke collapsed to emit its 2×10^9 negative electromagnetic photons which in time will also inseminate that many oversized hydron embryos.

Eventually and in both signs, each of the 2×10^9 positive photons from such splits score a direct hit on some oversized hydron in the premier galaxy. As negative polarity and photons entered the field a bit later they reach fewer candidates for conversion; let's make positive photon on the prowl a few hundred times more numerous. When any of these photons mono-polarize an embryo, that Unetron will eventually collapse to split 2×10^9 neutral mattrons in its subservient sign; aiming the subservient electromagnetic twin away while keeping the dominant twin.

This is a slow moving chain reaction in full flight that can only gradually disturb a local population of oversized embryos. Of the gargantuan number of hydrons embryos in a premier galaxy, most would likely to remain 10^{900}är wide for extreme periods without aid from passing photons that help in their compression. The vast embryos cannot be polarized when hit and they can continue to collect new photons in both signs for a long time with a certain a small advantage in the number of positive photons.

Adverse emission from photons passing between smaller embryos is more effective than between larger embryos but the hydrons longest collapse time is from its biggest size. When the first photons pass and emit boosted virgins from within the leverage point upon many hydrons, these virgins cut very shallowly into that vast hydron radius. Opposed already within its top layer they becomes ineffective as center wise compression upon the 10^{900}är wide embryo across a 10^{904}är average grid distance.

The overwhelming virgin majority from a photon is harvested by a vast nearby hydron but the depth of a virgins orbital cut is related to its 10^{50}¢ grid boost and I consider its extremely restricted center-wise compression upon an embryo null and void with only linear push in full working order. The closer a photon passes, the weaker is the temporary delivered surface compression. It is first from outside the leverage point of the 10^{900}ãr wide embryo at roughly 10^{919}ãr distance that the $3,16 \times 10^{57}$¢ orbital gain allows all the virgins hit without any cut to potential compression input (before the 10^{-38} center-wise cut).

In this gradual chain reaction, the small original number advantage of mono-polarized positive Unetrons above that of negatively mono-polarized Unetrons is permanently cemented. From the central galactic field, a few hundred batches of positive photons continue for a brief period to seed the galaxy (each batch adding 2×10^9 mono-polarized Unetrons); inseminating large hydron embryos in the positive sign.

Somewhere within the galactic center, a negative mono-polarized Unetron, also polarized by fluke, will almost shortly collapse and set off its own chain reaction with 2×10^9 negative photons); inseminating hydron embryos in the negative sign.

With hydrons in both signs collapsing, a brief period of positive hegemony has been broken. The number difference is not huge within the premier galaxy. With both signs represented in the field, I posit that out of every 480 embryos that collapse and emit electromagnetic photons, a single Unetron is negatively polarized and 479 are positively polarized. Take this trivial difference as an illustration of a system.

It is only initially that the distribution of hydron embryos within the galaxy allows merely a single hit that turns oversized embryos into mono-polarized Unetrons. As a rising number of collapsing hydrons is providing bursts of 2×10^9 photons in both signs and soon the overwhelming number of galactic embryos are being inseminated in both signs. The 480 positive advantage holds for the number of delivered photons and Unetrons are fast becoming an increasingly rare breed; even a standard neutral Unetron cannot survive when hit and polarized by a single photon.

As the harvest of electromagnetic photons of opposite polarity preserves the 480 number advantage of positive photons on the surface of most hydron embryos, this will result in internally garbled polarization when they collapse below radius 10^{881}ãr.

Galactic polarization

From multiple points at the center of a premier galaxy, photons of both signs are now heading out from collapsing embryos in all directions at the exceedingly slow speed of light. As each hits or passes by a set of oversized embryos, it raises the compression upon them but this virgin compression is local and does not penetrate far; a photon usually affects one or two hydrons at the time on passing between the embryos in 10^{1184}ãt (10^{-10} second). This does not lead to a sustained collapse but is likely to quicken the ongoing corralling and compression by way of mutual virgin exchanges.

To begin with, in a say 10^{927}år wide premier galaxy built of 10^{69} oversized 10^{900}år wide embryos, only 10^{-13} of all photons heading out from the galactic center avoid hitting a hydron. The minority that escapes the premier galaxy has traversed it for a hundred thousand years. When all the galactic hydrons have collapsed to standard size, the galactic transparency has increased so that only 10^{-81} of the photons hit a standard hydron before the majority leaves the galaxy after a hundred thousand years.

Until the chain reaction reaches its crescendo at a time when the when 2×10^9 photons stream out of a galactic center from 10^{69} collapsing hydrons, large embryos usurp all photons that hit them. This is a sizeable part of the 2×10^{77} positive electromagnetic photons and $4{,}17 \times 10^{74}$ negative photons that are ultimately split. How many of the average dose of 2×10^9 positive photons or $4{,}17 \times 10^6$ negative photons each embryo in the premier galaxy can collect is not a provision for maintaining the permanent positive advantage of 479 positive photons for every negative photon.

The first electromagnetic photon from the center of this premier galaxy can pass out of it after 10^{1207}åt in roughly a hundred thousand years. At first only 10^{-13} of all photons avoid hitting an embryo but a few initial thousand exit the galaxy, leaving minuscule trails in their wake where their adverse emission has caused embryos to shrink slightly at every point where they passed by. The majority blockage cannot prevent an electromagnetic photon minority from escaping the premier galaxy.

Earlier, I multiplied the polarization period for a single hydron made up of 10^{228} mattrons by the square root of mass units; extending the polarization time of an entire hydron 10^{114} fold. Conversion of a large number of hydrons takes more time than conversion of a single one but conversion of embryonic sizes to standard sizes within a galaxy is not directly tied to polarization in the premier galaxy although the two are closely related.

In the same general manner the structural collapse of 10^{69} hydrons takes more time than the collapse and emission response by a single hydron. Both these conversions are slowest at the start and fastest and most massive at the end. To use $\sqrt{}$ mass units to multiply a collapse time for a single embryo to account for the conversion speed of all 10^{69} hydrons in a galaxy from start to finish would make for a galactic conversion times that is $3{,}16 \times 10^{34}$ fold that of the average individual embryo.

When the first few photons exit the galaxy, only a small fraction of the innermost center hydrons in the premier galaxy have collapsed from 10^{900}år to 10^{852}år standard size and few of the large embryos in its massive outer layers are affected by purely local compression from a few passing photons.

In the collapse from 10^{900}ãr size, all balances change step by step down to 10^{852}ãr standard size. A shrinking hydrons drive for compression weakens while its expansion increases; the longest period being at largest size. If the collapse takes place in less than 10^{1192}ãt (the creation constant), a 10^{900}ãr embryo with its highly compressed mattrons has no time to allow mattrons to replenish their 10^9 times smaller primal mass whereas incoming photon pressure remains unchanged and effective upon the smaller masses.

If a collapse under grid pressure takes more than 10^{1192}ãt, growing mass resistance needs not be considered but if longer, the compression against repulsion must be reduced at each point down the chain of 10^{48} fold size difference by cutting compression at each point by mass lack compared to 10^{912} standard mass.

It is hard to tell how much average time it takes for a single hydron to collapse from 10^{900}ãr to 10^{852}ãr standard size under the onslaught of 2×10^9 or more electromagnetic photons from each nearby burst of a collapsing hydrons. It takes photons only 10^{1184}ãt to move out of local contact but this does not hold due to local exchange. The local photon presence has grown to such extent that there can easily on average be 2×10^9 photons present for a period that is longer than 10^{1184}ãt (or fewer photons for a longer period).

The primal mass of 2×10^9 traversing local photons grows swiftly to 10^{912} with 2×10^{-271} adverse virgin emission that with so hydron tight grids allows few of these virgins to slip by the nearest embryo. *This is why photons entering from another galaxy that later become the usual trigger of hydron collapses and galactic conversion are already of standard mass.*

Virgins split at 10^{904}ãr outside a 10^{900}ãr embryos 10^{919}ãr leverage point arrive on locally linear paths across 10^{19} fold its aural distance and the span is only 100 fold. My earlier idea for standard hydrons was to allow incoming virgin boost be cut as compression by the inverse leverage distance times the split distance. If used for a collapsing hydron embryo, virgin pressure input at 10^{50}¢ is modified by $(10^{-19}\times10^2)$ with temporary 10^{33}¢ compression.

With an embryos surface already rapidly shrinking under the onslaught of earlier photon emission, the question how 'temporary' compression input works over time after virgin coordinate paths are blocked in its top layer. The steady shrinkage means that the delivered orbits become ever closer to the leverage point. At this point for an embryo, an orbiting virgin is blocked by deeper denser mass and while the embryo continues to shrink, 'temporary' compression is being made permanent.

Virgins boosted to 10^{50}¢ may here arrive from far within its leverage point but as the shrinkage does not ease but increases instead, the compression is turned permanent so I allow the 10^{50}¢ work fully as compression input that is downgraded as center-wise compression to 10^{-57}; the collapse speed in an unbroken collapse is set in the embryos rotation period (here 10^{1131}ãt before it turns the other cheek) or else the embryo would expand again.

The pressure come from a great number of photons of various compressive drives after a slower initial stable start across the greatest radii. A lighter pressure can lasts for a longer period if photon presence is such that when the first photon batch leaves the local scene, other passing photons take their place.

When these vast hydrons collapse within 10^{1192}ãt, their electromagnetic photons hold 10^{-9} of standard photon mass and must therefore pass by 10^8 embryos before growing to full size and emission power. This weakens the initial compression ability in the premier galaxy but not so much in all later galactic conversions where collapses are driven by a greater stock of full bodies photons arriving from another galaxy. Weaker compression does of course push upon smaller mass which cancels the effect of the deficit.

One can set this up in the 10^{900}ãr wide embryos 10^{1131}ãt rotation time while under compression from 2×10^9 fully stocked photons upon its 10^{1131} primal mass as (2×10^{-271} virgins)$\times10^{50}$¢$\times10^{-57}$ which yields 2×10^{-278}pu and moves it to collapse at 2×10^{-278}¢ speed across the widest distances which takes 5×10^{1177}ãt to standard size. The electromagnetic photons take a longer 10^{1177}ãt to leave the local premises but due to the continuous exchange of photons within the grid, the pressure can continues for much longer.

This shorthand in a text boxes is to remind myself of values in the ordinary text. Photons cross galaxy in 10^5 years ($3,15\times10^7$sec)=$9,45\times10^{1206}$ãt. $1,58\times10^{1212}$ãt÷3×10^{1194}ãt=$5,27\times10^{17}$sec=$1,67\times10^{10}$years!

The next galaxy is not far away. The spread of a photon front takes a small bite out of the numbers with perhaps a thousand part of the photons hitting the nearby galaxy and the timeframe for all following galactic conversions is shorter. It is not up to gradual embryonic collapses or Unetrons that were mono-polarized. The following conversions relay on fully stocked incoming photons from outside the galaxy that arrive in full numbers after $1,67\times10^{10}$ years in numbers barely shy of the photon number eventually split by the new galaxy after implanting mismatched numbers that validate the original fluke polarization ratio.

The conversion of a galaxy simply indicates the collapse of all its embryos to standard hydrons. If it takes the oversized 10^{900}ãr individual embryo an average 5×10^{1177}ãt period to collapse to standard size and it takes $3,16\times10^{34}$ fold that period to convert the entire galaxy, the whole thing is over in $1,58\times10^{1212}$ãt after the ball starts rolling. This longer period of $5,27\times10^{17}$ seconds equals $1,67\times10^{10}$ Earth years.

Away from a converted galaxy, the first exchanged effect across to nearby galaxies is not slow electromagnetic photons but whatever of their highly boosted adverse virgin emission that avoids activation across to the next galaxy. This virgin and flux effect comes in gradual stages from the time the first few photons reached out of a converted galaxy till the last greatest burst of photon arrives midways to the next galaxy after $1{,}67 \times 10^{10}$ years where their effect peeks.

Before a rising number of electromagnetic photons start to bathe faraway galactic fields of developing hydrons in showers of singular photons and to embed potential polarities in their image, their adverse emission baths the galactic fields in virgins and flux but these fields are initially weak and lead to lesser intergalactic activation with most virgins delivered into galaxies.

The Galactic Conversion

Upon a pristine galaxy of Winter Gate proportion that is fairly close to a converting galaxy one can gauge the gradual photon influence. The galactic embryos are large and have taken a long time to shrink from formation to 10^{900}ãr when electromagnetic photons start passing into the 10^{927}ãr wide galaxy that contains 10^{69} hydron embryos that have hitherto merely been able to shrink from the 10^{904}ãr inter-embryo grid distances.

The photon number exiting the premier galaxy continues to rise until it reaches a climax after $1{,}67 \times 10^{10}$ years (or whatever) when the number of arriving photons tapers off rapidly in the next hundred thousand years. The conversion of a new galaxy takes roughly thousand times longer than for a photon to reach that next galaxy which they cross in 10^{1207}ãt.

It is first during the last stages of galactic conversion that inter-galactic photon number delivers the full blast of electromagnetic photons with virgin and flux compression across the inter-galactic void. Up until that, smaller fractions of the ultimate 2×10^{77} positive photons and $4{,}17 \times 10^{74}$ negative photons are dosed out and something like a ten thousandth part hit the nearby galaxy where the photons pass into the outer galactic surface and in among its less developed 10^{900}ãr wide hydron embryos.

The first photons hit a nearby galaxy already after several million years; if one ignores that photons from an earlier galactic conversion further back in the chain may already have started to affect the galaxy from further away. The first arriving photons are few and far between and 10^{900}ãr wide embryos continue to block them as they hit in both signs at a gradually rising rate.

In the inter-galactic void, as the input rises, photons become numerous enough for their adverse emitted virgin majority to be activated. The virgin input in the two signs is unequal virgin but this is corrected at the delivery end by unequal activation by inter-galactic flux that evens up the score. The surviving virgins then deliver the same pressure upon hydrons in both signs; not that this pressure plays a significant role. The flux spawned by unequal photon input does however deliver unequal flux pressure into the galaxy and this will later play an important role when galactic embryos have shrunk enough to sustain a garbled polarity.

At the end of the conversion era, the motherload of 10^{78} positive photons emit virgins across the inter-galactic distance with an activated majority. The flux from 10^{726} inter-galactic virgins gains 10^{464}¢ orbital drive during 10^{698}ãt consummation (10^{900}ãr hydrons block flux at around 10^{230}¢). This 10^{1888}ãr flux presence combined with 10^{990}ãr virgin presence or 10^{2878}ãr in 10^{2784}ãr inter galactic volume allows 10^{-94} of the negative virgins avoid activation and to hit the 10^{900}ãr wide embryos at 10^{62}¢ while the majority flux rips primal mass out embryos at a rate and eventual pressure that in a lumpy galaxy is not specific to any local mass collections.

Traversing photons set the end-game for galactic transformation in more ways than one. As hydron embryos in a galaxy collapse under local virgin emission, eventually from 2×10^9 positive photons, they become polarized and gain a garbled polarity of a neutral mattron stock in proportions of photons harvested by their surfaces. Adverse flux spawned by inter-galactic photons arrives highly boosted into these hydrons and internal backhand to this flux pressure will over time effectively shift mattrons of garbled polarity at a different rate within the hydrons.

When a hydron crosses the 10^{881}ãr polarity threshold, it suffices for it to contain one singular mattron (or one mattron polarized by fluke) to have its entire mattron grid polarized in that image. With a fair share of two billion positive and over six million negative singular mattrons; a brewing polarity runs like northern lights out from every singular mattron within an embryo and shapes pockets of a new polarity that is heavily garbled from the start.

For most mattrons within a hydron, the initial distribution of a polarized population is holistically asymmetric but locally more symmetric. However, during extensive periods after polarization, the high speed inter-galactic flux that is harvested by mattrons core flux will work continually to shift the position of polarized areas within any hydron endowed with garbled polarity. The more powerful negative inter-galactic flux is collected by cores of positive mattrons that are more easily carried deeper within the hydron embryos.

Generated by a disparate crowd of inter-galactic photons; negative highly boosted flux from 2×10^9 positive photons and positive flux from $4{,}17 \times 10^6$ negative photons will in the end bear upon every mattron core within each galactic hydron. This totally penetrating center-wise thrust is blocked within hydrons and yields an internal local backhand shadows between mattrons that in time proves quite effective.

Thus the Universal model does not solely come about by actual physical hits of 479 positive to single negative photon upon oversized embryos; it is in time also shaped by emission from photons on route between galaxies as the adverse virgins rapidly transform into flux in the void, aided by the parental boost from those split in precursor halos. This power of this collective primal flux orbiting holistically through a galaxy exceeds all local flux exchanges.

A surviving virgin sweeping into a new galaxy can hit one of 10^{69} embryos but all virgins are harvested before sweeping 10^{62} times through the galaxy; they can however do so only 10^{15} times before hitting an embryo and they cannot at first penetrate deep into the galaxy. An 10^{928}ãr virgin orbit at 10^{62}¢ takes 10^{866}ãt and widens by merely 10^{881}ãr in 10^{15} passes or 10^{-46} of the galactic radius; a shallow galactic surface area that may hold 10^{46} embryos and this would be the most concentrated virgin pressure from outside upon galactic embryos.

This is therefore not a powerful compression; even at the peak of photon emission around the galaxy, some 2×10^{929} negative virgins are emitted in a 10^{900}ãr embryos 10^{1131}ãt rotation period and 10^{-94} or 2×10^{835} hit an embryo. Of these, 2×10^{789} hit each the 10^{46} surface embryos at 10^{62}¢ and deliver 2×10^{851}pu virgin pressure that downsized to 10^{-38} delivers 2×10^{813}pu compression upon an embryos 10^{1131} mass; driving it inwards at 2×10^{-318}¢ against the resistance by the 10^{900}ãr embryos more lethargic $5,62 \times 10^{-350}$¢ expansion that falls far short of the compression.

If this were the only available virgin effect, it would start to shrink the 10^{900}ãr embryo in 5×10^{1217}ãt which may be the longest period in galactic creation but compression cannot be harvested by all galactic hydrons at the same rate but starts by embryos in its outermost layers. The hydron embryos have from early been gathered extremely slowly by peeking virgins corralling their outermost mattrons and the inter-galactic virgins add a weak voice to this withdrawal. It is unlikely that a galaxy, before being smitten by photons, has begun to coalesce into areas of greater density surrounded by less populated areas. It is first when corporeal photons start to physically pass into the galactic grid that hydron collapses start off in an accelerating progress.

Each large embryo harvests its share of passing photons in both signs at the 480 ratio and after a photon hits an embryo, its ability to compress or affect any other hydron seizes immediately and completely. Adverse photon emission is marginally activated in close encounters with embryos and a few and far between photons passing into a nearby galaxy are ill equipped to compress large close embryos it passes before hitting a large embryo slightly deeper within the galaxy. The uniform one to 479 disparity in photon signs is not a determining factor in virgin compression as virgins are harvested by hydrons as surface pressure irrespective of sign.

A growing frequency of trespassing photons from far off yields a short-lived but gradually increasing pressure and the embryos easily expand again. By the nature of things, new local splits of less fully stocked neutral mattrons have a weaker initial influence than these incoming fully grown photons. It takes photons a hundred thousand years to pass through a galaxy and the overwhelming majority soon hits an embryo.

As the galaxy is so opaque, photons that don't hit an embryo must pass between 10^{23} hydron pairs, one after the other, before exiting the galaxy on the other side. Unlike virgins, the electromagnetic photons enter the galactic surface linearly at 10^{-280}¢ and cannot penetrate far before hitting embryos. The outer layer embryos are the first to be affected as a photon hits already after passing between merely 10^8 embryonic pairs.

The galaxy is already a roughly globular structure when its embryos start to shrink throughout the outer layers; initially on a thin skin of surface embryos. The following newcomer photons will be able to penetrate a bit deeper between the shrinking embryos before hitting a large embryo and being taken out of circulation. As the electromagnetic number rises and the conversion continues across deeper than 10^8 grid distances between more than 10^8 embryos, the effect is aided by way of local collapses and rapidly proliferating locally split electromagnetic photons.

The first tier collapses of surface embryos are more gradual and take longer as there are fewer photons around. For photon pressure upon a local embryo to continue, an incoming photon must be replaced by a new photon within 10^{1184}ãt when the first passes out of contact. The first tier embryos take longer to collapse and eventually emit its own batch of photons.

The galactic conversion is eventually also self generated as multiple outer layers collapse and emit their own photons inwards and outwards. The local collapses provide close quarter electromagnetic emissions that are no longer in a supporting role but drive the galactic conversion inwards from the outer layers where large embryos continue to block photons for ten billion years.

The flux fronts that orbit into a galaxy from many directions are slowed and eventually stopped as primal mass is ripped out of mattrons. Galactic flux at 10^{463}¢ in orbits is first harvested at around 10^{230}¢ and must rip 10^{233} atoms out of 10^{69} hydrons to slow it down. The flux cannon fodder is quickly re-deposited into galactic hydrons as local coordinates focus the flux to local areas when the tighter boosts spread the effect until harvest is possible. All independent coordinate associations of an atom become part of local mass.

There is no simple illustration of a converting galaxy from start to finish as each snapshot will see embryonic sizes ranging from 10^{900}ãr to 10^{852}ãr but one helpful things is that non-polarized 10^{900}är embryos and polarized 10^{852}är standard hydrons block flux at roughly the same 10^{230}¢ speed.

During one second, every primal atom of a hydrons mass can be re-shifted innumerable times from within a hydron out to semi local distances. Before harvest in variable consummation periods, say 10^{698}ät, the emission of 10^{427} newborns by 10^9 photons for every hydron rips out 10^{234} flux atoms after reaching 10^{464}¢. The (standard) hydron loses 10^{661} atoms of 10^{1140} mass and during one second, this loss grows 10^{496} fold so that in principle a hydron can lose 10^{1157} atoms within a second; *exchanging and reinstating an all-new hydron mass 10^{17} times.*

It stands to reason that some long distance universal flux can move flux out from between galaxies into larger orbits with corresponding boosts and consummation but this may be marginal as most blocking mass resides within nearby galaxies. Galactic boost coordinates are not lost on hydrons, only divided upon an immense number of local primal flux atoms that focus the flux storm to local area where coordinates are modified and downgraded.

The massive flux generated by each activated inter-galactic virgin is shared fairly equally and deposited into a vast array of hydrons, outside and within the galaxy, where the end game of each single flux atom is to deliver all tiny coordinate snippets gained by it in collisions from myriad inputs. These minuscule primal parts act collectively upon the mass of a receiving hydron; even the original coordinate drive that set the flux storm linearly moving is part of the package. Both linear and orbital coordinate inputs from outside the galaxy blend with the coordinates assimilated within a galaxy.

After $1,67 \times 10^{10}$ years with its internal composition altered, the galaxy is built of compressed standard hydrons. A single incoming flux atom at 10^{464}¢ is slowed by 10^{234} local atoms and new flux coordinates from 2×10^{78} photons around the galaxy in all universal directions are delivered into every hydron. Orbital flux backhand does compress the hydron mass by internal backhand but the large scale galactic mass is would be more affected by the coherent linear snippets of inwards going virgin coordinates that are left by the same flux. This embedded linear drive is pushing the galactic mass inwards from all directions, directly and as backhand shadows.

An embryos harvest of photons prior to its collapse will stay lodged on its surface, lost among its 10^{152} neutral top layer mattrons, their coronal or core sizes set by local polarity. Neither flux nor virgin compression can push a singular photon deeper from a top layer position into a mattron lattice; and singular mattrons receive the weakest inwards push of all.

Ultimately, photons escaping a galaxy reach vast distances and may meet up with unstructured mattron fields on the Universal outskirts at distances way beyond our comprehension. An amorphous mattron field does not need to undergo a quantum split into corporeal hydron structures for these long distance electromagnetic photons to implant their potential future polarity into that mattron field.

Dark Matter

The mass estimate for the Milky Way galaxy of $8,38 \times 10^{68}$ baryons is said to be compromised by a need for 5 times more mass in the unexplained form of gravity producing 'Dark Matter'. This must be accounted for in some way if the inconsistency indeed exists because it indicates greater mass.

The Primal Code theory allows the fleeting existence of many oversized hydron embryos that cannot split neutral mattrons (i.e. emit light) or reflect photons and are therefore by definition 'dark matter'. If gravity were an inherent property of corporeal galactic mass, the presence of 'dark matter' could possibly affect the large scale effects. However as gravity is provided by linear emission delivered by flux from inter-galactic photons, greater gravity cannot be adjusted by merely adding mass into the galaxy; you must add the corresponding number of photons into the inter-galactic void.

Large hydron embryos do of course have a dark matter character before shrinking into the light. They do block galactic flux but even small polarized rapidly collapsing embryos wider than 10^{860}ãr cannot split or reflect mattrons so all hydrons above them in the chain are in effect 'dark matter'.

Not all galactic embryos in less populated areas may not have collapsed into the light while photons from fully converted galaxies provide gravity input from passing photons that is usurped by newly created galactic embryos. This however does not provide extra photon emission to bring on greater gravity effect. Dark matter can still be materializing in less populated galactic areas as diffuse hydron gas. That the presence of 'dark matter' is a steady effect over time need not mean that the collective effect is caused by the same mass. Any harvest by large dark matter embryos set by stoppable flux speed means that more effective blocking ability would also quicken the consummation time for inter-galactic flux at the same rate.

An odd embryo in less populated areas shrinks to 10^{881}ãr and is polarized where it blocks negative flux 10^{17} times faster at $4,1 \times 10^{247}$¢ (than a standard hydron that blocks it at $5,62 \times 10^{230}$¢). As an illustration, say that 10^{52} new galactic embryos harvest gravity producing flux while each collapses into the light after a fleeting moment of 10^{1197}ãt or thousand seconds. By collapsing so slowly across 10^{881}ãr, the newly created mass affects the galactic harvest of linear gravity by collecting flux 10^{17} times more effectively and therefore at an equal rate as all the 10^{69} standard galactic hydrons.

This equals a galactic production of 10^{49} new hydrons per second and they have the same blocking power as all the ordinary galactic mass. A little later, each new collapsing hydron adds its 2×10^9 newly split photons to the pot but this cannot account for 'dark matter'. The creation of 10^{49} new embryos per second within a galaxy is enough new mass to form a single sun every year but this is just one conceivable explanation of this concept; whether or not that explanation is needed.

Garbled Hydron Polarity

When a collapsing embryo crosses the 10^{881}ãr disparity threshold by way of incoming virgin compression with myriad singular mattrons on its surface in each sign, the polarization of its neutral mattrons runs swiftly through the internal grid and builds polarized domains of unequal proportions. Globular domain pulses head away from each singular surface mattron in all available directions as myriad polarization pulses claim the hydrons entire mattron population swiftly within 10^{1026}ãt.

Among a hydrons 10^{152} neutral surface mattrons, less than 2×10^9 positive singular mattrons and less than $4,17\times10^6$ negative mattrons are found in the random top layer positions where they originally. Each singular mattron can now launch a hemisphere polarity pulse (half globular) that blooms sideways and inwards away from every singular surface mattron.

Assume that incoming virgin pressure has evened out a lattice grid with 10^{152} top layer surface mattrons 10^{805}ãr apart. Before each of the propagated hemisphere pulses meets another polarity pulse in either sign, each of the 2×10^9 singular positive surface mattrons and $4,17\times10^6$ negative mattrons claim 5×10^{142} neutral surface mattrons whereupon the sideways spread of the surface domain pulse comes to a stop.

Where more numerous positive domain pulses meet one of its kind, the two polarity areas join up and lock rank sideways. As the fewer singular negative mattrons get to propagate a domain pulse for the same period of time before all the pulses meet up, the collective positive surface domains areas are correspondingly larger after locking ranks sideways.

At the same time, all the hemispheric domain pulses also spread inwards at the same speed towards the hydron center. After joining up sideways on the surface, each such pulse from a singular mattron will by necessity take the form of an inwards narrowing cone.

This allows each polarity (named after its coronal sign) to claim scattered mattron domains in direct relationship to the actual photon number in each sign. In narrowing cones that reach from surface to its center, the negative and positive domain pulses should reach the embryos center in 10^{1026}ãt. The polarization is therefore set by the proportions of singular photons and for the on average population of 10^{228} neutral mattrons some $9,998 \times 10^{227}$ turn positively polarized and $2,083 \times 10^{225}$ mattrons turn negatively polarized.

Importantly, the negative polarized mattrons are largely blocked out of the perfect center where the fewer narrower cones of negative mattrons become increasingly crowded by the broader positive cones and therefore prevented from reaching all the way to a center where positive cones from the hydron opposite side meet up with them and lock ranks.

The 10^{881}är wide hydron embryo is now solely built of polarized mattrons but these are arranged in myriad separated adverse domains if the hydron continues to shrink under a relentless virgin compression from outside. The allied emission in 10^{881}ät by $3,66 \times 10^{1134}$ mass is $3,66 \times 10^{823}$ subservient virgins of this, some $1,43 \times 10^{-28}$ or $5,23 \times 10^{795}$ sub virgins escape and each gains $5,05 \times 10^{31}$¢ with a $2,64 \times 10^{913}$ã presence in 10^{881}ät radius time among central subservient flux with $3,16 \times 10^{402}$¢ boost and $1,43 \times 10^{777}$ät consummation. The $2,36 \times 10^{1975}$ã collective positive flux presence combines to $2,64 \times 10^{913}$ã virgin presence as $6,23 \times 10^{2888}$ã within 10^{2643}ã volume and $1,61 \times 10^{-246}$ or $5,23 \times 10^{795}$ allied positive virgins escape; yielding $8,42 \times 10^{549}$pu expansion in the 10^{881}ät radius time or $8,42 \times 10^{-332}$pu per änna.

Walls of adverse domains within a newly polarized hydron do not spread or meet at random; the polarized population is 'perfectly' arranged in billions of positive cones and millions of negative cones (doubtless with some potholes). Dropping inter-mattron flux boost turns internal mattrons less compressed within a fast diminishing hydron volume. The rising backhand to internal inter-galactic flux to which all hydrons are transparent and that reaches all their

polarized mattrons gains in center-wise compression while the internal inter mattronal distances shorten and the mutual flux shadows darken.

After Polarization

Domains in both signs harvest inter-galactic flux by way of exchanged flux between mattron cores and there is no question which polarized mattrons in cones are moved most speedily inwards. Positive polarized cones that hold 479 times more mattrons that do negative cones are populated with the only mattrons that have negative cores and their domain has already joined up to claim the hydrons perfect center.

The internal flux backhand is the average of core and coronal mattron shadows across an average distance. The 479 fold inter-galactic negative flux is harvested and shared by 479 times more numerous positively polarized mattrons (little negative flux is harvested at slower speed by fewer negative coronas). The lesser inter-galactic positive flux is harvested and shared by the fewer negatively polarized mattrons so that the flux shadow input is much the same for polarized mattrons in both signs.

The actual shadow effect is different as it depends on the average distance apart for all affected mattrons in each sign. The hydrons central areas have been claimed by positive polarity and the backhand to this central mass is more effective pulling positively polarized mattrons inwards in their broader cones of positive polarity. The narrower cones of negative polarity are more scattered and further from the hydron center so the effective shadow across to the other side yields a weaker backhand.

Even if there had been no center dominance and that 10^{228} mattrons were evenly dispersed in the two signs in disparate numbers within a hydron, the average distances of 479 times more numerous positively polarized hydrons yields shorter shadows and greater center wise backhand.

With $9,98 \times 10^{227}$ positively and $2,083 \times 10^{225}$ negatively polarized mattrons, there are also two different center-wise drives (that would affect compression if the polarized mattrons were evenly distributed) set by the inverse square root of mass in both signs. For negatively polarized mattrons the center wise drive is ordinarily downsized by $2,19 \times 10^{-113}$ versus 10^{-114} for the positively polarized lot but even this full 21,9 fold center-wise fraction is not enough to outweigh a greater distance dilution.

The more numerous positively polarized mattrons gradually sink within the embryo ahead of the fewer negatively polarized mattrons as the hydron continues to shrink.

In a newly polarized embryo, a positive singular surface photon on top of its allied polarized cone can hold its surface position while allied positively polarized mattrons sink from under it and the original surrounding sea of allied coronas is gradually claimed by negative coronas. A negative singular surface photon on top of its allied cone has no problems either in staying put on a hydron surface in its original position while the entire hydron collapses under intense virgin pressure in both signs.

As a surface phenomenon, less bound singular mattrons are stationed furthest out of all mattrons and both signs keep their top places. Even one singular surface mattron that single handedly transformed a gargantuan positive polarity domain remains on the top layer surface while the mass of its subservient cone sinks below it to be replaced by neutral mattrons of the dominant polarity.

That singular positive surface mattron is soon surrounded sideways and downwards by negatively polarized neutral mattrons over which it holds little sway while they gradually claim the hydron surface and its positive polarity kin sinks deeper as their positions are taken over by encroaching negatively polarized mattrons. The positively polarized mattron majority has negative cores that are pushed inwards by more massive negative inter-galactic flux while fewer neutral negatively polarized mattrons harvest of weaker positive inter-galactic flux.

All singular mattrons of both signs gathered by oversized hydron embryos stay put on its surface for a long time, before and after polarization and in the random places where they hit.

The Quantum Charge

The migration of singular mattrons within a hydron exposed to different charges in penetrating flux pressure leads to an ultimate internal separation of the charges in two layers; at center and surface. Although mattron mists out of which hydrons embryos form are for the most part evenly distributed, differences in collapsing mass would not be uncommon. This which raises the question whether an embryo can gather so much excessive mass that is forms multiple quantum domains or layers of adverse internal polarity.

Positive subservient virgin emission out from a globular center does not push all disparately compressed neutral mattrons away from the central globe with equal input. Positive virgins that reach out of a center in widening orbits are most effectively activated and harvested along with their outwards coordinates by positive dominant core flux in negatively polarized cones.

This enhances the general shift of negatively polarized mattrons outwards from a positively occupied center, consolidating its initial standing within a stratosphere of the forming embryo. The hydron cannot return to its former homogenous state as changes in internal composition permanently assert themselves. Once created, the hydron center is effective in gradually flushing adversely polarized mattrons further away from the center in all directions.

From more massive than standard center, the directional repulsion works swiftly to clear the spaces surrounding the center of negatively polarized mattrons that are best swayed by its virgin messengers. The process meets less initial resistance from interior gridlock while all mattrons are situated in allied cones. But persistent expansive field lines signal a faster movement away from the center of negatively polarized mattrons and as they retreat their cones broaden towards the inner middle layers of the embryo. This broadening encroaches on the positively polarized cones that are losing mattrons inwards.

This positional shift can continues up to a point. A parallel response may be ongoing from the soon negatively polarized surface by way of orbital virgin pressure that cements the structural shift into adverse polarized layers for the same embryo; herding the positively polarized mattrons more effectively inwards vis-à-vis negatively polarized mattrons being pushed outwards.

What happens if the embryo is exceptionally massive is fairly simple. The expanding spherical wave of negatively polarized mattrons that are herded outward from the hydron center is mainly pushing upon the ends of millions of negatively polarized cones in a grid that contains one negatively polarized mattron to every 481 positively polarized mattrons which; given these forces, is not that great a difference.

As a spherical wave of outwards moving negatively polarized mattrons is sifted out through the mattron grid it can continue until the shift has gained them such numbers that it closes off all avenues for positive polarized mattrons to the center before a new negative layer reaches the middle layers of the massive embryo. The second layer negatively polarized mattrons now form a corporeal sphere within the grid and from this layer there emanates a negative collective subservient emission that is more effective upon positively polarized mattrons, both inwards and outwards within the hydron; and this drive is a new cohesive local influence.

Positively polarized mattrons outside the excessive impenetrable mass of a new layer of negatively polarized mattrons can no longer slip through its cohesive emission mesh where negative emission is more easily harvested by positive cores. Again, this repulsion is not merely a collective force that does not allow positive singular mattrons on the outside to pass through towards the center but it also increasingly repels a new spherical material wave of positively polarized mattrons on the outside of it.

For a while, a negative material wave consolidates as positively polarized mattrons are herded outward while collecting negatively polarized mattrons while a similar weaker effect is seen inwards from the embryos surface.

If hydron mass is excessive, on the outside of the first negatively polarized wave, positively polarized mattrons are herded away in the same fashion, leaving negatively polarized mattrons behind, foreboding another spherical consolidating layer of polarized mattrons. Through a massive hydron, this chain reaction can ripple from center outward and from the surface inward. The massive embryo can be split into quantum waves of alternating polarity although this might be in different proportions than the 479 fold difference in total numbers with a skin of neutrally compressed mattrons as a lining between the adversely polarized layers.

Primordial nucleosynthesis of the Standard Theory during early phases of the Universe is considered responsible for the production of light hydrogen isotopes and even the formation of heavy stable isotopes of hydrogen, helium and lithium. If one removes the notion that all of this took place 'in moments after the Big Bang', the formation of quantum layers in hydron embryos may perhaps fit that bill, albeit in a radically different timeframe; an intriguing phenomenon with lots of serious complications, all of which must be put aside for now.

The Double Layered Embryo

With a greater center-wise drive, the cones of positively polarized mattrons that claimed most of the embryos center now cement their central influence. The centers subservient positive emission is captured by cores of negatively polarized mattrons and the tail ends of jagged cones of negatively polarized mattrons are pushed more effectively away. The positively polarized center gains stature as negatively polarized mattrons are pushed into the middle layers and from there outwards to form a thin surface layer.

As the embryo shrinks, the positively polarized mattrons that were more numerous on top of their surface cones from the start soon find the surface sinking beneath them while the more elevated negatively polarized mattrons eventually claim the hydrons entire surface.

After swift polarization in 10^{1026}ãt, the shrinking embryo forms two layers and with all the cones gone, the massive positively polarized central globe is soon encased in a thin negatively polarized mantle. On the hydron scale, the two polarized layers are asymmetric while each has a perfect internal small scale symmetry of single malt polarity.

As all high speed flux compression is slowed by core flux and harvested by mattron cores, mattrons in the two opposite layers do not exchange mutually effective flux shadows. The virgin pressure continues to compress the hydron but there are no overarching forces available here to forge the thin negatively polarized outer layer to blend fully or partly back into the central domain in the face of mutually effective virgin exchanges. The layered hydron has one totally unified inherent structure spin and the outer layer remains firmly in place; much in the form of a thin evenly distributed cover all around the globular core.

Of every 480 mattrons within the joint embryo, the central globe has 479 for every mattron in the outer layer. The volume of the negative outer domain that encloses the larger center is set by $2,083 \times 10^{225}$ mattrons with similar inter-mattronal distances set by hydron size. The positive central domain claims a globular volume set by $9,98 \times 10^{227}$ mattrons.

Between the two adversely mono-polarized layers, as earlier between any adverse cones, there will by necessity form a thin buffer zone of neutrally compressed mattrons.

Emission from both layers across a buffer zone delivers neutrally blended emission that mid between the immense opposite domains provides neutral polarization of buffer zone mattrons. Away from the middle zone, mattrons turn more polarized and assume at some point full polarization by the closer domain.

10^{876}ãr Layered Snapshot

Intermittent photons passing by at various distances and occasionally even hitting is causing an hydron embryo to rapidly shrink and this is a snapshot of a layered embryo on reaching an 10^{876}ãr wide radius. 10^{228} compressed but polarized mattrons have here separate into two opposite polarity layers and the compressed mattrons and the 10^{800}ãr inter-mattronal distances afford mattrons $3,16 \times 10^4$ times smaller than standard primal masses.

$9,998 \times 10^{227}$ positively polarized mattrons at equal distances occupy the center and claim $9,998 \times 10^{2627}$ã volume (479 parts out of 480) with the same share of the hydron mass within a $9,998 \times 10^{875}$ãr central globe radius. The 'thin' negatively polarized outer layer takes up $2,083 \times 10^{2625}$ã volume (1 part out of 480) and encases the center domain. The top layer is $2,083 \times 10^{873}$ãr 'thick' and contains $2,083 \times 10^{225}$ negatively polarized mattrons.

Without an actual center, the outer layer has a different character than it would as an independent hydron. One cannot use a top layers $2,083 \times 10^{873}$ãr 'thickness' to gauge its engagement with locally orbital virgins or to set the presence of linear foreign flux by twice the thickness of the top layer when a flux atom cuts through the entire embryo without local boost.

Local flux is mostly blocked between communicating mattrons in each domain; a negatively polarized outer layer generates a very opaque positive (dominant) flux presence while a central layer generates negative (dominant) flux presence. The few virgins that escape each domain out of the layered hydron do so in widening outward orbits that pass through both domains and their flux fields myriad times. Their activation and escape rates are therefore set by how the individual virgin presence combines with its allied collective flux presence on the wing in the two opposite domains.

Virgins are quickly orbital within both layers and exit the central domain in circular orbits, shallow or center-wide, that widen on its way out through the surface layer. In reality most such orbits peek out of the central domain to pass through the greater adverse presence of local flux, some all the way out to the hydrons surface in average orbits from the perfect center.

From within the outer layer, orbits of virgins are never hydron wide from the start. They 'peek' inwards out of the outer layer and orbit back into it, cutting through the inner domain myriad times before taking leave of the hydron aura as free virgin emission. In each widening orbit through different parts of both layers, they encounter different fractions of flux presence.

More numerous virgins are emitted *out of polarized mattron coronas* in domains where they are adverse emission; a positively polarized central domain emits far more negative virgins that fuel its dominant negative flux. The outer layer releases vastly fewer virgins out of its mattrons in that (subservient) sign. A positive (subservient) virgin escapes the inner domain more easily to face the music of dominant positive flux within the outer domain. A negative virgin that escapes more easily out through the outer domain must plow repeatedly through dominant negative flux within the bigger and more massive central domain before it can eventually escape clean of the hydron.

Conversely, the central layers greater but more heavily activated dominant input of orbital negative virgins is not as extensively activated when passing through the outer layer among the allied subservient flux in its sign. The same goes for the outer layers positive virgins as they pass through a central domains subservient flux.

My simple method to gauge emission is to consider each domain as an independent hydron. It emits mostly dominant virgins that survive the flux of its layer and therefore escape even the layered hydron as the other domain puts up only subservient flux in this sign. There are doubtless more accurate approximations but I am fond of these shorthand illustrations.

A virgin from $5,62 \times 10^{313}$ãr wide parental orbit at $1,78 \times 10^{-251}$¢ exits the positively polarized (almost) 10^{876}ãr wide inner domain with a $2,37 \times 10^{30}$¢ sideways sweep and $2,36 \times 10^{906}$ã presence that grows marginally in widening orbits through the thin outer layer to $2,37 \times 10^{906}$ã presence. Before an outer layer virgin escapes the hydron after gaining $2,37 \times 10^{906}$ã presence, it must do so by crisscrossing in and out of the inner domain where 479 out of 480 of its local presence is housed.

Central negative emission out of 10^{876}ãr aura:

With the above method, the layered hydrons main emission is fairly simple as each virgin that escapes the hydron gains $2,37 \times 10^{906}$ã presence before it leaves it in 10^{876}ãt and its presence spent within each layer is according to relative volume (1 or 479 parts out of every 480).

Of the full $2,37 \times 10^{906}$ã virgin presence, a central virgin gains $2,36 \times 10^{906}$ã presence within the central domain and a top layer virgin gains $4,927 \times 10^{903}$ã presence within the top layer.

In 10^{876}ãt, the dominant negative flux in the center domain is driven by adverse emission of $3,15 \times 10^{819}$ virgins ($9,98 \times 10^{227}$)×($3,16 \times 10^{591}$). Their flux gains 10^{400}¢ inter-mattronal boost in $4,57 \times 10^{776}$ãt dominant consummation and puts up $1,44 \times 10^{1996}$ã collective flux presence within the hydron and the central domain.

Combining the $2,37 \times 10^{906}$ã presence of a dominant central virgin within the central domain gives $3,41 \times 10^{2902}$ã within a central $9,998 \times 10^{2627}$ã volume and allows $2,93 \times 10^{-275}$ of the adverse $3,15 \times 10^{819}$ negative virgins released in 10^{876}ãt to escape out of the central layer and thus also through subservient outer layer flux out of the layered hydron; giving the hydron an external $9,2 \times 10^{544}$ positive virgin emission in its 10^{876}ãt radius time.

Top layers positive emission out of 10^{876}ãr aura:

All positive virgins from the top domain that leave a hydron in 10^{876}ãt are boosted to $2,37 \times 10^{30}$¢ across its 10^{876}ãr radius to gain $2,37 \times 10^{906}$ã presence across that linear distance. Only one in 480 of that presence or $4,94 \times 10^{903}$ã is within the top layer itself.

In 10^{876}ãt, dominant positive outer layer flux is driven by adverse emission of $6,58 \times 10^{816}$ positive virgins $(2,083 \times 10^{225}) \times (3,16 \times 10^{591})$ where flux also gains 10^{400}¢ inter-mattronal boost in $4,57 \times 10^{776}$ãt dominant consummation and puts up $3,01 \times 10^{1993}$ã collective flux presence that stays in the top layer.

Combined with $4,94 \times 10^{903}$ã allied presence of a positive outer layer virgin, $1,48 \times 10^{2897}$ã presence within $2,079 \times 10^{2625}$ã volume would allow $1,4 \times 10^{-272}$ of the $6,58 \times 10^{816}$ adverse negative virgins released to escape out of the hydron or **$9,2 \times 10^{544}$ negative virgins** in 10^{876}ãt radius time.

A layered hydron distinguishes itself by emitting the same number of virgins in both signs; i.e. less emission at the dominant level than were it built as a singular domain with more escaping subservient virgins in one sign.

10^{876}ãr internal expansion and compression:

Having said that, there is a distinction to be made between the external emission and the different expansion drive upon the two layer of opposite polarity. One might assume that $9,2 \times 10^{544}$ virgins that escape the layered hydron in both signs are a unified expansion drive, but there is a twist.

The globular domain at its heart a layered hydron is in reality already a polarized hydron that emits more of its subservient virgins due to less allied flux within the central layer and more of which are activated and harvested within the adversely polarized outer layer with more linear virgin push being deposited into the top layer; pushing the outer mass layer outwards.

We can go through the motions; allied central emission does push the top layer away with an expansion drive depicted by $1,78 \times 10^{790}$ positive virgins $10^{228} \times (3,16 \times 10^{591}) \times (5,62 \times 10^{-30})$ in 10^{876}ã that each gain $2,37 \times 10^{906}$ã presence in 10^{876}ãt radius time among central subservient flux with 10^{400}¢ inter-mattronal boost and $5,62 \times 10^{770}$ãt subservient consummation. The 10^{1961}ã collective positive flux presence combines to $2,37 \times 10^{906}$ã presence of a positive virgin as $2,37 \times 10^{2867}$ã within 10^{2628}ã volume and $4,22 \times 10^{-240}$ of the $1,78 \times 10^{790}$ allied positive virgins escape; yielding roughly $7,51 \times 10^{550}$pu expansion drive upon the outer layer in 10^{876}ãt.

This is compounded by all peeking virgins in less than mattron wide orbits that are also activated in the outer layer and work as additional expansive element. These subservient virgins are in contradictory orbits and leave only their linear away drive when activated in the outer domain. The $2,37\times10^{30}$¢ effective orbital drive that at aural escape peeks out of the top layer surface indicates an average square root $1,54\times10^{15}$ fold additional expansion drive on top of the $7,51\times10^{550}$pu general expansion drive and leave it with quite a powerful $1,16\times10^{566}$pu expansion drive outwards upon the outer layer in 10^{876}ãt.

It would however take more than this level of incoming virgin pressure to compress a outer domain up against the central domain in the face of its effective expansion. If incoming virgin compression is unable to outgun this drive, it leads to a quickly widening breach as the central domain expands more slowly.

There is some minor shave off from harvested virgin pressure that arrives beyond the Neutrons leverage point to work upon the relatively thin Electron skin as compression pushing it up against the Protons buffer zone. I simply use the inverse 480 difference of the two layers and allow one of 480 of the virgin harvest to hold down the top layer, which brings the required $1,16\times10^{566}$pu virgin harvest to $5,57\times10^{568}$pu.

10^{876}ãr compression by external pressures:

The inter-galactic flux casts a backhand gravity shadow between internal mattrons but is not all that effective within a large 10^{876}ãr hydron. Internal shadows are least diluted between individual mattrons at 10^{800}ãr apart and most diluted across 10^{876}ãr wide hydron radius with an average backhand distance of 10^{838}ãr.

The backhand distance for the outer layer as center-wise flux compression is quite insignificant and can be ignored across to its allied mass on the other side of the full 10^{876}ãr Neutron width.

In the 10^{876}ãr wide central domain, the average shadow dilution between $3,16\times10^{627}$ãr mattron coronas across 10^{838}ãr is by the inverse of $3,16\times10^{-211}$ squared or 10^{-421} subservient shadow backhand while dominant shadows between $4,87\times10^{615}$ãr cores are diluted to inverse of $4,87\times10^{-223}$ squared or $2,37\times10^{-445}$. The average coronal and core backhand is then $1,54\times10^{-433}$ average within the center. Positively polarized mattrons harvesting negative inter-galactic flux, there is then the additional 10^{-114} (almost) center-wise cut that increases the central domains center-wise dilution to $1,54\times10^{-547}$.

Even in full numbers upon each such polarized embryo, 2×10^9 positive inter galactic photons would deliver flux from 2×10^{605} negative virgins to every Milky Way hydron in 10^{876}ãr. A 10^{463}¢ boost delivers 2×10^{1068}pu negative pressure that downsized to $1,54\times10^{-547}$ delivers $3,08\times10^{521}$pu compression in 10^{876}ãr radius time within the central (Proton) domain and delivers merely $3,35 \times 10^{-24}$ of the $9,2 \times 10^{544}$pu central expansion drive and are of no use in holding the hydron embryo together.

Two layered 10^{876}ãr hydrons with $9,2 \times 10^{544}$ virgins radius time emission in both signs can be subjected to mutual virgin compression by their exchange but say the $9,2 \times 10^{562}$pu across 10^{903}ãr grid distance cannot counter neither the central expansion downsized to 10^{-38} ($9,2 \times 10^{524}$pu against $9,2 \times 10^{544}$pu), nor the top layer expansion when downsized by 480 ($1,92 \times 10^{560}$pu against $1,16 \times 10^{566}$pu); not unless the two are paired at closer distances from where each hydron can deliver more mutual pressure than inter-galactic virgins.

The third compressive option at the peak of photon emission from around the galaxy where peak power comes late in the day; 2×10^{674} negative virgins are emitted of which 10^{-94} avoid flux and arrive in 10^{876}ãr radius time. How many of the 2×10^{580} virgins hit a 10^{876}ãr polarized neutral embryo depends on average sizes but if we assume that all the 10^{69} galactic hydrons are of the same 10^{876}ãr size, 2×10^{511} virgins hit a 10^{876}ãr neutral embryo at 10^{62}¢, and the 2×10^{573}pu virgin pressure cannot counter the central expansion by 10^{-38} (2×10^{535}pu against $9,2 \times 10^{544}$pu). Even if this holistic pressure cannot keep the embryo together, it can keep the central Proton from rejecting the Electron layer if downsized by 480 ($4,17 \times 10^{570}$pu against $1,16 \times 10^{566}$pu).

Of course, with a massive photon presence around the galaxy, photons are already passing en force through the galaxy with far greater individual effect. Peak photon emission from around a galaxy is likely to compress the larger gathering hydrons and secure buffer zones from widening into a breach until at sizes where virgins start to miss hydrons while meeting increased internal expansion at smaller sizes.

The greatest pressure upon polarized neutral embryos is undoubtedly from photons passing into the galactic surface and in among galactic embryos to become the main instrument of occasional relatively more rapid collapses that set the ball of galactic conversion rolling. To drive mass below 10^{855}ãr in the time a single photon passes out of contact midways is impossible but it speeds up the process at closer passes. A shrinking hydron would however harvest a rapidly diminishing part of the photon emission while its central expansion rapidly gains ground due to the smaller size.

While the pressure is upheld upon a layered hydron as a unit at sizes down below 10^{860}ãr, its buffer zone does develop into a breach. A layered hydron at some time almost invariably experiences pockets in compression caused by passing photons. If no other photon arrives to take the place of a former photon, the pressure is withdrawn and the layered hydron expands. Any fluctuation in its compression drive may upset its precarious balance but this rarely leads to a split of a large layered hydron until its width is close to the electromagnetic threshold.

Local hydron instability stems not only from a chance number of photons between a hydron pair at each time but even more from how close they pass. Compression can be powerful upon a small hydron but as the photon will pass more quickly out of maximum delivery, it dampens the effect.

Again, while it may take less than a second for a Neutron to collapse and split, to complete an aspiring chain reaction of splitting Neutrons across an entire galaxy takes another 10^{10} years.

10^{876}år blocking speed of incoming flux:

If we look at a 10^{876}år layered hydrons blocking power of flux, the negative engagement is exclusively in the central domain and set by its collective sphere (almost the whole hydron). The value for the central domain is shown by its escape rate of virgins by applying the $2,37 \times 10^{30}$¢ internal virgin boost to its $2,93 \times 10^{-275}$ negative escape rate; the central domain of this large layered hydron does not engage negative flux passing through at speeds above $1,44 \times 10^{244}$¢.

Although positive flux is blocked by the outer layer by passing through it twice on a linear path through the layered hydron, the end result of how the flux engages within the hydron is in the volume relation of the two signs. Logic says that if there were only one positively polarized mattron, it would not matter if situated at the hydron center or on its surface. Crisscrossing flux that passes through a polarized entourage will on average be in accord with that inter-mattronal flux exchange volume as the mattron exchange volumes are identical in both domains.

In other words, the inter-mattronal flux that interferes with traversing flux atoms in the top layer does so in accord with its 480 times smaller collective volume. The top layer therefore blocks positive flux at 480 times lower speed than the center ($1,44 \times 10^{244}$); which is an average speed below 3×10^{241}¢.

Part 6: The Unstable Layered Hydron

Buffer Zone and Breach

Direct emission of intact virgins across a buffer zone between opposite domains is a tiny minority and effect-wise rather insignificant. The buffer zone between two polarized domains contains a relatively small population of interactively dependant neutral mattrons polarized to various degrees. What makes it interesting is its potential to become a 'breach'.

The number of neutral buffer zone mattrons with compressed coronas is set by how many mattrons are needed to slow the orbital coordinate centers of two equal adverse flux winds to a full stop across the buffer zone between the two wall linings of totally polarized mattrons on opposite sides of it.

Were we to string a line of neutral mattrons at average grid distance across a buffer zone from one fully disparate coordination mattron in each lattice wall and across the buffer zone, we would see adverse pressure from each side turn ever less overbearing till it leaves only the local exchanges.

Average compression across each buffer zone area is first set when a linear orbiting flux pulse from myriad mattrons is brought to halt by an opposite linear flux, i.e. not the adverse flux emission between closer coordination mattrons. The effect away from each domain is similar as it would be away from a singular photon that is unable to re-polarize an already adversely polarized grid. In this case of course there are as yet no singular mattrons as all of these are found on the outside surface of the layered hydron.

Eventually all orbital flux in opposite linear directions from both sides has its orbital centers stopped somewhere in the buffer zone and all mattrons across that entire width become equally and more neutrally compressed by flux boosted across a greater distance. Calculated from this lattice point in the buffer zone, the disparity grows step by step to full polarity towards each fully polarized domain where mattron coronas can also keep somewhat more primal mass.

In this case the flux wind generated it is not from the virgin emission of a singular photon but a lattice wall from some of 10^{152} fully polarized standard mattrons that line each domain wall. Progress from complete disparity of the mattron population at a polarized wall to the neutrally balanced part of the buffer zone shows gradually dwindling disparity and lightening masses.

For illustration only, I give the initial buffer zone between the domain walls a population of 10^{36} mattrons where disparity ranges from full polarity at the two domain walls to neutral grid compression at a buffer zone balance point across a relatively narrow stratum. With 10^{152} fully polarized mattrons lining each domain wall, the buffer zones initial population between the adverse domains would be 10^{188} mattrons.

These 10^{188} affected mattrons are 10^{-40} of the hydrons standard population and the buffer zone layer therefore 10^{-40} of its radius. Relatively speaking, this is less than it sounds; a similar layer within Our Sun that spans 10^{-40} of its radius would be thinner than the width of one hydrogen atom.

If a buffer zone reaches say 10^{850}år (the aural size of a minimum hydron) in an expanding Neutron, average grid distances between 10^{188} mattrons of 100 times lesser mass would be 10^{788}år (same as the average in 10^{864}år wide embryos). The buffer zone distribution in a stretched out zone is not average but runs from domain grid distances (10^{779}år in 10^{855}år wide hydrons) out to the widest neutral part of the buffer zone grid. I use the square root of the 10^9 difference between 10^{788}år and 10^{779}år to set the largest neutral grid distances at $3{,}16 \times 10^{792}$år (just above that of average grids in 10^{868}år wide embryos where neutral mattrons are thousand times lighter).

Long range traversing flux from the two purely polarized main domains crosses the original 10^{-40} skin at right angles across 10^{36} mattron domains where mattrons of diverse disparity exchange and block lesser flux. The long range flux crossing the buffer zone is therefore rarely harvested within the zone but rather in the domain where allied core flux blocks it at high speed. This exchange is also instrumental in blending new inherent spin impulses within the two layers for unified structure spin.

Buffer zone mattrons are thus not as well equipped to follow the holistic hydron expansion or compression on their own volition. The move more by proxy when herded before a shrinking or an expanding domain wall.

In order to follow any shift in hydron size, the buffer zone mattrons will be pushed more 'physically' by one domain wall or the other as they do not individually heed the drift communicated by internal winds at quite the same rate as the fully polarized layers. As interactive units, they do adjust the buffer zone hydron grid to the reigning circumstances, albeit belatedly.

During hydron compression, the subservient virgins mutually exchanged between layers also forward an outer layers inwards movement to a central domain. Peeking subservient virgins that avoid flux activation at close hand and head across a buffer zone in orbits out of their domain wall can pass repeatedly in and out of its original domain and the other domain. More subservient virgins in all orbital directions are captured with intact linear thrust on passing through the greater core flux in an opposite domain.

A 10^{855}år wide collapsing hydron that can in principle split electromagnetic photons in a breach away from the inner sphere between polarized domains but cannot do so in a buffer zone grid environment. The pivotal condition to for an electromagnetic breach split is for wayward mattron within the breach to gather enough linear virgins to shift its mass into a split. To do so, it must first form a precursor halo which requires an extensive breach and extensive withdrawal of buffer zone mattrons or a blend of both. It requires a mattron to leave its grid environment and that is no easy task in a hydron grid.

The situation in the buffer zone does not change as long as the hydron is being compressed from outside. While local virgin pressure upon a hydron surface is forging a smaller hydron or keeping its size stable, the neutral stratum is a permanent fixture sandwiched between two adverse domains. This balance is first put into jeopardy if pressure upon the layered hydron eases and its body starts to expand again from a size well below 10^{855}ãr.

In an expanding hydron, the outer layer moves faster outwards than the central globe as both are driven outwards by subservient peeking central virgins that reach through the buffer zone. The central domain with greater escaping virgin main body emission cannot keep up with the 480 fold lighter outer domain that is moving more briskly outwards. As a layered hydron expands, grid distances between buffer zone mattrons start to widen and the neutral area moves gradually closer to the outer layer; neutral mattrons near the inner domain become more polarized as the inner domain encroaches on neutrally compressed buffer zone mattrons.

Distances in the direction of the inner domain tend to shorten during decompression as the buffer zone population is caused to 'drift' against the inner domain wall although distances are continually adjusted. The buffer zone mattrons that come in full close contact with the inner domain lining simply join it to become fully polarized grid mattrons of that sign. The odds are in favor of the inner domain claiming more neutral buffer zone mattrons if the buffer zone expands; which also means that the buffer zone would be rapidly losing of its disparate mattron population.

While the two layers are broader than the buffer zone, the local virgins exchanges are more equal though different from the broader inner virgin domain pushing upon the entire smaller outer domain. It is first after a breach forms in an extensive buffer zone that its population drops faster. The positive linear virgin input from a negatively polarized outer domain that becomes activated flux would be robust enough to bring the flux winds to a full stop at the mid buffer zone (they are equal while the two layers are broader than the buffer zone), if it not for the outer domains faster flight.

The central disparity will therefore edge out amongst the mattrons in the buffer zone and the neutrally compressed band shifts closer towards the departing outer layer. As the polarity adjustment continues, the neutrally compressed mattrons thin out to rapidly diminishing numbers. The buffer zone is now about to widen into a 'breach' between two more independent layers of altered exchange of linear push and inherent spin coordinates.

The two layers of the expanding hydron are parting ways linearly in all directions, mainly outwards and even collective spin has started to deviate. At first the altered spin is rather the segregation of coordinates signaled to the mattrons within each layer but soon the layers start to slowly orbit more structurally in opposite directions. In this respect, the mattrons in the mid buffer zone are unique. They harvest and consummating opposite spin coordinates and are not universally orbiting although that is of minor structural importance.

In say a 10^{850}år wide buffer zone where mattron population has dwindled to 10^{140} remaining mattrons of smaller masses, the average grid distances is 10^{805}år but this does not represent all the increasingly neutral transparent buffer zone.

If the zone population has dwindled to say 10^{167} lighter mattrons with average grid distances at 10^{796}år and domain wall distances at 10^{779}år, the mid zone average distance is greater by the square root and be for neutrally compressed mattrons just under the 10^{805}år disparity threshold. If mattrons are disparate at $3{,}16 \times 10^{804}$år, their cores would be dispersed by mutual flux exchanges and they would lose their disparity anyway.

It is first at 10^{824}år that neutral mattrons (of 10^{903} primal mass) cannot be sustained in a mattron grid; ripped asunder in mutual flux exchanges. If for example the population in a 10^{850}år fast widening buffer zone drops to 10^{129} mattrons with $2{,}16 \times 10^{809}$år average, the $1{,}47 \times 10^{824}$år grid distance within the neutral zone would exceed the 10^{824}år structural stability threshold.

The photosphere condenses towards respective domain walls on either side with diverging polarity as neutrally compressed mattrons in the mid buffer zone disperse into flux. As the ensuing breach widens with unraveling of mattrons, dispersed primal mass in opening orbits is blocked by mattrons closer to each domain wall. The primal mass disappears from the breach as a photosphere breaks into two photospheres that become increasingly polarized without much of an upheaval.

A widening of a breach is likely hard to stop (like an electromagnetic split when the neutral mattron mass has been hardwired into a split) whereupon the breach would rapidly widen due to the central domains quite effective repulsion of the outer negative layer.

Local virgin exchanges across a breach cease when the distance between two photospheres does not allow local orbiting virgins to cross. Long range virgin exchanges between two domains continue to orbit across a buffer zone into both structural elements but only the central globe sends off repulsive main body virgins that are already free. Repulsive Proton emission presses the smaller outer photosphere into its domain while peeking orbital virgins from the outer layer cannot push back the inner photosphere that withdraws more slowly towards its domain by other inputs. The Protons massive main body subservient emission literally blows the Electron layer off the Proton.

An expanding Neutron will inevitably enter an upheaval that separates the two layers that would have already started to diverge in spin directions about their common center. When separated, the central globe would come to spin independently within the thinning top layer domain that is departing away in a differential retrograde orbit.

Our perception often dovetails to conventional parameters of our senses; a ball throws across a field and suchlike. It is good to keep in mind that a mattron (say 10^{629}ãr wide) is presented with a daunting distance if pushed across an 10^{850}ãr wide buffer zone that is 10^{221} fold its radius. Even an electromagnetic photon must travel across 10^{98} fold its 10^{752}ãr precursor radius. It may be helpful to compare this to the insignificance of a mere 10^{18} fold trek measured in Solar radii which is the relative distance to the edge of the observable Universe. For a mattron, both the breach and the buffer zone are inconceivably vast.

The Neutron Split

The inner polarized hydron center that is positively polarized is the **Proton** part. The smaller positively polarized outer layer is the **Electron** part. As two cohabiting layers within a single hydron, they form a **Neutron**.

Positive virgin pressure from intergalactic photons or mutually exchanged virgins across short grid distances cannot deliver compression that can forge smaller Neutron embryos. The main pressure in alleviating their problems of instability comes after polarization with a rising number of passing photons in a chain reaction of shrinking Neutrons. These splits need not start with hydrons furthest along in their formation cycle; it can be set off anywhere as passing photon showers begin to permeate an entire galaxy.

The average distance of a photon passing between two hydrons is half their vast distance apart but they often pass closer at various distances to deliver powerful local pressure upon a layered Neutron that compresses it forcefully as a single unit. The outer layer and central domain shrink in step and keep the buffer zone firmly to a minimum. And while the inevitable subsequent pressure drop when the close photon passes on, the slack must be remedied by newcomer photons and if these do no arrive in time, the photon emission drops off rapidly; forcing the buffer zone to widen.

An individual Neutron split would generally be an oversized Neutron that becomes highly compressed, enough for its central mass to be locked up by internal galactic flux shadows, only to experience a destabilizing pressure drop. All Neutron splits are the same, whether standard 10^{852}ãr or larger 10^{855}ãr radius; the split of a 10^{860}ãr maximum hydrons with a part lockup of central mass is probably more messy but largely the same. Expansions in dropping pressure are unlikely to lead to splits for larger embryos shrinking from 10^{881}ãr polarization threshold down to around 10^{860}ãr.

On and off drop in pressure upon a large layered embryo is not a defining moment as consolidation can continue between mild temporary expansions. Larger Neutron embryos harvest a great deal more virgins than smaller ones and have a lesser expansion drive. The smaller sizes are also less stable as they cannot gather sufficient virgin pressure to compress layers as a unit in tandem without driving a buffer zone into a breach by rising expansion and widening away from a central domain locked up by galactic flux.

When virgin harvest temporarily drops and allows the buffer zone between Neutron layers to widen, chances are that the duration does not suffice to cause a breach and rarely at all above 10^{860}ãr. Repulsion by collective main body hydron emission between the layered domains needs continue long enough to destabilize the Neutron before another passing photons can pick up the slack. Stabilization after a rapid compression during the galactic conversion period depends usually on virgin pressure from external photons. Nevertheless, it is a common occurrence for a Neutron to shrink intact down to standard size.

The Electron Shell

After a severe rise with a subsequent drop in the jacked up virgin pressure, the widening buffer zone allows main Proton body emission an upper hand to accelerate the linear Omni-directional repulsion of the outer Electron shell which this virgin emission is unable to penetrate. Viewed from outside, the Neutron is to all appearances simply expanding under insufficient virgin pressure; its outer surface initially neither unraveling nor heavily distressed.

The expanding Neutron surface still holds its old original harvest of $2{\times}10^9$ positive singular mattrons, and old collection of $4{,}17{\times}10^6$ negative singular coronal mattrons (electromagnetic photons gathered earlier by its embryo). Up until now, local virgin pressure has been slowly consolidating the Neutron that without a broad buffer zone has no need for massive virgin compression to hold down the top layer.

During local grid passage of a traversing photon, an oversized Neutron experiences a sharp compression to a smaller size but the photon will soon depart from the local grid and the buffer zone starts to widen and a breach opens as the compressed Neutron is forced into rapid expansion.

After a Neutrons buffer zone eventually becomes a widening breach, the Proton globes powerful main body virgin emission pushes its Electron shell away from the center at accelerating speed in all directions. The outer shell responds in kind for considerable time by pushing the central Proton embryo inwards by its collective emission. If mattrons were billiard balls, which they are not, the outgoing main body repulsion across a breach would act as a compressive drive upon the inner lining of the Electron mass orb where layer upon layer of polarized mattrons would then be forced closer while being pushed outwards.

As interactive entities, mattrons react when moved closer in a grid layer as their exchanges work to maintain their internal grid distances. While the expansion pushes a thinning layer outwards, its drive does not significantly compact a cohesive Electron shell of $2,083 \times 10^{225}$ neutral mattrons that is expanding out into the void. The disparate mattrons on a widening shell area suggest a thinning mass orb although the mattrons are in no danger of losing polarity (requires 10^{29} fold grid distance). External emission by an expanding Neutrons become tainted by rising negative emission out of the Electron shell but such changes are soon overtaken by more drastic effects.

The process is slightly different according to circumstances. For a Neutron in a primeval converting galaxy, the trigger pressure would usually come from a passing electromagnetic photon that soon traverses out of reach from where it leaves little incoming virgin pressure from across a grid of enlarged hydron embryos; leading to rapid expansion and consequent collapse to a size at which it can be split (or not).

In a fully converted galaxy, a standard negative outer layer must be held up against the positive layer by pressure supplied by a closely bonded Proton. In either case, when local incoming pressure is abruptly lifted it causes an expanding buffer zone that can eventually develop into a breach.

Primeval Neutrons during galactic conversions stand for the overwhelming number of splits. After each individual split, depending on virgin winds, it will take considerable time in that heavy pressure environment to allow singular mattrons of either sign to rise from hydron surfaces and form precursors.

In the present day environment of advanced grids, a Neutron must have formed a stable pair relation with a Proton where it can harvest the incoming pressure supplied by its closely bonded companion. If not, or if the Neutron is pushed out from under the protecting umbrella of close Protons emission, it will split.

Virgins are overwhelmingly activated before hitting top layer mattrons. The coordinates delivered into mattron mass by the resulting top layer flux will therefore affect top layer mattrons differently. Flux from negative incoming virgins is harvested directly by negative coronas in a Neutrons negative top layer. The sprinkle of $4,17 \times 10^6$ negative singular top layer mattrons on the Neutron surface that are allied to its grid polarity at the same large coronal sizes are held up against the Neutron surface by this negative incoming virgin pressure delivered by local flux. This negative effect (within the top layer) is marginally more compressive upon a singular mattron corona than upon a neutral disparate mattron of the same mass that carries a lighter core baggage along with it.

The 2×10^9 positive singular mattrons on a negative Neutron surface are adverse to its grid polarity and compressed to core sizes where a death clot turns their mass temporarily lighter ($5,62 \times 10^6$ times at standard size and $1,62 \times 10^7$ times lighter at 10^{855}är size). Light cores housed within a mattron corona are easily carried along by local negative pressure but the coronas drag on the linear inwards push. On their own, light positive singular cores are effectively pushed deeper (within a top layer) by positive pressure from incoming virgins and held up to the Neutron surface.

All of this is also true in reverse for singular surface mattrons on positively polarized surfaces where allied singular mattrons are pushed more effectively inwards within the top layer. This does not tell us what singular mattrons are first to lift from a hydron surface. The outcome also rests on the balance between the two signs of incoming and outgoing virgin winds that allow the mattrons to rise from (standard) hydron surface which heavily complicates the end result (subservient main body repulsion are also around 10^7 fold the dominant virgin emission). Generally the singular photons stay effectively put on a hydron surface while it remains larger than standard size.

Note that before a hydron has been compressed to standard size, neutral top layer mattrons facing insufficient virgin pressure expand away from the surface as they are still not tied down by inter-galactic flux. This means that all singular mattrons surface mattrons find it hard to rise from the Neutron surface as the outer layers are not tied down in the grip of inter-galactic flux and the entire neutrally polarized surface simply expands under insufficient virgin pressure along with the singular mattron population.

The greatest effect upon negatively polarized grid mattrons on the Electron shell is not the linear subservient virgin repulsion streaming out of the Protons main body. It is the immensely boosted parental sweep of the same virgins as they arrive in large orbits from opposite directions that now imitate the shape of the Electron shell. This creates a vastly more powerful sideways backhand shadow that drives all mass towards the densest point on the Electron shell.

It is easy to understand why the collapse of the Electron subcloud is faster than this outwards rejection; the virgin side sweep along the orb is from the start roughly **10^{24} fold** its linear drive outside an aura and rises from there; on reaching 10^{854}är from the Proton, the sideways sweep of the shell is 10^{25} fold. The drive is without significant activation until the virgins enter in between mattrons in the polarized outer layer where they are activated; the drive is there delivered by locally exchanged mattron flux. The Protons positive orbital main body virgin emission is also driving the densest part of a collapsing Electron mass in a retrograde orbit relatively slowly outwards from the breach.

The mattrons on the globular Electron shell are moved 10^{24} times more efficiently towards the densest point on that shell which is reshaping the Electron shell after a distance of a few or a few hundred extra Neutron radii. Inevitably, the violent strength of opposite sideways orbital sweep carries parts of its thinning shell lattice towards the single spot that casts the orbs darkest backhand shadow.

As the thinnest part of the shell opens up and before much else can happen, the collective mass of the negative polarized neutral mattron lattice withdraws fairly cohesively towards a single orb spot where it forms a globular rotating Electron.

In this upheaval, a latticed congregation of $2,083 \times 10^{225}$ negatively polarized top layer mattrons is not extensively broken up as it gathers like a violently remodeled solid into an immense globular hydron; a fully functional standard Electron. The old 2×10^9 positive singular and $4,17 \times 10^6$ negative singular mattrons on the Electron shell were to begin with situated furthest away on the outer layer but these are differently affected by the sideways drive as the relatively thin backhand shadow causes the layer to collapse.

Electromagnetic Splits in the Breach

While the Protons main body emission is starting to batter the shell mass into a new structure, sparse emission from the shell has somewhat compacted the oversized central Proton while also pushing remaining buffer zone mattrons into a thin photosphere deep within the breach, closer around the oversized Proton. This is before the Electron shell on the Proton orb has extensively collapsed sideways into a globular Electron.

As the untied Proton photosphere of disparate mattrons is swiftly joining the Proton surface, this is not a stable environment. There are always a few last mattron stragglers left behind and these neutral photosphere mattrons can drop out of mutual flux contact close to the collapsing Proton surface. Due to 10^7 times more massive Proton repulsion, this is less likely to happen in a meager more evenly pushed photosphere up against the inner Electron shell. When neutral mattrons drop out of contact to hoist precursor sails for stability, even a small precursor may block enough virgins in the lower aura to hardwire a neutral mattron into an electromagnetic split.

For the outermost mattrons in the Protons low flying photosphere, the last buffer zone mattrons are special inasmuch as they were originally trapped between two layers moving in opposite inherent orbits with a mass majority void of inherent structural spin coordinates. They have now for a long time been adapting to opposite spins and have gained inherent mass coordinates that follow neither of the two layers. The split of these neutral mattrons into electromagnetic photons would be relatively devoid of the Neutrons inherent common coordinates and the resulting photon twins would be fairly straight line; albeit less so than electromagnetic photons without a trace of inherent structural coordinates.

The Proton cannot stave off the impending splits in its fast but unevenly shrinking photosphere but is instead instrumental in aiming the resulting twins.

These electromagnetic splits do not cause the Neutrons structural jeopardy that was achieved earlier, driven by repulsion by a central Protons main body virgins. But after such splits, the Electron shell cannot be easily persuaded by any local forces to rejoin the previously closely knit bond.

Photon pairs at close quarters around both hydrons see swift acceptance of inwards aiming electromagnetic twins opposite to hydron polarity. Allied twin are aimed outwards and depart at light speed in opposite directions in accordance with their number. Massive electromagnetic emission all around the hydron compress it to standard size. As the same photons pass across its leverage point, they can in time compress it to below standard size. Close to the surface in the Protons lower aura, little adverse virgin emission from the positive electromagnetic photons escapes entrapment by its main body and the free subservient main body repulsion upon the Electron shell will remain a more powerful component for some time.

These electromagnetic splits after a Neutron split are not intermittent. They are a simultaneous burst of billions of electromagnetic photon twins outside a soon standard Proton surface. Soon, all the negative electromagnetic twins aimed into the Proton have been blocked and only positive photons deliver a linear push upon the relatively close Electron by the increasing fraction of freely escaping negative virgins in numbers that will soon come to exceed the Protons collective positive main body emission. Adverse repulsion increases as photons continue outwards from the Proton at light speed behind the now rapidly expanding and sideway collapsing Electron shell.

As a single pair of adverse photons pours out exponentially more virgins close to the hydron surface, one may think that their presence in the lower aura, for example outside a Proton photosphere, could serve as a catalyst to additional electromagnetic splits. This does not happen; the initial neutrally blended massive output from an outermost split photosphere twins pushes all the other left behind stragglers up against a hydron surface by massive neutral virgin emission that cannot split them.

The initial electromagnetic burst is not a catalyst to multiple electromagnetic splits but limits them instead. The orbits of the adverse virgins reach across a small part of a Protons curved surface and eventually across its aural volume. As noted earlier, I assume that the average number of electromagnetic splits achievable within a standard Protons aural volume is 2×10^9 photon twins.

At the same time, as yet unperturbed by the electromagnetic photons deep within the Proton aura, the globular Electron is reaching the final chapter of consolidation. Of mattrons hurling sideways through space in many opposite directions on that polarized layer of negative mattron coronas, the sideways drive delivered by the Proton is by activated positive flux in the last stages between mattrons. Even so, this drive will effectively push neutral disparate mattron mass along when trapped by its light positive cores; the sideways movement communicated in exchanges with their heavier negative coronas.

The 2×10^9 old positive photons, compressed to core sizes and without the extra baggage of great coronal mass are nonetheless driven sideways at full greater speeds but they are surface material; they cannot be driven deeper than one mattron layer because the sideways virgins at 10^{24}¢ do not survive a top layer and the singular cores cannot harvest such local flux beyond the top layer.

The $4,17 \times 10^6$ old negative photons of coronal sizes are barely affected by the 10^{24} fold sideways virgin drive as they have no cores to communicate the ongoing movement to them. This is of course the Protons 10^7 times weaker dominant main body emission but the likely outcome is that the $4,17 \times 10^6$ old negative photons are lost to the Electron as they drop out of mattron contact and form precursors.

As the Electron collapses, even here there will be left behind stragglers of neutral mattrons in a shrinking photosphere of neutral mattrons dropping out of mutual flux contact when drawing closer to the forming globular Electron surface where a small precursor can block enough subservient negative virgins to hardwire its mass into an electromagnetic split.

The free negative wind from the Protons 2×10^9 electromagnetic photons in its lower aura is not yet massive enough to reverse a subservient main body negative virgin repulsion by an Electron that aims negative electromagnetic twins away from the Electron at light speed. The Protons positive main body subservient virgins can do no more than its virgin orb aids the Electrons acceptance of the newly split positive twins. The departing $4,17 \times 10^6$ negative photons that were lost from the Electron shell when it collapsed will help aim positive neutral mass of splitting mattrons into the Electron surface.

The Protons standard 10^{852}âr radius is 7,83 times the Electrons standard 1,277×10^{851}âr radius and the (7,83)³ difference of aural volumes (480) allows the Electron to split 480 times fewer electromagnetic photons or 4,17×10⁶.

Among 10^{152} polarized mattrons on the Electrons negatively polarized top layer, there now floats an infinitesimal share of old 2×10^9 singular positive mattrons and the new sprinkle of $4,17 \times 10^6$ positive electromagnetic twins sent inwards from its lower aura is a minor addition. Whereas I expected the Electron to keep a surface population of $4,17 \times 10^6$ singular positive mattrons, the logic is having none of it (as far as I can see).

Positive virgins from the $4,17 \times 10^6$ Electron batch of new electromagnetic negative photons and the 2×10^9 Proton batch of positive photons are soon battering both the reshaped Proton and the globular Electron (and eventually another nearby Neutron) from their leverage points. This massive temporary virgin pressure fades relatively quickly but compacts all three into standard costumes from within their leverage points.

In close vicinity, all three hydrons are now so relatively diminutive and their internal flux shadows so dark that inter-galactic flux backhand enters hydron evolution in a decisive manner. The flux can now hold the mattrons to standard internal 10^{776}ãr inter-mattron grid distances that provide rock bottom standard stability and secures their sizes to 10^{632}ãr Protons and $1,277 \times 10^{851}$ãr wide Electrons (or else the hydrons would expand again).

The only exception to the ability of flux shadows to hold these sizes by way of internal inter-galactic flux shadows is the outer negative layer of a nearby Neutron. Its central positive domain is a functional Proton while its negative outer layer has so diluted shadows of lesser input in the core sign that it must from thereon in rely on incoming virgin pressure from the companion Proton to stay intact; or else split.

For example, the electromagnetic blitz hammers a Proton into a 'standard' costume from its $3,96 \times 10^{861}$ãr leverage point by delivering $1,26 \times 10^{-242}$pu per ãnna orbital pressure that downsized to 10^{-38} yields $1,26 \times 10^{-280}$pu actual compression from 2×10^9 photons; outgunning its standard 10^{-300}pu per ãnna collective expansion (excessive compression is resisted by internal mattron halos). After electromagnetic photons of both signs pass beyond the relevant leverage points, the local virgin pressure starts to fade again.

The hydron majority is created at a time when the field of galactic photons is fully stocked. On the other hand, the first galactic Neutrons that are split receive lesser pressure which may not suffice to lock up the mattrons within a standard grid. These first new unbalanced hydrons must therefore expand again to find balance without opening a breach and without an overly broad buffer zone. They may be put out of balance multiple times before reaching a smaller size that allows a split and a batch of electromagnetic photons. After a breach opens and an outer layer is reshaped, the emitted electromagnetic photons will deliver the greatest virgin pressure that either domain will ever experience again. After that, electromagnetic splits by fully formed hydrons become more anemic and mostly in response to a local disturbances.

During a Neutron split, the outwards coordinates (harvested up until the electromagnetic burst in its breach) causes shell mass expansion effective in all linear directions by inherent coordinates. Overtaken by more neutered electromagnetic breach photons in fewer directions, the more linear stronger electromagnetic repulsion pushes the mass outwards. This push becomes increasingly mono linear as the orbiting shell mass gathers to a single universal direction. The shell mattrons slower already gained linear speed in all opposite directions is still effective in that new collective direction.

As the Electron shell opens up and its negatively polarized lattice starts to consolidate to a globular standard size, the main departing Electron mass is orbiting faster relative to the Proton due to inherent coordinates. Its orbiting becomes an effective blend of the Omni-directional main body push and the single (for each universal direction) drive by newly implanted linear virgins from new outgoing electromagnetic photons in the widening Proton breach.

The Electrons new path of departure is more marginally curved as dictated by the inherent coordinates within the Neutron and the new more powerful but less inherent drive. The flight of an Electron away from a Proton caused by the nearest powerful close quarter kick on a relative retrograde path will continue to be linearly corrupted but this common rotation based spin is not very effective in capping its linear drive by orbital neutralization in a faraway differential orbit at significant distance, even at increased inherent speed.

The fast electromagnetic photons will soon outreach the departing Electron and thereby allow the two adversely polarized hydrons to cast powerful virgin orbs between them when their mutual orbits allow. This will further slow the Electrons departure and allow it to settle into an orbit some distance away.

If there is another nearby Neutron companion, it would be rapidly orbiting the now split Neutron; a surviving Neutron companion is close to a Proton in fast gravity flux driven orbits while the consolidating Electron leaves the scene more linearly. Having few inherent coordinates in common with the Proton, a closely orbiting Neutron neutralizes weak linear electromagnetic repulsion to some extent but if it does not end up under the protective umbrella of the Proton, it too will eventually split.

Neutrinos

On the outside of the Electron mass shell as it breaks up, there can be smaller debris from the collapsing negatively polarized mattron grid that has moved outside the darkest shadow on the Proton backhand orb where orbital push here can push it sideways but not into the main Electron consolidating in the darker maelstrom of overwhelmingly orbital virgin winds.

Once such 'small' chunks of shell mass become situated outside a sphere set by the thickness of a forming Electron, much of the cohesive orbital backhand aiming towards the Electron disappears. After the shell opens up, its collapse is nonetheless 10^{24} times faster than its outwards speed as the Electron shell consolidates into a widening rotating globular Electron and little of it is left on the orb that lies outside the main Electron body.

The fewer virgins further out would deliver orbital thrust upon some lighter debris outside the darkest shadow that may miss its main orbital backhand and be prevented from joining the main body Electron. Even mattron debris on the outermost surface of the Electron shell would be keeping more or less the grid lattice from earlier. These smaller chunks of mass would consolidate slower and under lighter shadows anywhere on the Proton generated orb; originally outside the Electron. Such smaller dwarf hydron(s) of negatively polarized mattrons that form later than the Electron are also being repelled by the Proton.

The Neutrinos are affected by rising virgin emission on highly diverse orbs from all the recent electromagnetic photons that are likely to hamper their individual build by pushing shell mass towards the middle of the Proton orb which helps to deposit such chunks into the Electron. This robs Neutrinos of potential grid mass and deposits mattron stragglers in these outer areas back into the main mass centers. The featherweight Neutrino(s) are furthest from the central Proton and are actively repelled by electromagnetic photons. Much like the heavier Electron, they gain a slightly curving path by earlier inherent coordinates. Neutrinos can in spite of small spheres collect photons that hit them by they are likely bereft of electromagnetic splits of their own.

It is hard to give small chunks fashioned into globular dwarf hydrons any specific mass. It would be radically smaller than Electron mass; not unlike a sun losing out on fringe mass in the form of a few relatively tiny planets or a vast planet losing mass from its fringes into tiny moons. There would hardly be more than one or a couple of substantial Neutrinos during a split as they would join up on the original Proton orb if at the same distance; unless two are (unlikely enough) situated on precisely opposite sides of the orb.

To have an average Neutrino mass to work with, it seems 'reasonable' to put it at somewhere less than a millionth part of Electron mass or roughly 3×10^{218} negatively polarized mattrons. A conceptual illustration of this mass relation would be Mars versus Our Sun although it would be wise to recall that these tiny dwarf hydrons are anything but minuscule; they are truly gargantuan structures.

Evolving Spinfields

A hydron is a large collection of mattrons, each is moving within its own supreme coordinate system and for consummation in a larger groups of mattrons, 10^{948}ãt mattron consummation multiplied by the mattron number yield 10^{1176}ãt consummation of inherent spin within a standard Proton.

Some parts of a hydron are being pushed this way and other parts that way and the backhand to incoming strays or virgins will also continually influence hydron parts. The hydrons apparent spin of the moment, just as any submist that orbits it can also be a result of conflicting forces that have nothing to do with inherent spin. And, as the spinfield does not fuel all the hydrons apparent spin, neither is all that spin projected away from a hydron as inherent spin coordinates by its emitted virgins which are also a tiny part of the receiving mass which overwhelms that inherent message by its own inherent coordinates.

The Proton is the more massive hydron of an opposite pair. Each repulsion kick or backhand attraction between the gargantuan structures pushes distant mass in a universal line drawn at each momentary position through the center of the affected masses. At each moment, they implant linear motion in a specific universal direction. It does not really matter if this relative linear virgin push works upon an Electron or a loose swarm of inter-hydronal mattrons.

We know that if a submist has been moved sideways or formed any type of orbit by enforces means, each new linear implant is increasingly stymied by shifting focus; the implanted movement can be neutralized by precisely the right orbital speed. If orbital motion is insufficient to neuter the induced directional push every time a submist is hit by a virgin along a new universal line, the already implanted directional coordinates are still part of its coordinates and will force a shift in the submist position away or towards the Proton.

Every repulsive virgin emitted by a Proton carried the coordinates of its average main body rotation which, although a weak force, usually guides the orbital direction of other nearby submists. The Electron orbit is of course already imbued by the inherent spinfield of the original Neutron, albeit in an retrograde fashion, relatively speaking.

Hydron spinfield evolve; there is nothing to say that a hydron must rotate with perfect differential field of spin; albeit certainly more likely than to see a hydron keeping a single plane of spin for all mattrons orbiting its barycenter; somewhere in between may well be common. The spread of orbital planes of mattrons around a hydrons barycenter, inherent or enforced constricted by a grid may be greater than for orbital planes of planets within a solar system wobbling around an invariable plane at varying inclinations.

To what extent the Electron or a faraway mattron orbits the Proton twice before a correct antidote comes around in a differential orbit is not clear but its repulsion or attraction is confronted and neutralized or else the distance will be forced to change. Acceleration in an orbital field cannot be realized in orbits where repulsion or gravity are at regular intervals pitted against their mirror image but is easily without any effort achieved in inherent spinfields.

In the case of a rotating Proton, the Electron faces varying effects of repulsion or attraction bondage out to far distances where bonds are weak and purely orbital at slow speeds while antidotes are most effective at close quarters in a plane(s) of spin.

An interesting picture emerges when you look at a Proton with less than perfect differential spin; the enforcing effects of its repulsive emission and attractive backhand shadows upon external mass, especially mattrons far outside its aura. With perfectly differential rotation, where a momentary spin plane itself orbits the Proton, there is no notable difference for exterior mattrons on harvesting emitted virgins; they are repelled or attracted at the same rate in all available universal directions away from the Proton.

The more interesting illustrative is what happens when hydron spin is in a single plane which would be a rare occurrence but a familiar picture. With multiple orbital planes constricted to a broad invariable plane at varying or shifting inclinations, a similar but less definite enforced effect will be seen. Look at repulsion upon a loose exterior mattrons with precursors that have been orbiting a Proton in a single plane of spin (or a broad invariable plane). Assume that these mattrons are evenly spread all around the Proton and look at what happens to enforced repulsion as virgins on its orbital path away from parents realize a linear universal direction on hitting any submist that blocks them.

The point is that enforced repulsion has a different character out along a rotation axis where there is no orbital push to neutralize its outwards drive. The further you move from the rotation axis towards the plane of spin, the slower the delivered orbital push component becomes. At the rotation axis, all hitting virgins are perfectly linear and a mattron receiving the repulsion on the axis line falls freely away from the Proton with its flight unimpeded by the collective single plane spinfield (far distances show increasing orbital tendencies around an axis line with an effect weakened by dilution).

Conversely, the closer to the plane of spin an orbiting mattron is situated when hit by outgoing virgins, the more efficiently its faster collective orbit neutralizes that repulsion. The altering result from a rotation axis towards the plane of spin is not an all or nothing outcome but rather a gradual loss of linear repulsion that becomes ever less effective closer to the single plane.

If there were an exterior cloud of precursored mattrons, this different effect would shape that cloud. The spaces out from the Protons rotation axis would soon be cleared of nearby neutral mattrons that have been pushed further afield and out of tangible contact in polar directions. At a distance from out from the Protons rotation axis where a distant mattron aide by its orbital movement reaches balance between repulsion and backhand attraction, the mattron does not orbit the Proton; it orbits its rotation axis at that distance. Only virgins that hit a mattron in a Protons 'equatorial' or invariable plane of spin hit mattrons that actually orbit the Protons main corporeal body.

We can see the repulsion drive upon a mattron situated some way from the Protons rotation axis at this distance is by linear virgins that enter it along lines that on average always cut the distantly rotating hydron. At the same distance from a Proton, the mattron will in time gain orbital speeds that are slower than in the plane of spin and orbits around the rotation axis are also shorter.

In stable systems the repulsion would be matched by the linear backhand of enforced attraction. If the mattron gains a balance between repulsion and attraction, the inwards or outwards balance is achieved by angled virgin push from opposite directions. This means that part of the axis repulsion or backhand attraction is periodically brought into opposition where antidotes neutralize a mattrons departure or approach towards the rotation axis but this is only part of the linear enforced effect in either direction.

If a balance between enforced repulsion and 'attraction' changes, say in favor of repulsion, and distantly orbiting mattrons are being repelled along a single axis, they fall away out and around its rotation axis along a tube-like path. As directional nudges away from a rotation axis are annulled, a mattron is pushed away from a repulsive rotating Proton while remaining trapped in orbit around an axis from which it cannot easily change its distance. The closer the mattron is to the Proton, the greater is the orbital drive as part of the linear repulsion.

Polar bonds are always weaker and during periods of expansion when an exterior submit receives greater emission; instead of being forced outward in all directions at the same rate, all mass outside the plane of spin is more easily moved away from that plane (a cylinder like orbit around the rotation axis). This outward progress forms a seamless path that carries it partly away from the main central body and partly away from its plane of spin.

Faraway mattron at considerable distance from the axis that moves closer to a constricted plane of spin under influence of gravity backhand, also do it along a tube-like path (staying on the wall of a wide axis tube). If a balance is not reached after the outside body is accelerated towards the plane of spin, the 'falling' (in this case) mattron is forced to pass through that spin plane to enter a decelerating recession phase brought on by diametrically opposed implants; also towards the Protons plane of spin.

Mass under the influence of any large mass barycenter can be caused to fall back through the plane of spin again and again, its acceleration turning into deceleration. However, in the long run, if push or pull is not neutralized towards the plane of spin, the reach of these oscillations out of and back through that plane diminishes successively until the mass is eventually brought to heal in a stable orbit within a single plane of spin in a direction set by the spin of a large corporeal central body.

This development has striking similarities to solar, planetary or galactic systems and even planetary rings and is probably more relevant outside macro submists that are less prone to upheavals. Although distances for planetary structures are relatively small and the forces weaker, the reason for most of the mass to be found in a plane of spin may not be because it was originally situated there. The system may well have started out as a globular structure that due to an evolving barycenter spinfield brings hydron mass around a larger globular solar center to drift more effectively towards its single invariable plane of spin than towards the dominating central body; causing planet embryos to gather and form in the systems plane of spin

This phenomenon may be stable enough if there were a single invariable plane of spin around a Proton to shape the structural reformation of nearby inter-hydronal mattrons but the processes on the hydron scale are unlikely to allow these effects. Standard hydron bodies and bonds of are formed by other processes and oversized hydron embryos are unlikely to have single planes of spin. The electromagnetic photons from a collapsing Proton after a Neutron split have great globular symmetry and their massive effect outgun any repulsion and its minuscule relation to a Protons rotation axis.

For a mental two dimensional cross section picture of a Protons plane of spin, imagine an immense hourglass shape where the Proton is a tiny free-floating corn of sand centered in the empty neck of the hourglass which the single plane of spin cuts horizontally. Outside the neck, the presence of any surrounding mattrons would be densest and closest and such a mist would thin out from the neck in either directions while the hourglass mist broadens from the plane of spin at further distance.

A better three dimensional picture would lines of a solar magnetic field with a constant rotation axis. With differential spin that picture would remain similar except the rotating plane of spin would at each time be repelling mass out along a rotation axis that shifts in relation to the rest of the Universe as its axis would be orbiting the globular structure like the eye of an emitting quasar body.

Part 7: The Galactic Fields

Milky Way Hydrons and Photons

The wayward photons with precursor halos on a continuing trip through the Universe remain stable in the inter-galactic void and the majority of their adverse primal emission is already activated outside galaxies. The generated flux is however overwhelmingly ripped out of a nearest galactic mass field of more closely knitted hydrons that have adjusted to balanced standard sizes and darker flux fields.

Every galaxy is made up of three main singularly polarized hydron types; the Proton and the Electron that harvest high speed flux according to their core signs, positive or negative and a standard Neutrons that harvest flux at the same speed as the other two in the respective signs.

The Andromeda and the Milky Way galaxies lie within a distance of 10^{929}år but they are not alone. It is not easy to set apart an inter-galactic void to be assigned as private empty space around a galactic mass field. The Milky Way is not merely a barred spiral galaxy with a 10^{926}ãr wide globular halo of star clusters and a galactic disc surrounded by a vast hot globular gaseous halo reaching $1,5 \times 10^{926}$ãr or 160.000 light years towards the Magellanic clouds; a halo of unstructured mass that almost equals that of its corporeal mass.

An illustration of a Milky Way galaxy stipulates $8,38 \times 10^{68}$ baryons that with 2×10^9 positive and $4,17 \times 10^6$ negative for each photon passing through inter-galactic spaces in the original proportions. The Milky Way mass would be matched by $1,68 \times 10^{78}$ positive photons and $3,492 \times 10^{75}$ negative photons around the galaxy.

This illustration is probably too restricted. Giving the Milky Way a private space of $3,3 \times 10^{2786}$ã inter-galactic volume with corresponding $6,92 \times 10^{928}$ãr 'radius' derived from size of the Virgo Supercluster includes the surrounding inter-galactic spaces of exponential volume of freely roaming photons seems more reasonable where the Milky Way has a 'private' average contingent of these photons according to its hydron mass. Although the mental picture is clearer without such mixed flux where a virgin majority shared between more galaxies at variable sizes, this seems a more logical average.

I illustrate therefore with a Milky Way that harvests and shares the effects of adverse emission from its 'private' inter-galactic photons that pass within the $6,92 \times 10^{928}$ãr distance of this larger space.

When the nearest galaxy is the Milky Way that harvests this flux effect, it is affected by its share of the inter-galactic photons that pass within the $3,3 \times 10^{2786}$ã inter-galactic Virgo volume. The ripped out flux receives boost across various inter hydronal distances prior to harvest and is repeatedly chipped across the $6,92 \times 10^{928}$ãr inter-galactic distances at $2,63 \times 10^{464}$¢ but most of the boost is eventually harvested within the corporeal Milky Way galaxy.

Most of inter-galactic newborns in both signs would enter the main body galaxy already as activated flux. One reason why average flux harvest time is important is to estimate how much flux serves to activate virgins from within and outside the galaxy. The survival rate of virgins across various distances sets linear and orbital virgin backhand shadows as the inherent coordinates create an effective backhand as they collide into submits. Stopped in their tracks, they are even easily blocked by precursored mattrons of their sign.

When a long distance virgin is activated by a flux atom in say a galactic sized orbit, the resulting flux pair is a composite of influential coordinates, great many of which can cross the galactic void $2,63 \times 10^{464}$¢ faster along a highly convoluted path. This can bring the pair back to where it started and back again across the private galactic space within $1,32 \times 10^{464}$ãt while the average virgin covers that distance first in $6,92 \times 10^{928}$ãt.

There is a marked difference in how virgins and activated flux progress across a void. A virgin aims away from a parent anywhere within or outside the galaxy and traverses the void at birthspeed in an opening differential orbit with boosted parental sweep in strict accordance with inherent and unaltered coordinates until the newborn is activated.

Newborns from main body photons are boosted from 10^{-244}¢ in 10^{316}ãr wide parental orbits to $2,63 \times 10^{62}$¢ across the inter-galactic regions if they survive that trek as virgins. Photon precursor mass emits 10^{24} times fewer virgins but these gain $4,89 \times 10^{212}$¢ sweep across $6,92 \times 10^{928}$ãt. This great presence serves as a robust activating agent across the vast emptier regions outside a galaxy where virgins are rapidly activated but it also decimates the delivery of precursor virgins while any survivors are harvested by the first hydron they hit.

The overwhelming virgin majority is activated at different rates by variable flux within either the emptier volume of the private space of the Milky Way or the denser galactic volumes within it. The mixture makes for a more dodgy approximation of the average activation rates and actual activation within lesser local volumes when galactic flux becomes increasingly focused into and around the greatest mass density in the galaxy.

Galactic Flux and Virgin Pressure

It is clear with more than a billion inter-galactic photons for every hydron and activated flux boosted to 10^{16} times faster orbital sweep between two galaxies than between two hydrons, the galactic flux drive wins hands down against the drive of locally exchanged flux.

Flux across any larger distances is rarely blocked at subservient speeds. Protons, Neutron and independent Electrons block flux at high speeds in the respective dominant signs and this sets inter-galactic and inter-hydronal flux consummation periods. The Protons block negative flux at **$5,62 \times 10^{230}$¢** and Electrons of more meager positive flux at a slower **$9,13 \times 10^{228}$¢**.

The problem of illustrating an average of 'last leg' distances prior to harvest is that these are widely different with corresponding consummation periods. The spatial presence of diminishing long distance flux beyond a first galactic run and only a tiny part reaches inter-galactic voids. Flux in widening orbits is continually slowed linearly and orbitally but snippets of uncompromised coordinates are transferred to more flux carriers. While fewer long distance flux atoms carry local flux out from among hydrons, stars, galaxies and even across the universe, the majority becomes locally focused by mass when this flux is on its last leg orbits during the last average inter-hydron runs.

One also needs to estimate how virgins and flux come into conflict with the corresponding inter-galactic, inter-stellar, and solar flux presence as that ratio can seriously affect virgin exchanges within more local fields. Looking at the broad picture, virgin survivors across a linear $6,92 \times 10^{928}$ãr from outside the Milky Way orbit across that distance to engage with galactic hydrons. Proton and Neutron spheres at 10^{1704}ã and $1,631 \times 10^{1702}$ã Electron spheres block with their different areas while free precursored mattrons or photons with 10^{1504}ã halos can block virgins at speeds up to 10^{72}¢.

The **negative** inter-galactic <u>flux</u> presence from $1,676 \times 10^{78}$ positive photons in $6,92 \times 10^{928}$ãr that is takes a virgin to move across that distance and reach $2,63 \times 10^{62}$¢; is driven by $1,16 \times 10^{727}$ virgins. Having accepted that the focus of the central galaxy sets the average boost according to its average $3,22 \times 10^{926}$ãr 'radius', even inter-galactic flux in the 10^{906}ãr wide grid among $3,3 \times 10^{68}$ baryons is boosted to $1,79 \times 10^{463}$¢ before harvest at $5,62 \times 10^{230}$¢.

Within a sun, focused down to the last 10^{894}ãr inter-hydronal distance, the average consummation period for flux on the wing is $1,78 \times 10^{663}$ãt. The consummation period in inter-stellar space in a 10^{904}ãr grid is $1,78 \times 10^{673}$ãt and the inter-galactic harvest across 10^{906}ãr is $1,78 \times 10^{675}$ãt.

Within $3,3 \times 10^{2779}$ã galactic volume where flux majority is harvested mostly within suns, the collective $3,7 \times 10^{1853}$ã negative inter-galactic flux presence combines here with the $1,82 \times 10^{984}$ã presence there of a single inter-galactic virgin (10^{-7} of its $1,82 \times 10^{991}$ã inter-galactic presence) as $6,73 \times 10^{2837}$ã. This means that $4,9 \times 10^{-59}$ of all negative virgins from positive inter-galactic photons stay intact before they orbit out through and around the galaxy on the other side.

On crossing the larger private galactic space, few surviving virgins chance to hit a standard hydron and the majority passes through without a direct hit. Out of the $4,79 \times 10^{1857}$ã private Milky Way sphere, the collective spheres of $8,38 \times 10^{68}$ Protons and Neutrons collectively block an $8,38 \times 10^{1772}$ã area. A virgin has $1,75 \times 10^{-85}$ chance of hitting a baryon in each full orbit but during $6,92 \times 10^{928}$ãt it makes $2,63 \times 10^{62}$ orbits before exiting the galactic domain at $2,63 \times 10^{62}$¢. These orbital passes allow $4,6 \times 10^{-23}$ of the surviving virgins to actually hit a nucleon before continuing across to the nearest other galaxy after completing a first galactic trek. This loss comes on top of the $4,9 \times 10^{-59}$ loss by activation and builds the total loss of active virgins to **$2,25 \times 10^{-81}$**.

The non-hitting survivors continue to travel across the Universe where a virgin gains in collision ability with hydrons (distance spheres do not grow at the same rate as volumes of corresponding mass) by more orbits due to greater boost. Increased boost and time raises activation at a matching rate as effects cancel out and the tiny minority of lost virgins are returned by proxy with marginal extra pressure that I drop from the wider Universe.

The **positive** inter-galactic flux presence driven by $3,492 \times 10^{75}$ negative photons across the Milky Way in $6,92 \times 10^{928}$ãt counts $2,42 \times 10^{724}$ virgins that initially reach $2,63 \times 10^{62}$¢ sweep. Flux at $1,79 \times 10^{463}$¢ is harvested at $9,13 \times 10^{228}$¢ by inter-galactic Electrons across 10^{894}ãr during $1,095 \times 10^{665}$ãt. This collective positive inter-galactic flux presence is $4,74 \times 10^{1852}$ã that with 10^{-7} of its $1,82 \times 10^{991}$α single inter-galactic virgin presence or $1,82 \times 10^{984}$ã combines as $8,62 \times 10^{2836}$ã positive presence within the $3,3 \times 10^{2779}$ã Milky Way volume where $3,83 \times 10^{-58}$ of all positive virgins from negative inter-galactic photons stay intact at $2,63 \times 10^{62}$¢ and have a shot at a submist.

The positive virgins have the same $1,75 \times 10^{-85}$ chance as negative virgins of hitting a baryon in each orbit and $4,6 \times 10^{-23}$ of the surviving positive virgins hit a nucleon; a positive loss of **$2,11 \times 10^{-80}$**.

When counting delivered pressure after $2,25 \times 10^{-81}$ downsizing but with a $2,63 \times 10^{62}$¢ orbital boost (and 1pu Universally linear drive), virgin pressure upon a standard hydron would still fall short of resisting hydron expansion. For a galaxy built of mainly larger Neutrons, the hitting rates are larger and even the activation rate is lower due to shorter flux consummation and they capture more of virgin pressure. While a standard galactic baryon in a fully converted galaxy is on average hit by $1,18 \times 10^{-289}$pu per ãnna from the more numerous 2×10^9 positive photons assigned to each. Incoming virgin pressure is downsized to 10^{-38} and $1,18 \times 10^{-327}$pu compression cannot keep a Neutron breach from widening against a central 10^{-300}pu expansion but a galaxy of larger Neutron sizes facilitates virgin compression and balance.

Main body hydrons or precursored mattrons (or photons according to sign) cast local backhand shadows by blocking virgin pressure with physical spheres.

Short Distance Flux Shadows

There is an important factor to be addressed concerning the transfer of push from photons outside the galaxy by fast orbital flux at $5,62 \times 10^{230}$¢ in any conceivable orbital direction through all allied hydrons. The perfectly contradictory orbital primal coordinates of a pair of split twins are imbedded in flux and deposited into every galactic hydron. This highly convoluted orbital push cannot cast a linear gravity shadow onto another nearby hydron (except when blocked at extremely close distances).

The path of a flux atom from the last hydron it rips through (without being blocked) has a convoluted orbit that opens at $5,62 \times 10^{230}$¢ as a coordinate center moves non-linearly out of the next to last hydron at that speed. Ripe for harvesting after engaging a spinning or orbiting atom its new widening gyrating path is locally boosted with great spatial presence. The flux may not end up in the hydron that its next to last trajectory appeared to aim at. Before this slowdown it aimed and ended up in all conceivable directions in changing orbits and is as often as not chipped right across the galaxy.

The independent primal spin of two split adverse newborns emitted by a blend of inter-galactic photons mostly in perfectly opposite directions. The original coordinates of primal boost do not change but as they end up evenly spread among all hydrons within the galaxy, all the orbital coordinates are matched and offset by flux in opposite orbital directions.

The orbital coordinates embedded in flux atoms do indeed cast a shadow with an effective backhand between hydrons within a galaxy when harvested. but this is always an orbital shadow cast in both opposite directions. When this shadow falls upon another hydron on any distance orb, it falls upon both sides of the same hydron with an equal backhand in both directions. These orbital coordinates cast a linear shadow with an effective backhand within hydrons when harvested (between the internal mattrons) but whether born of primal or parental spin; after ending up in all galactic hydrons cause in opposite direction, no linear push upon galactic hydrons is experienced.

Aside from orbital coordinates, there is another set of coordinates inherent to both twin newborns born of inter-galactic photons that are not opposite when dispersed into flux; this is their movement away from a parent.

All flux is blocked by galactic hydrons and hydrons do cast a shadow within the galaxy in accordance with the harvested number of flux atoms. The linear effect of a virgin (and its eventual flux) is effective in whatever linear direction incumbent upon its orbit. This effect carried by flux within a galaxy is countered linearly from the other side of the galaxy but this linear push away from a parent must be cancelled by opposite linear collisions.

The smallest input from $3,492 \times 10^{75}$ to $1,676 \times 10^{78}$ photons outside a galaxy is the relatively tiny 1¢ linear newborn drive away from a parent but when this coordinate push enters the galaxy imbedded in primal flux, it is broadly directional on the galactic scale as the thrust works in any direction away from the original parent.

This part of the cast shadow has a different backhand because each flux atom has a snippet of a linear push imbedded and the same message is cast between all galactic hydrons; not orbitally but linearly. All flux atom push weakly in the original broad direction where the presence of another hydron prevents a tiny linear push fragment from being delivered, thereby causing a linear backhand.

Note that an inter-galactic virgin that hits a galactic hydron generates an internal flux shower through which all its coordinates are communicated to the rest of the hydron mass and its push in not orbital within the galaxy where this cast a local shadow with a linear backhand upon another hydron that stands in its path. A local virgin with a smaller than galactic orbit cannot cast a galactic shadow except by linear means but it can certainly cast local shadow across distances smaller than its orbit.

Mattron Stability in Galactic Space

Hitherto, I have regarded mattrons as immune to passing high speed flux. Nevertheless, the wear and tear of flux affects haloed mattrons that traverse the void, neutral or singular and it is necessary to confirm that the flux does not affect a mattrons inter-submist stability when it interacts within the galaxy.

With average sizes of important players; Protons, Electrons, Neutrons, photon emission and activation levels largely in hand, flux effects on mass presence between hydrons, suns and galaxies can be broadly approximated.

In a tight mattron grid, if mattron mass is dispersed in one sign or both, it has nowhere to go except to condense back into mattrons when the local field density allows. In a greater grid, dispersed mass of whatever sign does not affect a local strayfield for long. Long distance flux is mostly and swiftly harvested by a gargantuan hydron at dominant flux speeds. First look at a virgin crossing a say 10^{894}ãr inter-hydron distance with 10^{45}¢ sweep and 10^{939}ã presence within that 10^{2682}ã local volume.

In the 10^{894}ãt time for survivors to escape a local volume $3,492 \times 10^{689}$ **positive virgins** are emitted by $3,492 \times 10^{75}$ negative inter-galactic photons. Look at flux that has been focused within a $3,22 \times 10^{926}$ãr galaxy and boosted to $1,79 \times 10^{463}$¢. Harvested by Electrons at $9,13 \times 10^{228}$¢ in $1,095 \times 10^{665}$ãt between hydrons. Before harvest, it gains $6,84 \times 10^{1817}$ã activated presence within the $3,33 \times 10^{2779}$ã galactic volume. Of all this flux presence in 10^{894}ãt, only 3×10^{-98} or $2,05 \times 10^{1719}$ã is present within the inter-hydron volume. With 10^{939}ã presence combining at $2,05 \times 10^{2658}$ã in the 10^{2682}ã local volume; $2,05 \times 10^{-24}$ of positive virgins is activated by galactic flux on the first trek before leaving the local volume.

Similarly in 10^{894}ãt, some $1,68 \times 10^{692}$ **negative virgins** are emitted by the $1,676 \times 10^{78}$ positive inter-galactic photons and focused within and boosted to $1,79 \times 10^{463}$¢ across a $3,22 \times 10^{926}$ãr galaxy to be harvested by Protons at $5,62 \times 10^{230}$¢ with $1,78 \times 10^{663}$ãt consummation. It has gained $5,35 \times 10^{1818}$ã negative presence within the $3,33 \times 10^{2779}$ã galactic volume in 10^{894}ãt but of this, only 3×10^{-98} or $1,61 \times 10^{1721}$ã is found within the local volume. With 10^{939}ã negative presence combining at $1,61 \times 10^{2660}$ã within this larger 10^{2682}ã volume; only $1,61 \times 10^{-22}$ negative virgins will be activated by galactic flux before the virgin leaves the inter-hydron volume. (The vastly faster precursor virgins of both signs in 10^{376}ãr wide parental orbits with 10^{-64}¢ parental sweep reach 10^{195}¢ and cover 10^{1089}ã spatial points across 10^{894}ã and all would be activated before reaching a close by hydron).

Now follow a negative virgin split within a positive inter-galactic photon on a trek using the same premises with its $1,79 \times 10^{463}$¢ boost across the $3,22 \times 10^{926}$ãr galaxy until it has interacted with galactic mass and chipped out roughly 10^{232}

atoms needed to slow its orbital drive before a resulting flux shower can be blocked by hydrons and harvested by mattrons during 10^{663}ãt consumption.

Slowed and kicked across the galaxy, all the flux atoms become gradually more focused upon galactic mass in repeated collisions towards ever greater mass populations near the galactic center in tightening orbits; the rising dominance of local coordinates in the slowing more numerous atoms on passage through smaller hydron grids will continuously focus the flux to tighter active volumes.

While 10^{461} other virgins join in from 10^{78} photon points around the galaxy, all slowed to stoppable speed, they form a locally oriented flux field that consists of 10^{693} newborn atoms that pass at 10^{230}¢ through a few last inter-hydronal exchange domains on the last leg before harvest. Pick one such atom passing its next to last hydron domain on a 10^{894}ãr inter-hydron path and ask the question; what presence in this last hydron domain is this passing atom most likely to interact with before harvest?

Although galactic flux has the greatest large scale presence and best odds for further hits, its atoms are relatively few locally and the interactions in a local setting demand time. For example, locally exchanged hydron flux may need 10^{664}ãt before harvest at 10^{230}¢ while galactic flux may need 10^{696}ãt. In both cases, the flux atoms interact with whatever local mass that has the greatest field presence per ãnna. The following comparison of local presence between hydrons shows which particles in a local field are predominantly hit on passage where those with the highest presence have the highest hitting rate.

1. An activated dominant inter-mattron newborn in a mattron exchange within a standard hydron has 10^{388}ã presence per ãnna and covers 10^{1051}α spatial points during 10^{663}ãt local consumption. The 10^{228} mattrons in the inter-hydronal field emit 10^{611} dominant atoms in that time and they have the greatest collective presence of **10^{1662}ã**.

   Each rips 10^{359} dominant atoms out of mattron cores during roughly 10^{747}ãt consummation or 10^{270} atoms in 10^{663}ãt, ripping 10^{881} atoms into the field in that time (with the same original 10^{1662}ã collective presence) supplying all the cannon fodder required.

2. A mass atom in a standard mattron in parental orbit at 10^{-244}¢ covers 10^{419}ã spatial points during 10^{663}ãt local consumption. A single inter hydronal domain has a sustainable stock of in excess of 10^{1140} such atoms with a collective presence of **10^{1559}ã**.

3. An activated inter-hydron newborn gains a great 10^{447}α presence per ãnna and during 10^{663}ãt a hydron emits a maximum of roughly 10^{392} virgins that have a collective presence of **10^{1502}ã**.

300

This locally exchanged flux has so much presence that even 10^{-160} less than for exchanged core flux are chipped away, 10^{451} atoms are not to be found and most of the 10^{392} available atoms with flux are kicked out by galactic flux; confirming the wider dispersal of locally activated newborns.

4. A precursor halo mass atom in 10^{376}ãr wide parental orbit with 10^{-64}¢ sweep covers 10^{599}ã spatial points during 10^{663}ãt local consummation. With at most 2×10^9 photons in the inter-hydronal field, roughly 10^{899} precursor mass atoms (10^{-22}) have a collective presence of **10^{1498}ã**.

As this mass is not replenished, a precursor will eventually be eroded and must by nature be intermittent during more extensive intervals.

5. A dominant inter-mattron virgin without sweep covers 10^{663}ã spatial points during 10^{663}ãt local consummation if it stays intact. With on average roughly 10^{-79} surviving the 10^{776}ãr trek in 10^{776}ãt, the average presence of each emitted virgin is merely 10^{584}ã. The 10^{228} mattrons in the inter-hydronal field emit 10^{611} dominant virgins in 10^{663}ãt and the stable field has a collective average presence of **10^{1195}ã**.

6. An inter-hydron virgin with 10^{45}¢ sweep covers 10^{708}ã spatial points during 10^{663}ãt local consummation. As all seem to survive the first run and a hydron emits a maximum of roughly 10^{392} virgins in 10^{663}ãt, the secure virgins gain a collective presence of **10^{1100}ã**.

It is the mattronal mass that fuels the slowdown of galactic flux and other fields need not be tapped into although they fractionally are in this zero sum game where the same number of ripped out mass is re-deposited. Precursor mass is not rapidly eroded by galactic flux and the inter-submist stability of mattrons in voids is relatively stable in the 10^{1192}ãt it takes a standard mattron to double its 10^{912} mass.

The outcome indicates that mattrons that traverse for extensive periods through voids will have lots of atoms chipped out into greater orbits by fast long distance flux but their main bodies are in no danger of being ripped apart and precursor stability holds intermittently firm in local fields between hydrons, stars or galaxies.

Part 8: Stable Hydrons

The Proton

I am using marginally less mass than for my chains 10^{228} mattrons and I'll drop the fractions for a Proton of permanent positive polarity that gives the rock bottom 'standard' stability of 10^{776}ãr inter-mattronal distances. This mattron mass must be held together by negative flux or be forced to expand.

During 10^{852}ãt radius time, internal mattrons emit 10^{800} negative dominant and 10^{764} positive subservient virgins. Those not activated, escape the surface at birthspeed with 10^{24}¢ orbital sweep.

The 10^{800} negative dominant virgins activated between mattrons gain 10^{388}¢ boost during $5,62 \times 10^{746}$ã dominant consummation. The $5,62 \times 10^{1934}$ã dominant flux presence combines with the 10^{876}ã virgin presence as $5,62 \times 10^{2810}$ã within a 10^{2556}ã Proton volume which lets $1,78 \times 10^{-255}$ or $1,78 \times 10^{545}$ negative virgins escape its surface at birthspeed.

The 10^{764} positive virgins gain 10^{388}¢ inter-mattronal boost during 10^{740}ãt subservient consummation as flux sweeps 10^{1892}ã collectively within the Proton. An orbital virgins 10^{876}ã presence combines to a 10^{2768}ã presence within the 10^{2556}ã volume where 10^{-212} of the positive virgins escape the 'standard' proton main center untouched and 10^{552} virgins that also account for its 10^{552}pu expansion.

$$10^{24} \times (1,78 \times 10^{-255}) = 1,78 \times 10^{-231} = 5,62 \times 10^{230}$$

Protons harvest little positive flux at 10^{188}¢ but blocks negative flux at up to **$5,62 \times 10^{230}$¢**.

Negative inter-galactic flux from 2×10^9 positive photons in 10^{852}ãt is driven by 2×10^{581} virgins at average $1,79 \times 10^{463}$¢ galactic boost and delivers $3,58 \times 10^{1044}$pu negative inter-galactic pressure to every galactic Proton in radius time.

The average inter-mattronal shadows is between 10^{776}ãr and 10^{852}ãr Proton radius at 10^{814}ãr mean distance and standard coronal mattron shadows fade to 10^{-364} and core shadows to 10^{-391} with a $3,16 \times 10^{-378}$ average that with the additional inverse 10^{-114} square root of mass yields $3,16 \times 10^{-492}$ dilution of the $3,58 \times 10^{1044}$pu incoming flux to $1,13 \times 10^{553}$pu compression versus the Protons 10^{552}pu expansion. This keeps the standard hydron just within its 10^{852}ãr radius unless inter-galactic flux adjusts by distances or by altered virgin population; neither of which are constant values.

If compressed beyond 'standard' size, inter-mattronal boost can no longer uphold the balance of mattron coronas and primal halos step in to pick up the slack. I put the smallest stable hydron at a balanced 10^{850}år radius with a small part of coronal mass dissolved into halos. The pressure upon the standard Proton, given these inputs does not signal such circumstances; its surface that in star terms would seem diamond smooth is almost perfectly balanced by the negative galactic flux.

The Electron

Even the Electron is likely to get its $2,083 \times 10^{225}$ central mattrons of 10^{912} atoms at 10^{776}år inter-mattron distances permanently locked into the grip of the weaker positive inter-galactic flux to gain standard stability.

A standard Electron would be held to quarters within a $2,083 \times 10^{2553}$ã central volume of $1,277 \times 10^{851}$år radius and contains $2,083 \times 10^{225}$ mattrons of permanent negative polarity; also in the iron grip of inter-galactic flux.

During $1,277 \times 10^{851}$åt radius time, its mattrons emit $2,66 \times 10^{760}$ negative subservient and $2,66 \times 10^{796}$ positive dominant virgins that activated gain 10^{388}¢ boost during $5,62 \times 10^{746}$åt adverse and 10^{740}åt allied consummation. The $1,49 \times 10^{1931}$ã dominant flux presence combines with a virgin that with $3,57 \times 10^{23}$¢ sweep gains $4,559 \times 10^{874}$ã presence. The combined $6,79 \times 10^{2805}$ã within the $2,083 \times 10^{2553}$ã Electron volume lets $3,068 \times 10^{-253}$ or **$8,16 \times 10^{543}$ positive virgins** escape its surface at birthspeed.

The $2,66 \times 10^{1888}$ã subservient flux presence combines with $4,559 \times 10^{874}$ã virgin presence as $1,213 \times 10^{2763}$ã within the $2,083 \times 10^{2553}$ã Electron volume and lets $1,717 \times 10^{-210}$ or $4,567 \times 10^{550}$ negative virgins escape its surface at birthspeed and these also represent the **$4,57 \times 10^{550}$pu Electron expansion**.

$$3,57 \times 10^{23} \times 3,068 \times 10^{-253} = 1,095 \times 10^{-229} = 9,13 \times 10^{228}¢$$

This Electron can block negative flux at a slow $2,08 \times 10^{185}$¢ but receives little of it as most has been blocked elsewhere at dominant speeds. It blocks positive flux at a robust **$9,13 \times 10^{228}$¢** (same as a standard Neutron).

Positive inter-galactic flux from $4,17 \times 10^{6}$ negative photons during its $1,277 \times 10^{851}$åt radius time has a $1,79 \times 10^{463}$¢ drive by $5,32 \times 10^{577}$ virgins that deliver $9,52 \times 10^{1040}$pu positive inter-galactic pressure to every galactic Electron in radius time, whether free or bound by Neutrons.

Looking for an average internal inter-mattronal shadows we find it between 10^{776}år and $1,277 \times 10^{851}$år Electron radius at $3,574 \times 10^{813}$år mean distance. The internal standard shadows are $7,831 \times 10^{-364}$ for standard coronas but for the mattron cores $7,831 \times 10^{-391}$ with an average of $2,48 \times 10^{-377}$. With a smaller mass based $4,564 \times 10^{-112}$ dilution on top, the incoming flux is downsized as active center-wise compression to **$1,132 \times 10^{-488}$**.

The $9,52 \times 10^{1040}$pu positive pressure brings **$1,08 \times 10^{553}$pu** compression that is $235,87$ fold its $4,57 \times 10^{550}$pu radius time expansion. The field has no problem in holding Electrons in the stable grip of positive inter-galactic flux with a surface that in star terms would be of diamond smooth quality.

If the Electron is slightly compressed, it remains stable as the exchanged internal mattron virgin halos make up for dropping pressure exchanges and the consequently increased expansion renders a new balance.

The extra internal emission adds marginally to internal core sign flux which blocking speed and shadow dilution is marginally diminished. Shorter inter-mattronal distances do lead to faster consummation of positive flux (which cannot be substituted for by lesser boost). This causes an Electron to block positive flux at slower speeds but does not change its average pressure harvest.

Thus, unwilling to complicate my standard illustration for this small discrepancy, I allow its internal mattron coronas to resist the compression full out with halo exchanges to afford the Electron $1,08 \times 10^{553}$pu virgin expansion versus its $1,08 \times 10^{553}$pu radius time flux compression.

The other alternative to expand it to a new standard shell lattice of slightly larger than standard 10^{776}ãr inter-mattronal distances and a larger standard Electron radius which is a more accurate method. This upsets the aforesaid in various ways and leads to greater inter-mattron boosts and smaller mattrons with all kinds of other complications. For the continued expediency of this illustration, I stick to the standard image.

The Neutrino

The Neutrino is likely to get its illustrative 3×10^{218} central mattrons, each of 10^{912} atoms, permanently locked in the grip of positive inter-galactic flux and thus gain stability when its internal inter-mattron distances stabilize at around 10^{776}ãr standard range. A Neutrino would then be held to quarters within 3×10^{2546}ã volume which yields $1,73 \times 10^{849}$ãr radius with 3×10^{218} mattrons of permanent negative polarity; in the grip of inter-galactic flux.

$$10^{776}ãr = 10^{2328}ã \quad (3 \times 10^{218}) = 3 \times 10^{2546}ã = 1,73 \times 10^{849}ãr$$

During $1,73 \times 10^{849}$ãt radius time, its 3×10^{218} negatively polarized mattrons of 10^{912} primal mass emit $5,19 \times 10^{787}$ positive dominant virgins and 10^{-36} negative virgins escape mattron coronas as $5,19 \times 10^{751}$ subservient virgins. The activated exchanged virgins gain 10^{388}¢ boost during a $5,62 \times 10^{746}$ãt dominant and 10^{740}ãt subservient consummation.

The $2{,}92 \times 10^{1921}$ã dominant positive flux presence combines with a virgin that gains $4{,}16 \times 10^{22}$¢ sweep and $7{,}2 \times 10^{871}$ã presence. Within the 3×10^{2546}ã Neutrino volume the combined $2{,}1 \times 10^{2793}$ã allows $1{,}43 \times 10^{-247}$ or $7{,}42 \times 10^{540}$ positive virgins escape its surface at birthspeed or $4{,}29 \times 10^{-309}$pu per ãnna.

The $5{,}19 \times 10^{1879}$ã subservient flux presence combines with $7{,}2 \times 10^{871}$ã virgin presence as $3{,}74 \times 10^{2751}$ã within such a 3×10^{2546}ã Neutrino volume and lets $8{,}02 \times 10^{-206}$ or $4{,}16 \times 10^{545}$ negative virgins escape its surface at birthspeed or $2{,}4 \times 10^{-304}$pu per ãnna which also represent negative virgin expansion in its negative subservient sign or $4{,}15 \times 10^{545}$pu in radius time.

With $1{,}73 \times 10^{849}$ãr radius and 3×10^{1130} primal mass, a single newborn at a Neutrino center gives it $5{,}77 \times 10^{-282}$¢ inherent surface rotation in 3×10^{1130}ãt. The 10^{752}ãr wide mattron precursor within the Neutrinos $1{,}73 \times 10^{849}$ãr aura would block $3{,}34 \times 10^{-195}$ of the small hydrons collective emission and the effective harvest by a mattron precursor in the aura is $1{,}43 \times 10^{-503}$ positive and $8{,}02 \times 10^{-499}$ negative virgins per ãnna.

The Neutrino can block positive flux at $1{,}68 \times 10^{224}$¢ while Electrons block most inter-galactic flux at $9{,}13 \times 10^{228}$¢ which would seem to leave only $1{,}84 \times 10^{-5}$ of positive galactic flux for Neutrinos to block.

Positive flux from $4{,}17 \times 10^{6}$ inter-galactic negative photons during its $1{,}73 \times 10^{849}$ãt radius time has a $1{,}79 \times 10^{463}$¢ drive by $7{,}21 \times 10^{575}$ virgins that downsized to $1{,}84 \times 10^{-5}$ because of the lesser ability to block positive flux delivers $2{,}37 \times 10^{1034}$pu inter-galactic positive pressure to every Neutrino in its radius time.

Looking for an average internal inter-mattronal shadows we find it between 10^{776}ãr and $1{,}73 \times 10^{849}$ãr Neutrino radius at $4{,}16 \times 10^{812}$ãr mean distance. The internal standard shadows are $5{,}76 \times 10^{-362}$ for 10^{632}ãr coronas and $5{,}77 \times 10^{-389}$ for $3{,}16 \times 10^{618}$ãr sized mattron cores with an average of $1{,}82 \times 10^{-375}$. With a smaller mass based $1{,}73 \times 10^{-109}$ dilution on top, incoming flux harvest is downsized as active compression to $3{,}15 \times 10^{-484}$ per ãnna.

The $2{,}37 \times 10^{1034}$pu positive pressure brings $7{,}47 \times 10^{550}$pu compression versus $4{,}15 \times 10^{545}$pu radius time expansion. In this case the field is overly compressive by a factor of $1{,}8 \times 10^{5}$ upon a Neutrino structure. This is much larger than for a standard Electron and requires considerable adjustment and diminished size for a Neutrino to be perfectly balanced and to keep a standard internal mattron grid.

The difference is supplied by mattrons that let loose more primal emission as precursor virgins from mattron surfaces as free primal mass out between the grid mattrons which raises subservient emission and hydron expansion. The Neutrinos standard nature is permanently upset as its mattrons can no longer remain at standard sizes. The Neutrino adjusts at smaller size but it warrants a better illustration and as they only affect other hydrons across close distances and have less grid influence, I leave them for now.

Surface Retinue

In order to make sense of overall hydron structures, I must understand what surface retinue of mattrons is left after hydrons have been hammered into a standard suit by local virgin pressure. This could be anything outside the hydrons top layer after its main mattron population enters the lockup by inter-galactic flux. Fluctuations in an inter-galactic flux field are too small to affect a mattron population under lock-up more than marginally.

Although a hydrons resistance to compression emanates from the center, the standard state locks all its neutral mattrons in place throughout the hydron. The local virgin photon pressure that forces a 'standard state' from the surface is not solely center-wise and the lock-up of neutral hydron mass is therefore holistic; not center specific.

If local virgin pressure that was instrumental in herding neutral surface mattrons inwards and evened up internal inter-mattron distances in the hydrons outer layers while inter-galactic lockup got a grip, were totally lifted, it might at most cause a fractional surface expansion. It seems possible that minor compressions and expansions of the hydron surface take place from time to time but rarely enough to loosen grip of inter-galactic flux. When a rare maximum Neutron is split, this may entail some unbound mass of its central domain but this mass would join the center during the split.

I cannot see anything that can constitute an 'unbound' retinue except for singular mattrons. A standard hydrons top layer is not a boiling cauldron of local effects like a solar surface but a tranquil lattice. A neutral mattron finds it near impossible to be pried out of a hydrons surface. If it for any reason strays outside its 'diamond smooth' top layer and loses local flux exchanges, it forms a neutral precursor and is split electromagnetically deep in the lower hydron aura; destined to enter inter-galactic space in a hundred thousand years or so.

The only notable exception to 'standard' lock-up are the singular surface mattrons of any polarity that float on a hydron top layer surface.

Singular surface mattrons are, relatively speaking, few and far between and all of them are at various times exempt from the standard grip. The fact that all of them were initially exclusively in a hydrons dominant sign and adverse to its polarity, this changes quickly with time as photons continue to hit the standard hydrons in both signs.

A singular surface mattron in a hydrons *subservient* sign that wanders outside its top layer during weak incoming virgin pressure is repelled to light speed out through its aura by subservient emission; a singular *dominant* mattron is repelled by 10^7 times weaker emission. Both can be fully or partly endowed with inherent hydron coordinates (set by how long it has been part of the hydron); the local phenomenon of curved orbital paths due to inherent structural coordinates will be dealt with when relevant.

The Neutron

Through a long creation process from great sizes, embryos form loosely bonded grid pairs with feeble mutual repulsion and virgin backhand. Close hydrons communicate best with a nearest neighbor and are therefore often paired in doubles that have been pushed closer over time. Both hydrons of each neutral pair are unlikely to remain stable in a galactic grid unless if a neutral companion splits within a distance that fits the bill for the other neutral companions survival, in which case one of the pair can remain a Neutron while the other splits.

At the point in galactic evolution when local virgin pressure upon Neutrons is forever decreasing, while often also locally increasing and decreasing due to passing photons, splits are the norm. Unlike early hydron consolidation with gradual shrinkage and in step, collapses and massive local pressure beds for more individual local development.

When a pair of Neutron embryos has shrunk to a size that can set off electromagnetic splits in a breach, and they again face dropping pressure, they are unlikely to complete that cycle in exactly the same period of time. The more progressed Neutron is likely to lose its footing and blow its cover with great outpouring of photons before the instability of its Neutron companion has reached that critical stage.

The split of the first Neutron bathes the nearby Neutron in boosted virgin pressure from photons of both signs; the closer, the greater the pressure. Almost immediately, the un-split more unhurried companion Neutron will harvest an authoritative inwards virgin push that stabilizes its widening buffer zone or developing breach. If the Neutron companion is too far away, this virgin outpouring will merely give a temporary stability until the photons have departed whereupon the other Neutron is again forced to expand.

The massive photon emission would soon help move a Neutron and its Proton companion closer by casting massive temporary virgin backhand between them. Passing photons from a splitting neighbor and their emission from outside a leverage point compress the other Neutron to standard size and lock up its positive central domain to harness negative inter-galactic flux shadows. Although standard central state is achieved, a negatively polarized Electron shell will remain free of the bond forged by negative core shadows. Not only are the backhand shadows cast by the outer layer those of weaker positive inter-galactic flux but they are cast across the entire Neutron and become extensively diluted.

When the upheaval after its formation fades away and the local pressure subsides, the only recourse for a stable Neutron is for its outer layer to be held up against its central domain by virgin pressure alone.

There is no way to keep the Neutron unit intact by available flux pressure; it must be locally compressed by massive virgin pressure. Any major drop in that pressure destabilizes a Neutron companion and causes a split. Whether the virgins are negative or positive does not matter as most are harvested by the polarized mattrons in the outermost top layer. Without the lockup, the Neutron companion must be held at standard size by local virgin pressure from a close Proton companion, or else it will split.

After a long time of bonded association, the Electron shell can be blown off from a widening breach if its distance to the Proton is altered enough for its harvest of local virgin pressure to drop. How close the hydron pair needs be in order for a Neutron surface to be held in check by boosted virgin emission from a Proton companion should become clear when all facts are at hand.

The distance must allow steady local balance that prevents the composite structure from splitting after the galactic field settles. This is likely to provide a narrow stability window in close Proton vicinity where pressure upon its surface fits the bill. If it is close enough to obtain stability, it is likely that an already consolidated Electron passes the Neutron companion by in opening orbit to be banished to a larger stable orbit far beyond the baryon pair.

Of the three a hydrons paired in a local field, a Neutron generates weaker external emission of dominant strength in both sign. It is not that the other two hydrons behave as if the Neutron were not there but they are mainly 'aware' of it through complications that must be dealt with later.

A stable 10^{852}ãr Neutron is built of (almost) 10^{228} positively polarized mattrons at (almost) radius 10^{852}ãr with mattrons of both polarizations containing standard 10^{912} atoms. The $2{,}079 \times 10^{225}$ negatively polarized mattrons are held as a relatively thin top layer at standard distances solely by incoming virgin pressure.

Neutron emission in both signs out of its aura is identical to the negative dominant Proton emission with **$1{,}78 \times 10^{545}$ negative and positive** virgins escaping its surface during 10^{852}ãt radius time. The Neutrons ability to block high speed negative flux is equal to the Protons at **$5{,}62 \times 10^{230}$¢.**

The path of a passing positive flux atom through the average central point must cut twice through an outer layer which $2{,}079 \times 10^{225}$ mattrons make up its average blocking power of positive inter-galactic flux in accordance with the domain volume. Like a standard Electron, a standard Neutron blocks positive flux at **$9{,}13 \times 10^{228}$¢.**

The globular domain at the heart of a Neutron is a Proton that collectively emits allied positive virgins that are easily harvested by the positive adverse core flux fields in the negatively polarized outer layer. Their outwards linear push also serves as an expansion drive that is similar upon the top layer as upon the central Proton and the holistic Neutron.

For the Neutron as a whole, the 10^{-38} cut in harvested virgin pressure for center-wise compression is still valid. Like any hydron, it can be compressed holistically by greater virgin pressure or expanded by the lack of it. While the incoming pressure is sufficient, its layered quality is preserved with the same internal distances as between mattron within a Proton. However, expansion by the central Proton upon the outer layer gains additional push on top of its 10^{552}pu general repulsion drive and I use the square root of 10^{24} orbits by peeking virgins to account for this; namely 10^{12} fold to an effective 10^{564}pu radius time expansion drive.

To assess the compression required, upon the relatively thin surface skin of adverse polarity that can prevent a buffer zone in widening into a breach needs detailed analysis. There is a shave off from harvested virgin pressure that arrives from beyond the Neutrons leverage point works upon a relatively thin Electron skin; I earlier used the inverse of the 480 layer difference to shave off the virgin harvest that holds down top layer mattrons. A balanced standard Neutron that requires 10^{564}pu to resist its general expansion would then require $4,08 \times 10^{566}$pu incoming virgin pressure from beyond its leverage point or **$4,08 \times 10^{-286}$pu** pressure per ãnna in order to keep its buffer zone intact and prevent it from widening into a breach.

Already bound up by inter-galactic flux backhand, the Proton part of the standard Neutron is fully balanced. Below the thin negative Electron shell, the Proton would otherwise require holistic 10^{-300}pu central compression supplied by 10^{-38} of incoming virgin pressure harvested by the Neutrons top layer surface from beyond the Neutrons leverage point or a virgin harvest of 10^{-262}pu per ãnna.

Part 9: Gravity

Galactic Gravity

The root to Gravity is long distance flux from outside a galaxy; specifically the linear part of the original newborns drive away from its parent.

Galactic hydrons are subjected to the mutual backhand of linear snippets of imbedded linear push. Nothing is carried out of line or diluted. This tiny push is always linearly active in a hydron where it is eventually lodged and when flux goes missing in an opposite linear direction (from the other side of the galaxy), an affected hydron can be spatially moved. Gravity reveals itself when a large collection of hydrons prevents other hydrons from receiving this linear push by standing in the way.

A virgin activated outside the galaxy has a single original linear drive from outside the galaxy. The linear drive is divided upon over 10^{232} flux atoms that reach over 10^{463}¢ and the linear drive is geared down by interacting with primal mass in all galactic hydrons. This immense primal mass shower is focused into ever tighter orbits all over the galaxy and the 10^{232} flux atoms are blocked and divided upon all of the $8{,}38 \times 10^{68}$ galactic hydrons.

Of all these convoluted flux paths, only the linear drive can cast a locally coherent linear shadow as each tiny snippet remains faithful to its share of that original linear push away from a parent. Whereas all the orbital push by flux is annulled, the embedded collective linear push is not.

The actual shadow cast between all spheres of galactic hydrons (harvested by mattron cores) sets the 'dose' of linear backhand push prevented by every hydron of allied polarity in any direction upon all others and sets a stable linear flux shadow between galactic hydrons that is diluted by the inverse of the distance squared.

As hydrons of adverse polarity block the jumbled flux with 10^{42} speed difference, their mutual shadows create ineffective backhand snippets between them; embedded subservient linear push is not blocked at speeds that are blocked by the other hydron as dominant flux. The disappearance of flux blocked in the dominant sign does not register as lost backhand upon an adverse hydrons that is barely aware of the existence of such flux.

At standstill (without orbiting), a non-boosted backhand push can at some distance be balanced by virgin repulsion exchanged between hydrons with a presence augmented by $\sqrt{r}$ boost. And if either of the contrary winds has a stronger point effect, a specific orbital motion can bridge the unbalance at a given distance (although orbital bonds are usually formed by a single point attraction). With submists of solar or galactic sizes, escaping virgins are of course relatively few and external repulsion less significant across the vast distances, not least compared to the greater mass to be moved.

In the Primal Code, gravity (not repulsion) is a galactic mass shadow where bodies experience a backhand measured by mass 'm' that casts a relatively constant linear shadow upon any given distance sphere. Any mass caught in such a gravity bond is driven by flux backhand in a free-fall neutralized on an orbital path around a massive submist system.

A galaxies 'm' gravity effect upon an orbiting sun on some sphere at any distance from its average mass center fashions a backhand where the total sphere effect must be diluted by the distance sphere measured in hydron radii. To describe its mass by its gravity shadow, a sun during one whole orbit around the galaxy harvests a gravity backhand equal to $\sqrt{m}$.

The effective gravity pull by a large gathering of hydrons (such as a sun) would also be $\sqrt{m}$ upon an orbiting submist, whether this is a hydron or a planet. The submists point speed during its 'r' sized orbit will be continually hampered by the shifting focus which eventually neutralizes this backhand push by supplying a directly opposite push.

The average linear backhand drive can therefore at each point be set as the square root of the orbital period; so that at whatever distance, the speed of orbital displacement is set by $\sqrt{r}$. The net point speed in orbit (dropping all constants) that balances the actual backhand push will therefore alter with both distance and mass in accordance with $\sqrt{m} \div \sqrt{r}$.

The Gravitational Constant

The Gravitational Constant is a measure of an 'inherent force of attraction' working between two masses where 'G' equals $6,67 \times 10^{-11}$nt. The Newton unit of force is the force required to accelerate one kilogram of mass at the rate of one meter per second.

With the numbers of my ideal chain in hand, it is simple to estimate the value of the Gravitational Constant within the framework of the Primal Code, seeing that all such effects is this conceptual theory are derived merely from mass, speed and distance.

One gram on Earth weighs roughly $5,976 \times 10^{26}$ nuclei (hydrons) and each of these is on average built of 10^{228} standard mattrons. A kilogram therefore contains $5,976 \times 10^{254}$ mattrons and if each is built of 10^{912} primal atoms, an Earthly kilogram holds a primal population of $5,976 \times 10^{1166}$ atoms.

The force required to accelerate $5,976 \times 10^{1166}$ primal atoms across a distance of 10^{906}ār during a period of 3×10^{1194}āt demands a speed of equal to $3,33 \times 10^{-289}$¢. Such a speed can be achieved by colliding 2×10^{878} primal atoms at 1¢ birthspeed into a mass of $5,976 \times 10^{1166}$ primal atoms, which then equals the 'force' of one Newton. The Primal Code uses 'pus' or pressure units per second.

Accordingly, one Newton Force Unit equals the 'force' of 2×10^{878}pus.

For comparison, the linear non-boosted 3×10^{914}newborns that are emitted by each sign within a standard mattron per second equals $1,5 \times 10^{36}$nt in each sign.

The Gravitational Constant (G) of $(6{,}67\times10^{-11}\text{nt}) \times \text{meters}^2/\text{kg}^2$ quantifies the actually measured net gravity shadow at work between the mass of two one kilo masses at the Earth's surface. Here the effect is measured by two kilo masses, each built of $5{,}976\times10^{254}$ mattrons or $5{,}976\times10^{26}$ standard hydrons, placed one meter (10^{906}ãr) apart.

With G as $6{,}67\times10^{-11}$nt and one Newton as 2×10^{878}pus, G thus delivers $1{,}334\times10^{868}$pus (per kilo) '*backhand push*' across the distance of one meter.

As an opaque flux sphere of a certain blocking speed sets the hydrons gravity shadow, (not its mass), replace the mass with blocking value of the opaque hydron spheres within the one kilo weight; using a balanced 10^{1704}ã standard Proton or Neutron spheres.

The Nucleons harvest flux and casts exterior shadows with dilution shown by relating their spheres to the one meter sphere (10^{1812}ã divided by 10^{1704}ã) where each casts an individual gravity shadow that is diluted 10^{108} fold upon a Nucleon within the other kilogram mass one meter away.

This 10^{108} times weaker (than full shadow) is first multiplied by the number of hydrons casting the shadow and again by the number of hydrons harvesting the backhand push(missing out on harvesting the flux).

The Nucleon number casting the shadow in a one kilo mass is $5{,}976\times10^{26}$ and so is the number of Nucleons affected by the shadow at the receiving end in the other kilo with mass^2 value $3{,}571\times10^{53}$. This one kilo shadow, diluted by 10^{108}, is therefore $3{,}571\times10^{-55}$ part of one uncut hydron shadow.

The measured net effect of 'G' ($1{,}334\times10^{868}$pus) between two kilo masses is thus shown to equal $3{,}571\times10^{-55}$ of the full linear undiluted negative surface shadow that is cast by a single Nucleon.

The full original external linear shadow cast by each Nucleon that holds this one meter straight line on a hydron scale to be blocked by each of the other hydrons is therefore $3{,}736\times10^{922}$pus backhand pressure ($1{,}868\times10^{44}$nt since one Newton is 2×10^{878}pus). This is the shadow cast by each Nucleon before it spreads away from a hydron surface.

The virgin emission generated by 2×10^{9} inter-galactic photons per second counts 6×10^{923} virgins with a linear 6×10^{923}pus or rather $1{,}2\times10^{923}$pus push when reduced 5 times by the flux that dark matter harvests elsewhere before it reaches the one kilo masses. That this is twice the measured value does not overly bother me with all the reservations attached to this illustration.

The Gravitational Constant is simply the result of $5{,}976\times10^{26}$ full hydron shadows of $1{,}868\times10^{44}$nt each at source with total $1{,}116\times10^{71}$nt backhand to blocked stray pressure by each of $5{,}976\times10^{26}$ hydrons or a $6{,}67\times10^{97}$nt total backhand push diluted 10^{108} times across this distance.

The result is $6,67 \times 10^{-11}$nt, namely the value of 'G'.

Earth weights in at $5,97223 \times 10^{24}$kg with each kg containing $5,976 \times 10^{26}$ Nucleons or $3,569 \times 10^{51}$ Nucleons with an average 10^{1704}ã sphere. Earth's own radius is $6,372 \times 10^3$ kilometers, each measuring 10^{909}ãr and with $4,06 \times 10^{1825}$ã divided by 10^{1704}ã equaling $4,06 \times 10^{121}$, the average shadow that each Earth Nucleon casts upon another within the affected kilogram mass on Earth's surface is diluted $4,06 \times 10^{121}$ times.

To get the collective effect between the two masses, the end effect from a full hydron shadow is first multiplied by the number of hydrons casting this backhand value and then by the number of hydrons (not) harvesting it. Thus the $3,569 \times 10^{51}$ Earth Nucleons cast their $1,868 \times 10^{44}$nt shadow each, which gives Earth a full $6,666 \times 10^{95}$nt external average gravity shadow that is on average diluted by $4,06 \times 10^{121}$ across the Earth's radius to give it a backhand effect of $1,642 \times 10^{-26}$nt upon every Nucleon sized surface patch.

The result upon every one kilo mass situated at the Earth's surface is thus an effect of $1,642 \times 10^{-26}$nt harvested by the backhand of $5,976 \times 10^{26}$ hydrons (of the kilo mass); namely a net shadow of 9,8nt. It takes 9,8nt per second to support the mass of one kilo at the Earth's surface.

'G' must by necessity be measured by the greater negative inter-galactic flux input as blocked by the bigger dynamic spheres of opaque positively polarized hydron centers. The Electrons that block of the lesser positive inter-galactic flux with smaller spheres are paired in orbital bonds to heavier Protons and their lighter masses move along with them, making the Electron shadows much less relevant in measuring 'G'.

Nonetheless, a single free Electron at Earth surface is affected by all the collective Electron shadows of the positive flux these block within Earth. Neutrons block negative flux at the roughly same speed as Protons and positive flux at roughly same speed as Electrons.

Each of the $3,569 \times 10^{51}$ Earth Electrons (half free and half latent) cast a positive shadow by $1,631 \times 10^{1702}$ã dynamic spheres upon the $4,06 \times 10^{1825}$ã Earth sphere and an individual Electron on the Earth surface blocks the shadows from within that are diluted $2,489 \times 10^{123}$ fold.

One cannot directly derive the Electron share of positive intergalactic flux from 'G' but the inter-galactic input per second from $3,492 \times 10^{75}$ negative inter galactic photons upon $8,38 \times 10^{68}$ (free and latent) galactic Electrons or $4,17 \times 10^6$ for each supplies $1,25 \times 10^{921}$ positive virgins per second which at birthspeed deliver

1,25×10^{921}pus in linear backhand or 6,275×10^{42}nt to each of 3,569×10^{51} Earth Electrons. Their dynamic spheres cast a collective 2,24×10^{94}nt gravity shadow that is diluted by 2,489×10^{123} across the Earth's radius and leaves a 9×10^{-30}nt backhand push upon a single freely roaming Electron at the Earth's surface.

The individual 1,642×10^{-26}nt Proton backhand shadow is 3,284×10^{852}pus that is working upon its 10^{228} mattrons. The 9×10^{-30}nt Electron backhand shadow or 1,8×10^{849}pus working to push 2,083×10^{225} mattrons delivers a relative effect that is **1824** times weaker.

Gravity in the two signs does not affect primal mass in the same way; the perception that all 'mass' falls at the same rate in a gravity field is flawed. It affords a 479 times more massive Proton a backhand push that is roughly 1824 times more powerful than an Electrons backhand push which can be mistaken to mean that a Proton is 1824 times more massive when measured as primal mass falling in a gravity field.

An Electron harvests a total backhand divided upon all its mattrons but the assumption that the measured result depends on a mass 'number' we cannot see is flawed. It is the harvested backhand that sets the rate and the Proton is 1824 times 'heavier' in this respect only. The outcome may need some tweaking of numbers but to be consequent, I let this stand as a template illustration into which you can eventually fit the correct numbers.

Part 10: Auxiliary Photons and Coronas

Auxiliary Photons

All photons are singular mattrons. They have all been created through an electromagnetic split of a neutral mattron where the twins are sent off at light speed in opposite directions, one usually into space and the other into a hydron. At the time of the split, they carry all structural inherent coordinates of their ancestral neutral mattron and they can even carry all or part of the structural inherent coordinates of their ancestral hydron.

If the ancestral neutral mattron is part of an inherently spinning hydron, it will early on acquire a majority mass by exchanging flux that is inherently in tune with the hydrons common coordinates. On the other hand, if a newly arrived mattron of any type becomes part of a hydrons top layer, it gains an increasing fraction of the hydrons inherent spin coordinates after residing on the hydron surface for long enough.

The singular mattron can be moved by an enforced linear newborn push of the moment where the orbital component of that push is inherent to hydron spin while the linear push component is all new and not part of any inherent coordinates.

If any mattron on a hydron surface can lift, it will be pushed away from the hydron by repulsive virgin wind; on average from the hydron center. As the mattron heads out through the hydron aura, the part of its mass carrying inherent structural hydron coordinates depends on the length of its stay on the hydron surface. Most neutral mattrons have been residents on a hydron surface longer than have incoming photons that recently hit the hydron and would therefore have fully consumed coordinates when split in the aura; and if incoming from outside the aura, a neutral mattron would be split without structural hydron coordinates.

When a singular mattron hits a hydron, even when highly boosted in vast orbits, its side sweep within an original inherent spinfield and its linear enforced movement come to an abrupt stop as its coordinates are countered by new virgin coordinates transferred by the flux from 10^{36} coordination mattrons and eventually up to 10^{228} mattrons within the hydron. A 'collision' brings a photons movement within its original inherent spinfield to full mutual stop much like when a primal particle hits other primal particle, which is what actually happens, and the slate of the incoming photons inherent spinfield relative to the Universe is mostly wiped free.

A clean slate does not mean that the blocked photons mass is immediately in possession of any new structural hydron coordinates. These are gathered by its primal mass through internal harvests of exchanged primal mass from all its mattrons until these inherent coordinates of a common Universal spinfield constitute a majority of its primal mass.

For a recently blocked singular mattron, inherent hydron coordinates have not always been fully transferred by way of shifting out its primal mass when it is re-launched as a singular mattron and its inherent orbital movement is then in accord with an already transferred fraction of the hydrons inherent spinfield. Being singular mattrons, their launch, even when subservient at roughly light speed is different from an electromagnetic launch.

I call all launched or re-launched singular mattrons 'auxiliary photons'.

In a primal split by a parent particle, the 'away' movement is incumbent upon all directions of its inherent orbit (as perceived by a universal observer) where the orbital speed and linear away speed are equal inherent properties. An auxiliary photon is launched by an entire hydron by escaping input from all its inherent mattrons moving in a common spinfield.

The coordinates of various contradictory orbits blend but, given time, their common coordinates are communicated after individually spinning parents from spinning mattrons within a spinning hydron come at an exterior photon in an aura in primal orbits that are first resolved later by the photon. These mixed messages may affect its individual spin but the inherent hydron spin coordinates common to all these primal particles.

The launching speed of auxiliary photons is set by the harvest of enforced repulsion by virgins from fully inherent parents and these linear directional system spin coordinates become inherent to the photon in 10^{948}ãt. The repulsion virgins that drive it away at birthspeed deliver an orbital push component from the hydrons coordinate center in all Universal directions. Although communicated by a relatively insignificant hydron rotation, the photon is at this time to some extent one with those hydron coordinates.

An auxiliary photon can be fully in tune with hydron spin or only fractionally tainted by its coordinates. If it is recently blocked and rapidly re-launched, the auxiliary photon follows whatever part of the hydron spinfield coordinates it could harvest during its period on the hydron surface (in convoluted ways that we come to later). After launch, a fully inherent auxiliary photon eventually brings the photon to form an inherent orbit around the hydrons coordinate center at the distance by that variable harvest and eventual resistance.

Photons are lasting structures that all carry some levels of inherent system coordinates away from a hydron. Their collective repulsion is not as linearly precise as repulsion by an adverse electromagnetic twin of opposite spin away from a common coordinate center while they are joined at the hip. For auxiliary photons that have less than fully adapted to the emitting hydrons coordinates take even more convoluted paths. An auxiliary photon can be repelled by a hydrons subservient aural winds at light speed or by dominant virgin winds at roughly million times slower speed.

Auxiliary paths taken by mass within a fully or partly compliant auxiliary photon, driven away from a hydron follow the inherent coordinates of both hydron and photon spinfield that extend to the edges of the Universe and the photon experiences gradually boosted sweep by the component of inherent coordinates harvested on the hydron surface (a photons independent spin coordinates driven by its own internal emission become fully inherent after 10^{948}ãt). The collective hydron coordinates are communicated to its entire mattron population after resolved conflicts and a surface photon can gain a full or fractional coordinates ranging from linear to significantly curving.

An initially insignificant departure sweep increases from the aura by a multiple of linearly traversed hydron radii in accordance with the inherent coordinate component. When the orbital component eventually comes to equal its launch speed at some distance, an auxiliary photon forms an orbit around the hydron.

The away aim of its original repulsion has by then long become one with its inherent structural coordinates of the hydron spinfield and its outwards aim has become Omni-directional to that extent; a photon is therefore not brought to a halt in a closed orbit on that account.

The repulsion of an auxiliary photon away from the hydron derives from its aural drive and its inherent orbit does not cancel this outwards drive when it forms an orbit around the hydron unless there are contrary virgin winds out there that can resist that original inherent outwards drive it by backhand push. Constant linear backhand shadow across an orbital distance yields a linear push in a single shifting direction; this is not the case for an auxiliary photon which wholly inherent orbit is expansive in all directions incumbent on its orbit.

A fully inherent auxiliary photon departing with inherent hydron coordinates has its orbital sweep boosted by the distance traveled as measured in hydron radii. It can therefore gain exponentially superluminal orbital sweep at greater distances which comes as something of an unexpected shock but I can find no logic to refute it.

In the aftermath of galactic conversion when photons start to hit all types of hydrons at superluminal speeds, their speed outguns even the standard hydrons ability to resist subservient photons at light speed. The gargantuan hydrons stop superluminal photons at any speed with minimal effect and the photons lose most of their former coordinates to the coordination mattrons on the surface they hit but also generate new spin by internal emission. They can also gain fast inherent hydron spin coordinates if they stay long enough to harvest enough common hydron coordinates before re-launch.

Auxiliary photons in vast ever opening orbits have no limits on potential orbital reach away from a hydron. A perfectly linear photon crosses a 10^{932}år observable universe would not find it easy to hit a single one of 10^{78} baryons of 10^{852}år width on that trek (in a space populated by one baryon per 10^{2718}å cubic meter); their collective discs block 10^{-82} of that distance sphere.

Again, for a perfectly linear photon, a similar baryon mass collected into roughly 10^{22} suns of 10^{914}år wide discs block out all but 10^{-14} collectively on the same trek across the observable universe. The massive flux fields and multiple other photons between hydrons in solar coronas or chromospheres make it unlikely that a photon at robust superluminal speed can pass deep into a sun; much less to emerge unscathed on the other side like a runaway star crossing a galaxy.

Consider that photons emitted in opening orbits gain speed in orbit with fully inherent hydron coordinates; a Protons 10^{-288}¢ surface sweep (10^{-8} light speed) allows a fully inherent auxiliary photon 10^{72} fold light speed across the observable universe which jacks the hitting rate up 10^{72} fold. Repeated orbital paths at superluminal speeds radically increases photon hitting rates across large distances. The majority of re-launched photons hit the hydrons in solar coronas far earlier that that at slower superluminal orbital speeds. Here their coordinate slate is wiped free and they are re-launched in due course after collecting variable levels of inherent hydron coordinates.

After hitting a hydron, an auxiliary photon is carried along by its inherent top layer rotation. For a long time, this drive is enforced, not inherent, and when a singular mattron has been on a surface long enough to consummate majority mass of the hydrons collective inherent coordinates, it has become fully endowed within its inherent coordinate system of spin.

Let's use a standard Proton to illustrate with its 10^{-288}¢ inherent orbital sweep around its coordinate center. My crude rate for consummation of inherent spin in a submist was based on the standard mattrons 10^{948}ãt consummation and the vast Proton of 10^{228} fold mattrons consummates new collective rotation after 10^{1176}ãt.

I assume (for illustration) that any photon blocked by any hydron has taken to heart all its inherent spin coordinates after a hydrons consummation period of new collective spin; a period that allows its internal mass to collect majority coordinates from the hydron. For a baryon this happens after 10^{1176}ãt and for an Electron after $2,08×0^{1173}$ãt.

Sooner or later, auxiliary photons on a hydron surface experience a period of insufficient incoming virgin wind and are re-emitted. At that time, the harvested mass share of the hydrons inherent structural spin coordinates can range from insignificant coordinate mass to majority mass. If a photon is blocked by a Proton and re-emitted first after 10^{1176}ãt, it has gained all the Protons inherent spin coordinates that will corrupt its otherwise linear aim; its outwards path will be gradually curving in tune with its earlier path while on the hydron surface.

With its mass fully endowed with structural coordinates, it soon achieves an orbital speed to match the speed of its auxiliary launch which is greater than the speed of hydron rotation; at which point its orbit will form at that distance point. Although this speed is faster than that of the hydron rotation speed at surface, the distance is greater by the same factor; a fully endowed auxiliary photon makes each differential orbit in the same period as the main body hydron takes for each differential rotation.

When the incoming virgin pressure drops enough in one sign and not in the other; singular surface mattrons in that sign move individually out from a hydron surface. Singular mattron are less bound to a hydrons top layer than neutral mattrons in the grip of inter-galactic flux and rise smoothly out of surface contact and out through the aura. Due to their mutually effective repulsion, the two signs are not launched at the same time and cannot form a neutral auxiliary swarm in widening orbit around the hydron.

A photon swarm can eventually run into a heavy backhand from incoming virgin winds that can slow or break its expansion and eventually recall the auxiliary photons to a hydron surface. Subservient au-photons are emitted earlier by greater subservient winds and are generally held down by more meager virgin winds from paired hydrons of adverse polarity. Their inherent coordinates reach larger orbits before they can turn orbital which makes it harder for diluted backhand shadows to break their stride and bring them back to the hydron that launched them. Subservient orbital bonds are fully operational but they are more easily disturbed in ways different from their left behind hydron and the distant subservient auxiliary photons are mostly lost as they face exterior virgin winds.

Other rarer birds could be neutral mattrons that hit a hydron at boosted sweeps and are split inside or outside a standard aura. If the speed of approach is lower than light speed but fast enough to bring a precursor close enough for subservient headwinds to exceed the electromagnetic threshold, the neutral mattron is split. The dominant twin is (in principle) aimed into a standard hydron where all coordinate sweeps are neutralized while the subservient twin is repelled linearly at light speed.

As no approaching mattron can be stopped and repelled if its side sweep speed is greater than light speed, even incoming neutral mattrons at such speeds would, after an aural split, see both the split twins enter the hydron surface where their former neutered coordinates render a pair of auxiliary photons of opposite signs.

If a boosted speed of a neutral mattron are extreme, standard subservient hydron emission may not even be able to deliver the required repulsion in $10^{912}\alpha t$ to hardwire its neutral mass into a split if it (unlikely enough) passes too quickly through the aura before it joins the hydron surface where it will be balanced as a disparate neutral mattron.

Hydron Coronas

The hydron corona is a major field of interest as it affects almost every aspect of mutual hydron communication but herding their many fluid effects into a coherent conceptual illustration is easier said than done.

The original singular <u>negative</u> surface mattrons on the Proton surface can form a dominant fairly close cloud of its 2×10^9 au-photons that individually orbit its main body hydron by inherent coordinates at a similar distance. This is the Protons 'standard' corona in both numbers and distance.

The original singular <u>positive</u> surface mattrons on an Electron surface form a dominant cloud of 2×10^9 au-photons individually orbiting its main body at shorter distances; the 'standard' Electron corona.

A stable Neutron under the patronage of a Proton companion keeps 2×10^9 original singular positive surface mattrons and $4,17\times10^6$ singular negative surface mattrons and can under special circumstances launch its old and new stocks of auxiliary photons but this must be dealt with separately.

The adverse virgin emission by a 'permanent' relatively close dominant corona hits a hydron in boosted orbits. This incoming subservient coronal pressure can outgun the hydrons subservient emission by a large margin and though there are holes in that coverage, generally, subservient surface au-photons cannot be launched unless a dominant corona has first been grounded.

A standard Proton emits $1,77\times10^{-307}$ negative (dominant) virgins per ãnna and during $2,38\times10^{1135}$ãt, a 10^{752}ãr wide dominant auxiliary precursor will therefore block 10^{-200} of $4,21\times10^{828}$ virgin repulsion with $4,21\times10^{628}$pu repulsion upon 10^{912} atoms. The dominant auxiliary photon is driven out of the Proton aura by directional central virgins in $2,38\times10^{1135}$ãt at slow $4,21\times10^{-284}$¢ far below light speed. This speed is actually $2,10\times10^{-284}$¢ due to the half effect loss in the primal repulsion but to keep this all sizes purely relative with as few constants as possible, I also drop that one too.

The dominant corona has a great chance to become an enduring fixture when countered by incoming virgin winds (generated by a local Electron or other local entities) that create a backhand towards the Proton against its a soon diluted repulsion.

The $4,21\times10^4$ fold difference in $4,21\times10^{-284}$¢ speed repulsion speed gained in the aura by an auxiliary photon with full inherent coordinates compared to its earlier 10^{-288}¢ surface sweep reveals the distance to which it must travel in hydron radii before its inherent orbital speed becomes equal to its established $4,21\times10^{-284}$¢ Omni directional linear speed out of the aura.

For a standard Proton, the coordinates of dominant auxiliary photons fully adapted to inherent Proton coordinates after 10^{1176}ãt brings them to orbit the Proton after traveling $4,21\times10^4$ Proton radii to a lowest common coronal denominator at distance $4,21\times10^{856}$ãr and the trek out to that distance at $4,21\times10^{-284}$¢ takes 10^{1140}ãt same as the Protons rotation period. This does not mean that the 2×10^9 negative au-photons settle into non-widening orbits this far within the Protons $3,96\times10^{861}$ãr leverage point. Taming the widening corona on its outwards route thereafter can only be achieved by contrary virgin winds.

$$10^{852}ãr \div 4,21\times10^{856}ãr = 5,64\times10^{-10} \ (gain \ 4,87\times10^{21}) = 2,75\times10^{12} = 3,64\times10^{-13}$$

Across a stable dominant $4,21 \times 10^{856}$ãr wide corona, the $2,05 \times 10^{26}$¢ virgin boost with $4,87 \times 10^{21}$ orbital gain allows $3,64 \times 10^{-13}$ of the positive adverse coronal virgins from 2×10^9 au-photons to slip by its main body as external free positive emission of $\mathbf{7,28 \times 10^{-284}}$ virgins per ãnna (the Protons main body subservient emission counts $7,28 \times 10^{16}$ times fewer 10^{-300} virgins per ãnna).

For a standard Electron the inherent coordinates of dominant singular mattrons that have become fully adapted to inherent Electron coordinates after $2,08 \times 10^{1173}$ãt follow the same rules. The $1,277 \times 10^{851}$ãr wide Electron emits $6,39 \times 10^{-308}$ positive virgins per ãnna with $3,57 \times 10^{23}$¢ surface sweep where repulsion by $1,15 \times 10^{827}$ virgins during $1,8 \times 10^{1134}$ãt pushes a positive auxiliary precursor that harvests $6,132 \times 10^{-199}$ or $7,05 \times 10^{628}$ virgins on its way out through the Electron aura. The $7,05 \times 10^{628}$pu linear drive upon its mattrons 10^{912} atoms drives is on a slightly curving path at $7,05 \times 10^{-284}$¢. As in the Protons case, this dominant emission speed is far below light speed.

The $1,15 \times 10^3$ times difference between an Electrons $6,14 \times 10^{-287}$¢ rotation speed and the auxiliary photons $7,05 \times 10^{-284}$¢ emission speed, allows each to travel merely $1,15 \times 10^3$ Electron radii away to $\mathbf{1,47 \times 10^{854}}$ãr in $2,08 \times 10^{1137}$ãt (same as the Electrons rotation period) before the 2×10^9 au-photons form inherent orbits that continue to widen unless met by contrary virgin winds.

If this smaller coronal size seems insignificant; an atom orbiting Our Sun at over thousand solar radii is orbiting beyond Neptune.

$$1,277 \times 10^{851}\tilde{a}r \div 1,47 \times 10^{854}\tilde{a}r = 7,55 \times 10^{-7} \text{ gain } 1,05 \times 10^{22} = 7,93 \times 10^{15} = 1,26 \times 10^{-16}$$

Across a stable dominant $1,47 \times 10^{854}$ãr wide corona, the $1,21 \times 10^{25}$¢ virgin boost with $1,05 \times 10^{22}$ orbital gain allows $1,26 \times 10^{-16}$ of the negative adverse coronal virgins from 2×10^9 positive au-photons to slip by its main body as external free negative emission numbering $\mathbf{2,52 \times 10^{-287}}$ virgins per ãnna. The Electron main body negative subservient emission fields $7,06 \times 10^{13}$ times fewer $3,567 \times 10^{-301}$ virgins per ãnna.

A standard Neutrino that is originally without singular mattrons will later be hit by a few photons of both signs. These direct hits are rarer than for the larger hydrons as Neutrinos of variable mass could be million times smaller targets than baryons and, as they are often unpaired, they are less likely to cast heavy virgin backhand shadows.

Using 3×10^{218} mattrons at $1,73 \times 10^{849}$ãr standard size; some $4,29 \times 10^{-309}$ dominant positive virgins are emitted per ãnna of which an aural precursor harvesting $3,34 \times 10^{-195}$ or $1,43 \times 10^{-503}$pu. Dominant repulsion out of the Neutrinos $1,73 \times 10^{849}$ãr wide aura would take $1,1 \times 10^{1132}$ãt which allows an auxiliary positive photon precursor to harvest $1,57 \times 10^{629}$ virgins and generate a $\mathbf{1,57 \times 10^{-283}}$¢ drive that would ordinarily push the photon out of the aura in the same $1,1\times10^{1132}$ãt.

It is difficult for 'standard' Neutrinos to launch harvested au-photons with fully inherent mass (after 3×10^{1166}ãt); the Neutrinos $5,77\times10^{-282}$¢ inherent rotation speed at surface is already 36,75 times faster than its $1,57\times10^{-283}$¢ dominant auxiliary launching speed.

The inherent orbital coordinates of a fully compliant au-photon are already moving the photon faster sideways than enforced repulsion pushes it away. That the photon is already rapidly orbiting the Neutrino deep within its aura makes it more vulnerable but the less stable orbits can continue to widen. Little adverse auxiliary photon emission slips by the main body Neutrino as free external emission. Furthermore, as Neutrinos are unlikely to accompany other hydrons, they are probably unable to keep any coronas.

Generally about dominant hydron coronas when local grid pressures have subsided after Neutron splits, earlier mutual main body exchanges between hydrons are replaced by emission from 'standard' singular coronas that are adverse to hydron polarity. These dominant photons can be held at 'arms length' in differential orbits and they can often be recalled.

Again, for adversely polarized paired hydrons, adverse coronal exchanges cast backhand virgin shadows upon auxiliary photons in local orbit. These more numerous virgins that are adverse to a hydrons polarity but allied to local dominant auxiliary photons and pull at them more effectively than the subservient au-photons. A powerful freely escaping adverse coronal emission exchanged by hydrons causes darker backhand shadows where virgins work on coronas according to the auxiliary sign.

For hydrons of opposite polarity, mutual virgin exchanges push upon the other's auxiliary coronal mattrons while they leave allied coronas unaffected. Adverse coronal virgins that slip by a main body do not affect allied photons in the home corona or in auxiliary orbits around other local grid hydrons of the same polarity. They do however push auxiliary coronal photons orbiting local adverse hydrons linearly away by allied virgin winds; as they would the hydron itself.

As stable standard dominant coronas affected by virgins winds from an adverse bonded hydron can be orbiting far within their hydrons leverage point which causes the home hydrons main body to block the vast majority of such adverse virgin emission. A Proton and an Electron corona holds an equal number of dominant photons where each single auxiliary photon emits 10^{20} times the free subservient emission of an entire standard Proton; the coronal free virgin wind can be massive even when the virgin escape is heavily restricted by the main body hydron.

In the case of a surviving standard Neutron, after first being critically expanded and then heavily compressed when a nearby Neutron undergoes a split; it is normally paired to a Proton that has an Electron of its own. The collective emission out of a Neutron aura is equal in both signs at the level of a Protons subservient emission and its neutral virgin winds cannot cause electromagnetic splits of wayward neutral mattrons.

The compressed Neutron has had its positively polarized central (Proton) mass hammered into a standard suit and locked up by inter-galactic positive flux but its thin negatively polarized outer layer shell cannot be tied down by weaker positive inter-galactic flux as this backhand must reach through the main body Neutron mass and cast its shadows upon mattron cores on the Neutrons other surface side.

A standard Neutron, stabilized by a nearby Proton, remains in possession of its old harvest of 2×10^9 singular positive and $4,17 \times 10^6$ singular negative photons gathered as an embryo. The singular surface mattrons in its top layer were initially held up against the Neutron surface by general virgin pressure while local virgin winds from electromagnetic photons in both signs faded as they departed out of grid contact. A companion Proton is of course far more effective in blocking a Neutrons auxiliary launch in the positive sign than a Neutron can be in grounding auxiliary Protons launch in the negative sign (same as a Proton blocking Neutron launch in the negative sign).

The Neutron must ride out subsiding field pressure in close company of a Proton but can only do so if its thin negatively polarized outer surface receives in excess of $4,08 \times 10^{566}$pu incoming virgin pressure in 10^{852}ãt radius time or $4,08 \times 10^{-286}$pu per ãnna.

$$10^{852}\tilde{a}r \div 4,94 \times 10^{868}\tilde{a}r = 4,1 \times 10^{-34}(gain\ 4,49 \times 10^{15}) = 1,84 \times 10^{-18} \times 10^{-300}pu = 1,84 \times 10^{-318}pu \times 2,22 \times 10^{32} = 4,08 \times 10^{-286}pu$$

The massive **$4,08 \times 10^{-286}$pu** per ãnna virgin exchange required to stabilize a Neutrons top layer by a Protons boosted subservient main body emission is easily found at closer distances. At $4,94 \times 10^{868}$ãr distance, a Proton is within reach to stabilize the Neutrons top layer by its positive 10^{-300}pu main body emission boosted to $2,22 \times 10^{32}$¢ which delivers the pressure needed to keep it intact. Proton coronas can do this at greater distances but as coronas can be lost or recalled, main body emission is the only stable long term source. This does not mean that the two baryons can bond at $4,94 \times 10^{868}$ãr as there are myriad other effects like mutual repulsion that can disturb or prevent close bonds.

The Proton can later add its massive positive coronal emission on top of its main body input but all that must be dealt with when actual distances and bonds come into clearer focus.

Auxiliary Clusters

It is clear that, as chance would have it, that after continually harvesting auxiliary coronal mattrons they could at times happen to cluster in closer proximity before being re-launched. It is hard to know whether clustering is entirely down to chance as it depends on the considerable difference in how smoothly auxiliary photons are re-launched by a main body hydron from 'polar' areas of varying sizes or recalled under weak or massive virgin orbs.

At both ends of a coronal run there are parameters that can to some extent influence how auxiliary photons become distributed across the surface of a rotating hydron. Each end of a coronal run may 'take out' or cancel potential mutual effects. Under a powerful orb, a hydrons polar area may be heavily restricted with few auxiliary photons emitted at longer intervals as it takes the hydron longer to rotate them onto an small active re-launching area. On the other hand, those emitted under these circumstances are delivered more swiftly back into a hydron surface which hardly bring them closer together after a coronal pod run. I cannot see clustering as a significant effect while it should not be rejected out of hand

Conversely under a weak incremental virgin gravity orb recalls, especially when all orbs are weak, the hydrons polar areas are larger and rotation is quicker in putting auxiliary photons onto bigger unshielded areas where they can be re-launched. The intervals of launch are shorter with more auxiliary photons simultaneously in flight. On the other hand, they are delivered back into the surface of a rotating hydron after myriad orbits on a lengthy coronal run and could be more evenly distributed across the hydron surface.

Subservient Coronas

An allied singular mattrons of hydron polarity is not a disturbing element on a hydron surface. Marginally bound to its surface, singular allied mass in the subservient sign and hydron polarity keeps coronal size and has no core; just one among 10^{152} identical subservient coronas in a hydrons top layer. A potential auxiliary photon is lost in a sea of its coronal kind and perfectly at ease with its disparate coordination mattrons.

As most Protons are paired to an Electron with opposite sign exchanges, each hydron harvests more numerous dominant virgins while emitting more subservient virgins. This makes it easier to launch subservient au-photons by main body subservient virgin emission that outguns dominant repulsion ten million times. Rising into an aura when virgin winds cannot hold it down under a subservient launch window, the auxiliary photon forms a precursor; its smooth launch from an allied hydron surface causes no disturbance. As auxiliary launches in the two signs rarely coincide, there is no coronal blend of emission but the two auxiliary signs can alternate in forming coronas.

Although coronal compression is mostly more than sufficient to ground an auxiliary photon in the other sign, it covers incomplete parts of the hydrons surface; especially low flying standard coronas. Each au-photon can cover a tiny fraction with adverse emission of a hydron surface at each time. This spotty pressure can nevertheless deliver powerful average resistance upon a surface of a rotating hydron which can resist the launch from temporarily deficient surface patches. If there are any opposite orbiting coronal photons in orbit, a powerful adverse emission may blow them away but generally the compression from a singular corona is instrumental in resisting the launch of opposite singular mattrons unless that corona is first recalled.

Virgins boosted from close coronas cut too shallowly into the main body hydron to deliver much orbital compression; not least fast virgins from the 10^{888} primal precursors mass that exit its 10^{752}ãr borders with 10^{124}¢ exit sweep that is further boosted but delivers only 1¢ linear speed compression into the hydron surface.

The formation of Protons left them with a full pot of dominant negative singular surface mattrons and without any subservient singular mattrons. Their initial auxiliary coronas are dominant and it is first later after hydrons are hit by subservient auxiliary photons they may be unable to keep for long. Most subservient auxiliary photons are re-launched relatively swiftly during intermittent coronas after collecting fewer inherent Proton coordinates. On top of that, they are driven away at light speed by the 10^7 fold subservient emission (except on Neutron surfaces that repels by dominant emission in both signs).

Dominant coronas are generally built of auxiliary photons that have become fully endowed with inherent hydron spin coordinates. Subservient coronas are probably also built of more au-photons that to various degrees have had less time to gather inherent coordinates. In both cases, the way photons fractional harvests of hydron coordinates are repelled from hydrons is different from the fully endowed auxiliary photons and this will be dealt in due time. Fully endowed subservient coronas are repelled 10^7 times faster to further linear reach before their inherent sweep forms an orbit around the hydron. They are therefore also affected by highly diluted contrary virgin backhand that is mostly unable to slow their pace so that even fully endowed progressive subservient coronas are easily lost to the home hydron.

The subservient coronas are less cohesive and given to quick dispersal but, depending on distance, they keep up adverse freely escaping virgin pressure upon the home hydron for considerable time. Their highly boosted virgins can be blocked by multiple local hydrons and generate powerful backhand shadows. As these effects are extremely varied, I present no examples until specific circumstances of coronal communications are more firmly in hand.

The balancing acts between hydrons form a very sophisticated regulatory system that adjusts singular coronas by local and long distance exchanges in a Pandora's box of complicated intermittent and unstable balances that are difficult to fit into such a conceptual framework; pushing the envelope of my comfort zone in keeping this logical chain transparent.

A standard Proton emits 10^{-300} positive subservient virgins per ãnna of which a subservient aural precursor blocks 10^{-200}. During 10^{1132}ãt, the 10^{632} harvested virgins by a singular mattrons 10^{912} atoms drives it out from the Proton aura along a curving path at 10^{-280}¢ speed during 10^{1132}ãt. For a subservient auxiliary photon fully adapted to inherent Proton coordinates after over 10^{1176}ãt on its surface (a rare staying power), its inherent curvature comes to follow the Protons 10^{-288}¢ rotation speed at **10^{860}ãr** distance from its surface after heading outwards across 10^8 Proton radii.

The inherent orbital sweep at 10^8 Proton radii in orbit at 10^{860}ãr matches its light speed launching at **10^{-280}¢** and forms an inherent orbit at this distance where it clears one orbit at this and any distance thereafter in tune with the Protons 10^{-288}¢ surface rotation which takes 10^{1140}ãt.

$$10^{852}\tilde{a}r \div 10^{860}\tilde{a}r = 10^{-16} \ (gain \ 10^{20}) = 10^{-4}$$

Across a subservient 10^{860}ãr distance, the 10^{28}¢ boost of adverse virgins with 10^{20} orbital gain allows 10^{-4} of negative adverse virgins from each positive auxiliary photon to slip by the Protons main body as 10^{-284} free negative virgins per ãnna which may continue to increase. The fully inherent auxiliary photon forms a widening orbit that can only be stopped by contrary backhand virgin shadows.

One positive auxiliary photon delivers 10^{16} times more free negative virgins from this coronal distance than the 10^{-300} per ãnna positive virgins emitted by the main body Proton itself.

The 10^{860}ãr distance on forming the initial orbit is only just within the Protons $3{,}96 \times 10^{861}$ãr leverage point and the majority of adverse coronal emission is soon passing by the hydron as free field emission. Subservient coronas of a few auxiliary photons with full hydron coordinates can be firmly bound to their hydrons by way of inherent coordinates but as they were not ordinarily part of the hydron population for long, they form orbits at remote distances where the ability of backhand is seriously weaker and their orbits continue to widen at largely undiminished speed.

Subservient auxiliary photons can therefore rarely, even as a temporary phenomenon, be considered coronas; they mostly constitute a re-emission of a blocked photon (or multiple photons). They are also easily disturbed from below by dominant coronas after they travel too far to keep the dominant surface photons grounded.

A standard Electron emits $3,58 \times 10^{-301}$ negative subservient virgins per ãnna and a subservient aural precursor blocks $6,132 \times 10^{-199}$ or $2,2 \times 10^{-499}$. In $7,63 \times 10^{1130}$ãt, $1,68 \times 10^{632}$ harvested virgins upon its 10^{912} atoms drives it out of a $1,277 \times 10^{851}$ãr Electron aura along a curving path at $1,68 \times 10^{-280}$¢ speed in $7,6 \times 10^{1130}$ãt. For a subservient photon fully adapted to inherent Electron coordinates after $2,08 \times 0^{1173}$ãt on its surface (a rare staying power), the curvature brings it to orbit the Electron after $2,74 \times 10^{6}$ Electron radii at $\mathbf{3,5 \times 10^{857} \tilde{a}r}$ where its inherent orbital sweep from $6,14 \times 10^{-287}$¢ Electron rotation comes to equal its $1,68 \times 10^{-280}$¢ directional repulsion speed. It can of course only form stable non-widening orbits at this distance if held back by a contrary virgin wind backhand that can wear down the expansion.

$$1,277 \times 10^{851} \tilde{a}r \div 3,5 \times 10^{857} \tilde{a}r = 1,33 \times 10^{-13} \text{ gain } 2,16 \times 10^{20} = 2,87 \times 10^{7} = 3,48 \times 10^{-8}$$

Across a stable $3,5 \times 10^{857}$ãr wide coronal distance, the $5,92 \times 10^{26}$¢ virgin boost with $2,16 \times 10^{22}$ orbital gain allows $3,48 \times 10^{-8}$ of the negative coronal virgins from each subservient positive au-photon to slip by the Electrons main body as $\mathbf{3,48 \times 10^{-288}}$ free external negative virgins per ãnna. Again, the single negative au-photon delivers roughly the same number of free positive virgins at this distance as would a fully stocked dominant positive corona ($2,52 \times 10^{-287}$ per ãnna); it delivers $8,77 \times 10^{16}$ times more free positive virgins than a main body Electron collectively emits of subservient negative virgins ($3,567 \times 10^{-301}$ per ãnna).

Far within the Electrons $3,3 \times 10^{860}$ãr leverage point distance, a subservient fully endowed Electron corona faces more backhand than a subservient Proton corona but the conditions dictate whether an auxiliary subservient photon is recalled.

From a standard $1,73 \times 10^{849}$ãr wide Neutrino, $2,4 \times 10^{-304}$ subservient negative virgins are emitted per ãnna and an aural precursor will harvest $3,34 \times 10^{-195}$ of those or $8,02 \times 10^{-499}$. During $4,64 \times 10^{1129}$ãt, the $3,72 \times 10^{631}$ harvested virgins push upon 10^{912} atoms and drive a precursor out of a $1,73 \times 10^{849}$ãr wide aura at $3,72 \times 10^{-281}$¢ during $4,64 \times 10^{1129}$ãt.

The Neutrinos main body rotation period of 3×10^{1130}ãt is $6,46$ fold its launch period. The $6,46$ difference between a Neutrinos $5,77 \times 10^{-282}$¢ rotation speed and $3,72 \times 10^{-281}$¢ repulsion of fully adapted subservient auxiliary photon after over 3×10^{1166}ãt on its surface requires it to travel across $6,46$ Neutrino radii to $1,12 \times 10^{850}$ãr in $4,64 \times 10^{1129}$ãt where its auxiliary curvature normally forms a widening orbit at $3,72 \times 10^{-281}$¢.

$$1,73 \times 10^{849} \tilde{a}r \div 1,12 \times 10^{850} \tilde{a}r = 2,39 \times 10^{-2} \text{ gain } 1,64 \times 10^{22} = 3,92 \times 10^{20} = 2,55 \times 10^{-21}$$

Across a stable $1,12\times10^{850}$ãr wide 'corona', a $1,06\times10^{23}$¢ virgin boost with $1,64\times10^{22}$ orbital gain allows $2,55\times10^{-21}$ of positive adverse coronal virgins from each negative au-photon to slip by its main body as $\mathbf{2,55\times10^{-301}}$ virgins per ãnna in free positive emission. The $2,4\times10^{-304}$ negative virgins in main body subservient emission are $1,06\times10^{3}$ times fewer than from each positive subservient auxiliary photon the Neutrino blocks and launches (its creation left the Neutrino mostly without singular mattrons).

$1,73\times10^{849}$ãr$\div$ 10^{895}ãr$=2,99\times10^{-92}\times$ no aid $\times$ $(6,96\times10^{-279})= 2,08\times10^{-370}$ free virgins at $3,16\times10^{45}$¢ $=6,57\times10^{-325}$pu

For a subservient au-photons to travel $6,45$ radii out to $1,12\times10^{850}$ãr and orbit just outside the aura at $3,72\times10^{-281}$¢ demands virgin backhand that is too weak to pull it back which is likely in the average 10^{895}ãr hydron grid where small unpaired Neutrinos do not cast dark virgin shadows and are not likely to hold on to subservient auxiliary photons. The big grid contributors of negative backhand virgins are Electrons that deliver $6,57\times10^{-325}$pu to standard Neutrinos across a 10^{895}ãr grid which represents $2,74\times10^{-21}$ of their $2,4\times10^{-304}$ negative repulsion per ãnna.

If a Neutrino, against the odds, forms a subservient positive $1,12\times10^{850}$ãr wide corona, the $1,06\times10^{23}$¢ virgin boost with $1,64\times10^{22}$ orbital gain allows $2,55\times10^{-21}$ of its adverse positive virgins to slip by its main body and the $2,55\times10^{-301}$pu negative free emission from each of a few harvested auxiliary photons makes the Neutrinos external overall emission less charged as its negative main body emission is $2,4\times10^{-304}$ virgins per ãnna.

Given that Neutrinos can be of highly diverse sizes with varied coronas, there is little point in pursuing these less than mainstream entities and I forthwith drop them from this illustration.

A standard Neutron has no way to launch a subservient corona and can at best launch auxiliary surface photons of both signs into its smaller dominant corona.

Part 11: Hydron Exchanges

Virgin Gravity Orbs

I want to set up a couple of illustration to show the nature of exchanges between fully formed standard hydrons. Look at a hydron pair where local virgin backhand is sufficient to keep all singular surface mattrons from being pushed into auras by the dominant surface repulsion.

The main body Proton emits 10^{552} positive virgins during 10^{852}ãt radius time or 10^{-300} per ãnna. Setting an Electron and Proton at average 10^{894}ãr apart, the greatest exchange (without coronas) of subservient main body virgins sweeps at 10^{45}¢ for every linear step to gain $10^{939}\alpha$ presence before leaving that local 10^{2676}ã volume.

As the majority of the virgins survive in orbit across the inter-hydronal distance, main body Proton emission is $1{,}277 \times 10^{551}$ virgins in $1{,}277 \times 10^{851}$ãt Electron radius time. Though boosted to 10^{45}¢ across 10^{894}ãr the virgin only makes $1{,}277 \times 10^{2}$ orbits while it moves out across an Electron wide orb and this augments the linear $1{,}63 \times 10^{-86}$ hitting rate of $1{,}277 \times 10^{551}$ virgins to allow $2{,}08 \times 10^{-84}$ to hit the Electron with $2{,}66 \times 10^{467}$ positive virgins bearing upon it in radius time, delivering $2{,}66 \times 10^{512}$pu orbital pressure.

All the positive Proton main body emission harvested by an Electron works as linear repulsion in accordance with $2{,}66 \times 10^{467}$ virgins in Electron radius time. The virgins arrive in large inter-hydronal orbits at 10^{45}¢ to repel its main body linearly away from the Proton by $2{,}66 \times 10^{467}$pu ($2{,}08 \times 10^{-384}$pu repulsion per ãnna). The orbital push also generates a local positive virgin backhand shadow that cannot affect a negative Electron corona.

Virgin compression cannot counter the Electrons $4{,}567 \times 10^{550}$ radius time expansion but as the disparate mattron masses of all Electrons are held in the grip of positive galactic flux, the Electron needs no extra compression.

A collective $2{,}66 \times 10^{512}$pu Electron backhand is created by the $2{,}66 \times 10^{467}$ blocked positive virgins *that arrive exclusively sideways into the Electron on an Electron thick sphere membrane in vast inter-hydronal orbits at 10^{45}¢.*

A virgin gravity orb is a highly defined sphere membrane that a virgin slavishly follows; it is not random at all. It always arrives into the blocking hydron from all directions equally but these are all parallel to the virgin sweep. Correspondingly, the shadow to all the blocked virgins can cause no backhand except out along the parallel directions on a globular 'membrane' which has the thickness of the blocking hydron. It is only within that orb that each virgin is moved in orbit around the Proton in all directions at that distance (aside from its 1¢ linear speed).

The $2{,}66 \times 10^{512}$pu backhand shadow cast by harvested virgins follows the contours of blocked virgin coordinates; the virgin shadow reaches all around (and even beyond) the emitting Proton on an Electron thick orb. Its backhand is exclusively effective within and along this relatively thin globular shadow 'membrane'. This means that the dilution of the shadow is by simple radius (in hydron radii) and not by that radius squared.

The 1¢ 'away' mode of a virgin carries it outwards during each whole orbit divided by 10^{45}¢ boost and it moves outwards across 10^{849}ãr during each full orbit around the Proton across 10^{894}ãr. This widens the orb and dilutes its backhand shadow across the orb by extra $6{,}82 \times 10^{-3}$ of the $1{,}277 \times 10^{851}$ãr Electron radius. As virgin coordinates harvested by the Electron widen the potential backhand shadow in each virgin orbit by $6{,}82 \times 10^{-3}$ of the Electron radius, each virgin makes on average $1{,}277 \times 10^{2}$ orbital passes around the Proton along the Electron thick orb membrane between the time enters the orb until it has moved beyond that orb thickness.

As only those virgins that hit the Electron cast a backhand shadow away from it and $2{,}08 \times 10^{-84}$ of the virgins emitted by the Proton hit the Electron, few of the $127{,}7$ orbital passes by a virgin bring it even close to the Electron. Individual virgins that indeed hit it, do so on a path crossing the Electron while most other passes on the orb are not remotely close to it.

The $2{,}66 \times 10^{512}$pu collective shadow is cast sideways along the lines of the globular orb membrane away from the Electron which generates a linear **$2{,}08 \times 10^{-339}$pu** per ãnna surface backhand. This main body virgin shadow falls upon a positive singular mattron on the Electron surface when Electron rotation brings it under the orb. A positive au-photon in the Electron corona is exclusively affected when its coronal orbit brings it to cross that Electron wide orb at its relatively great coronal distance.

Due to the Electrons surface curvature, there is a difference between the collective gravity shadow cast by virgins arriving in a parallel manner on an orb and the surface *'attraction'* upon mattrons (by the same blocked virgins). The virgin that for example hits a surface mattron at a straight angle in a mid orb of Electron thickness and delivers full linear top layer pressure while blockage leaves a full backhand shadow. Another parallel virgin that hits a surface mattron at the inner or outer edge of the Electron thick orb has a tangential inwards thrust into the globular Electron surface which ultimately delivers no inwards pressure at the 'poles' while the blockage of the virgin still leaves a full backhand shadow out along the virgin orb.

The (spot) *surface pressure* by $2{,}66 \times 10^{512}$pu push of the of the collective $2{,}66 \times 10^{467}$ virgins ($2{,}08 \times 10^{-339}$pu collective attraction per ãnna) works fully in the middle of the virgin gravity orb but has dwindling inwards push closer to the inner or outer edges of the orb as virgins hit closer to the 'poles'. At right angles to the gravity orb, there is no inwards push at all. Again, the inwards push is different from the projected shadow by blocked effect.

On any spot under the mid gravity orb generated by the paired Proton, the Electrons $8,16\times10^{543}$ dominant positive radius time emission ($6,39\times10^{-308}$) is $3,07\times10^{31}$ times more powerful than the collective $2,08\times10^{-339}$pu 'attraction' for the entire surface (dropping all constants). The singular positive surface mattrons can therefore easily be repelled from anywhere on the surface and be sent off as auxiliary photons with precursor halos.

A parallel virgin to an Electron wide orb that is delivered tangentially into a surface mattron at the 'pole' pointing towards the Proton delivers only 1¢ linear speed as pressure into the hydron. At the opposite pole along the orb edges at the far side pointing from the Proton, it yields 'no' inwards pressure but pushes a mattron linearly instead away from the pole at 1¢. Between the 'poles and the equator', surface pressure delivered by blocked virgins yields a gradual center wise aim spanning from tangentially 1¢ to the full 10^{45}¢ boost; the most effective pressure loss being at the two 'poles'.

The collar of the circular virgin gravity backhand orb encircles the Electron where the orb plane meets its main body from all parallel directions. Gravity backhand to blocked virgins, whether under the orb or elsewhere, performs rather like a cardboard cutout that blocks out light.

The Electrons positive main body virgin emission resists the incoming positive surface backhand with repulsion that is equally effective anywhere on the Electron surface and not limited to the directions of the virgin gravity orb (when ignoring all other possible virgin gravity orbs).

During one rotation of the Electron, a singular surface mattron receives the greatest surface 'attraction' under the mid virgin orb where it receives full (spot) compression during part of the Electrons differential spin cycle until the rotation brings that surface area to a perfectly right angle to the gravity orb that its surface spot traverses for a short while.

This may seem academic under circumstances with a Proton 10^{894}ãr away where repulsion wins the surface tug of war between opposite virgin winds anyway with a $3,07\times10^{31}$ margin but adverse hydrons can move closer while robust launch of singular positive surface mattrons can continue as before at right angles to the main virgin gravity orb.

Negatively polarized neutral mattrons are all held to the standard surface in the iron grip of galactic flux. If an Electron surface holds negative singular mattrons, their fate depends on what negative virgins orbs are present to hold them down against the Electrons powerful subservient emission.

The lost polar effect between paired adverse hydrons does not mean that singular positive surface mattrons positioned at the perfect 'poles' receive no surface attraction; they receive virgin backhand from multiple orbs of both signs from other hydrons stationed a bit further away in the general grid from where fewer of their virgins arrive due to greater dilution.

Across an average 10^{895}är hydron grid and successively further away in the grid, bigger and ever more numerous orbs without activation by galactic flux hit a hydron from any direction on other virgin orbs than those of the paired hydrons at higher speed while greater dilution renders them weaker.

A Proton at 10^{895}är that emits $1,277\times10^{551}$ virgins in an Electrons $1,277\times10^{851}$ät radius time across ten times the distance is diluted 100 times but modified by the 3,16 fold boost to 31,6 fold dilution (and successively weaker on other orbs from further away). Singular positive surface mattrons are pushed into the aura on a linear path from a 'polar' area a bit more weakly than under the mid gravity orb.

In this tug of war between surface attraction generated by main body Proton emission across the grid and local Electron repulsion, the Electrons entire surface is without any area where attraction under the virgin gravity orb outguns its dominant surface repulsion.

It does not take much to jack up the number of virgins in a local gravity orb, for example if a Proton acquires a corona or by emission from a passing photon in the local field. Under heavy virgin orbs, singular surface mattrons run into heavy headwind and their launch into a dominant corona is prevented under the orb.

If two adverse hydrons draw much closer, their two coronas can easily recall an orbiting auxiliary corona from under the heavy perfectly defined orb that both mutually generate and keep it grounded (if not at right angles to it).

With a grounded corona, the hydron rotation brings a recalled auxiliary surface photon from under the powerful dominant main virgin gravity orb that recalled it; occasionally moving it onto a right angle polar position where it only meets weaker dominant emission orbs from across the hydron grid.

Here the au-photon may again face incoming dominant pressure that is insufficient to resist its launch. In other cases of heavier orbs, it may be brought to a spot of absolutely weakest resistance after multiple hydron rotations from where the auxiliary surface photon may at last be launched. While the incoming virgin winds in both signs eclipse hydron repulsion across its entire surface, no auxiliary photons can be launched.

Now turn this illustration around:

A main body Electron emits $3,567\times10^{-301}$ negative virgins per änna or $3,567\times10^{551}$ virgins during the Protons 10^{852}ät radius time that all survive the trek across maximum 10^{894}är to the paired Proton. Of these survivors, aided by 10^3 orbits, 10^{-81} or $3,567\times10^{470}$ virgins boosted to 10^{45}¢ hit the Proton sphere with

$3{,}567{\times}10^{515}$pu negative pressure and create a virgin backhand that is limited to a Proton thick orb.

A collective $3{,}567{\times}10^{515}$pu backhand is created by the $3{,}567{\times}10^{470}$ blocked negative virgins that arrive exclusively sideways on a Proton thick membrane at 10^{45}¢ and casts a $3{,}567{\times}10^{-337}$pu sideways shadow attraction per ãnna along globular orb membrane lines of Proton thickness around and away from the Electron surface. Such main body virgin shadows fall upon negative singular surface mattrons as Proton rotation brings them under the orb.

Incoming backhand surface attraction of **$3{,}567{\times}10^{-337}$pu** per ãnna works fully in the mid virgin gravity orb but delivers dwindling inwards push the closer it hits to the opposite orb edges at right angles (the 'poles'). In the mid gravity orb (from and around the paired Electron but cast by the Proton), a singular negative mattron would take wings by the Protons $1{,}77{\times}10^{-307}$pu dominant negative repulsion that is $4{,}96{\times}10^{29}$ times stronger than the orbs incoming thrust at surface.

Similar to the Electron, the Proton also harvests lesser virgin pressure from many more diluted orbs from across an average 10^{895}ãr hydron grid and further away. This polar surface pressure attraction is also $31{,}6$ times less than under the mid orb and therefore too weak to prevent the Proton from launching its singular surface mattrons as auxiliary photons.

It is clear that singular surface mattrons can be launched out through the aura and away from the surface on linearly curved paths from anywhere on the surface by both Electrons and Protons at 10^{894}ãr apart when taking only mutual main body hydron emission exchanges into account.

Virgin gravity orbs are not source dependant. For example, free photons of any kind generate the same kind of virgin gravity orbs that work in more or less exactly the same way for any hydron that blocks them.

Coronal Exchanges

Free adverse virgin emission represents a massive rise in the power of local virgin gravity orbs cast by grid hydrons. Multiple orbs of various but weaker strengths from further away in the average grid cover an unpaired hydrons surface in an overlapping pattern that cannot always completely cover the right angle launching areas with sufficient pressure to hold down singular surface mattrons in areas of least resistance at right angle launching sites.

The pressure of diluted orbs across a standard 10^{895}ãr grid generates weak virgin backhand that cannot prevent launch of auxiliary photons (launches of subservient photons are resisted even more weakly). Relatively stationary far off local virgin gravity orbs of varying strength will restrict the launch of auxiliary photons at right angles from rotating hydrons to variable degrees.

Standard dominant coronas which expansion beyond the standard distance has been stopped is the mainstay of hydron communication. I am using $4{,}21\times10^{856}$ãr wide dominant Proton coronas and $1{,}47\times10^{854}$ãr wide Electron coronas, each stocked with auxiliary photons that are adverse to the hydrons polarity and in fully inherent coronal orbits.

A dominant standard Proton corona of 2×10^9 negative auxiliary photons releases **$7{,}28\times10^{-284}$ free** positive virgins per ãnna from inside its leverage point and this emission is $7{,}28\times10^{16}$ fold the Protons collective and free main body emission of 10^{-300} positive subservient virgins.

A dominant standard Electron corona of 2×10^9 positive auxiliary photons releases **$2{,}52\times10^{-287}$** negative virgins per ãnna from well inside its leverage point or $7{,}04\times10^{13}$ fold the main body Electrons $3{,}58\times10^{-301}$ collective negative subservient emission.

Standard dominant coronas in differential orbit are perfectly globular structures. Coronal photons of fully inherent coordinates follows the hydrons inherent field coordinates at any distance. At the coronal distance, launched auxiliary photons pass first and then periodically through the backhand of a local virgin orb of main body hydron thickness which is by then a small part of its differential orbit.

As a photon passes first through a 'stationary' virgin gravity orb at launch speeds, it is affected under the orb by a backhand drive towards the main body hydron until it passes out of the hydron thick and continues in orbit. During the backhand to incoming harvests under virgin gravity orbs from locally paired adverse hydrons this 'attraction' towards a main body hydron saps the photons outgoing speed.

Depending on the orb power, a complete stop to coronal expansion can happen gradually across further distances or it can happen immediately at the standard distance. At the coronal distance, there is little other resistance to an inherent movement in widening orbit and as the auxiliary photon will pass repeatedly through the local virgin gravity orb after its outwards path is cancelled, it is gradually brought closer to the home hydron; eventually all the way back to its surface. Often such recalls take place in incremental steps during innumerable orb passes but closely paired hydrons can allow recall in a single great jump.

There is an important property of virgin gravity orbs at any distance that leaves a heavy footprint; repulsive virgin emission and gravity orb backhand shadows are diluted at a different rate.

Whereas main body virgin emission is evenly distributed upon a distance sphere the inverse square law in hydron radii, the backhand push under the orb loses effect by the inverse distance in hydron radii (non-squared). This delivers a vastly more focused backhand attraction from a small slice of that faraway sphere but none for the rest of the coronal orbit.

At any distance, the width of a two-dimensional orb shadow is like a paper cutout area of a hydron that casts it. A coronal au-photon follows its orbital circumference at multiple hydron radii; the dilution of the virgin orb shadow upon a distant coronal precursor during passage through an orb is by the simple inverse of the distance measured in hydron radii.

When an auxiliary coronal photon orbits its hydron (relative to the universe), it must pass periodically through the local stationary virgin gravity orb at great distance from its hydron. The less diluted backhand push towards the hydron can at that distance exceed the hydron repulsion under the orb (even if it does not at the hydron surface). After orbiting through the relatively thin orb at the coronal distance, the backhand will be absent for the rest of the orbit.

If the 'attraction' after each passage is massive enough, an auxiliary photon continues to move closer to the hydron; if the orb 'attraction' falls short of the launching speed, the auxiliary photon continues to move further away from the hydron in opening orbit. The outwards auxiliary photon speed is slowing while its inherent orbital speed is inherently accelerating.

Any new repulsion effect upon an auxiliary photon during its full orbit is directional in a great number of universal directions where each has its own counter directions that neutralizes that small effect. This differs from the two orb directions (each other's counter directions) as spatial shift of a photons orbital elements under the orb delivers radically bigger effect. Repulsion of an fully inherent auxiliary photon or superior backhand causes any shift from or towards the hydron to slow its inherent speed in orbit at closer distances and speed it up at further distances.

Unlike a planet falling round a sun against a gravity shadow (where orbital speed increases if it is to stay in permanent orbit), a push that brings an au-photon closer to the center of its inherent spinfield; slows its orbital speed without outside forces forcing it to do so. Additional virgin winds harvested on its way inwards are mostly unable to resist its approach.

Excessive backhand works under an orb linearly in a universal direction towards the hydron center on each passage through the orb. The inherent orbit brings an auxiliary photon in and out of the hydron thick orb at launch speed at the standard coronal distance but faster as the orbit widens. After the orbital expansion is brought to a stop, every new gained linear motion towards a main body hydron on passage through an orb by the implanted backhand 'attraction' comes gradually to aim further to the side of the main body hydron on leaving the other side of the virgin orb; consequently the realized attraction speed lasts longer than merely during the orb passage.

I assume that the average linear attraction towards the hydron continues to work as linear universal motion after each passover by a square root of the orbital coronal distance measured in hydron radii. This will bring the auxiliary photon closer to its hydron on each passage in longer incremental steps than can be achieved during the passover period alone. This linear drift is limited to a period shorter than the auxiliary orbital period (at superluminal speed).

When backhand attraction is greater under an orb, it cannot be adjusted by the more spread main body repulsion and the auxiliary orbit shrinks. An orbiting auxiliary photon is then pushed slightly closer to the main body hydron on each of intermittent passages through the orb and withdrawal is not neutralized by resistive repulsion. The orb thickness becomes an ever greater part of each orbit at a slower speed as the auxiliary photon orbits closer and each new withdrawal step is bigger that the last.

As inherent coordinates slow auxiliary photons in orbits on drawing closer by the full distance drop (multiple of hydrons radii), the orb passage slows at closer quarters and darker attraction works for greater part of a shortening orbit. The resistance by emitted main body virgins does however increase and dilution of attraction and repulsion turns more equal. But even when a constant diluted dominant main body repulsion upon an auxiliary precursor (under an orb) gains the upper hand upon an already lowered corona, the repulsion cannot re-expand the orbiting corona.

In coronal exchanges between bonded adverse hydrons, a smaller coronal size restricts the rate of free external emission that can slip by each hydrons main body. The closer an auxiliary corona orbits, the less free emission fuels these inter-hydron exchanges. When the orb from a paired adverse hydron is generated by similarly shrinking corona, paired hydrons deliver a mutually weakening backhand exchange of virgin gravity orbs.

Generally, auxiliary photons are more easily withdrawn to hydron surfaces when adversely polarized hydrons draw closer. Due to difference in coronal sizes, repulsion and orb backhands, dominant coronas of paired hydrons may not shrink in perfect sync. For example, a part-ways shrinkage of a small Electron corona may deliver orb drive to a larger Proton corona which marginally slows its outwards widening orbits. The first million orbits may leave almost unbridled dominant auxiliary launch speed but the continuing slowdown with more freely exchanged virgin emission from beyond leverage points would rarely allow expansion beyond hydron retrieval.

There is a correlation between the strength of a virgin recall orb that tells us that, in spite of lesser dilution under the orb, it can at best marginally break the outwards auxiliary pace when not significantly stronger than a hydrons repulsion at surface. It's backhand at the standard coronal distance must be able to match its surface launch speed after a first few orbits or the corona will continue to expand in inherent orbits until the collective slowdown brings it to heel.

There is a small possibility that a weak orb that is recalling an auxiliary photon is stronger than the hydron repulsion in orbit at standard coronal distance (due to the lesser orb dilution) while it is weaker than the hydron repulsion at the surface. This may occasionally cause coronal withdrawal to seize part-ways from a hydron and create a lower perpetual corona with less free emission but such an outcome would be exceptional.

If a massive virgin gravity orb is constant and non-proactive, like orbs that are generated by fixed emission from say a passing electromagnetic photon, a hydron corona can be recalled to its surface without going through any reciprocal adjustments. Under massive orbs, an auxiliary photon can be so heavily attracted from orbit on passing through its first thin virgin gravity orb that it is brought back to the hydron surface in a single jump before it can exit after crossing the hydron thick orb once.

When an auxiliary photon is pushed inwards across the event horizon into a surface lattice of disparate mattrons, they simply lose their precursors and take on either coronal or core size according to polarity. In these examples, the auxiliary photons are already in full possession of inherent hydron coordinates but if wanting, they continue to augment the stock of inherent coordinates gained during earlier surface visits while on the surface and subsequent re-launches are in full accord with their latest share of inherent coordinates.

Coronal Size Formation

Set aside that all hydrons appear mutually repulsive and assume than an wayward Proton and a wayward Electron are approaching within a hydron grid at lesser than the grids average distance while exchanging ever more free coronal virgins by darkening virgin orbs.

As Electrons are in reality split away from Protons, this approach would be the other way around from greater to closer mutual distance which induces a similar pattern of shifting exchanges in reverse. Keep in mind that this is not quite symmetrical and there are difference in approaching or departing adverse hydrons caused by initial inputs that affect the general outcome.

In the following illustration of a long distance approach, assume that the general grid pressure has long subsided and can no longer keep singular surface mattrons grounded. Electrons and Protons can launch auxiliary photons that form a local corona at some specific distance set by incoming orbs. The permanent potential coronas are initially singular in one sign only; positive for the Electron and negative for the Proton.

The distances between two exchanging hydrons dictates a great deal about coronal formation and stability. To make coronal exchanges conceptually clearer, a handful of snapshots of dominant coronas orbiting a Proton and an Electron at shrinking distances are given as examples. To show all the ongoing changes, a billions snapshots would not suffice; making do with several snapshots demands that a reader uses a mind's eye to link up the effects between the handful of presented stepping stones.

Remember always, and this is important, that two hydrons with coronas are two equipages separated in space by greater distances than we can easily imagine; the coronas are each hydrons local phenomenon. As entities, they affect each other's coronal emission by way of communicating by mutually harvesting virgins allied to a variable stock of auxiliary photons that are locally orbiting them at each time.

We start with an illustration of a wayward Proton and Electron that are approaching across 10^{894}ãr apart (one tenth of the average hydron grid distance) with complications that make this a less than a simple setup. Approach of unpaired hydrons from far distances is actually an abnormal setting but the handful of successively smaller distances will now be used as stepping stones down to 10^{870}ãr apart to exemplify inter hydron behavior.

The first difficulty is setting a starting point for their initial coronal reach but there is a general rule that governs this according to circumstances of immense variety. These dictate where an expanding Electron corona under a Proton generated orb actually stop expanding and start to be recalled again; and vice versa. As adverse coronal exchanges across enormous distances are reciprocal, one must set up a conceptual rate of local coronal reach as well as rules for the periods of incremental coronal shrinkage required after the expansion of the local inherent coronas has been brought to heel.

True to form, I use a extremely rough approximation for coronas that reach beyond standard size, including those that reach outside the leverage point from where auxiliary photons gain virgin emission that is fully free of the local main body hydron. This ballpark figure is a brutal arbitrary illustration where two expanding dominant local coronas expand at launch speed under the inter-hydron orb at the standard coronal distance against an insufficient resistance.

If the launch speed outguns the potential pullback speed induced by the local backhand, the auxiliary photon reach further across a distance 'β' that is a multiple of standard coronal radii. The expansion of the fully inherent coronas would continue outward across relatively extensive distances while a weak backhand to virgin exchanges chips at the original launch speed that is not replenished after that aural launch.

The coronal effect exchanged between the two hydrons of adverse polarity across 10^{894}ãr shows that the released dominant coronas will first move out across their respective auxiliary standard sizes and keep moving outwards from there due to the weak incoming orb resistance.

'β' stands for the difference is launch speed at the coronal distance and the potential counter speed at that standard distance where an auxiliary photon starts to orbit the hydron. This backhand resistance slows its outwards thrust on a first passage under the orb at the standard coronal distance and in every subsequent larger orbit thereafter at a dropping rate due to both dilution and the accelerating inherent speed of an auxiliary photon orbiting through the orb.

Remember; there is only one correct way to calculate coronal reach under various circumstances (and this is probably not it).

Subservient Auxiliary Launch

Before continuing, let us take a look at the launch of auxiliary photons that are **subservient** and launched by a hydrons greater subservient main body virgin emission. Take for example a subservient launch by an Electron at $\mathbf{10^{894}}$**ãr** from a Proton to explain the difficulty by negative auxiliary photons launched at $1,68 \times 10^{-280}$¢ to form a stable local coronas.

The Protons positive coronal emission is entirely impotent in resisting a negative auxiliary photon rising from the Electron surface and cannot deliver any effective orb backhand. However, the Protons negative main body emission of $1,77 \times 10^{-307}$ virgins per ãnna delivers a weak negative backhand orb across to the Electron equipage (ignoring that there can often be darker negative coronal emissions from across the average grid).

$1,277 \times 10^{851}$ãr $\div 10^{894}$ãr $= 1,63 \times 10^{-86}$ *(gain $1,277 \times 10^{2}$)* $= 2,08 \times 10^{-84} \times 1,77 \times 10^{-307} =$
$3,68 \times 10^{-391}$pu $\times 10^{45}$¢ $= 3,68 \times 10^{-346}$pu *vs* $6,39 \times 10^{-308}$pu $= 1,74 \times 10^{38}$ *fold repulsion*

Across 10^{894}ãr, these negative virgins hit a single Electron boosted to 10^{45}¢ and cast a $3,68 \times 10^{-346}$pu shadow backhand out from the Electron surface.

10^{752}ãr $\div 1,47 \times 10^{854}$ãr $= 4,63 \times 10^{-205} \times 1,15 \times 10^{3} = 5,32 \times 10^{-202}$

The $3,68 \times 10^{-346}$pu negative backhand shadow generated by the Protons virgin orb and cast from the Electron surface, diluted to $5,32 \times 10^{-202}$ at the Electrons standard coronal distance, delivers $1,96 \times 10^{-547}$pu negative orb backhand.

A fully inherent subservient auxiliary photon starts to orbit the Electron at $1,47\times10^{854}$ãr and passes at $7,05\times10^{-284}$¢ speed through the $1,277\times10^{851}$ãr Electron wide orb in $1,81\times10^{1134}$ãt where a weak negative $1,96\times10^{-547}$pu pull blocks $3,55\times10^{587}$pu negative virgins and could in principle drive the photon towards its Electron at $3,55\times10^{-325}$¢. Against a $1,68\times10^{-280}$¢ linear outwards speed, this falls $4,73\times10^{44}$ times short and the auxiliary photon continue orbiting outwards with almost no effective resistance.

In this case the reach of an auxiliary photon away from the Electron before its expansion could be stopped by weak rapidly fading backhand at β fold standard coronal measure is $4,73\times10^{44}$ fold $1,47\times10^{854}$ãr standard at **$6,95\times10^{898}$ãr**. No recall is to be had as this distance which is larger than the average hydron grid where myriad other and contrary backhand shadows are in play.

This example of a fully inherent subservient auxiliary photon shows there is no way for an ordinary subservient auxiliary photon from either Electron or Proton to have its outwards linear speed slowed more than marginally in local orbits and verifies that any subservient auxiliary photon that carries a lesser fraction than all of the hydrons inherent coordinates, has its outwards speed even less linearly impeded while orbiting outwards.

The launch of a subservient auxiliary photon can therefore essentially be viewed as the emission by a hydron of a single photon at light speed. It there are other subservient photons present on a hydron surface, the first launch may be followed by other subservient photons to follow the first (if they can be rotated across a polar area while the launch window remains open during the time at hand). Nevertheless, each launch is an unrelated occurrence, albeit tainted by partly common coordinates.

Subservient photons blocked by a hydron spend dissimilar periods on its surface and gain various parts of its inherent coordinates. With few common coordinates, they can be re-launched quickly but for example under active adverse coronas they can be kept on the surface long enough to harvest full inherent coordinates. Neither case is enough to break their stride after they are launched at light speed.

As coronas go, we are solely concerned with dominant photons of fully inherent hydron coordinates bound to their hydron after a re-launch at slow dominant speeds. Dominant photons are often blocked and re-launched with partial share of inherent coordinates but if they cannot be recalled to turn into permanent residents, recently blocked dominant photons escape further afield, often at superluminal orbital speeds, to be blocked by other hydrons.

The following comments on hydron coronas refer therefore exclusively to dominant auxiliary photons that are periodically recalled from local coronas and that gradually gain a full share of their inherent coordinates.

Coronal Starting Points

We must have some neutral starting point so this setup shows a wayward Proton and Electron on approach from afar with fully stocked extensive local coronas built of fully inherent dominant auxiliary photons. Let's assume for illustration only that the coronal reach is to respective leverage point for both hydrons which then start simultaneously to mutually communicate by fully free virgins; an arbitrary setup to be sure, but one that shows the general thrust of adverse exchanges between the different hydrons.

The Electron corona is therefore orbiting at its $3,3 \times 10^{860}$ãr leverage point from where roughly 10^{-271} negative virgins escape per ãnna while the Proton corona orbits at its $3,96 \times 10^{861}$ãr leverage point which allows roughly 10^{-271} positive virgins to freely slip by the main body Proton per ãnna.

$$1,277 \times 10^{851}ãr \div 10^{894}ãr = 1,63 \times 10^{-86} \ (gain \ 1,277 \times 10^2) = 2,08 \times 10^{-84} \times 10^{-271} =$$
$$2,08 \times 10^{-355}pu \times 10^{45}¢ = 2,08 \times 10^{-310}pu \ vs. \ 6,39 \times 10^{-308}pu = 3,07 \times 10^2 \ fold \ weaker$$

One can on bases of the Electrons local $3,3 \times 10^{860}$ãr wide coronal distance with terminated auxiliary expansion, estimate the local backhand speed and the period for the recall to the Electron surface of its dominant auxiliary photons under a the virgin orb generated by a distant Proton.

$$10^{752}ãr \div 3,3 \times 10^{860}ãr = 9,18 \times 10^{-218} \times 2,58 \times 10^9 = 2,37 \times 10^{-208}$$

Proton input 10^{-271}pu with $2,08 \times 10^{-310}$pu positive local backhand cast by the Electron which after local dilution to $2,37 \times 10^{-208}$ with $4,93 \times 10^{-518}$pu backhand at the $3,3 \times 10^{860}$ãr distance where each auxiliary photon has been brought to a linear stop but is orbiting at $1,58 \times 10^{-277}$¢ superluminal speed after $6,14 \times 10^{-287}$¢ inherent Electron surface rotation has accelerated $2,58 \times 10^9$ times outwards in fully inherent orbit.

At a constant $3,3 \times 10^{860}$ãr coronal distance, the $4,93 \times 10^{-518}$pu backhand pulls back at a precursored dominant auxiliary photon orbiting the Electron as it passes at $1,58 \times 10^{-277}$¢ superluminal speed through the $1,277 \times 10^{851}$ãr Electron wide orb in $8,08 \times 10^{1127}$ãt. A backhand by $3,98 \times 10^{610}$pu harvested positive virgins under way, drive it in the direction of the local Electron at $3,98 \times 10^{-302}$¢.

As the average linear attraction towards a hydron continues to work as linear universal motion after each passover by a square root of the orbital coronal distance measured in hydron radii, it should add $\sqrt{2,58 \times 10^9}$ extra linear working time after each passover to the active $8,08 \times 10^{1127}$ãt passage whereupon its $3,98 \times 10^{-302}$¢ linear drift lasts for $4,10 \times 10^{1132}$ãt. The backhand drift moves the auxiliary photon $1,63 \times 10^{831}$ãr closer to the local Electron on each passage or by

3,26×10^{831}ãr twice in each orbit which is 9,88×10^{-30} of its 3,3×10^{860}ãr wide local corona in each orbit. Using my shorthand method of furthest orbits, falling towards the Electron in 1,01×10^{29} orbits where each requires 2,08×10^{1137}ãt which makes for a relatively swift **2,10×10^{1166}ãt** recall period into the Electron surface (5,97×10^{-29} seconds).

Upon the Proton in the other direction, one can on bases of its local 3,96×10^{861}ãr wide coronal distance with terminated auxiliary expansion, estimate the local backhand speed and the period for recall to the Proton surface of dominant auxiliary photons under the Electron generated orb.

10^{852}ãr÷10^{894}ãr =10^{-84} (gain 10³) =10^{-81}×10^{-271} =10^{-352}pu×10^{45}¢=10^{-307}pu vs 1,77×10^{-307}pu. The orb is 1,77 times weaker than repulsion under the mid orb

As negative virgins reach across to the vast field to the Proton at 10^{894}ãr, boosted to 10^{45}¢, aided by 10³ orbits, some 10^{-81} hit or 10^{-352} virgins that generate 10^{-307}pu in negative surface pressure. This local orb casts a virgin backhand shadow focused out along a wedding band set by the areal cutout of the Proton on Proton thick globular orb reaching all around the Electron. Locally, this orb is 1,77 times weaker than repulsion from the Proton surface and negative auxiliary photons can rise from anywhere on its surface.

10^{752}ãr÷ 3,96×10^{861}ãr= 6,39×10^{-220}× 3,96×10^9 = 2,53×10^{-210}

An Electron generated input with 10^{-307}pu local negative backhand cast by the Proton is diluted locally to 2,53×10^{-210} with 2,53×10^{-517}pu backhand at the 3,96×10^{861}ãr distance where its corona is brought to linear stop. With a Protons 10^{-288}¢ inherent surface rotation the re-launched auxiliary photon is accelerated 3,96×10^9 times outwards to 3,96×10^{-279}¢ superluminal speed in its local fully inherent coronal orbit.

At the linearly stopped 3,96×10^{861}ãr coronal distance, the 2,53×10^{-517}pu backhand pulls back at the precursored dominant auxiliary photon locally orbiting the Proton as it passes at 3,96×10^{-279}¢ superluminal speed through the 10^{852}ãr wide Proton orb in 2,53×10^{1130}ãt with 6,4×10^{613}pu backhand harvest of negative virgins that drive towards the Proton at 6,4×10^{-299}¢.

As the average linear attraction towards a hydron continues to work as linear universal motion after each passover by a square root of the orbital coronal distance measured in hydron radii, it adds √3,96×10^9 extra linear working time after each passover to the 2,53×10^{1130}ãt passage and allows the 6,4×10^{-299}¢ linear drift to continue for 1,59×10^{1135}ãt. This moves an auxiliary photon 1,02×10^{837}ãr closer to the Proton on each passage or twice by 2,04×10^{837}ãr in each orbit which is 5,15×10^{-25} of its 3,96×10^{861}ãr wide corona in each orbit. Using my shorthand of furthest orbits, falling locally towards the Proton in $1,94 \times 10^{24}$ orbits where each requires 10^{1140}ãt leads to **$1,94 \times 10^{1164}$ãt** recall period ($6,47 \times 10^{-31}$ seconds) into the Proton surface.

The Electron corona is here recalled during $2,10 \times 10^{1166}$ãt or roughly hundred times slower than the Proton corona in $1,94 \times 10^{1164}$ãt; a result of arbitrarily placing the coronas at different leverage points. If, on the other hand, you start with two grounded coronas, the speed of dominant auxiliary launch from an Electron surface can bring the auxiliary photons out to its closer leverage point and full free emission 100 times faster than the Protons auxiliary re-launch can.

Coronal Formation by Adverse Hydrons
10^{894}ãr

There is no question here that the hen preceded the egg; that coronas are launched before they could be recalled. The myriad exchanges thereafter can put almost any coronal constellation on the table when two adverse hydrons meet and the outcome can turn out radically different. Opting for the hen rather than the egg, this illustration shows two hydrons as they enter closer from the average 10^{895}ãr grid distance and start simultaneously to launch dominant auxiliary photons as they approach across 10^{894}ãr.

There are other weak orbs incoming across the wider 10^{895}ãr grid and the weak main body emission of both hydrons across 10^{894}ãr but are looking at how the local coronas progress under each other's growing virgin input while they expand outwards in auxiliary orbits.

It is essential to use a mental picture that shows the relatively immense distance between the two hydrons equipages and not to confuse this with the local reach of their coronas. If we were in our world to consider a hydron of football size set by the reach of its corona and equal in width to its leverage point, the approaching football of the other hydron would be outside the observable universe.

It complicates coronal formation when using the rules for a steady virgin orb input that the exchanges are mutual and rising. When auxiliary photons reach standard sizes, the exchanged virgin orbs have not reached a potential orb strength that ultimately shapes their corona.

It is therefore necessary here to use a two step approximation where β at their standard coronal reach (first by the faster Electron) sets the potential reach for both hydrons. The mutual coronal orb input from the standard reach is weaker than the input from a final reach against the weak emission. For a better picture, use the average of these two orb inputs as a steady orb at the standard reach for both hydrons which yields the ultimate reach β that sets their ultimate coronas.

The output of free virgins by auxiliary photons that can rise anywhere from both surfaces is heavily restricted on reaching standard coronal reach. The faster expanding Electron corona reaches $1,47 \times 10^{854}$ãr standard coronal size in $2,08 \times 10^{1137}$ãt at $7,05 \times 10^{-284}$¢ while the local corona in the Protons distant equipage at $4,21 \times 10^{-284}$¢ lags behind time wise as its auxiliary photons have by then reached out across $8,76 \times 10^{853}$ãr or $2,08 \times 10^{-3}$ of its $4,21 \times 10^{856}$ãr standard coronal reach.

The free emission of the expanding Electron corona at the standard reach is $2,52 \times 10^{-287}$ virgins per ãnna while free virgin emission out of the Protons immature $8,76 \times 10^{853}$ãr corona is limited to $1,44 \times 10^{-291}$ virgins per ãnna.

The first step in finding the Electrons new coronal size are the first effects of the $1,44 \times 10^{-291}$ virgin input from the Proton, This cannot stop the coronal expansion of its newly launched auxiliary photons at the standard coronal distance which expansion continues with increasing free coronal emission.

$$1,277 \times 10^{851}ãr \div 10^{894}ãr = 1,63 \times 10^{-86} \ (gain\ 1,277 \times 10^{2}) = 2,08 \times 10^{-84} \times (1,44 \times 10^{-291})$$
$$= 3 \times 10^{-375}pu \times 10^{45}¢ = 3 \times 10^{-330}pu \ vs \ 6,39 \times 10^{-308}pu = 2,13 \times 10^{22} \ fold\ weaker$$

The positive virgins hit the single Electron boosted to 10^{45}¢ across the vast empty space of 10^{894}ãr and, aided by $1,277 \times 10^{2}$ orbits cause $2,08 \times 10^{-84}$ of them to hit and the 3×10^{-375} blocked positive virgins cast 3×10^{-330}pu shadow per ãnna out from the Electron surface; an area-wide backhand wedding band curving as an inconceivably vast globular orb all around the Proton.

The outwards pace of the Electrons dominant photons is marginally slowed as they orbit through the orb at the $1,47 \times 10^{854}$ãr standard distance. Other orbs from beyond 10^{895}ãr may deliver greater pressure inputs from different angles but these are ignored in this illustration.

$$10^{752}ãr \div 1,47 \times 10^{854}ãr = 4,63 \times 10^{-205} \times 1,15 \times 10^{3} = 5,32 \times 10^{-202}$$

The 3×10^{-330}pu positive backhand shadow generated by the Protons limited virgin orb and cast from an Electron surface is diluted to $5,32 \times 10^{-202}$ at the Electrons standard coronal distance and delivers $1,6 \times 10^{-531}$pu orb backhand upon each auxiliary precursor.

A fully inherent dominant auxiliary photon at $1,47 \times 10^{854}$ãr has started to orbit the Electron, passing at $7,05 \times 10^{-284}$¢ speed through the $1,277 \times 10^{851}$ãr Electron wide orb in $1,81 \times 10^{1134}$ãt where the $1,6 \times 10^{-531}$pu pull blocks $2,9 \times 10^{603}$pu positive virgins under way which drives it in the direction of the main Electron at speed $2,9 \times 10^{-309}$¢. This cannot stop its $7,05 \times 10^{-284}$¢ expansion speed that meets it head as a resistance speed that falls $2,43 \times 10^{25}$ times short and allows its auxiliary photon to continue orbiting outwards.

The $7,05 \times 10^{-284}$¢ launch speed is $2,43 \times 10^{25}$ fold the potential $2,9 \times 10^{-309}$¢ inwards speed generated on its first passage under the local orb. According to my approximation with β as $2,43 \times 10^{25}$, the expansion reach of auxiliary photons is brought to a full stop and starts to be reversed after $2,43 \times 10^{25}$ standard $1,47 \times 10^{854}$ãr wide coronal measures at distance $\mathbf{3,57 \times 10^{879}}$**ãr**. This vast local coronal reach is far outside its $3,3 \times 10^{860}$ãr leverage point with 2×10^{-271} free coronal emission instead of its $\mathbf{2,52 \times 10^{-287}}$ virgins per ãnna from the standard distance or $7,94 \times 10^{15}$ times greater input.

The problem with this is that the Proton coronas virgin input has also been increasing and, as a consequence, does not allow this wide expansion of the Electron corona. We can now look at the situation when the Proton corona reaches standard $4,21 \times 10^{856}$ãr which it does in 10^{1140}ãt. At that point the Electron corona has expanded at $7,05 \times 10^{-284}$¢ to merely $7,05 \times 10^{856}$ãr and its free emission input has risen to $\mathbf{2,64 \times 10^{-283}}$**pu**.

As to the distant Proton equipage across 10^{894}ãr this $2,64 \times 10^{-283}$pu virgin input per ãnna is the first step in the Protons coronal development.

10^{852}ãr ÷ 10^{894}ãr = 10^{-84} *(gain 10^3)* = $10^{-81} \times 2,64 \times 10^{-283}$ = $2,64 \times 10^{-364}$pu × 10^{45}¢ = $2,64 \times 10^{-319}$pu vs $1,77 \times 10^{-307}$pu. *The orb is $8,85 \times 10^{11}$ times weaker*

As these negative virgins reach across to the vast field to the Proton at 10^{894}ãr, boosted to 10^{45}¢, aided by 10^3 orbits, some 10^{-81} of them hit it or $2,64 \times 10^{-364}$ virgins that generate $2,64 \times 10^{-319}$pu negative surface pressure.

10^{752}ãr ÷ $4,21 \times 10^{856}$ãr = $5,64 \times 10^{-210} \times 4,21 \times 10^4 = 2,37 \times 10^{-205}$

The $2,64 \times 10^{-319}$pu negative backhand shadow generated by the Electron orb and cast by the Proton surface is diluted locally to $2,37 \times 10^{-205}$ out to the standard coronal distance and delivers $6,26 \times 10^{-524}$pu orb backhand.

When a re-launched fully inherent dominant auxiliary photon starts to locally orbit the Proton at $4,21 \times 10^{856}$ãr, it passes at $4,21 \times 10^{-284}$¢ through its 10^{852}ãr wide orb in $2,38 \times 10^{1135}$ãt. During that time the $6,26 \times 10^{-524}$pu pull allows it to block $1,49 \times 10^{612}$pu negative virgins under way which drives it in the direction of the home Proton at $1,49 \times 10^{-300}$¢. This local backhand cannot stop the head on $4,21 \times 10^{-284}$¢ expansion speed with a resistance that falls $2,83 \times 10^{16}$ times short and an auxiliary photon continues to orbit outwards.

The Proton corona launch at $4,21 \times 10^{-284}$¢ is $2,83 \times 10^{16}$ fold the resistance by a potential $1,49 \times 10^{-300}$¢ inwards speed generated on first passage under the orb.

With β as $2,83 \times 10^{16}$, the reach of the re-launched auxiliary photon is terminated and reversed after $2,83 \times 10^{16}$ standard coronal measures of $4,21 \times 10^{856}$ãr at a distance of **$1,19 \times 10^{873}$ãr**. This distance is also far outside the Protons $3,96 \times 10^{861}$ãr leverage point and would potentially allow full free coronal emission in return across the vast field towards the Electron.

The problem is that the Electrons coronas increased virgin input stood in the way of this vast β expansion of the Proton corona. We can look at the situation when the Proton corona reaches across its $3,96 \times 10^{861}$ãr leverage point which it does at $4,21 \times 10^{-284}$¢ in $9,41 \times 10^{1144}$ãt. At that point the Electron corona has expanded at $7,05 \times 10^{-284}$¢ to $6,63 \times 10^{861}$ãr and its free emission input has risen to the coronal majority of **2×10^{-271}pu** and both hydron equipages are now emitting at their greatest possible input.

The rough method used here is to take the average of the two respective coronal orbs (standard reach and fully outside leverage point inputs) and use that value to limit the initial value at respective standard reach.

Upon the Proton, the average virgin input is $(2 \times 10^{-271}\text{pu}) \div (2,64 \times 10^{-283}\text{pu})$ or $7,58 \times 10^{11}$ times darker $2,3 \times 10^{-277}$pu orb. The Protons initial β $2,83 \times 10^{16}$ is thereby downsized to β $3,73 \times 10^{4}$ times its standard $4,21 \times 10^{856}$ãr corona and to an actual $1,57 \times 10^{861}$ãr corona reaching to 0,4 of its $3,96 \times 10^{861}$ãr leverage point with marginally downsized free emission of $1,96 \times 10^{-272}$pu.

> 10^{852}ãr $\div$ $1,57 \times 10^{861}$ãr $= 4,06 \times 10^{-19}$ *(gain $2,52 \times 10^{19}$) =one in 10,23 of 2×10^{-271}*
> *$=1,96 \times 10^{-272}$pu*

Upon the Electron, the average virgin input is $(1,44 \times 10^{-291}) \div (2 \times 10^{-271}\text{pu})$ or $1,39 \times 10^{19}$ times darker $2,88 \times 10^{-281}$pu orb and the initial β $2,43 \times 10^{25}$ is thereby downsized to β $1,75 \times 10^{6}$ times its standard $1,47 \times 10^{854}$ãr corona to an actual $2,57 \times 10^{860}$ãr size that is 0,78 of its $3,3 \times 10^{860}$ãr leverage point with slightly less than full free emission of $1,02 \times 10^{-271}$ virgins per ãnna.

> *$1,277 \times 10^{851}$ãr $\div$ $2,57 \times 10^{860}$ãr $= 2,47 \times 10^{-19}$ (gain $7,96 \times 10^{18}$) =one in 1,97*
> *of 2×10^{-271} $= 1,02 \times 10^{-271}$pu*

There is no way to make holistic illustrations; to show the final outcome in a single picture. The outcome is woven of many strands that must be explained separately to gain an understanding of what is happening. So keep in mind that these numbered snapshots of adverse hydrons passing in total isolation are shown with other important effects notwithstanding. The mutual effects of other strands will later be shown to radically alter these outcomes in different ways; so consider these strands less than the whole truth.

Mismatching Recall Periods
10^{894}ãr apart

So here we are with two hydrons mutually approaching across 10^{894}ãr as the two large local coronas are close to fielding full free 2×10^{-271} emission; the Proton $1{,}96\times10^{-272}$ virgins and the Electron $1{,}02\times10^{-271}$ virgins per ãnna.

> $1{,}277\times10^{851}$ãr$\div10^{894}$ãr $= 1{,}63\times10^{-86}$ (gain $1{,}277\times10^{2}$) $= 2{,}08\times10^{-84}\times 1{,}96\times10^{-272} =$
> $4{,}08\times10^{-356}$pu $\times 10^{45}$¢ $= 4{,}08\times10^{-311}$pu vs $6{,}39\times10^{-308}$pu$=$
> $1{,}57\times10^{3}$ times weaker medial orb than surface repulsion!

One can on bases of the Electrons local **$2{,}57\times10^{860}$ãr** wide coronal distance with terminated auxiliary expansion, estimate the local backhand speed and the period for the recall to the Electron surface of dominant auxiliary photons under a the virgin orb generated by a distant Proton.

> 10^{752}ãr$\div 2{,}57\times10^{860}$ãr$= 1{,}51\times10^{-217}\times 2{,}01\times10^{9} = 3{,}04\times10^{-208}$

The Protons $1{,}96\times10^{-272}$pu input creates a $4{,}08\times10^{-311}$pu positive local backhand cast by the Electron which after local dilution to $3{,}04\times10^{-208}$ with $1{,}24\times10^{-518}$pu backhand at the $2{,}57\times10^{860}$ãr distance where each auxiliary photon has been brought to a linear stop but is orbiting at $1{,}23\times10^{-277}$¢ superluminal speed after $6{,}14\times10^{-287}$¢ inherent Electron surface rotation has accelerated $2{,}01\times10^{9}$ times outwards in fully inherent orbit.

At a constant $2{,}57\times10^{860}$ãr coronal distance, the $1{,}24\times10^{-518}$pu backhand pulls back at a precursored dominant auxiliary photon orbiting the Electron as it passes at $1{,}23\times10^{-277}$¢ superluminal speed through the $1{,}277\times10^{851}$ãr Electron wide orb in $1{,}04\times10^{1128}$ãt. A backhand by $1{,}29\times10^{610}$pu harvested positive virgins under way, drive it in the direction of the local Electron at $1{,}29\times10^{-302}$¢.

As the average linear attraction towards a hydron continues to work as linear universal motion after each passover by a square root of the orbital coronal distance measured in hydron radii, it should add $\sqrt{2{,}01\times10^{9}}$ extra linear working time after each passover to the active $1{,}04\times10^{1128}$ãt passage whereupon its $1{,}29\times10^{-302}$¢ linear drift lasts for $4{,}66\times10^{1132}$ãt. The backhand drift moves the auxiliary photon $6{,}01\times10^{830}$ãr closer to the local Electron on each passage or by $1{,}2\times10^{831}$ãr twice in each orbit which is $4{,}68\times10^{-30}$ of its $2{,}57\times10^{860}$ãr wide local corona in each orbit. Using my shorthand method of furthest orbits, falling towards the Electron in $2{,}14\times10^{29}$ orbits where each requires $2{,}08\times10^{1137}$ãt which makes for an extensive $4,45 \times 10^{1166}$ãt recall period into the Electron surface ($1,48 \times 10^{-28}$ seconds).

As the corona cannot be recalled all the way to the Electron surface where the orb is $1,57 \times 10^3$ times weaker than the main body repulsion, the outcome of that must be dealt with. However, what governs that outcome is which corona is recalled faster; so we must first look at the Proton recall.

Upon the Proton in the other direction, one can on bases of its local $1,57 \times 10^{861}$ãr wide coronal distance with terminated auxiliary expansion, estimate the local backhand speed and the period for recall to the Proton surface of dominant auxiliary photons under the full $1,02 \times 10^{-271}$ input into the Electron generated orb.

10^{852}ãr$\div 10^{894}$ãr $=10^{-84}$ (gain 10^3) $=10^{-81} \times 1,02 \times 10^{-271} = 1,02 \times 10^-$ ^{352}pu$\times 10^{45}$¢$=1,02 \times 10^{-307}$pu vs. $1,77 \times 10^{-307}$pu. The medial surface orb is weaker than the main body repulsion that outguns it 1,74 times

---

As negative virgins reach across to the vast field to a Proton at 10^{894}ãr, boosted to 10^{45}¢, aided by 10^3 orbits, some 10^{-81} of the $1,02 \times 10^{-271}$ hit or $1,02 \times 10^{-352}$ virgins that generate $1,02 \times 10^{-307}$pu in negative surface pressure. This local orb casts a virgin backhand that is 1,74 times weaker than the main body surface repulsion from the Proton. All negative auxiliary photons can rise from anywhere on the surface except but they cannot be recalled all the way to the Proton surface under the medial orb.

10^{752}ãr$\div$ $1,57 \times 10^{861}$ãr$= 4,06 \times 10^{-219} \times 1,57 \times 10^9 = 3,37 \times 10^{-210}$

Initially, the Electron generated input with $1,02 \times 10^{-307}$pu local negative backhand cast by the Proton and locally diluted to $3,37 \times 10^{-210}$ generates a $3,44 \times 10^{-517}$pu backhand at the $1,57 \times 10^{861}$ãr distance. With a Protons 10^{-288}¢ inherent surface rotation the re-launched auxiliary photon is accelerated $1,57 \times 10^9$ times to $1,57 \times 10^{-279}$¢ superluminal speed in inherent coronal orbit.

At the linearly stopped $1,57 \times 10^{861}$ãr coronal distance, the $3,44 \times 10^{-517}$pu backhand pulls back at the precursored dominant auxiliary photon locally orbiting the Proton as it passes at $1,57 \times 10^{-279}$¢ superluminal speed through the 10^{852}ãr wide Proton orb in $6,37 \times 10^{1130}$ãt with $2,19 \times 10^{614}$pu backhand harvest of negative virgins that drive towards the Proton at $2,19 \times 10^{-298}$¢.

As the average linear attraction towards a hydron continues to work as linear universal motion after each passover by a square root of the orbital coronal distance measured in hydron radii, it adds $\sqrt{1,57 \times 10^9}$ extra linear working time after each passover to the $6,37 \times 10^{1130}$ãt passage and allows the $2,19 \times 10^{-298}$¢

linear drift to continue for $2,52 \times 10^{1135}$ãt. This moves an auxiliary photon $5,52 \times 10^{837}$ãr closer to the Proton on each passage or twice by $1,1 \times 10^{838}$ãr in each orbit which is $7,01 \times 10^{-24}$ of its $1,57 \times 10^{861}$ãr corona in each orbit. Using my shorthand of furthest orbits, falling locally towards the Proton in $1,43 \times 10^{23}$ orbits where each requires 10^{1140}ãt with **$1,43 \times 10^{1163}$ãt** recall period ($4,77 \times 10^{-32}$ seconds).

This recall cannot quite recall an auxiliary photon all the way into the Proton surface as the repulsion balances at $1,74$ times outside the 10^{852}ãr Proton at a short distance of $1,74 \times 10^{852}$ãr. The Proton is recalled to this distance in $1,43 \times 10^{1163}$ãt which is **$3,11 \times 10^3$ times faster** than the Electrons potential coronal recall in $4,45 \times 10^{1166}$ãt. This is an important difference as the Electron corona can be marginally recalled across **$3,22 \times 10^{-4}$** of its width; during the recall of the Proton corona.

$$by 10^{852}ãr \div 1,74 \times 10^{852}ãr = 3,3 \times 10^{-1} \, (gain \; 7,59 \times 10^{23}) = 2,5 \times 10^{23} = 4 \times 10^{-24}$$
$$of \; 2 \times 10^{-271} = 8 \times 10^{-295}$$

This leaves the Electron corona largely unchanged with its roughly 10^{-271} free virgin emission while the Proton coronas free virgin emission is heavily circumcised to 4×10^{-24} of the 2×10^{-271} coronal emission or to 8×10^{-295} which practically terminates its free emission back upon the Electron and prevents any further significant recall of the Electron corona.

Before continuing, the time has come to look at the important strand in adverse communication I mentioned earlier. This mutual effect cannot be ignored any longer in a holistic illustration as it can radically affect what is happening across these distances and is becoming a governing factor.

Magnetic Reversal

According to popular belief, adverse nuclei display a powerful mutual attraction instead of the massive repulsion that the Primal Code has hitherto shown to exist through orbitally aided repulsion between all hydrons. The saving grace is an effect that I call 'Magnetic Reversal'. This is one of the main strands that affects outcomes shown earlier in snapshots between two passing adverse hydrons.

A similar effect by repulsion in a local field by emitted virgins was touched on earlier when disparate virgin winds formed 'bulges' on mattrons where a certain threshold shift gave a side effect that could hardwire a mattron into an electromagnetic split. In hydrons we see how virgin repulsion upon main body hydrons blocks all virgins at the receiving end while auxiliary coronal photons that orbit the hydrons block the exchanged repulsive virgins in one sign only.

Furthermore, the mutual long distance virgin emission exchanged between hydrons affords a hydrons main body and a precursored auxiliary photon in its corona unequal virgin harvest. The linear long distance push displaces mass at the receiving end can therefore alter the mutual spatial relationship between the main body hydron and its orbiting corona.

'Magnetic Reversal' is born of the disparity in linear coronal shift versus main body linear shift. If an adverse hydron is pushed more slowly than the inherent auxiliary precursors in its corona, more of the corona is brought to and comes to emit from the far side of a main body hydron that is no longer at it coronal center. The mismatching main body and coronal shifts cause the adverse coronal emission to deliver more of its repulsion from the far side of the hydron after receiving the distant catalyst push and becomes therefore counter wise to the catalyst wind.

Whereas magnetic reversal causes 'attraction' between adversely paired hydrons, there is an opposite 'magnetic' shift where a distant main body hydron is instead shifted further than its local coronal precursors. This yields an opposite effect as the coronal emission from the windward side enhances the linear repulsion already taking place by the catalyst virgin wind upon the main body hydron; augmenting the repulsion between two allied hydrons.

'Magnetic Transfer' is used for clarity to represent enhanced repulsion between allied hydrons. In either case, the resultant coronal shift in relation to the main body hydron is the primary cause of a subsequent coronal push. In magnetic transfer the far off repulsion is greater upon the main body hydron than upon orbiting auxiliary photons in its corona. Magnetic transfer relies on repulsion shifting a hydrons main body instead of its corona; here the far off virgin repulsion is greater upon orbiting auxiliary precursors than upon the hydrons main body.

In both types of magnetic shift, the hydrons main body is after a period of time no longer found at the perfect center of its coronal swarm.

'G' gravity by inter-galactic flux, even between allied hydrons, is rarely a factor in magnetic transfer. At far distances, the gravity backhand is set by the inverse-square law but as hydrons draw closer, repulsive virgin harvests are aided by virgin orbits to thereby gain edge compared to the gravity backhand. At very close quarters where 'G' backhand becomes a significant force between allied hydrons, mutual virgin repulsion is massively aided by virgin orbits that prevent 'G' from overwhelming the mutual repulsion except at very short distances.

Between two isolated hydron of adverse polarity, displacement of coronal mattrons by mutually effective auxiliary repulsion is far greater upon one expanded coronal mattron of precursor size than upon each of the tightly packed mattrons within the main body hydron of standard sizes where the main body collective harvest of the linear push is shared by all of them.

One decisive parameter in any shift is the recall of auxiliary photons when they pass under a mutual virgin gravity orb. A coronal shift is for example impossible under a dark virgin orb with the one jump recall of a pod corona where re-launched auxiliary photons never reach the hydrons other side. The buildup of positional shift requires complete orbits by auxiliary photons; whether these are intermittently recalled or are without a short term coronal recall. After being recalled to a hydron surface and until re-launch, there is no positional shift of an auxiliary surface photon while that lasts.

Note that although a pod corona cannot be shifted to the far side of a main body hydron by virgins generating a local one step recall orb, a third weaker orb at right angle to the main local orb can push an auxiliary photon in each pod linearly further to the far side of the main body hydron and cause a weaker magnetic reversal towards the source of linear repulsion arriving at correct angle.

Generally speaking, auxiliary precursors spend more time on the far side of a main body after being pushed there by the catalyst virgin wind and from where they generate an effective tailwind. If the main body is instead pushed beyond the inherent coronal center during magnetic transfer, the orbiting corona lags behind if the singular repulsion does not affect its auxiliary sign. These photons will then generate an effective tailwind from the side closer to the repulsion source.

While a hydron stays at the perfect center of its coronal swarm, the linear part of Omni-directional coronal repulsion harvested by the main hydron has its universal directional drive largely neutralized.

In lieu of mathematics, magnetic reversal is by a catalyst virgin wind that shifts auxiliary photon orbits to the far side of the main body hydron during some working period (whether the period of adjacent passage by two hydrons or a recall period). Two hydrons affecting each other's auxiliary photons may for example form unrecalled continual coronas over long lasting periods.

When the accumulation of repulsive virgins shifts an auxiliary coronal precursor to the far side of a distant hydron, it is really a repositioning of its entire orbit. The entire orbit is therefore shifted fairly evenly in space since the linear drive pushes the auxiliary photon away at every point in its orbit until the hydron is no longer at the perfect center of that circular orbit. The linear one sided repulsion shift may not initially alter the perfect circle shape of the photons fast local orbit to a more elliptical shape but this can happen later when coronal recall becomes more effective.

Auxiliary speed in orbit is set by way of hydron rotation; this is an inherent orbital speed set by coronal distance and it is in no way enforced (it should be emphasized that although a dominant captured auxiliary photon with less than a full pot of inherent coordinates can be re-launched by a hydron to greater distances, that orbit lends itself to different rules). If intermittently recalled, the photon loses none of the inherent coordinates it has gained on the hydron surface but can continue to collect more mass coordinates every time the auxiliary photon lingers on its surface; the tendency is towards the lowest common coordinator of a fully standard corona.

Photons also orbit its hydron in differential planes parallel to far off adverse hydrons and there is therefore an average blend of various orbital paths that allows auxiliary photons to stay longer or shorter under an orb; consider therefore this rough illustration a conceptual exercise.

The linear effect that moves an auxiliary photon closer to a hydrons main body on the windward side is not only parallel to the hydron at the closest part in orbit but even to the sides; like a comet would pass closer to the sun at the sides, even in a perfectly circular orbit. The distance is shorter to the sides and a slower inherent photon takes longer to pass through a hydron wide orb at closer points which forces a faster recall. Even if a powerful catalyst wind is driving magnetic reversal, the ultimate recall of the auxiliary photons will usually come before any of them are lost into a hydron tail but this is rarely applicable for less that fully adapted photons.

There are various effects playing a part as the shift pushes a whole corona to the leeward side of a hydron. Pushing an auxiliary orbit far beyond a main body hydron on one side during a long working period leads to a growing counter push of magnetic reversal up until the leverage point which is also the point where a coronal orbit can be pushed beyond the hydrons (and its own) coordinate center.

Usually it is the time of adjacent hydron passage that calls the shots in the effects of both magnetic reversal and transfer. Under a dark orb, a fast recall can make for a shorter initial working period (unless for pod coronas without magnetic reversal) but after an auxiliary photon is re-launched, the magnetic reversal becomes active again and the already generated speed of the hydron gathers the same add-on effect with a longer period of passage working as if it were accelerated by accumulation during all that time.

The center of a precursors inherent coordinate system remains the hydron center and when its orbit is forcefully pushed towards that coordinate center, it causes the entire auxiliary orbit to shrink.

When the orbital shift to the far side comes to equal the coronal width, an auxiliary photon in orbit at the near side can hit the main body hydron (if circumstances allow) and be recalled that way instead of making an orbital escape. Either way it happens, intermittent recall cannot prevent re-launch and further magnetic reversal. To explain the system at work (simply and inaccurately), I use the full period of passage as the working period (unless paired hydrons can neutralize linear push in mutual high speed orbits).

An inherent auxiliary photon in orbit across the further distance on the far leeward side speeds up and makes the trek in the same time as across the windward side. Each photon in continuous differential orbit will eventually cover all spatial points on a coronal sphere but the speed difference cannot cover the squared number of points. The emitting corona will spend more time on the larger coronal sphere on the leeward side than on the smaller part of the coronal sphere on a windward side.

Two virgin winds affect the aim of magnetic reversal or transfer across the distances they work; one is the catalyst virgin wind harvested by a distant hydron or orbiting photon body. The other virgin wind works from auxiliary photons that have been shifted from a coronal center by that catalyst wind. This local emission works upon a local main body hydron across the coronal distance and there are significant cuts in the effect driving the local hydron by the resultant magnetic effect or transfer; especially local linear virgin aim.

The greatest downsizing of magnetic reversal or transfer woven by these two virgin winds stems from the aim of its local tailwind drive from its own orbiting auxiliary photons; in a direction in which a local main body hydron is pushed by repulsive virgins. that arrive from its corona in many opposite directions. The closer the corona, the more virgins will arrive sideways from auxiliary photons causing great linear neutralization. My way of illustrating this is to divide the distance across which the coronal wind works, measured in local main body radii, into the distance across which the catalyst wind works and use the net result to downsize the linear local push of magnetic reversal or magnetic transfer.

If a strong catalyst wind shifts the corona or main body across its entire width, the downsized reversal or transfer works full out (until then). The ability of a distant catalyst virgin wind to shift an auxiliary orbit in a working period has a bearing on magnetic reversal where two coronal halves weigh in against each other, ranging from total to minor shifts. If a weak catalyst wind shifts an auxiliary orbit at the receiving end across smaller parts of the coronal reach during a working period, there will be downsizing by that fraction of coronal shift on top of the linear downsizing of magnetic reversal or transfer.

On top of the above, a substantial downsizing of magnetic reversal and magnetic transfer is caused if there are fewer simultaneously active photons in a corona. Coronas around two very close hydrons may come to sport only one occasional photon in flight to deliver the tailwind (if that corona is not a pod corona). The numbers can vary from none to average 2×10^9 photons.

And lastly but not least, a large corona that reaches far outside its leverage point will see much of its virgin emission slip by the local hydron with fewer virgins able to affect magnetic reversal or transfer as they do not hit the local main body. In that direction at least, all that free escape is lost as a counter wise magnetic reversal.

An important aspect of this is that, if a circular auxiliary orbit is pushed clean beyond the center of a main body hydron (its own coordinate center) the unbending rules of inherent behavior take over. An fully endowed auxiliary photon is in coherent inherent orbit around its own coordinate center at the center of its hydron at faster speed than that of coronal drift.

Should its orbit drift by enforced means beyond its inherent coordinate center, the corporeal photon remains in inherent orbit (within which it has been suddenly shifted). Its orbit is no longer slow on a closer side or fast on a leeside; it is a full orbit around the coordinate center; set at the distance (and thereby the speed) it found itself at when its circular orbit passed by its own coordinate center. The fully endowed photon remains one with its center at the center of this new circular orbit by following its the hydron spin at whatever distance at the moment it finds itself passing beyond the hydron center in orbit.

When an orbits passes beyond the coordinate center, it is as likely to be at its furthest reach as at its closest reach; the average being roughly half of the orbital width. Its new halved orbit would then in time widen from that point across that exponential distance at the drift speed and, if it is again driven beyond the hydron center, its inherent orbital speed is once again unerringly set by the hydrons inherent coordinates.

A fractionally endowed photon with less than full pot of hydron coordinates is less affected by the hydrons coordinate center than its ongoing repulsion. It is the photons own inherent coordinate center that is pushed beyond the hydron center but the photon is no longer at the hydrons coordinate center.

Please take some of the foresaid with a grain of salt. The Primal Code itself is rigid in its governance, allowing but one specific outcome but this cannot be said of my various blunt illustrations. Far greater attention is needed to find the accurate field values (and correct average values for the universal chain).

Magnetic Reversal by Adverse Hydrons
10^{894}är apart

There are another element than the orb push embedded in the exchanged emission of coronal virgins that affects potential coronal recall by hydrons much earlier and speeds up the process.

The behavior of an auxiliary photon in its inherent orbit is guided by two main drifts; the catalyst push that works upon every point of its orbit and causes the universally linear away drift of the entire orbit and thus a hydron corona. And then there is a linear sideways push towards the hydron under the orb from two points in each orbit by the same but boosted virgins each time the inherent auxiliary photon crosses the local orb and is pushed closer to the center of its own coordinate system.

The linear drive away from the respective hydrons upon coronal photons orbiting the other hydron works to shift the orbits of these photons further away beyond the leeward side of the home hydron, leading them to orbit closer to the home hydron on at windward side.

Long before the two adverse hydrons engage across 10^{895}ãr in their private hydron domain, there is a hinder for a locally orbiting auxiliary photon to approach close to either hydron; their mutually generated orbs are too weak to allow the photons to join hydron surfaces against greater surface repulsion.

$1,277 \times 10^{851}$ãr $\div 10^{894}$ãr $= 1,63 \times 10^{-86}$ (gain $1,277 \times 10^2$) $= 2,08 \times 10^{-84} \times 1,96 \times 10^{-272} = 4,077 \times 10^{-356}$pu $\times 10^{45}$¢ $= 4,077 \times 10^{-311}$pu Vs. $6,39 \times 10^{-308}$pu surface emission = the orb is 1567,33 times weaker than the repulsion at the surface

---

Taking our cue from the earlier snapshot of the two adverse hydrons advancing across the first leg of 10^{894}ãr at 10^{-283}¢, start with two full adverse coronas. A full number of auxiliary photons orbiting at **$2,57 \times 10^{860}$ãr** around the Electron with 10^{-271} free negative virgins per ãnna and **$1,57 \times 10^{861}$ãr** corona around the Proton with $1,96 \times 10^{-272}$ free positive virgins per ãnna. The time for magnetic reversal to work with these inputs is **$1,43 \times 10^{1163}$ãt** before the Proton corona is recalled to lower orbits with its free virgins decimated during the remaining period of passage.

From the Proton during this active time, its corona delivers $2,08 \times 10^{-84}$ or $4,08 \times 10^{-356}$pu linear repulsion by positive virgins upon the distant main body Electron across 10^{894}ãr. The harvest by each of $2,08 \times 10^{225}$ standard mattrons is **$1,96 \times 10^{-581}$pu** per ãnna as perfectly linear push.

The catalyst wind from the Proton upon each auxiliary precursor in orbit around the Electron is 10^{-284} across 10^{894}ãr of the $1,96 \times 10^{-272}$ positive virgins. This **$1,96 \times 10^{-556}$pu** unfocused harvest upon each auxiliary precursor pushes the coronal auxiliary photon 10^{25} times faster than the main body Electron.

On approach across their new domain at 10^{-283}¢, during a fraction of the 10^{1177}ãt working time, the orbits of the auxiliary photons come to spend more time on the Electrons leeside. During the Proton coronas $1,43 \times 10^{1163}$ãt active period, the $1,96 \times 10^{-556}$pu focused repulsion drives each circular orbit by $2,8 \times 10^{607}$ virgins at $2,8 \times 10^{-305}$¢, pushing each across 4×10^{858}ãr which is $1,56 \times 10^{-2}$ of the Electrons **$2,57 \times 10^{860}$ãr** coronal reach that is well within its $3,3 \times 10^{860}$ãr leverage point.

A counter wind of magnetic reversal upon the Electrons main body is by $1,56 \times 10^{-2}$ of the Electron coronas 2×10^{-271} virgin input per ãnna that all hit its main body with a one sided thrust aims in the face a linear catalyst wind from the Proton. The magnetic reversal wind is heavily downsized to by the difference in catalyst distance and coronal radii to $2,57 \times 10^{-34}$ for a perfect aim $(2,57 \times 10^{860} \text{ãr} \div 10^{894} \text{ãr})$ and delivers $8,02 \times 10^{-307}$pu counter wind shared by its $2,08 \times 10^{225}$ standard mattrons as $3,86 \times 10^{-532}$pu per ãnna.

During $1,43 \times 10^{1163}$ãt, a main body Electron mattron gathers $5,52 \times 10^{631}$ virgins of magnetic reversal that can bring the main body at $\mathbf{5,52 \times 10^{-281} \text{¢}}$ with a bulls eye aim towards the distant Proton and the $1,96 \times 10^{-581}$pu direct repulsion $(1,97 \times 10^{49}$ times weaker) is insignificant in comparison.

The Electrons main body does not leave its corona behind. Although it is being pushed at $5,52 \times 10^{-281}$¢ across its own corona in an opposite direction to the windward side, able in principle to push it across $2,57 \times 10^{860}$ãr already during a swift $4,66 \times 10^{1140}$ãt period.

Should orbital drift by enforced means push an entire circular auxiliary photon orbit beyond its inherent coordinate center, the photon remains one with its coordinate system, The photon continues to follow inherent Electron spin at the distance point and moment when its orbit drifted beyond the hydron center (the corporeal photon can find itself almost anywhere on that orbit but on average at roughly half the distance). The thing that thereafter unerringly sets a new perfectly circular differential orbit for the auxiliary photons is its actual distance from the Electrons coordinate center that totally governs its inherent 'rotation'. This also means an new corona that without any recall would be slightly lower with less free virgin emission.

There is a reason for why this enforced drive cannot push the main body Electron across its own coronal width in that brief period; namely that the auxiliary photons orbit the coordinate center at the Electron center at much faster speed than the new magnetic reversal speed. The inherent photons in its $2,57 \times 10^{860}$ãr corona in the same $2,08 \times 10^{1137}$ãr period as the Electron main body rotates and this $1,4 \times 10^{-277}$¢ superluminal speed is much faster than $5,52 \times 10^{-281}$¢ speed induced by magnetic reversal. A weak Electron orb and greater repulsion simply carry its corona along, pushing it effectively along in the same direction at the windward side in every orbit.

Upon the Proton in the other direction across 10^{894}ãr; the Electrons $2,57 \times 10^{860}$ãr corona is close to its leverage point and delivers $1,02 \times 10^{-271}$ or back to the Proton during its $1,43 \times 10^{1163}$ãt initial stage before coronal recall. To compare the reaction of the hydron coronas we can look at the exchange during the $1,43 \times 10^{1163}$ãt period.

$$10^{852}\text{ãr} \div 10^{894}\text{ãr} = 10^{-84} \ (g10^3) = 10^{-8} \times 1,02 \times 10^{-271} = 1,02 \times 10^{-352}\text{pu} \times 10^{45}\text{¢} =$$

> $1,02 \times 10^{-307}$pu v $1,77 \times 10^{-307}$pu$=1,74$ more repulsion

This orb delivers a $1,07 \times 10^{-307}$pu versus the $1,77 \times 10^{-307}$pu repulsion and although the repulsion still carries the day, it is merely 1,74 fold and does not allow inherent photons to enter its surface, even though the can reach reaching extremely close to $1,74 \times 10^{852}$ãr at 1,74 times the Protons width.

The adverse Electron coronas emission repels negative precursors in the Proton corona by a catalyst wind that delivers a 10^{-284} of $1,02 \times 10^{-271}$ virgins upon each auxiliary photon or **$1,02 \times 10^{-555}$pu** linear push per ãnna. During the Protons $1,43 \times 10^{1163}$ãt working period, each auxiliary photon collects $1,46 \times 10^{608}$ virgins and is carried at $1,46 \times 10^{-304}$¢ across $2,09 \times 10^{859}$ãr of the Protons $1,57 \times 10^{861}$ãr corona ($1,33 \times 10^{-2}$).

As the coronal shift nears its consummation, the magnetic reversal wind by $1,33 \times 10^{-2}$ of the 2×10^{-271} coronal emission is downsized in effect upon the main body Proton to $1,57 \times 10^{-33}$ by the difference in catalyst distance and coronal radii ($1,57 \times 10^{861}$ãr$\div 10^{894}$ãr) where the $4,18 \times 10^{-306}$pu contra wind is shared by 10^{228} mattrons that yields $4,18 \times 10^{-534}$pu per ãnna magnetic reversal drive. You recall that the Electron was at that point accelerated by roughly 92,34 times more $3,86 \times 10^{-532}$pu per ãnna. During $1,43 \times 10^{1163}$ãt, a main body Proton mattron gathers $5,98 \times 10^{629}$ virgins that can drive its main body at $5,98 \times 10^{-283}$¢ through magnetic reversal and the Electron is moving faster towards the Proton at $5,52 \times 10^{-281}$¢.

Electron closing in on a Proton
7×10^{893}ãr apart

A pertinent fact is that as the two hydrons move closer, at first without significant changes; after drawing 1,43 times closer across 10^{894}ãr distance to **7×10^{893}ãr**, the Electron generated orb has become strong enough to recall the Proton corona all the way to its surface.

> 10^{852}ãr$\div 7 \times 10^{893}$ãr$=2,04 \times 10^{-84}$ ($g.1,2 \times 10^{3}$)$=2,45 \times 10^{-81} \times 10^{-271}=$
> $2,45 \times 10^{-352}$pu$\times 8,37 \times 10^{44}$¢$=2,05 \times 10^{-307}$pu Vs. $1,77 \times 10^{-307}$pu $=$ *The medial surface orb is now 1,16 times stronger than the main body repulsion*

> 10^{752}ãr$\div 4,21 \times 10^{856}$ãr$= 5,64 \times 10^{-210} \times 4,21 \times 10^{4} = 2,37 \times 10^{-205}$

The $2,05 \times 10^{-307}$pu negative backhand shadow generated by the Electron orb and cast by the Proton surface is diluted locally to $2,37 \times 10^{-205}$ out to the standard coronal distance and delivers $4,86 \times 10^{-512}$pu orb backhand.

When a re-launched fully inherent dominant auxiliary photon starts to locally orbit the Proton at $4,21 \times 10^{856}$ãr, it passes at $4,21 \times 10^{-284}$¢ through its 10^{852}ãr wide orb in $2,38 \times 10^{1135}$ãt. During that time the $4,86 \times 10^{-512}$pu pull allows it to block $1,16 \times 10^{624}$pu negative virgins under way which drives it in the direction of the home Proton at $1,16 \times 10^{-288}$¢. This local backhand cannot stop the head on $4,21 \times 10^{-284}$¢ expansion speed with a resistance that falls $3,63 \times 10^4$ times short and an auxiliary photon continues to orbit outwards.

The Proton corona launch at $4,21 \times 10^{-284}$¢ is $3,63 \times 10^4$ fold the resistance by $1,16 \times 10^{-288}$¢ inwards speed generated on first passage under the orb. With β as $3,63 \times 10^4$, the reach of a re-launched auxiliary photon is reversed after $3,63 \times 10^4$ standard coronal Proton measures of $4,21 \times 10^{856}$ãr at a distance of **$1,53 \times 10^{861}$ãr**.

The orbits of the Protons newly launched auxiliary photons are now just a tad lower that their earlier $1,57 \times 10^{861}$ãr and allows marginally fewer free virgins to escape across the vast field towards the Electron. The Proton corona will now be re-launched myriad times during the 7×10^{1176}ãt passage at 10^{-283}¢ across **7×10^{893}ãr** while the perpetual Electron corona remains largely unchanged. This does not mean that is hegemony over the Proton corona is upheld. The Protons longest coronal period is its recall period and its free coronal emission can be considered constant and roughly equal to earlier at 10^{894}ãr with $1,96 \times 10^{-272}$ virgins per ãnna. This will cause the Electron corona to be recalled to its surface in **$4,45 \times 10^{1166}$ãt** and beds for a completely different situation of oscillating mutual exchanges during the **$1,27 \times 10^{1173}$ãt** passage at **$5,52 \times 10^{-281}$¢** (under the drive of magnetic reversal gained across 10^{894}ãr).

It does not take longer than $4,45 \times 10^{1166}$ãt for the Electron corona to begin losing some free virgin emission in lowering orbits. The consequence of this first reaction is not really that intermittent Proton re-launch of auxiliary photons gains a greater coronal beach but rather that a faster Proton corona can only be recalled partially towards the Proton surface.

When the Electron generated orb is no longer enough to allow total recall of the Proton corona, free coronal emission upon the Electron is turned off. The backhand of the Electron corona is no longer fast enough for significant recall during the **$1,27 \times 10^{1173}$ãt** passage and it keeps hegemony over the Proton corona until the hydrons draw closer where recall of the Electron corona again allows re-launch and another period of free emission with a slowly lowering corona until repeat performances are possible at ever closer distances.

Magnetic reversal across 7×10^{893}ãr

At 10^{894}ãr after the Protons 'almost' recall in $1,43 \times 10^{1163}$αt, auxiliary photons at **7×10^{893}ãr** were recalled and re-launched into $1,53 \times 10^{861}$ãr orbits. At this distance, the development beds for a different situation for magnetic reversal by these oscillating mutual exchanges during passage across 7×10^{893}ãr which takes $1,27 \times 10^{1174}$ãt at $5,52 \times 10^{-281}$¢.

The pertinent fact is that Proton corona with largely same free emission needs to be recalled across a similar distance as before in identical working periods for magnetic reversal before the Proton generated orb is largely shut off. The Electrons local corona has the same 2×10^{-271} tailwind input but a slightly greater neutralization of aim from slightly closer orbits.

The magnetic reversal during $1,27 \times 10^{1174}$ãt will therefore not accelerate the already achieved $5,52 \times 10^{-281}$¢ tailwind drive of the earlier magnetic reversal. For clarification, one should point out that magnetic reversal across further distances, say 10^{895}ãr, could have brought the Electron to higher speeds than $5,52 \times 10^{-281}$¢ but that is not part of this illustration.

We therefore see mainly the Electron hurling across 7×10^{893}ãr at a speed of at least $5,52 \times 10^{-281}$¢ with a marginally lower corona and crossing to lesser distances with a bulls eye aim at the Proton.

Other Starting Points

The above starting point for an Electron and Proton entering a grid domain from opposite directions without coronas is just a simplified illustration and there are millions of other starting points where already established coronas start to be recalled by way of different mutual exchanges as the two hydrons draw closer. Before the two adverse hydron draw close enough to form their smaller private mutual grid exchange domain, each would have a corona that is a product of exchanges in the last neighboring domain they crossed.

A smaller Electron corona than in the above example would lead to faster recall of an Electron corona instead of the Proton corona; which would turn the outcome on its head with a subdued Electron corona and a full Proton corona. Depending on coronal sizes, input from neighboring domains across further distances can affect local coronas more than another closer hydron. Coronas are always likely to shrink or expand out of sync whatever is their starting point and this leads to different tipping points in the interplay of the local adverse hydrons.

Hydron equipages are brought seriously out of sync if either corona sinks below its leverage point, tipping the balance into recalling either corona faster than the other where the more slowly recalled corona gains hegemony over the re-launch of the other local corona that is then re-launched against a more robust orb an even smaller more rapidly recalled corona.

While the Electron corona is not even recalled, it hardly matters in the ebb and flow of coronal evolution that its re-launch launch across the leverage point to full free emission is 100 times faster than the launch of the Proton corona. Inability to move in sync in the various phases of auxiliary launches or coronal recalls seems in the long run through myriad repeated oscillations to strengthen the hand of the less easily recalled Electron corona. In the long run an Electron corona is destined to gain hegemony in any lasting bond.

The Inactive Surface Period

After launch and recall periods, the remaining time in a photons auxiliary cycle is its inactive surface period. A Proton makes one differential rotation at 10^{-288}¢ in 10^{1140}ãt or twice that in differential orbit but its rotation is not the main component in setting inactive surface time after a swift auxiliary recall and until the surface photons eventual re-launch from a Proton polar area.

A photon that is recalled to the Proton surface must be rotated onto a right angle re-launch polar area which size is here set by the Electron generated orb. The robust local orb sets how many rotations the Proton must make before it can move a surface photon to and across a potential re-launch spot. These polar areas of variable sizes can be relatively tiny and shifting parts of the hydron surface in the time it takes differential hydron rotation to bring each of the individual surface photons across one of the two polar areas where they can be re-launched becomes 'time out' from the photons active participation in free coronal emission.

The stronger the local dominant orb, the smaller the dominant launching areas and the longer the wait for auxiliary re-launch. A longer inactive period under a dark orb diminishes the number of auxiliary photons that are active in the corona at each time and thereby decrease the level of free emission that can slip by the main body hydron from each auxiliary photon at the achieved coronal reach but, first of course, the corona must be recalled.

To illustrate the inactive period and thereby the part of the auxiliary photon stock active in the hydron corona, I use the ratio by which the backhand of the main virgin gravity orb eclipses the hydrons dominant main body repulsion at surface; this ratio is used to multiply the hydrons rotation period. The rate at which an inactive surface period exceeds active periods of auxiliary photons during launch or recall renders an inactive surface period; and the part of the coronal photon stock that can partake in free external emission.

Electron at 10^{890}ãr apart

Jumping to ten thousandth part of the earlier distance means ignoring ten thousand snapshots of coronal changes and, for illustration ignoring what has taken place across the greater distances in between. One can probably assume that the Electron is moving closer at $5,52 \times 10^{-281}$¢ and crossing this 10^{890}ãr distance in a short $1,81×10^{1170}$ãt where somewhere along the way, it is likely to lose its hegemony over the Proton corona.

We start this illustration by ignoring what has taken place from 10^{894}ãr to 10^{890}ãr and arbitrarily assume that the Proton corona is in the cycle where it is first recalled part ways to the surface with lights out until the orb from the Electron grows dark enough to recall auxiliary photons all the way. Although this may have happened earlier than at 10^{890}ãr, the changes are of similar nature. At that point before Proton recall and re-launch at 10^{890}ãr, we can ascertain the size of the perpetual Electron corona required in order to recall the Proton corona in this illustration.

10^{852}ãr$÷10^{890}$ãr $=10^{-76}$ (gain 10^5) $=10^{-71}×1,8×10^{-279} = 1,8×10^{-350}pu×10^{43}$¢$=$ $1,8×10^{-307}$pu vs. $1,77×10^{-307}$pu; a slightly darker orb.

When the Electron generated orb across 10^{890}ãr is fuelled by an Electron coronas $1,8×10^{-279}$ free virgin emission per ãnna, the mid Proton surface orb turns marginally stronger than its main body repulsion and the $1,8×10^{-307}$pu backhand at the Proton will in a period that is shorter than the period of passage allow a recall of its corona all the way to the surface.

$1,277×10^{851}$ãr$÷ 2,04×10^{857}$ãr$=3,92×10^{-13}$ (g. $2,83×10^{20}$)$=1,11×10^8=$ $8,59×10^{-9}×2×10^{-271}=1,8×10^{-279}$ free virgins.

To meet this low level of free virgin emission, the Electron corona must in preceding steps have shrunk from $2,57×10^{860}$ãr to $2,04×10^{857}$ãt where it emits this minimum requirement of $1,8×10^{-279}$ free virgins per ãnna for this illustration. That said, the original $1,57×10^{861}$ãr corona around the Proton with $1,96×10^{-272}$ free positive virgins per ãnna is still more slowly recalled than is the Proton corona and we can look at some complications.

Note that although the Electrons $1,8×10^{-307}$pu orb diminishes or totally disappears, the Proton corona does not widen. Should the Proton corona for some other reason be recalled and re-launched at 10^{890}ãr apart, its negative backhand shadow generated by the Electron orb and cast by the Proton surface is diluted to $3,84×10^{-205}$ and delivers $6,91×10^{-512}$pu orb backhand and we can look at its coronal re-launch.

When a re-launched fully inherent dominant auxiliary photon starts to locally orbit the Proton at $4,21×10^{856}$ãr, it passes at $4,21×10^{-284}$¢ through its 10^{852}ãr wide orb in $2,38×10^{1135}$ãt. During that time the $6,91×10^{-512}$pu pull allows it to block $1,64×10^{624}$pu negative virgins under way which drives it in the direction of the home Proton at $1,64×10^{-288}$¢. This local backhand cannot stop the head on

$4,21\times10^{-284}$¢ expansion speed with a resistance that falls $2,57\times10^4$ times short and an auxiliary photon continues to orbit outwards.

The Proton corona launch at $4,21\times10^{-284}$¢ is $2,57\times10^4$ fold the resistance by a potential $1,64\times10^{-288}$¢ inwards speed generated on first passage under the orb. With β as $2,57\times10^4$, the reach of the re-launched auxiliary photon is terminated and reversed after as many $4,21\times10^{856}$ãr standard coronal measures at distance **$1,08\times10^{861}$ãr**. This distance is just a little less than its original **$1,57\times10^{861}$ãr** corona within the Protons $3,96\times10^{861}$ãr leverage point. It would allow somewhat less than its earlier free coronal emission in return across the field to the Electron but not to change what is happening.

Even giving us a new linearly stopped $1,08\times10^{861}$ãr coronal distance, the $6,91\times10^{-512}$pu backhand pulls back at the precursored dominant auxiliary photon locally orbiting the Proton as it passes at $1,09\times10^{-279}$¢ superluminal speed through the 10^{852}αr wide Proton orb in $9,17\times10^{1130}$ãt with $6,34\times10^{619}$pu backhand harvest of negative virgins that drive it towards the main Proton at $6,34\times10^{-293}$¢.

As the average linear attraction towards a hydron continues to work as linear universal motion after each passover by a square root of the orbital coronal distance measured in hydron radii, it adds $\sqrt{1,08}\times10^9$ extra linear working time after each passover to the $9,17\times10^{1130}$ãt passage and allows the $6,34\times10^{-293}$¢ linear drift to continue for $3,01\times10^{1135}$ãt. This moves an auxiliary photon $1,91\times10^{843}$ãr closer to the Proton on each passage or twice by $3,82\times10^{843}$ãr in each orbit which is $3,54\times10^{-18}$ of this example of a $1,08\times10^{861}$ãr corona in each orbit. Using my shorthand of furthest orbits, falling locally towards the Proton in $2,82\times10^{17}$ orbits each requiring 10^{1140}ãt leads to **$2,82\times10^{1157}$ãt** recall (or $9,4\times10^{-38}$ seconds) to the Proton surface.

What we are now seeing clearly at 10^{890}ãr is how the relative sizes of two adverse coronas have been changing. The Electron corona has continued to shrink much faster than the Proton corona that then cannot be recalled until at much closer distances which requires a longer $1,81\times10^{1170}$ãt period of passage to change. The Electrons dropping free emission delivers less resistance to the Proton re-launch (until itself is recalled and re-launched). If the Electron corona is recalled earlier than the Proton corona, its re-launch with new coronal reach is set at that point.

So let us look at the Electron coronas recall with the extreme changes that meet it at 10^{890}ãr instead of the gradual changes in in-between steps as it approaches from 10^{894}ãr.

10^{852}ãr÷ $1,08\times10^{861}$ãr=$8,57\times10^{-19}$ (g.$3,04\times10^{18}$)=one in $2,61\times$ 2×10^{-271}= $7,66\times10^{-272}$ *free virgins from the Proton corona*

Let's use the new $1,08×10^{861}$ãr wide Proton corona (for illustration only) that during $2,82×10^{1157}$ãt recall period emits $7,66×10^{-272}$ free virgins per ãnna; a widely more powerful emission that the $1,8×10^{-279}$ free virgins the new shrunken $2,04×10^{857}$ãr wide Electron corona emits. Even with a slightly smaller Proton corona, the changes in this a jump start illustration are clear.

$1,277×10^{851}$ãr÷10^{890}ãr = $1,63×10^{-78}$ (gain $1,277×10^4$) = $2,08×10^{-74}×$ $7,66×10^{-272}$ = $1,59×10^{-345}$pu × 10^{43}¢ = $1,59×10^{-302}$pu vs. $6,39×10^{-308}$pu= $2,49×10^5$ times darker medial orb than surface repulsion!

10^{752}ãr÷ $2,04×10^{857}$ãr= $2,4×10^{-211}×$ $1,6×10^6$ = $3,84×10^{-205}$

The Protons $7,66×10^{-272}$pu input creates a $1,59×10^{-302}$pu positive local backhand cast by the Electron which after local dilution to $3,84×10^{-205}$ with $6,11×10^{-507}$pu backhand at the $2,04×10^{857}$ãr distance where each auxiliary photon has been brought to a linear stop but is orbiting at $9,82×10^{-281}$¢ superluminal speed after $6,14×10^{-287}$¢ inherent Electron surface rotation has accelerated $1,6×10^6$ times outwards in fully inherent orbit.

The Protons $7,66×10^{-272}$pu input creates a $1,59×10^{-302}$pu positive local backhand cast by the Electron which after local dilution to $3,84×10^{-205}$ with $6,11×10^{-507}$pu backhand at the $2,04×10^{857}$ãr distance where each auxiliary photon has been brought to a linear stop but is orbiting at $9,82×10^{-281}$¢ superluminal speed after $6,14×10^{-287}$¢ inherent Electron surface rotation has accelerated $1,6×10^6$ times outwards in fully inherent orbit.

At a constant $2,04×10^{857}$ãr coronal distance, the $6,11×10^{-507}$pu backhand pulls back at a precursored dominant auxiliary photon orbiting the Electron as it passes at $9,82×10^{-281}$¢ speed through the $1,277×10^{851}$ãr Electron wide orb in $1,3×10^{1131}$ãt. A backhand by $7,94×10^{624}$pu harvested positive virgins under way, drive it in the direction of the local Electron at $7,94×10^{-288}$¢.

As the average linear attraction towards a hydron continues to work as linear universal motion after each passover by a square root of the orbital coronal distance measured in hydron radii, it should add $\sqrt{1,6×10^6}$ extra linear working time after each passover to the active $1,3×10^{1131}$ãt passage whereupon its linear $7,94×10^{-288}$¢ drift lasts for $1,64×10^{1134}$ãt. The backhand drift moves the auxiliary photon $1,3×10^{847}$ãr closer to the local Electron on each passage or by $2,6×10^{847}$ãr twice in each orbit which is $1,27×10^{-10}$ of its $2,04×10^{857}$ãr wide local corona in each orbit. Using my shorthand method of furthest orbits, falling towards the Electron in $7,87×10^9$ orbits where each requires $2,08×10^{1137}$ãt which makes for a **$1,64×10^{1147}$ãt** recall period into the Electron surface ($5,47×10^{-48}$ seconds).

That the inactive period on the Electron surface is $5,18\times10^{1142}$ãt because of an orb that is $2,49\times10^5$ times darker than surface repulsion, this does not hamper free coronal emission that lasts during $1,64\times10^{1147}$ãt. What governs that outcome is that the Electrons $1,64\times10^{1147}$ãt coronal recall is **$1,72\times10^{10}$ times shorter** than the Proton coronas $2,82\times10^{1157}$ãt recall period at this distance; needless to say, all to be calculated when correct numbers are eventually agreed upon.

Finally after the first recall of the Electron corona (by way of this dodgy illustration), the first re-launch of its new corona brings it out to a different and a very much shorter coronal reach and we can look at that.

$1,277\times10^{851}$ãr$\div10^{890}$ãr $= 1,63\times10^{-78}$ (gain $1,277\times10^4$) $= 2,08\times10^{-74}\times 7,66\times10^{-272} =$ $1,59\times10^{-345}$pu $\times 10^{43}$¢ $= 1,59\times10^{-302}$pu vs $6,39\times10^{-308}$pu$= 2,49\times10^5$ times darker medial orb than surface repulsion!

10^{752}ãr$\div 1,47\times10^{854}$ãr standard $= 4,63\times10^{-205}\times 1,15\times10^3 = 5,32\times10^{-202}$

The $1,59\times10^{-302}$pu positive backhand shadow generated by the Protons limited virgin orb and cast from an Electron surface is diluted to $5,32\times10^{-202}$ at the Electrons standard coronal distance and delivers $8,46\times10^{-504}$pu orb backhand upon each auxiliary precursor at the standard distance where the auxiliary photon completes its first orbit.

A fully inherent dominant auxiliary photon at $1,47\times10^{854}$ãr has started to orbit the Electron, passing at $7,05\times10^{-284}$¢ speed through the $1,277\times10^{851}$ãr Electron wide orb in $1,81\times10^{1134}$ãt where the $8,46\times10^{-504}$pu pull blocks $1,53\times10^{631}$pu positive virgins under way which drives it in the direction of the main body Electron at **$1,53\times10^{-281}$¢.**

This terminates the Electron coronas $7,05\times10^{-284}$¢ expansion and allows no further widening of standard coronal reach. It also means that the next Electron coronal recall from $1,47\times10^{854}$ãr after the $1,81\times10^{1134}$ãt period of crossing the orb adds $\sqrt{1,15\times10^3}$ extra $6,14\times10^{1135}$ãt working time as the $1,53\times10^{-281}$¢ backhand drive towards the main body Electron brings it closer by $9,39\times10^{854}$ãr after a single passage. This powerful recall of an auxiliary photon requires less than half an orbit for final recall already within **$9,61\times10^{1134}$ãt.**

$1,277\times10^{851}$ãr$\div$ **$1,47\times10^{854}$ãr** $= 7,55\times10^{-7}$ (gain $1,05\times10^{22}$) $= 7,93\times10^{15}=$ $1,26\times10^{-16}\times 2\times10^{-271} = 2,52\times10^{-287}$ free virgins

The $5,17\times10^{1142}$ãt inactive period on the Electron surface owing to an orb that is $2,49\times10^5$ times darker than its surface repulsion does hamper its free coronal emission as it is now the longest coronal period. $2,49\times10^5$ times longer than the

$12,08\times10^{1137}$ãt re-launch period and $9,61\times10^{1134}$ãt recall, it cuts the number of auxiliary photons simultaneously out in the Electron corona from 2×10^9 to $8,03\times10^3$ orbiting photons. Whereas a full standard corona is restricted to $2,52\times10^{-287}$ free emission, this now drops to **$1,01\times10^{-292}$** free virgins per ãnna.

$$10^{852}\text{ãr}\div10^{890}\text{ãr}=10^{-76} \text{ (gain } 10^5)=10^{-71}\times1,01\times10^{-292}=1,01\times10^{-363}pu\times10^{43}¢=$$
$$1,01\times10^{-320}pu \text{ vs. } 1,77\times10^{-307}pu =1,75\times10^{13} \text{ fold}$$

Setting my trepidation aside, the shoe is clearly on the other foot with the Proton having gained hegemony over the Electron corona at 10^{890}ãr apart and it does not seem likely to lose that hegemony any time soon. A much weaker Electron generated orb cannot recall the Proton corona closer to its surface. The new Proton corona is neither compressed or expanded but left in place for the remainder of the period of passage for the two hydrons.

The hydron coronas recall must await a shorter distance apart that allows more harvest and darker orbs to remedy the situation and allow the Proton corona to be gradually lowered until it too can be totally recalled.

Pod Coronas

Before we look further at closer coronal exchanges, I want to point out the important change in coronal behavior that (in this case) happened to the Electron at around 10^{890}ãr under the greater Proton generated orb. Similar swift recalls can take place from coronas at other distances and this single step execution of coronal recalls has important side effects.

*In this case the Proton corona with its hegemony will usher in a different coronal state for the Electron corona where auxiliary photons are recalled so quickly on forming their first orbit and entering a dark virgin gravity orb at the standard coronal distance that they are recalled all the way to the Proton surface happens in a single jump during a single passage under the orb. This creates an 'immediate recall collar' that spans the Protons entire circumference and divides its emission field into two separate independent emission fields at right angles to the local virgin orb; the two coronal **pods** on either side of the powerful orb plane become totally separated.*

An auxiliary photon that orbits into the orb at the coronal distance cannot pass beyond that collar plane but is recalled all the way to a hydron surface. Although an auxiliary photon cannot therefore enter the coronal pod on the hydrons other side, the recalled auxiliary photon can be re-launched from either of its two respective 'polar areas' into either of its two coronal pod.

The image of two magnetic poles comes to mind except that the identical separated fields are in the same sign. The collar where a virgin orb meets the main body hydron spans its entire circumference; separating the right-angle emission poles by a hydron thick orb plane between them. This recall collar reaches out from its surface all around the generating hydron; not only to the coronal reach where this separation collar is (relatively) razor sharp at that distance in the division of two immense coronal pods.

In a pod corona, auxiliary coronal paths are no longer orbital; they cannot reach around the hydron but are isolated into two launching pods on either side of a single recall orb. Unaffected by a hydrons differential rotation, one of its coronal pods and 'polar area' would always point towards the nearest paired hydron of adverse polarity that is generating the dark virgin gravity orb and the other pod would always point away from it.

During a hydrons rotation period, the plane of this local orb that separates the pods cannot be notably altered; it usually takes longer to move a nearest hydron far enough across say 10^{890}ãr to alter the position of it generated orb plane enough to count; compared to local effects.

Although each standard re-launch with half orbit curvature is followed by a swift one step harvest of the same auxiliary photon, external free emission continues; albeit downsized by longer inactive surface periods under a dark orb. A corona that is divided into two continuous pod fields has the same reach as a perpetual spherical corona but continual re-launch would cut its perpetual freely escaping emission additionally by half (constant). The larger fully orbital Proton corona would under these circumstances be in practical terms, a perpetual corona that cannot be recalled; even in tiny incremental steps under the weak Electron generated orb (ignoring all other inputs from the general local grid).

Magnetic Reversal Across 10^{890}ãr

The approach at the two adverse hydrons continues from across 10^{894}ãr to smaller distances at $5,52 \times 10^{-281}$¢, so we are now ten thousand times closer and looking at the two hydrons **10^{890}ãr** apart which they can traverse during $1,81 \times 10^{1170}$ãt. There are is only one main magnetic reversal drive by orb inputs here during different time periods; the $1,47 \times 10^{854}$ãr highly compressed Electron corona is a pod corona that cannot utilize magnetic reversal and works only during the first $1,64 \times 10^{1147}$ãt recall and is too weak to matter.

The main period of magnetic reversal takes over when the Electron has been has been subdued by a probably original $1,57 \times 10^{861}$ãr wide Proton corona. The weak catalyst wind across 10^{890}ãr from the Electron upon an auxiliary precursor in orbit around the Proton is 10^{-276} of the Electrons $1,01 \times 10^{-292}$ free negative virgins with $1,01 \times 10^{-568}$pu harvest for each auxiliary photon but the drive would lasts during the remaining **$1,81 \times 10^{1170}$ãt** period of passage; except that it is likely drive faster and therefore covers the distance in less time..

During $1,81\times10^{1170}$ãt, the $1,01\times10^{-568}$pu repulsion drives each auxiliary orbit linearly by $1,83\times10^{602}$ virgins at $1,83\times10^{-310}$¢ and across $3,31\times10^{860}$ãr or 4,74 of the Protons $1,57\times10^{861}$ãr corona. With 4,74 limiting magnetic the reversal from a continual 2×10^{-271} tailwind downsized to $1,57\times10^{-29}$ or ($1,57\times10^{861}$ãr÷10^{890}ãr) with $6,62\times10^{-301}$pu magnetic reversal shared by all of its 10^{228} standard mattrons as $6,62\times10^{-529}$pu per ãnna magnetic reversal.

This active counter wind drives the Protons main body against the catalyst wind by a harvest of $1,2\times10^{642}$ virgins $1,2\times10^{-270}$¢ and the ongoing $5,52\times10^{-281}$¢ speed of its already earlier generated bulls eye approach pales beside such a powerful magnetic transfer drive. Due to the high speed, the $1,81\times10^{1170}$ãt working period is no longer actual as the Proton potentially crosses this 10^{890}ãr distance at $1,2\times10^{-270}$¢ which would then only take $8,33\times10^{1159}$ãt.

The actual working period is the $1,23\times10^{1165}$ãt average during which the $6,62\times10^{-529}$pu magnetic reversal blocks $8,14\times10^{636}$ virgins and gains an actual $8,14\times10^{-276}$¢ superluminal speed that can bring the Proton towards the Electron across 10^{890}ãr in $1,23\times10^{1165}$ãt. This superluminal speed seriously outguns the earlier approach speed of the two hydrons with a new bulls eye approach during a shorter period of passage.

$$1,277\times10^{851}\text{ãr}\div10^{890}\text{ãr} = 1,63\times10^{-78}\,(gain\ 1,277\times10^4) = 2,08\times10^{-74}\times 7,66\times10^{-272} = 1,59\times10^{-345}pu$$

As to the Protons direct repulsion upon the Electrons main body during the $1,23\times10^{1165}$ãt period delivers most repulsive virgins by its $7,66\times10^{-272}$ input, the $1,59\times10^{-345}$pu push delivers $1,96\times10^{820}$ virgins upon the Electrons $2,08\times10^{1137}$ primal mass and generates $9,42\times10^{-318}$¢ repulsion speed. It requires much closer distances to break the suggested $8,16\times10^{-276}$¢ speed of the Proton towards the Electron. If it eventually meets enough repulsive resistance to slow the approach, the boosted orbs of the same virgins would be so dark that it affect coronal recall and emission to a degree that could even leave only their smaller steady main body emission.

There is a wide choice of possible setups to these illustrations and I am trying show the complex strands of an ever changing system at work; how the hydron equipages mutually affect each other. The examples are not specific to whatever distance is selected or the original linear speed; or induced speeds suggested upon one hydron or another driven by magnetic reversal.

Closer Encounters at 10^{880}ãr

For a new illustration, jump to 10^{-10} part of the earlier distance while ignoring 10^{-10} snapshots of coronal changes in between and whatever has taken place during that passage. Assume that the Electron corona remains heavily subdued while the Proton corona is less restrained and the oncoming Electron is crossing 10^{880}ãr at $1,12\times10^{-275}$¢ during $8,9\times10^{1154}$ãt.

Assume that the Proton corona is in the cycle where auxiliary photons are brought to the surface with such backhand force that their free emission of has fallen as close to the Protons 10^{-300} per ãnna main body virgin emission as it can. For this to happen, the Proton corona needs to be almost a pod corona facing an orb of enough free Electron emission to make this happen. This is not a stable exchange and we can then look at the first few steps of where this oscillation is heading.

Step 1; take for example a standard Electron pod corona that delivers free coronal emission downsized because of inactive surface periods at **3×10^{-293}** per ãnna and look at the oscillating balance at that point.

$$10^{852}ãr \div 10^{880}ãr = 10^{-56} (g.\ 10^{10}) = 10^{-46}\times3\times10^{-293} = 3\times10^{-339}pu\times10^{38}¢ = 3\times10^{-301}pu$$
$$vs.\ 1,77\times10^{-307}pu = 1,69\times10^{6}fold$$

The 3×10^{-301}pu backhand at the Proton surface can recall its corona all the way to its surface and re-launch it against the orb to standard distance where its auxiliary photons start to orbit.

$$10^{752}ãr \div 4,21\times10^{856}ãr = 5,64\times10^{-210} \times 4,21\times10^{4} = 2,37\times10^{-205}$$

At 10^{880}ãr apart, the 3×10^{-301}pu negative backhand shadow generated by the Electron orb and cast by the Proton surface is diluted to $2,37\times10^{-205}$ at the standard coronal distance and delivers $7,11\times10^{-506}$pu orb backhand. When a re-launched fully inherent photon starts to locally orbit a Proton at $4,21\times10^{856}$ãr standard distance, it passes at $4,21\times10^{-284}$¢ in $2,38\times10^{1135}$ãt through a 10^{852}ãr wide orb. The $7,11\times10^{-506}$pu pull allows it to block $1,69\times10^{630}$pu negative virgins under way which drives it in the direction of the home Proton at $1,69\times10^{-282}$¢. The working time is longer by $\sqrt{4,21\times10^{4}}$ or $4,88\times10^{1137}$ãt but it would take $2,49\times10^{1138}$ãt to recall the photon all the way to the surface. This Proton corona is still roughly 5 orbits from becoming a one step pod corona.

The orb at the Proton surface is $1,69\times10^{6}$ fold and the Protons 10^{1140}ãt rotation and launch period yields a $1,69\times10^{1146}$ãt inactive surface period. In practice, this indicates a less frequent launch of auxiliary photons that are recalled in one jump; leaving the Proton with fewer auxiliary photon in its corona. Of the 2×10^{9} auxiliary stock, only $1,18\times10^{3}$ photons are on average active in the corona at the same time.

A standard emission by a fully stocked Proton corona of 2×10^{9} negative auxiliary photons releases $7,28\times10^{-284}$ free positive virgins per ãnna. In this corona of standard reach, only $5,92\times10^{-7}$ of the coronal input is available as free coronal emission or $\mathbf{4,31\times10^{-290}}$ positive virgins per ãnna. This is still $4,31\times10^{10}$ times more than is provided by the Protons main body emission of 10^{-300} positive subservient virgins.

$1,277\times10^{851}$ãr$\div10^{880}$ãr $= 1,63\times10^{-58}$ (gain $1,277\times10^9$) $= 2,08\times10^{-49}\times 4,31\times10^{-290} =$
$8,96\times10^{-339}$pu $\times 10^{38}$¢ $= 8,96\times10^{-301}$pu vs $6,39\times10^{-308}$pu $=$ the orb backhand
outguns the surface repulsion $1,4\times10^7$ fold

10^{752}ãr$\div 1,47\times10^{854}$ãr standard $= 4,63\times10^{-205}\times 1,15\times10^3 = 5,32\times10^{-202}$

The $8,96\times10^{-301}$pu positive backhand shadow generated by the Protons limited virgin orb and cast from an Electron surface is diluted to $5,32\times10^{-202}$ at the Electrons standard coronal distance and delivers $4,77\times10^{-502}$pu orb backhand upon a auxiliary precursor at the standard distance on completing its first orbit.

A fully inherent dominant auxiliary photon at $1,47\times10^{854}$ãr has started to orbit the Electron, passing at $7,05\times10^{-284}$¢ speed through the $1,277\times10^{851}$ãr Electron wide orb in $1,81\times10^{1134}$ãt where the $4,77\times10^{-502}$pu pull blocks $8,63\times10^{632}$pu positive virgins under way which drives it in the direction of the Electron at $\mathbf{8,63\times10^{-280}}$¢.

This terminates the Electron coronas $7,05\times10^{-284}$¢ expansion and allows no further widening of that standard coronal reach. It also means that the Electron coronal recall from $1,47\times10^{854}$ãr has no need for $\sqrt{1,15\times10^3}$ extra $6,14\times10^{1135}$ãt working time after an orb crossing as the Electron now has a **pod corona** that is recalled in a single jump already within $\mathbf{1,7\times10^{1133}}$**ãt**.

The $2,91\times10^{1144}$ãt inactive period on the Electron surface ($1,4\times10^7$ fold orb versus surface repulsion times the $2,08\times10^{1137}$ãt rotation period) is now the longest coronal period. It is of course longer than the $1,7\times10^{1133}$ãt recall but it is also $1,4\times10^7$ times longer than the $2,08\times10^{1137}$ãt re-launch period. This cuts the number of auxiliary photons that are simultaneously in the corona from 2×10^9 to $1,43\times10^2$ orbiting photons and the $2,52\times10^{-287}$ free emission from a full corona drops to $\mathbf{1,8\times10^{-294}}$ free virgins per ãnna.

The Electron now emits 6 times fewer virgins compared to the 3×10^{-293} that was compressing the Proton in this exercise; with still fewer free virgins from the Electron corona against the powerful Proton orb, there is no stability at this point to solve the ongoing imbalance.

Step 2; having downsized the free emission of the standard Electron pod corona to $\mathbf{1,8\times10^{-294}}$ per ãnna eases pressure upon the Proton and allows its return pressure upon the Electron to grow which elicits a response of further dropping free Electron emission which accelerates the imbalance.

$$10^{852}ãr \div 10^{880}ãr = 10^{-56} (g.\ 10^{10}) = 10^{-46} \times 1,8 \times 10^{-294} = 1,8 \times 10^{-340}pu \times 10^{38}¢ =$$
$$1,8 \times 10^{-302}pu\ vs\ 1,77 \times 10^{-307}pu = 1,02 \times 10^5\ fold$$

The Electron generated new $1,8 \times 10^{-302}$pu backhand at the Proton surface will continue to recall the Proton corona its surface and re-launch it against a weakening orb to standard distance where its auxiliary photons start to orbit.

$$10^{752}ãr \div 4,21 \times 10^{856}ãr = 5,64 \times 10^{-210} \times 4,21 \times 10^4 = 2,37 \times 10^{-205}$$

At 10^{880}ãr apart, the $1,8 \times 10^{-302}$pu negative backhand shadow generated by the Electron orb and cast by the Proton surface is diluted to $2,37 \times 10^{-205}$ at the standard coronal distance and delivers $4,27 \times 10^{-507}$pu orb backhand. When a re-launched fully inherent photon starts to locally orbit the Proton at a standard distance of $4,21 \times 10^{856}$ãr, it passes at $4,21 \times 10^{-284}$¢ through its 10^{852}ãr wide orb in $2,38 \times 10^{1135}$ãt. The $4,27 \times 10^{-507}$pu pull allows it to block $1,02 \times 10^{629}$pu negative virgins under way which drives it in the direction of the home Proton at a speed of $1,02 \times 10^{-283}$¢.

This beats back auxiliary photons re-launched at $4,21 \times 10^{-284}$¢ when they reach the standard corona distance and prevent the Proton corona from expanding. The working time is longer by $\sqrt{4,21 \times 10^4}$ or $4,88 \times 10^{1137}$ãt but it would take $4,13 \times 10^{1139}$ãt to recall the photon all the way to the surface. This Proton corona is a further 84,63 orbits from becoming a pod corona.

The orb at the Proton surface is $1,02 \times 10^5$ fold and the Protons 10^{1140}ãt rotation and launch period yields a $1,02 \times 10^{1145}$ãt inactive surface period. In practice, this indicates a less frequent launch of auxiliary photons that are recalled in one jump; leaving the Proton with fewer auxiliary photon in its corona. Of the 2×10^9 auxiliary stock, only $1,96 \times 10^4$ photons are on average active in the corona at the same time.

The standard emission by a fully stocked Proton corona of 2×10^9 negative auxiliary photons releases $7,28 \times 10^{-284}$ free positive virgins per ãnna. Though only $9,8 \times 10^{-6}$ is available as free coronal emission or $\mathbf{7,13 \times 10^{-289}}$ positive virgins per ãnna it is still $7,13 \times 10^{11}$ times more than is provided by the main body Protons 10^{-300} positive subservient virgin main body emission.

$$1,277 \times 10^{851}ãr \div 10^{880}ãr = 1,63 \times 10^{-58} (gain\ 1,277 \times 10^9) = 2,08 \times 10^{-49} \times 7,13 \times 10^{-289} =$$
$$1,48 \times 10^{-337}pu \times 10^{38}¢ = 1,48 \times 10^{-299}pu\ vs\ 6,39 \times 10^{-308}pu = the\ orb\ backhand$$
$$outguns\ the\ surface\ repulsion\ 2,32 \times 10^8\ fold$$

$$10^{752}\tilde{a}r \div 1,47 \times 10^{854}\tilde{a}r \; standard = 4,63 \times 10^{-205} \times 1,15 \times 10^3 = 5,32 \times 10^{-202}$$

The $1,48 \times 10^{-299}$pu positive backhand shadow generated by the Protons limited virgin orb and cast from an Electron surface is diluted to $5,32 \times 10^{-202}$ at the Electrons standard coronal distance and delivers $7,87 \times 10^{-501}$pu orb backhand upon each auxiliary precursor at the standard distance where the auxiliary photon completes its first orbit.

A fully inherent dominant auxiliary photon at $1,47 \times 10^{854}$ãr has started to orbit the Electron, passing at $7,05 \times 10^{-284}$¢ speed through the $1,277 \times 10^{851}$ãr Electron wide orb in $1,81 \times 10^{1134}$ãt where the $7,87 \times 10^{-501}$pu pull blocks $1,42 \times 10^{634}$pu positive virgins under way which drives it in the direction of the Electron at **$1,42 \times 10^{-278}$¢** with increasing orb pressure.

This terminates the Electron coronas $7,05 \times 10^{-284}$¢ expansion and allows no further widening of that standard coronal reach. It also means that the Electron coronal recall from $1,47 \times 10^{854}$αr has no need for $\sqrt{1,15 \times 10^3}$ extra $6,14 \times 10^{1135}$ãt working time as the Electron has an even firmer pod corona that is recalled in a single jump already within **$1,04 \times 10^{1132}$ãt.**

The $4,64 \times 10^{1145}$ãt inactive period on the Electron surface ($2,32 \times 10^8$ fold orb versus surface repulsion times the $2,08 \times 10^{1137}$ãt rotation period) is now the longest coronal period. It is of course longer than the $1,04 \times 10^{1132}$ãt recall but it is also $2,32 \times 10^8$ times longer than the $2,08 \times 10^{1137}$ãt re-launch period which cuts the number of auxiliary photons that are simultaneously in the corona from 2×10^9 to $8,62$ orbiting photons. The $2,52 \times 10^{-287}$ free emission from a full corona now drops to **$1,09 \times 10^{-295}$** free virgins per ãnna.

The outcome from this strange exercise of the Electrons gradually dropping free emission shown through steps from 3×10^{-293} to $1,09 \times 10^{-295}$ to $275,23$ times fewer virgins from the start. The upshot of this deficiency is that the Electron pressure upon the Proton corona is decreasing and the Proton input of more free coronal emission upon the Electron is causing this imbalance to continue. The Proton gains greater hegemony and its corona soon breaks out from standard size which in turn affects both hydrons in this altering exchange.

Step 3; When the dropping free Electron emission reaches **10^{-296}** virgins per ãnna, this begins to affect the Protons coronal size.

$$10^{852}\tilde{a}r \div 10^{880}\tilde{a}r = 10^{-56} \; (g. \; 10^{10}) = 10^{-46} \times 10^{-296} = 10^{-342}pu \times 10^{38}¢ = 10^{-304}pu \; vs.$$
$$1,77 \times 10^{-307}pu = 5,65 \times 10^2 \; fold$$

The 10^{-304}pu backhand at the Proton surface recall its corona to its surface and re-launch it against the orb to standard distance where its auxiliary photons start to orbit.

10^{752}ãr÷ $4,21×10^{856}$ãr= $5,64×10^{-210}×$ $4,21×10^4 = 2,37×10^{-205}$

At 10^{880}ãr apart, the 10^{-304}pu negative backhand shadow generated by the Electron orb and cast by the Proton surface is diluted to $2,37×10^{-205}$ at the standard coronal distance and delivers $2,37×10^{-509}$pu orb backhand. When a re-launched fully inherent photon starts to locally orbit the Proton at a standard distance of $4,21×10^{856}$ãr, it passes at $4,21×10^{-284}$¢ through its 10^{852}ãr wide orb in $2,38×10^{1135}$ãt. The $2,37×10^{-509}$pu pull allows it to block $6,64×10^{626}$pu negative virgins under way which drives it in the direction of the home Proton at $6,64×10^{-286}$¢. With β 63,4 the standard Proton corona expands to **$2,67×10^{858}$ãr**.

When an inherent photon orbits the Proton at $2,67×10^{858}$ãr, it passes at $2,67×10^{-282}$¢ through its 10^{852}ãr wide orb in $3,75×10^{1133}$ãt where the orb backhand of $2,37×10^{-509}$pu pull allows it to block $8,89×10^{624}$pu negative virgins under way which drives it in the direction of the home Proton at $8,89×10^{-288}$¢. The working time is longer by $\sqrt{2,67×10^6}$ or $6,12×10^{1136}$ãt but it would take $3×10^{1145}$ãt to recall an photon all the way to the surface so this is now far from being a pod corona. Due to the longer recall, the presence of simultaneous photons in the corona is not restricted by the $5,65×10^2$ fold surface orb.

10^{852}ãr÷ $2,67×10^{858}$ãr=$1,4×10^{-13}$ (gain $6,12×10^{20}$)=$8,57×10^7$= $1,17×10^{-8}$
×**$2×10^{-271}$**=$2,34×10^{-279}$ free coronal virgins

The robust kickback of $2,34×10^{-279}$ free coronal virgins from the Proton corona upon an already heavily limited Electron corona becomes another game changer.

$1,277×10^{851}$ãr÷10^{880}ãr = $1,63×10^{-58}$ (gain $1,277×10^9$) = $2,08×10^{-49}×$ $2,34×10^{-279}$ = $4,87×10^{-328}$pu × 10^{38}¢ = $4,87×10^{-290}$pu vs $6,39×10^{-308}$pu = the orb backhand outguns the surface repulsion $7,62×10^{17}$fold

10^{752}ãr÷ $1,47×10^{854}$ãr standard = $4,63×10^{-205}×$ $1,15×10^3 = 5,32×10^{-202}$

The massive $4,87×10^{-290}$pu positive backhand pulse generated by the Protons revitalized virgin orb and cast from an Electron surface is diluted to $5,32×10^{-202}$ at the Electrons standard distance and delivers $2,59×10^{-491}$pu orb backhand upon each auxiliary precursor as an auxiliary photon makes its first orbit.

A fully inherent dominant auxiliary photon at $1,47 \times 10^{854}$ãr has started to orbit the Electron, passing at $7,05 \times 10^{-284}$¢ speed through the $1,277 \times 10^{851}$ãr Electron wide orb in $1,81 \times 10^{1134}$αt where the $2,59 \times 10^{-491}$pu pull blocks $4,69 \times 10^{643}$pu positive virgins under way which drives it in the direction of the Electron at **$4,69 \times 10^{-269}$¢** superluminal speed.

The Electrons coronal recall from $1,47 \times 10^{854}$ãr in this pod corona is in a single brief jump already in **$3,13 \times 10^{1122}$ãt.**

The $1,58 \times 10^{1155}$ãt inactive period on the Electron surface ($7,62 \times 10^{17}$ fold orb versus surface repulsion times $2,08 \times 10^{1137}$ãt rotation period) is by far the longest coronal period. It cuts the auxiliary photon number on the wing in its corona at the same time from 2×10^9 to average $2,62 \times 10^{-9}$ orbiting photons. The $2,52 \times 10^{-287}$ per ãnna free emission from a full corona drops to **$3,31 \times 10^{-305}$** free virgins.

However, this full drop in the Electron generated orb upon the Proton never comes into play because the Electron continues to deliver $3,567 \times 10^{-301}$ virgins per ãnna in steady main body virgin emission as it has done all along. This main body emission is now $1,08 \times 10^3$ times greater than the coronal emission.

Step 4; When the free Electron emission drops to **$3,57 \times 10^{-301}$** virgins per ãnna, this steady main body output bring stability of sorts to the Proton corona. A different predicament that cannot be altered by Proton responses comes to the fore under this steady Electron generated orb. This weak orb at this hydron distance apart cannot recall the Proton corona all the way to its surface; it can lower the Proton corona over time and put into closer orbits that decimates the Protons free coronal emission.

$$10^{852}\text{ãr} \div 10^{880}\text{ãr} = 10^{-56} \ (g10^{10}) = 10^{-46} \times 3,57 \times 10^{-301} = 3,57 \times 10^{-347}pu \times 10^{38}¢ = 3,57 \times 10^{-309}pu \ vs. \ 1,77 \times 10^{-307}pu = 49,58 \ fold \ repulsion$$

The $3,57 \times 10^{-309}$pu backhand at the Proton surface can recall its corona to $4,96 \times 10^{853}$ãr from its surface ($49,58$ times its main body) under the mid orb but it cannot recall its auxiliary stock all the way; and no re-launch leaves the Proton corona in this low limbo.

There are all kinds of fringe possibilities in this scenario, like that some of the first photons can be recalled and therefore re-launched to even greater than $2,67 \times 10^{858}$ãr distance from the Proton with greater free virgin emission while a majority or a part of the auxiliary stock is left in lower limbo; I take the easy way out by using the toggle all out.

$$10^{752}\text{ãr} \div 2,67 \times 10^{858}\text{ãr} = 1,4 \times 10^{-213} \times 2,67 \times 10^6 = 3,74 \times 10^{-207}$$

The question is how long it takes the Electron generated $3,57×10^{-309}$pu backhand orb to recall the Proton corona part ways from the $2,67×10^{858}$ãr distance it had reached earlier. At 10^{880}ãr apart, the backhand is diluted to $3,74×10^{-207}$ at the coronal distance and delivers $1,34×10^{-515}$pu orb backhand. A fully inherent photon orbits the Proton at $2,67×10^{858}$ãr and crosses its 10^{852}ãr wide orb at $2,67×10^{-282}$¢ in $3,75×10^{1133}$ãt and the $1,34×10^{-515}$pu pull allows it to block $5,03×10^{620}$pu negative virgins under way which drives it in the direction of the home Proton at $5,03×10^{-292}$¢. Although the working time is longer by $\sqrt{2,67×10^6}$ or $6,13×10^{1136}$ãt, it takes far longer or $\mathbf{5,31×10^{1149}ãt}$ to recall the corona **part ways** to $4,96×10^{853}$ãr from its surface.

10^{852}ãr$÷$ $4,96×10^{853}$ãr$=4,07×10^{-4}$ (gain $1,42×10^{22}$)$=5,78×10^{18}= 1,73×10^{-19}$
$×\mathbf{2×10^{-271}}=3,46×10^{-290}$ free coronal virgins

With **'step 4'** we have therefore come a full circle to another game changer; namely, we are back to **'step 1'** where the Proton emission was $\mathbf{4,31×10^{-290}}$ where the Proton had just started off this oscillation phase by increased free emission and eventual widening its own orbital corona.

When the orb generated by the gradually subdued Electron pod corona is eventually unable to recall the Proton corona to its surface, the balance tips over into allowing more auxiliary photon stock on the Electron surface to leave its surface which takes us back into a new oscillation cycle.

Magnetic Reversal Across 10^{880}ãr

At 10^{880}ãr, the Electron pod corona is not affected by magnetic reversal and the main magnetic reversal is that of the less restricted Proton corona. The longest recurring cycle is part-ways recalling of the $2,67×10^{858}$ãr Proton corona in $\mathbf{5,31×10^{1149}ãt}$ but the recurring oscillation cycle lasts during the full working period of $8,9×10^{1154}$ãt passage while the Electron approaches at $1,12×10^{-275}$¢ while mostly delivering $3,57×10^{-301}$ free main body virgins.

The catalyst wind across 10^{880}ãr upon an auxiliary precursor orbiting the Electron is 10^{-256} and allows it to harvest $3,57×10^{-557}$ virgins. During $8,9×10^{1154}$ãt time of passage, the $3,57×10^{-557}$pu repulsion drives each auxiliary orbit by $3,18×10^{598}$ virgins at $3,18×10^{-314}$¢ and across $2,83×10^{841}$ãr or $1,06×10^{-17}$ of the $2,67×10^{858}$ãr corona and the magnetic reversal from a continual $2×10^{-271}$ tailwind is cut by $1,06×10^{-17}$ and further to $2,67×10^{-22}$ due to ($2,67×10^{858}$ãr$÷10^{880}$ãr) with the resulting $5,66×10^{-310}$pu magnetic reversal shared by its 10^{228} standard mattrons as $5,66×10^{-538}$pu per ãnna magnetic reversal.

During the collective $8,9×10^{1154}$ãt active period of passage, this counter wind across 10^{880}ãr drives the Protons main body against the catalyst wind by a maximum harvest of $5,04×10^{617}$ virgins and the generated $5,04×10^{-295}$¢ speed pales beside the ongoing $1,12 \times 10^{-275}$¢ bulls eye speed generated upon the Proton at 10^{890}ãr. One can safely conclude that the magnetic reversal drive upon the Proton at 10^{880}ãr is not a decisive factor (other field effects, foreign to this private exchange, notwithstanding).

Closer still at 10^{878}ãr

This time we jump just a hundred times closer with the same setup as the two 'attractive' adverse hydrons reach to 10^{878}ãr from 10^{880}ãr and cross the smaller distance at $1,12 \times 10^{-275}$¢ during $8,9 \times 10^{1152}$ãt. At 10^{880}ãr, the Electron generated orb of $3,57 \times 10^{-301}$ virgins per ãnna in main body emission was not enough to bring the Protons auxiliary corona all the way back to its surface. This resulted in free emission from a lowered orbit that intermittently eased pressure upon the Electron and allowed rising emission from its heavily subdued corona that in turn helped to completely recall the Proton corona with a consequent re-launch.

We can now look in on these two adverse hydrons at 10^{878}ãr at the point where the Electrons coronal emission drops below $3,57 \times 10^{-301}$ virgins in main body emission per ãnna. As more of the steady weak virgin orb input is harvested at this distance what changes is that the Electron generated orb can recall the Proton corona all the way to its surface.

10^{852}ãr$\div 10^{878}$ãr$=10^{-52}(g.10^{11})=10^{-41} \times 3,57 \times 10^{-301}=3,57 \times 10^{-342}pu \times 10^{37}$¢
$=3,57 \times 10^{-305}pu$ v. $1,77 \times 10^{-307}pu=201,69$ fold orb

The $3,57 \times 10^{-305}pu$ backhand cast at the Proton surface can easily recall its corona all the way to its surface which allows the re-launch of its auxiliary stock to greater distances against the weak Electron generated orb.

10^{752}ãr$\div 4,21 \times 10^{856}$ãr$= 5,64 \times 10^{-210} \times 4,21 \times 10^4 = 2,37 \times 10^{-205}$

At 10^{878}ãr apart, the $3,57 \times 10^{-305}pu$ negative backhand shadow generated by the Electrons main body generated orb and cast by the Proton surface is diluted to $2,37 \times 10^{-205}$ at the standard distance and delivers $8,46 \times 10^{-510}pu$ backhand. When a re-launched fully inherent photon starts to locally orbit the Proton at a standard distance of $4,21 \times 10^{856}$ãr, it passes at $4,21 \times 10^{-284}$¢ through its 10^{852}ãr wide orb in $2,38 \times 10^{1135}$ãt. The $8,47 \times 10^{-510}pu$ pull allows it to block $2,02 \times 10^{626}pu$ negative virgins under way which drives it in the direction of the Proton at $2,02 \times 10^{-286}$¢. The standard Proton corona will be expanded by β 47,98 to **$2,02 \times 10^{858}$ãr** with a marginally (1,32 times) shorter reach than earlier.

10^{752}ãr$\div 2,02 \times 10^{858}$ãr$= 2,45 \times 10^{-213} \times 2,02 \times 10^6 = 4,95 \times 10^{-207}$

When an inherent photon orbits the Proton at $2,02×10^{858}$ãr, it passes at $2,02×10^{-282}$¢ through its 10^{852}ãr wide orb in $4,95×10^{1133}$ãt where the orb backhand of $1,77×10^{-511}$pu pull allows it to block $8,77×10^{622}$pu negative virgins under way which drives it in the direction of the home Proton at $8,77×10^{-290}$¢. The working time is longer by $\sqrt{2,02×10^6}$ or $7,04×10^{1136}$ãt but it would take $2,3×10^{1147}$ãt to recall an photon all the way to the surface so this is not a pod corona. Due to the longer recall, the presence of photons in the corona is not restricted by the $2,02×10^2$ fold surface orb.

10^{852}ãr ÷ $2,02×10^{858}$ãr=$2,45×10^{-13}$ (gain $7,04×10^{20}$)=$1,72×10^8$= $5,81×10^{-9}×2×10^{-271}$=$1,16×10^{-279}$ free coronal virgins

In this case a robust kickback of $1,16×10^{-279}$ free coronal virgins from the Proton corona upon the already heavily limited Electron corona is no longer a game changer because the Electrons constant main body emission can now recall the Proton corona. The exchange has entered phase with almost perpetual Proton corona that will from here continue to shrink at smaller distances apart while the Electrons pod corona is put ever more out of commission while keeping constant main body virgin emission.

$1,277×10^{851}$ãr÷10^{878}ãr = $1,63×10^{-54}$ (gain $1,277×10^{10}$) = $2,08×10^{-44}×1,16×10^{-279}$ = $2,41×10^{-323}$pu

As to the Proton direct repulsion upon the Electron body in $8,9×10^{1152}$ãt period of passage, the $2,41×10^{-323}$pu push delivers $2,15×10^{830}$ virgins upon the Electrons $2,08×10^{1137}$ primal mass generating $1,03×10^{-307}$¢ repulsion speed and it clearly requires closer distances for the $1,12×10^{-275}$¢ speed of approach to be broken.

Magnetic Reversal Across 10^{878}ãr

At 10^{878}ãr, the Electron pod corona is not affected by magnetic reversal and during the $8,9×10^{1152}$ãt period of passage at $1,11×10^{-275}$¢, all magnetic reversal is from the Proton corona while the Electron delivers $3,57×10^{-301}$ free virgins per ãnna main body emission.

The catalyst wind across 10^{878}ãr upon an auxiliary precursor in orbit around the Electron is 10^{-252} and allows each to harvest $3,57×10^{-553}$ virgins. During the $8,9×10^{1154}$ãt time of passage, the $3,18×10^{-602}$pu repulsion drives each auxiliary orbit at $3,18×10^{-310}$¢ and across $2,83×10^{845}$ãr or $1,4×10^{-13}$ of the $2,02×10^{858}$ãr corona and the magnetic reversal effect from a continual $2×10^{-271}$ tailwind is downsized by $1,4×10^{-14}$ and further to $2,02×10^{-20}$ due to ($2,02×10^{858}$ãr÷10^{878}ãr) with the resulting $5,66×10^{-304}$pu magnetic reversal shared by its 10^{228} standard mattrons as $5,66×10^{-532}$pu per ãnna magnetic reversal.

During the collective $8,9\times10^{1152}$ãt active period of passage, this counter wind across 10^{878}ãr drives the Protons main body against the catalyst wind by a maximum harvest of $5,04\times10^{621}$ virgins and the generated $5,04\times10^{-291}$¢ speed pales beside the ongoing $1,12\times10^{-275}$¢ bulls eye speed generated upon the Proton already at 10^{890}ãr. One can again safely conclude that magnetic reversal drive upon the Proton at 10^{878}ãr (albeit more than $5,66\times10^{-538}$pu at 10^{880}ãr) is not a decisive factor.

Extremely close 10^{874}ãr

In this illustration, jump to 10^{-4} part of the earlier distance and assume that the two 'attractive' adverse hydrons are crossing 10^{874}ãr at $1,12\times10^{-275}$¢ during $8,9\times10^{1148}$ãt where the Proton faces the Electron generated steady main body emission orb of $3,57\times10^{-301}$ virgins per ãnna.

$$10^{852}\tilde{a}r \div 10^{874}\tilde{a}r = 10^{-44}(g.10^{13}) = 10^{-31}\times3,57\times10^{-301} = 3,57\times10^{-332}pu\times10^{35}¢ =$$
$$3,57\times10^{-297}pu\ v\ 1,77\times10^{-307}pu = 2,02\times10^{10}fold\ orb$$

$$10^{752}\tilde{a}r \div 4,21\times10^{856}\tilde{a}r = 5,64\times10^{-210}\times 4,21\times10^{4} = 2,37\times10^{-205}$$

At 10^{874}ãr apart, the $3,57\times10^{-297}$pu negative backhand shadow generated by the Electrons main body generated orb and cast by the Proton surface is diluted to $2,37\times10^{-205}$ at the standard distance and delivers $8,46\times10^{-502}$pu backhand. When a re-launched fully inherent photon starts to locally orbit the Proton at a standard distance of $4,21\times10^{856}$ãr, it passes at $4,21\times10^{-284}$¢ through its 10^{852}ãr wide orb in $2,38\times10^{1135}$ãt. The $8,46\times10^{-502}$pu pull allows it to block $2,01\times10^{634}$pu negative virgins under way and drive it in the direction of the home Proton at $2,01\times10^{-278}$¢ superluminal speed which prevents the corona from expanding further and recalls the auxiliary photon to the surface in $2,09\times10^{1134}$ãt before it can cross the orb.

Even the Proton corona has become a pod corona and the time for magnetic reversal is over for both hydrons.

The $2,02\times10^{10}$ fold surface orb limits the number of auxiliary photons in the Proton corona from 2×10^{9} to a 0,1 photon average. This means that every period with one single auxiliary photon in flight, is followed with a ten times longer period while there is no photon on the wing.

The standard emission by a fully stocked Proton corona of 2×10^{9} negative auxiliary photons releases $7,28\times10^{-284}$ free positive virgins per ãnna. Though only $2,02\times10^{-10}$ is available as free coronal emission or **$1,47\times10^{-293}$** positive virgins per ãnna, this still remains $1,47\times10^{7}$ times more than is provided by its 10^{-300} positive subservient virgin main body emission.

$1,277 \times 10^{851} \tilde{a}r \div 10^{874} \tilde{a}r = 1,63 \times 10^{-46}$ *(gain $1,277 \times 10^{12}$)* $= 2,08 \times 10^{-34} \times 1,47 \times 10^{-293}$
$= 3,06 \times 10^{-327} pu \times 10^{35} ¢ = 3,06 \times 10^{-292} pu$ *vs* $6,39 \times 10^{-308} pu =$ *the orb backhand*
outguns the surface repulsion $4,79 \times 10^{15}$ fold

$10^{752} \tilde{a}r \div 1,47 \times 10^{854} \tilde{a}r$ *standard* $= 4,63 \times 10^{-205} \times 1,15 \times 10^{3} = 5,32 \times 10^{-202}$

The massive $3,06 \times 10^{-292} pu$ positive backhand generated by the Proton orb and cast from an Electron surface is diluted to $5,32 \times 10^{-202}$ at the Electrons standard distance and delivers $1,63 \times 10^{-493} pu$ orb backhand upon each auxiliary precursor as an auxiliary photon makes its first orbit.

A fully inherent dominant auxiliary photon at $1,47 \times 10^{854} \tilde{a}r$ has started to orbit the Electron, passing at $7,05 \times 10^{-284} ¢$ speed through the $1,277 \times 10^{851} \tilde{a}r$ Electron wide orb in $1,81 \times 10^{1134} \tilde{a}t$ where the $1,63 \times 10^{-493} pu$ pull blocks $2,95 \times 10^{641} pu$ positive virgins under way. Even the restricted Proton input at this close distance drives it in the direction of the Electron at $2,95 \times 10^{-271} ¢$ superluminal speed. It bears repeating that the Electrons coronal recall from $1,47 \times 10^{854} \tilde{a}r$ is still making this corona a pod corona that is recalled to the surface in a single jump within **$4,98 \times 10^{1124} \tilde{a}t$**.

The **$9,96 \times 10^{1152} \tilde{a}t$** inactive period on the Electron surface ($4,79 \times 10^{15}$ fold orb versus surface repulsion times the $2,08 \times 10^{1137} \tilde{a}t$ rotation period) is now by far the longest coronal period. It cuts the number of auxiliary photons simultaneously in the corona from 2×10^{9} to on average $4,18 \times 10^{-7}$ orbiting photons and the free $2,52 \times 10^{-287}$ emission from a fully stocked corona drops to **$5,26 \times 10^{-303}$** virgins per ãnna. In spite of somewhat lesser pressure from the Proton, the Electrons $3,57 \times 10^{-301}$ main body emission still outguns its own coronal emission.

Under other circumstances, the passage across $10^{874} \tilde{a}r$ at $1,12 \times 10^{-275} ¢$ in $8,9 \times 10^{1148} \tilde{a}t$ could conflict with the Electrons $9,96 \times 10^{1152} \tilde{a}t$ inactive surface period but as the main body virgin input is not delivered by its corona, it can be ignored. The Protons $2,02 \times 10^{1150} \tilde{a}t$ inactive surface period is 22,7 times longer than the period of passage but does not clash with the averages.

$1,277 \times 10^{851} \tilde{a}r \div 10^{874} \tilde{a}r = 1,63 \times 10^{-46}$ *(gain $1,277 \times 10^{12}$)* $= 2,08 \times 10^{-34} \times 1,47 \times 10^{-293}$
$= 3,06 \times 10^{-327} pu$

As to the Proton direct repulsion upon the Electrons main body during the $8,9 \times 10^{1148} \tilde{a}t$ period of passage, the $3,06 \times 10^{-327} pu$ push delivers $2,72 \times 10^{822}$ virgins upon the Electrons $2,08 \times 10^{1137}$ primal mass generating $1,31 \times 10^{-315} ¢$ repulsion speed that is far from stopping a $1,12 \times 10^{-275} ¢$ speed of approach.

Words of warning

It is unlikely that adverse hydrons ever get this close. Aim at $1,12\times10^{-275}$¢ speed across 10^{880}ãr at a moving Electron target is $1,63\times10^{-58}$ of the Electron distance sphere. This is similar to sending a spacecraft to hit a sun across 10^{11} observable universes without once adjusting its initial aim. At some distance the Proton will almost certainly pass by the Electron (or vice verse) and thereafter continue to depart the premises.

Such a close encounter, at whatever distance it comes, is of course not the end of their acquaintance. Much like under a gravity spell on approach, the departing coronas can soon again launch their auxiliary photons with a further reach and the hydrons will again be subjected to magnetic reversal from another direction. They can (in principle) form a lasting orbital bond between them as they fall in increasingly orbital fashion away from and towards each other under the influence of magnetic reversal.

To top it off, all these steps during approach can be altered by the simple intrusion of a passing photon (or another hydron) at a distance where their free emission can ground the corona of either of these two adverse hydrons.

$$10^{852}\tilde{a}r \div 10^{889}\tilde{a}r = 10^{-74}\,(3,16\times10^6) = 3,16\times10^{-68}\times10^{-280} = 3,16\times10^{-348}pu\times10^{42}¢$$
$$= 3,16\times10^{-306}pu\; 1,77\times10^{-307}pu = 17,85\,fold\;orb$$

$$1,277\times10^{851}\tilde{a}r \div 10^{888}\tilde{a}r = 1,63\times10^{-74}\,(g.1,277\times10^5) = 2,08\times10^{-69}\times\,10^{-280}\times10^{42}¢$$
$$= 2,08\times10^{-307}pu\;vs.\;6,39\times10^{-308}pu = 3,26\,fold$$

A passing positive photon grounds a Proton corona from roughly 10^{889}ãt away and a negative photon grounds an Electron corona from over 10^{888}ãt away. A corona that cannot launch its auxiliary photons allows the other hydron to launch its corona to vast distances and to gain at least a temporary hegemony while the hydron with the grounded corona fields a smaller input of free main body virgins; thus a grounded corona puts all illustrations out of business.

How balances come about or change in innumerable alternating steps is best dealt with by calculations that are better suited to the task than conceptual illustrations. Nonetheless, the mechanisms of mutual influences have become clearer against the backdrop of these rough illustrations and can therefore be more perceptively studied.

Adversely Bonded Hydrons

It is likely that both hydrons have already formed mutual enforced orbits around each other as they build up mutual speed on drawing closer whether they do this equally or one hydrons chases the other until it may become the others turn to put up the chase before a bond is realized where both adverse hydron often form pod coronas.

A Proton with coronal pods keeps robust coronal repulsion and may repel the Electron faster than it is pushed in return. When the Proton has pushed the Electron across more than a certain distance, the Electron generated orb has weakened to such extent that is allows the Protons auxiliary photons to make a few orbits.

Losing a pod corona, magnetic reversal is re-activated upon the Proton that closes the gap again. This continuous instability forms a stable local instability bond that is extremely hard to break. It periodically leaves enough coronal repulsion to repel the Electron until the Protons magnetic reversal after halfheartedly chasing the orbiting Electron will intermittently turn into a pod corona which is successful in terminating the Electrons flight. Paired to a single Electron, these shifts continue to ad infinitum, causing a small orbital wobble around a central point in a double system.

When linear virgins from either paired hydron are blocked by an auxiliary photon in a pod corona, they cannot push it further to the leeside orbit of its hydron since the virgin aim is at right angles to the mutual recall planes.

It must be emphasized that an auxiliary photon can block virgins from other entities that arrive at other angles parallel to the closest orb plane. Such linear push can drive an auxiliary photon in a pod orbit further to the leeside of its hydron and create magnetic reversal for that individual hydron (not the adverse hydron pair). In tighter hydron groups, these less decisive more complicated bond patterns are rarely as powerful as between isolated adverse hydrons.

Where hydrons of opposite polarity enter a special bond of coronal pods, the mutual main body repulsion that periodically works to push them apart has a drive far superior to any backhand attraction from inter-galactic flux. The 'G' backhand is also cast in different signs between adverse hydrons and delivers therefore a feeble shadow with negligible force between a Proton and an Electron.

This, in a nutshell, is a bond fashioned by adverse magnetic reversal, or by the lack of it, between two adverse hydrons; a simple pirates map to where the treasure is buried. These interacting effects need detailed foundation and analyses for every one of all the various possible setups that can be imagined in this situation.

A word may be needed on why a Proton does not collect more Electrons. When a new further off Electron passes within say 10^{888}är parallel to bonded Proton Electron pair loosely bonded the new approaching Electron emits the most free negative coronal virgins and orb input upon the three players. The bonded Proton is already limited by its Electron to a pod corona with few free virgins per ãnna; the bonded Electron emits only main body emission.

With almost no magnetic reversal to drive the intruding Electron closer to the established pair, its established linear speed and aim are important (it has a 10^{-7} chance of passing this close in the average grid. The new Electron is almost always at a different angle from the Proton than is its bonded but the coronal emission of the new Electron can easily ground a Proton corona when the bonded Electron has strayed too far.

The main change is that the balance of the bonded hydrogen pair is upset. The bond is forged by the intermittent mutual repulsion that brought them out to a distance where the Proton corona grew strong enough for magnetic transfer to be triggered under the Electrons meager main body emission. As it is the stronger component, this reaffirms the bond but this cannot happen while the Proton corona remains grounded.

As the bonded hydrons mutually repel each other, the bonded Electron gains increased ability to launch auxiliary photons into a corona which will increase its repulsion of the Proton; gradually driving the two apart across considerable distances. Not only does this accelerate the Electrons repulsion of its bonded Proton but it allows is new pod corona to generate magnetic transfer upon the new faraway Electron passing at 10^{888}ãr. This one sided repulsion can be massive upon a faraway Electron with its less restricted corona and more robust negative coronal emission from the bonded Electron repels the far-off Electrons main body by way of magnetic transfer.

With the reflected faraway Electron moving away and no longer able to ground the Proton corona, that corona is partly released and the earlier bond restored, albeit at a greater distance. After a large discontinuous transition jump, some magnetic reversal will help to bring the bonded hydrons together again as the passing Electron passes away in the grid.

Haven been kicked asunder to a different state that is only stable for a short while bears a striking resemblance to the familiar quantum leap of an Electron in an atom as it suddenly shifts from one energy level to another. In this case it is clearly not caused by the emission and absorption of a photon but rather by the same passing photon.

If the rejected Electron is lingering about in a hydron grid with nowhere to go after being pushed weakly away, the consequence is that it may soon be able to ground the Proton corona again only to be repelled again by magnetic transfer from the bonded Electron. It is unlikely that an extra Electron can be held weakly on either side of the 10^{888}ãr wide globular shell in a dynamic alternating bond and distant Electrons on the outer fringes could form weak and less tangible, easily disturbed bonds that are prone to easy swaps.

In relative terms, the bonded Electron jumped into a wider orbit during the approach of the intruding Electron. When the pair is again attracted by way of magnetic reversal; they jump back into a smaller stable bonded orbit.

The above coronal grounding and interactions are symmetric; a bonded hydrogen atom can of course also be approached by a free negative photon (or Proton) that instead disturbs the bonded pair by grounding the Electron corona from a distance with an adverse but quite similar outcomes.

Protons Magnetic Transfer 10^{894}ãr

The gravity flux backhand 'G' with $3,736 \times 10^{922}$pus or $1,25 \times 10^{-272}$pu per ãnna is based on inter-galactic flux backhand between two Protons that block negative flux at the same speed with identical mutual backhand fully operative between them by the inverse square law. The 'G' backhand yields too weak attraction between adverse hydrons to have a meaningful effect in local exchanges except at extremely close quarters but it can be used as a comparison 'constant' against direct repulsion and magnetic transfer.

To illustrate *Magnetic Transfer*, one has to look at an 'otherwise isolated' pair of polarized standard hydrons because almost all hydrons form grids of diverse sizes. I must therefore illustrate within an 10^{895}ãr wide grid where isolation is impossible because of the significant adverse inputs that can affect the situation from multiple adverse hydrons in the surrounding grid.

At 10^{894}ãr inter-hydron distance for two 'isolated' hydrons within a 10^{895}ãr grid as shown in magnetic reversal between two adverse hydrons, _**replace the Electron with another Proton**_ and see the radically different magnetic transfer at work between the two allied hydrons instead of magnetic reversal. Here the magnetic transfer works by way of free exchange of positive adverse coronal virgins that repel main bodies but can't interact with allied coronas that would be launched to vast distances if they were ever recalled.

Closeness is not in their nature. If forced closer, mutual Proton repulsion, coronal or main body, continues during their approach and work to push any two isolated Protons apart with accelerated magnetic transfer.

To set some coronal reach for the Proton, one has to select some input of negative virgins from all surrounding domains, adjacent across the 10^{895}ãr grid or from further away. This could for example be from an adverse hydron pair at 10^{890}ãr apart where an Electron emits $1,8 \times 10^{-279}$ free coronal virgins per ãnna that allows the recall of the Protons corona; or it could be a loose Electron emitting 2×10^{-271} free virgins. To have some ground to build on, one can pick some average emission, let's use 10^{-276} free virgins from multiple Electrons at higher or lower emission levels from closer or further to have shaped the coronal reach around the two Protons; as an illustration only.

10^{852}ãr ÷ 10^{895}ãr = 10^{-86} *(gain 3,16×10²)=3,16×10⁻⁸⁴×10⁻²⁷⁶=3,16×10⁻³⁶⁰pu* $×3,16×10^{45}$¢=10^{-314}pu vs. $1,77×10^{-307}$pu. *The main body repulsion is now 1,77×10⁷ times stronger than orb backhand generated by other domains*

10^{752}ãr ÷ $4,21×10^{856}$ãr= $5,64×10^{-210}$× $4,21×10^{4} = 2,37×10^{-205}$

The 10^{-314}pu negative backhand shadow generated by the strongest grid orb and cast by the Proton surface is diluted locally to $2,37×10^{-205}$ out to the standard coronal distance and delivers $2,37×10^{-519}$pu orb backhand.

When a re-launched fully inherent dominant auxiliary photon starts to locally orbit the Proton at $4,21×10^{856}$ãr, it passes at $4,21×10^{-284}$¢ through its 10^{852}ãr wide orb in $2,38×10^{1135}$ât. During that time the $2,37×10^{-519}$pu pull allows it to block $5,64×10^{616}$pu negative virgins under way which drives it in the direction of the home Proton at $5,64×10^{-296}$¢. This local backhand cannot stop the head on $4,21×10^{-284}$¢ expansion speed with a resistance that falls $7,46×10^{11}$ times short and auxiliary photon continue to orbit outwards. With β as $7,46×10^{11}$ standard coronal Proton measures of $4,21×10^{856}$ãr, the reach of a launched auxiliary photon was reversed at distance **$3,14×10^{868}$ãr**.

So let's pick **10^{868}ãr** wide coronas around both Protons with a perpetual $2×10^9$ stock of auxiliary photons (until recalled) with $2×10^{-271}$ per ãnna free positive virgins emission. These coronal virgins do not have any mutual coronal effect but the 10^{-300} per ãnna negative main body virgins from both the allied hydrons affect both negative coronas with steady emission.

10^{852}ãr÷10^{894}ãr=10^{-84}*(gain 10³)= 10⁻⁸¹ × 10⁻³⁰⁰ = 10⁻³⁸¹pu × 10⁴⁵¢ = 10⁻³³⁶pu vs.* $1,77×10^{-307}$pu = $1,77×10^{29}$*deficiency*

The 10^{868}ãr wide coronas cannot be lowered by the weak 10^{-336}pu negative main body generated orb across their 10^{894}ãr distance apart. This orb level would allow far greater coronas and the two 10^{868}ãr coronas that enter this exchange from other domains are not recalled or lowered and stay therefore with us throughout most of this illustration.

10^{852}ãr ÷ 10^{868}ãr=10^{-32} *(gain 10¹⁶)= 10⁻¹⁶= all 2×10⁻²⁷¹coronal virgins escape and* $2×10^{-287}$ *virgins hit their own Proton*

A **Proton** corona that emits 2×10^{-271} positive free virgins that reach across 10^{894}ãr to the other Proton boosted to 10^{45}¢ with 10^{-81} hitting, push the other Protons main body by 2×10^{-352}pu linear virgin repulsion that does not affect negative auxiliary photons orbiting the allied hydron. Shared by 10^{-228} main body mattrons, the virgins work to push each main body mattron linearly away by a weak **2×10^{-580}pu** per ãnna main body catalyst wind.

During the 10^{1177}ãt period of passage at 10^{-283}¢, the 2×10^{597}pu harvest drives the Proton bodies to 2×10^{-315}¢ across 2×10^{862}ãr, moving the main body Proton across 2×10^{-6} of 10^{868}ãr coronal width and downsizes magnetic transfer on top of the 2×10^{-287} magnetic transfer wind is downsized to by the difference in catalyst and coronal distances to 10^{-26} in aim (10^{868}ãr$\div10^{894}$ãr) and delivers 4×10^{-319}pu repulsion wind shared by 10^{228} standard mattrons as 4×10^{-547}pu per ãnna.

During 10^{1177}ãt, each main body mattron gathers 4×10^{630} virgins of magnetic transfer that can bring the main bodies mutually to **4×10^{-282}¢** away from each Proton. This repulsion speed it greater than the **10^{-283}¢** speed of approach which cannot therefore continue.

The upshot is that the Protons are so mutually repulsive by the magnetic transfer that they will not allow continued approach against the massive headwind but are instead reversed or reflected into departing before reaching 10^{894}ãr apart when approaching at 10^{-283}¢. At higher speeds than outside enforced 4×10^{-282}¢, the two Protons can of course continue to approach.

If we compare this **4×10^{-319}pu** magnetic transfer upon the Proton with 'G' of $1,25\times10^{-272}$pu per ãnna diluted to 10^{-84} across 10^{894}ãr or $1,25\times10^{-356}$pu between the two Protons at this distance, the magnetic transfer is **$3,2\times10^{37}$** times greater than 'G'. As they draw closer, a Protons main body repulsion increases in delivery, aided by virgin boost unlike the inverse-squared $3,736\times10^{922}$pus shadow backhand that 'G' casts by each Proton with its constant 'G' input of $1,25\times10^{-272}$pu per ãnna.

Protons 10^{890}ãr apart

10^{852}ãr$\div10^{890}$ãr$=10^{-76}$ $(g.10^5)= 10^{-71} \times 2\times10^{-271}= 2\times10^{-342}pu\times10^{43}$¢ $= 2\times10^{-298}$pu

A **Proton** corona that emits 2×10^{-271} positive free virgins that reach across 10^{890}ãr to the other Proton boosted to 10^{43}¢ with 10^{-71} hitting, push the other Protons main body by 2×10^{-342}pu linear virgin repulsion which does not affect negative auxiliary photons orbiting the allied hydron. Shared by 10^{228} main body mattrons, the virgins push each linearly away by a weak **2×10^{-570}pu** per ãnna catalyst wind upon the main body Proton.

During the 10^{1171}ãt period of passage at a more penetrating 10^{-281}¢ speed, the 2×10^{601}pu harvest repels Proton bodies at 2×10^{-311}¢ across 2×10^{860}ãr, moving a main body Proton across 2×10^{-8} of its 10^{868}ãr coronal width that leads to a downsizing of magnetic transfer to 2×10^{-8}. The 2×10^{-287} magnetic transfer wind is downsized to by the difference in catalyst and coronal distances to 10^{-22} in aim (10^{868}ãr÷10^{890}ãr) and delivers 4×10^{-317}pu repulsion wind shared by 10^{228} standard mattrons as 4×10^{-545}pu per ãnna which is a much stronger mutual repulsion than the direct one.

During 10^{1171}ãt, each main body mattron gathers 4×10^{626} virgins of magnetic transfer that can repel the main bodies mutually at **4×10^{-286}¢** with a repulsion speed of less than the proposed **10^{-281}¢** speed of approach that can therefore continue unaided against the new headwind.

If we compare this **4×10^{-317}pu** magnetic transfer upon the Proton with 'G' of $1,25\times10^{-272}$pu per ãnna diluted to 10^{-76} across 10^{890}ãr or $1,25\times10^{-348}$pu between the two Protons at this distance, the magnetic transfer is **$3,2\times10^{31}$** times greater. As they draw closer, a Protons main body repulsion increases in strength, aided by virgin boost against inverse-squared $1,25\times10^{-272}$pu per ãnna shadow backhand of 'G' cast by each Proton.

Protons 10^{880}ãr apart

$$10^{852}\text{ãr}÷10^{880}\text{ãr} =10^{-56}\,(gain\ 10^{10})= 10^{-46} \times 2\times10^{-271}= 2\times10^{-317}pu$$

A Protons 10^{868}ãr corona that emits 2×10^{-271} positive free virgins reaching across 10^{880}ãr to the other Proton boosted to 10^{38}¢ with 10^{-46} hitting, push the other Protons main body by 2×10^{-317}pu linear virgin repulsion which does not affect negative auxiliary photons orbiting the allied hydron. Shared by 10^{-228} main body mattrons, the virgins push each main body mattron linearly away by a weak **2×10^{-545}pu** per ãnna main body catalyst wind.

During the 10^{1161}ãt period of continued passage at 10^{-281}¢, a 2×10^{616}pu harvest drives Proton bodies to 2×10^{-296}¢ across 2×10^{865}ãr, moving them across 2×10^{-3} of the 10^{868}ãr coronal width that leads to downsizing of magnetic transfer. The 2×10^{-287} magnetic transfer wind is downsized by the difference in catalyst and coronal distances to 10^{-12} in aim (10^{868}ãr÷10^{880}ãr) to deliver 4×10^{-302}pu repulsion shared by 10^{228} standard mattrons as a powerful 4×10^{-530}pu per ãnna.

During 10^{1161}ãt, each main body mattron gathers 4×10^{631} virgins of magnetic transfer that repels the main bodies mutually away at **4×10^{-281}¢**; a repulsion speed that is 4 times faster than the earlier penetrating **10^{-281}¢** speed of approach. The Protons are so mutually repulsive that the magnetic transfer does not allow continued approach against the massive headwind and they are reversed before reaching 10^{880}ãr apart unless they are forced by other means to approach at faster than 10^{-281}¢ speed.

If we compare this **$4×10^{-302}$pu** magnetic transfer upon the Proton with 'G' of $1,25×10^{-272}$pu per ãnna diluted to 10^{-56} across 10^{880}ãr or $1,25×10^{-328}$pu between the two Protons at this distance, the magnetic transfer has dropped to being merely **$3,2×10^{26}$** times greater. If they draw closer, the Protons main body repulsion continues to increase in strength, being more aided by virgin boost against inverse-squared $3,736×10^{922}$pus shadow backhand of 'G' cast by each Proton with an unaltered 'G' input of $1,25×10^{-272}$pu per ãnna.

10^{852}ãr $÷10^{880}$ãr $=10^{-56}$ (10^{10}) = $10^{-46}×2×10^{-271}=2×10^{-317}$pu *direct repulsion*

The direct repulsion of the two main bodies by free coronal repulsion is also gathering steam and can in 10^{1161}αt create a repulsion drive of its own by $2×10^{844}$ virgins. This can mutually push them asunder at **$2×10^{-296}$¢** which has a long way to go in order to put other forces out of play.

Protons 10^{874}ãr apart

10^{852}ãr$÷10^{874}$ãr $=10^{-44}$ *(gain* $10^{13})$= 10^{-31} × $2×10^{-271}$= $2×10^{-302}$pu

A Proton corona that emits $2×10^{-271}$ positive free virgins that reach across 10^{874}ãr to the other Proton boosted to 10^{35}¢ with 10^{-31} hitting, push the other Protons main body by $2×10^{-302}$pu linear virgin repulsion which does not affect negative auxiliary photons orbiting the allied hydron. Shared by 10^{-228} main body mattrons, the virgins push each main body mattron linearly away by **$2×10^{-530}$pu** per ãnna main body catalyst wind.

If we assume that the Protons were approaching at a penetrating 10^{-280}¢ light speed which makes for a 10^{1154}ãt working period of passage, the linear repulsion in that time can in principle drive a Proton body by $2×10^{852}$pu virgins to speed $2×10^{-288}$¢ and across $2×10^{866}$ãr which as $2×10^{-2}$ of its 10^{868}ãr corona downsizes magnetic transfer. The $2×10^{-287}$ active magnetic transfer wind is cut by 10^{-6} difference in catalyst and coronal distances (10^{868}ãr$÷10^{874}$ãr) whereupon the $4×10^{-295}$pu repulsion is shared by 10^{228} standard mattrons as $2×10^{-523}$pu magnetic transfer per ãnna.

During 10^{1154}ãt, each main body mattron gathers $2×10^{631}$virgins into magnetic transfer that can in principle repel a main body at **$2×10^{-281}$¢** away from the other Proton. This repulsive resistance by magnetic transfer speed is now merely 5 times weaker than the **10^{-280}¢** light speed approach which is not broken at 10^{874}ãr.

If we compare this 4×10^{-295}**pu** magnetic transfer upon the Proton with 'G' of $1,25 \times 10^{-272}$pu per ãnna diluted to 10^{-44} across 10^{874}ãr or $1,25 \times 10^{-316}$pu between the two Protons at this distance, the magnetic transfer, although more massive, has dropped to a lower ratio and is merely $\mathbf{3,2 \times 10^{21}}$ fold 'G'. As the Protons draw closer, their main body repulsion increases in strength and is aided by virgin boost against inverse-squared $3,736 \times 10^{922}$pus shadow backhand of 'G' cast by each Proton of steady 'G' input of $1,25 \times 10^{-272}$pu per ãnna.

$$10^{852}ãr \div 10^{874}ãr = 10^{-44} (10^{13}) = 10^{-31} \times 2 \times 10^{-271} = 2 \times 10^{-302}pu$$

Then there is also the direct repulsion of the two main bodies driven by free coronal repulsion; a force that during 10^{1154}ãt create a repulsion drive of its own by 2×10^{852} virgins that repels the Protons at $\mathbf{2 \times 10^{-288}}$¢. This does not put other forces out of play but we can see where this is heading as magnetic transfer at 2×10^{-281}¢ is now merely 10^7 fold the direct repulsion.

Protons 10^{872}ãr apart

$$10^{852}ãr \div 10^{868}ãr = 10^{-32} (gain\ 10^{16}) = 10^{-16} = all\ 2 \times 10^{-271} coronal\ virgins\ escape\ and$$
$$2 \times 10^{-287}\ virgins\ hit\ their\ own\ Proton$$

$$10^{852}ãr \div 10^{872}ãr = 10^{-40} (gain\ 10^{14}) = 10^{-26} \times 2 \times 10^{-271} = 2 \times 10^{-297}pu$$

A Proton corona that reaches across a hundredth part of the distance between the allied hydrons and emits 2×10^{-271} positive free virgins that reach across 10^{872}ãr boosted to 10^{34}¢ with 10^{-26} hitting, push the other Protons main body by 2×10^{-297}pu linear virgin repulsion but does not interact with coronas. Shared by 10^{-228} main body mattrons, the virgins push each main body mattron linearly away by a weak $\mathbf{2 \times 10^{-525}}$**pu** per ãnna main body catalyst wind.

During a 10^{1152}ãt working period at light speed approach, the 2×10^{627}pu harvest drives the Proton bodies to 2×10^{-285}¢ across 2×10^{867}ãr, moving the main body Proton across a fifth of the 10^{868}ãr coronal width (if it still holds) and leads to downsizing of the 2×10^{-287} magnetic transfer wind on top of the catalyst and coronal distance downsizing to 10^{-4} (10^{868}ãr $\div 10^{872}$ãr) which combined delivers 4×10^{-292}pu repulsion wind that, shared by 10^{228} standard mattrons, is a powerful 4×10^{-520}pu drive per ãnna.

If we compare this powerful $\mathbf{4 \times 10^{-292}}$**pu** magnetic repulsion with 'G' of $1,25 \times 10^{-272}$pu per ãnna diluted to 10^{-40} across 10^{872}ãr with $1,25 \times 10^{-312}$pu drive between the two Protons, the mutual attraction by 'G' is $\mathbf{3,2 \times 10^{20}}$ times weaker. And during 10^{1152}ãt, each main body mattron gathers 4×10^{632} virgins of this magnetic transfer that can in principle push the main body at 4×10^{-280}¢ or 4 times light speed away from the other Proton.

$$10^{852}\tilde{a}r \div 10^{872}\tilde{a}r = 10^{-40}(10^{14}) = 10^{-26}\times2\times10^{-271} = 2\times10^{-297}pu$$

The weaker direct 2×10^{-297}pu coronal repulsion drive upon the two main bodies can in principle repel the Protons apart during 10^{1152}ãt by 2×10^{855} virgins at 2×10^{-285}¢ but falls by the wayside because of the more powerful magnetic reversal.

As to the potential recall of the 10^{868}ãr wide Proton corona which is at a ten thousandth part of the distance apart, it is not very effective in moving auxiliary photons closer to the home Proton during the limited time at hand. With 2×10^9 photons in orbit, the orb projected from the other Proton surface would have become marginally more spread at the edges but from a source distance of 10^{16} times the orb width; even this a minor effect.

$$10^{852}\tilde{a}r \div 10^{872}\tilde{a}r = 10^{-40}(10^{14}) = 10^{-26}\times10^{-300} = 10^{-326}pu\times10^{34}¢ = 10^{-292}pu \text{ vs.}$$
$$1,77\times10^{-307}pu = 5,65\times10^{14} \text{ times more powerful orb}$$

As to the recall of the 10^{868}ãr wide Proton corona that is now 10^{-4} of their distance apart, the 10^{-292}pu negative main body orb is now $5,65\times10^{14}$ times more powerful at the surface than the negative dominant $1,77\times10^{-307}$ virgins in surface emission. This has started to move the auxiliary corona closer to the home Proton but the time is too short; only a drop in the extra enforcing wind can stop the Protons and give them a longer working period.

$$10^{852}\tilde{a}r \div 10^{872}\tilde{a}r = 10^{-40}(gain\ 10^{14}) = 10^{-26} \times 1,77\times10^{-307}pu \times 10^{34}¢ = 1,77\times\times10^{-299}pu$$
$$vs.\ 1,77\times10^{-307}pu$$

$$10^{752}\tilde{a}r \div 10^{868}\tilde{a}r = 10^{-232}\times 10^{16} = 10^{-216}$$

At the steady 10^{868}ãr coronal distance, the $1,77\times10^{-515}$pu backhand (versus $1,77\times10^{-531}$pu repulsion) pulls back at the precursored inherent dominant auxiliary photon orbiting the Proton as it passes at 10^{-272}¢ superluminal speed through the 10^{852}ãr wide Proton orb in 10^{1124}ãt with $1,77\times10^{609}$pu backhand harvest of negative virgins that drive towards the Proton at $1,77\times10^{-303}$¢.

As the average linear attraction towards a hydron continues to work as linear universal motion after each passover by a square root of the orbital coronal distance measured in hydron radii, it adds $\sqrt{10^{16}}$ extra linear working time after each passover to the 10^{1124}ãt passage and allows the $1,77\times10^{-303}$¢ linear drift to continue for 10^{1132}ãt. This moves an auxiliary photon $1,77\times10^{829}$ãr closer to the Proton on each passage or twice that by $3,54\times10^{829}$ãr in each orbit which is

3,54×10^{-39} of its 10^{868}ãr corona in each orbit. Using my shorthand of furthest orbits, falling locally towards the Proton in 2,82×10^{38} orbits each requiring 10^{1140}ãt leads to **2,82×10^{1178}ãt** recall to the Proton surface. This approach is coming closer to the end of the line of forcefully pushing Protons together.

Protons 10^{868}ãr apart

10^{852}ãr÷ 10^{868}ãr=10^{-32} (gain 10^{16})= 10^{-16}= $2×10^{-271}$coronal virgins escape and $2×10^{-287}$ virgins hit the other Proton

Across most of 10^{868}ãr, of 2×10^{-271} virgins emitted by a Proton corona and boosted to 10^{32}¢ some 10^{-16} hit and push the other Proton by linear 2×10^{-287}pu virgin repulsion that, shared by 10^{228} main body mattrons, driving each of its internal mattrons away by a weak **2×10^{-515}pu** per ãnna. As the two Protons are at this point unlikely to orbit each other effectively on their bulls eye path, this linear repulsion push does not suffer extensive neutralization by contrary directions.

Having said that, the virgins across at least half the 10^{868}ãr that hit each Proton push the bodies apart by direct repulsion and by magnetic transfer due to mutual shifts of main bodies. The coronas still keep more auxiliary photons between the Protons with a transfer drive pushing them apart. Half the auxiliary photons are inherent to one of the Protons but not the other.

10^{852}ãr ÷10^{868}ãr =10^{-32} (10^{16}) = 10^{-16}× $2×10^{-271}$ = $2×10^{-287}$pu

During roughly 2,5×10^{1147}ãr working period at new 4×10^{-280}¢, this magnetic transfer peaks at the mid 10^{868}ãr. The 5×10^{860}pu harvest drives the Proton bodies to 5×10^{-280}¢ across 1,25×10^{868}ãr, moving the main body Proton to the edge of its coronal span where it is held. This does not lead to downsizing of the 2×10^{-287} magnetic transfer wind which is also uncut by the catalyst and coronal distance (10^{868}ãr÷10^{868}ãr) and delivers **2×10^{-287}pu** per ãnna repulsion wind that, shared by 10^{228} standard mattrons, is **2×10^{-515}pu**. During 2,5×10^{1147}ãr, each main body mattron would gather 5×10^{632} virgins of magnetic transfer that push the main bodies mutually apart at **5×10^{-280}¢** or 5 fold light speed.

If we compare this powerful **2×10^{-287}pu** magnetic transfer repulsion with 'G' of 1,25×10^{-272}pu per ãnna diluted to 10^{-32} across mid 10^{868}ãr where it delivers **1,25×10^{-304}pu** drive between the two Protons, 'G' attraction is still **1,6×10^{17}** times weaker.

10^{852}ãr ÷10^{868}ãr =10^{-32} (10^{16}) = 10^{-16}× 10^{-300} = 10^{-316}pu

If we compare the 10^{-316}pu direct repulsion drive of the two main bodies to the mutual $1,25 \times 10^{-304}$pu 'G' backhand attraction from mid 10^{868}ãr, the 'G' attraction is **$1,25 \times 10^{17}$ times stronger** were it not for the magnetic transfer.

A new change materializes on Proton approach across the mid 10^{868}ãr when the orbiting coronas moves beyond the other main body Protons that are being forced closer. Each Proton casts a negative main body generated orb from the other Proton around itself away. During a brief period across 10^{852}ãr wide part of that mid 10^{868}ãr trek, mutually cast orb backhands by negative main body emission coincide exactly with the spheres of the auxiliary photons orbiting at 10^{-272}¢. Although these orbs are briefly occupied by all the auxiliary photons, the chance of hitting a Proton is minimal.

As auxiliary photons are thereafter orbiting outside the inter-Proton distance, no part of the auxiliary coronas are now found between the Protons and there can no longer be any shift of their main bodies to the far side of each corona to cause greater coronal virgin emission to emanate from there. Instead, if the two Protons continue to move closer, the coronal overlay becomes closer in form and outside coronal shift becomes more fractional. The magnetic transfer is still there but, out of the blue and from now on, its powerful 2×10^{-287}pu starts to lose its punch. The further the Protons reach towards the common center of two coronas, there more the magnetic transfer repulsion drops until it leaves only the weaker direct main body repulsion which is weaker than 'G' attraction already at 10^{868}ãr.

Coronas can no longer be recalled by local orbs after the coronal distances move beyond the other main body Proton. Another new effect are backhand orb shadows generated by virgins boosted to 10^{32}¢ from auxiliary photons cast by each Protons; at this point restricted to 10^{-32} of the input. There is a small 10^{-32} change that this orb affects the other Proton along the first orb and the net 10^{-64} change makes this weak 2×10^{-319}pu push towards the center of the double system.

Protons 10^{864}ãr apart

After crossing the thin common negative auxiliary and orb sphere and after the Protons have moved a hundred million times closer to **10^{864}ãr**, a more precise overlay of the two coronas puts both Protons closer to one common coronal center. The rapidly approaching Protons are not orbiting on a rough bulls eye path but may by now be passing parallel to each other while linear repulsion and 'G' attraction is not being extensively neutralized.

10^{852}ãr ÷ 10^{864}ãr = 10^{-24} (gain 10^{18}) = 10^{-6} of 10^{-300} or 10^{-306} main body virgins repel the Protons mutually apart

---

The least of the repulsion effects is the direct main body Proton virgin that works to drive them mutually apart by **10^{-306}pu.**

The coronal distance is the same as before; far outside the two Protons distance apart. The auxiliary photons are still in mismatched numbers on either side of each Proton but no longer between the Protons. Nevertheless, each Proton still receives more repulsive virgins from the direction of the other Proton and is thereby repelled.

However, the relative shift that earlier set the level of applicable virgins to carry the repulsion while a different extra aim dilution from between Protons set the final effect as the product of these numbers, the inter-Proton distance will in this case set both the level and the aim; in other words the distance in Proton radii squared. With coronas so nearly in circular orbit and Protons so close at the coronal center, both the direct coronal repulsion and magnetic transfer is this time fractionally set by $(10^{868}$ãr$\div 10^{864}$ãr$)^2$ with 10^{-8} effect.

10^{852}ãr$\div 10^{868}$ãr$=10^{-32}$ (gain 10^{16})= 10^{-16}= 2×10^{-271}coronal virgins escape and 2×10^{-287} virgins repel each Proton

---

With the Protons approaching 10^{860}ãr at 5×10^{-280}¢ in 2×10^{1139}ãr the exchange of 2×10^{-287}pu virgins reaching each Proton and pushes it by 10^{-8} linear effect of **2×10^{-295}pu virgin repulsion and magnetic transfer** per ãnna; shared by 10^{228} internal main body mattrons.

From 10^{864}ãr apart, for this example of intact coronas, the full gravity flux backhand 'G' ($3,736\times10^{922}$pu per second) is diluted by the inverted-square law to 10^{-24} of its $1,25\times10^{-272}$pu per ãnna and works to pull the two Protons linearly closer by a **$1,25\times10^{-296}$pu** drive.

*The mutual repulsion is now merely **16 fold** the 'G' gravity attraction.*

The Strong Force

With the Protons approaching 10^{863}ãr with 2×10^{-287}pu virgin harvest per ãnna, the mutual repulsion is by 10^{-10} linear effect and **2×10^{-297}pu push.**

From 10^{863}ãr apart, for this example of intact coronas, the full gravity flux backhand 'G' ($3,736\times10^{922}$pu per second) is diluted by the inverted-square law to 10^{-22} of its $1,25\times10^{-272}$pu per ãnna and works to pull the two Protons linearly closer by a **$1,25\times10^{-294}$pu** drive.

We are looking at the first example of two standard Protons coming under the spell of 'G' flux gravity to an extent that overwhelms all other local effects. Gravity and the 'Strong Force' are considered the two remaining fundamental theories in nature outside the scope of unification; that these forces are one and the same may simplify the picture; the Strong Force is Gravity.

'G' gravity backhand is now a ruling local force at 10^{863}âr. The gravity attraction is $\mathbf{6,25 \times 10^2}$ times more powerful at than the mutual 2×10^{-297}pu repulsion between the isolated Proton pair. From one moment to the next between 10^{864}âr and 10^{863}âr, instead of resisting the approach forcing the Protons closer, the repulsion is suddenly lifted and they are sucked closer with more than 100 fold linear drive. An allied bond forms at far closer distances than adverse bonds.

What we are looking at is to all intents and purposes the formation of an old fashioned orbital gravity bond; two stars coming under each other's gravity spell and falling closer while gaining speed on corrupted paths as they start to orbit closer along initially more elliptical paths.

The above setup, given specific distances and coronas, demonstrates the problem with using simple illustrations. The normal thing for Protons is to have Electrons at say 10^{874}âr apart. Were we to follow two Protons being forced to close ranks with orbiting Electrons where mutual coronal recall is extremely influential, the distance would be different for 'G' to take over. Although the intermittent weak 'G' authority would be there, it is first at the 10^{863}âr distance (even if enforcing pressure were lifted, unless the entire input of inter-galactic 'G' weakens significantly, that the mutual gravity bond turns unbreakable.

At greater distances, two Proton coronas can be recalled and re-emitted against mutual main body exchanges to lower orbits than their distance apart. This would allow massive magnetic transfer to mutually repel them at great speed while diminishing 'G' gravity works between them by the inverse square law. Here the Protons would fly apart whereas a coronal recall within 10^{863}âr would lead to no alterations that can break the bond.

If the Protons are sucked together on an ongoing linear approach from 10^{863}âr to say a thousand times smaller 10^{860}âr distances to form elliptical orbits that through the later corona effects eventually turn more circular. These fast orbits add stability by increased neutralization of linear drives in any one universal direction. The mutual orbits around the incorporeal center of a double or multiple system balances would already be gaining motion. Slower orbital movement further afield does not allow shifts by such brutal forces.

If 10^{860}âr seems close, make no mistake, these standard baryons face a formidable exchange distance across 10^8 fold their radii, the relative orbiting distance between Our Sun and Proxima Centauri, the nearest galactic sun.

In nature, there is one obvious candidate (aside from human fiddling) for individual enforced system pressure that can forge double or larger groups within the average hydron grid into closer contact. This would be during the periods of galactic and solar conversion where oversized hydron embryos are without coronas during increasing 'G' flux backhand; forming gravity bonds where further approach is resisted by a slower balanced orbital effect.

There is very little 'G' gravity working between a bonded Proton Electron which leads to volatile reactions if for example a corona is grounded and one or both are pushed into vastly wider orbits. There are so many special cases in this dance that there is little point in serving up an endless row of illustrations; for example when two hydrogen atoms are forced closer, Proton coronas are grounded when the Electrons are intermittently at right angles to various degrees.

Two Electrons Forced Closer

If two of the lighter and flightier Electrons had left their Protons to roam free, they too could be forced closer by virgin winds and would also resist the move by mutual repulsion and magnetic transfer. The initial effects are similar to those between Protons around which most Electrons orbit. In the absence of other local effects than their own, after magnetic transfer dissipates, there is no point at which repulsion gains enough on G^e attraction to outgun it at close distances.

As to gravity, the robust negative inter-galactic gravity flux is harvested at higher speeds by Protons or Neutrons; an Electron cannot block this flux that is ineffective between Electrons. Nonetheless, the $4,18 \times 10^{-274}$pu per ãnna ($1,255 \times 10^{921}$pus) positive flux G^e gravity backhand is blocked by Electrons at high speeds and casts its backhand shadows between them.

An orbital bond of two Electrons must be forged by outside virgin winds that overwhelm mutual linear repulsion upon lighter $2,083 \times 10^{225}$ mattron masses. This requires relatively larger pressure than between two Protons that also harvest greater 'G' while forging a similar Proton bond. Before an orbital gravity bond is arrived at, there is a long history of mutual approach where various forces also resist such a further advance.

More to the point, an Electron has a vastly greater affinity of bonding with a Proton; if forcing two Electrons together is at all possible, a large local field must be free of Protons that repel Electrons at great distances. That a Proton comes into being during a Neutron split with its own Electron suggests Electron bonds may be possible in theory but remain a rare anomaly.

The Surviving Neutron

A very close Neutron pair in a primeval field after the local field pressure has subsided experiences expansion on facing diminishing local pressures. If their expansion drives were perfectly equal, both would split at the same time which is highly unlikely in an uneven pressure environment.

In the primeval field, neither Neutron is of standard size and neither can be stabilized by the others weak main body emission exchange during their extensive periods of expansion and one Neutron would almost certainly split before the other. As the fading local photon presence loses its punch, the buffer zones between two opposite layers gradually develop into the breach that can eventually cause a Proton Electron split as the larger Neutrons outer layer is being repelled at breakneck speed while consolidating at close quarters into a soon rejected Electron.

It takes a free Neutron $2,6 \times 10^{1197}$ãt to split in a depressurized environment but one must assume that primeval expansion periods during greater pressure were lengthy and uneven due to exchanges of frequently passing photons in local fields where many Neutrons are undergoing similar splits. As the linear photons in both signs have departed after Neutron splits, local pressure would eventually drops below the heavy **$4,08 \times 10^{-286}$pu** per ãnna incoming virgin pressure that is needed to support a Neutrons top layer.

The only way for a Neutron to survive in this depressurized field, after the split of a nearby companion, is staying extremely close to that Proton, as it would have been expanding for the same reason. Stabilized by the massive outpouring from electromagnetic photons heading out from two new nearby hydrons the remaining Neutron is hammered into a standard costume.

Close Neutrons would from the start be orbiting at high speed and remain in these close high speed orbits as a left behind Proton takes the place of the split Neutron companion. While a fast orbiting Neutron cannot be easily dislodged, a slower orbiting Electron continues on an outwards path away from its Proton to pass far beyond the closely orbiting Neutron survivor.

$$10^{852}\tilde{a}r \div 4,94 \times 10^{868}\tilde{a}r = 4,1 \times 10^{-34}(gain\ 4,49 \times 10^{15}) = 1,84 \times 10^{-18} \times 10^{-300}pu =$$
$$1,84 \times 10^{-318}pu \times 2,22 \times 10^{32} = 4,08 \times 10^{-286}pu$$

After the electromagnetic pressure fades, a standard Proton bonded to a close Neutron showers it with powerful 10^{-300} per ãnna positive subservient main body virgin emission. The needed $4,08 \times 10^{-286}$pu top layer compression is reached with a Proton **$4,94 \times 10^{868}$ãr** away with a formidable close quarter boosted virgin exchange in both signs.

The standard surface of a closely paired Neutron is in possession of its old harvest of photons from early expanded stages; 2×10^9 positive and $4,17 \times 10^6$ negative singular mattrons. The powerful backhand orb causes long inactive surface periods and coronas do not play a role at this point.

There is a reciprocal effect in this exchange; while a Proton emits 10^{-300}pu positive and $1,77 \times 10^{-307}$pu negative main body virgins, a companion Neutron emits $1,77 \times 10^{-307}$pu main body virgins per ãnna in both signs but even these deliver so powerful virgin orbs at this short distance between the hydrons that they can only be matched by closely passing photons.

The $4,94 \times 10^{868}$är distance from a Proton is the absolute outer parameter that provides pressure stability to a Proton-Neutron bond. There are other effects from within this outermost parameter that can destabilize this bond but these must be dealt with in due time.

Grid hydrons (not closer companions) can be brought together for many reasons. They are unlikely to be mutually orbiting in the lattice and certainly not at high enough speeds to neutralize the massive repulsion between baryons as local electromagnetic pressure subsides. A Protons main body repulsion would push Neutrons by dominant main body emission out from under its protective pressure umbrella at the parameter distance and turn exceptionally effective if the Proton is able to launch its corona.

External Hydron spinfields

Set aside all other movement within a hydron and every resident at any depth will pass a full orbit around its center in a non-collision orbit in the time it takes the circumference of that orbit to complete a perpendicular orbit around the hydron center. It is impossible to cancel coordinate motion (relative to an independent observer) that is inherent to every internal atom by way of internal emission.

The one component carried outwards by the newborns of a hydrons main body emission and therefore common to all its primal population is inherent collective spin coordinates. Although inherent collective spin is constantly affected by a virgin addition with new spinfields, the common denominator of the past is the collective rotation of the moment. The ongoing net coordinates of additional primal spin may be zero when eventually paired up in collisions but these influences are yet to be boiled down from an enormous variety of contrary movement.

Inherent speed that applies to all the primal population in differential orbit around a hydron center is embedded in its external emission. Born of primal parents with common coordinates, all primal emission that escapes a hydron is endowed with an identical inherent collective orbital sweep (in addition to all the other coordinate effects affecting coronal orbits).

In a standard 10^{852}är Proton, the fully developed inherent orbit of surface mattrons is at 10^{-288}¢. This inherent differential spinfield across the distance to the next hydron grows by the distance in hydron radii. However, as the effect is communicated by hitting virgins, distribution cuts this by r√r. Any enforced effects upon mattrons or parental particles within the hydron are not inherent and cannot be communicated as such away from the hydron.

The inherent coordinates communicate an unopposed nudge in the outside strayfield, brought to bear upon outside mass. Clashing parental motion like primal orbits opening at birthspeed do facilitate distribution of an inherent spinfield but these contrary orbital effects will broadly cancel out.

The hydrons external inherent spinfield may be an insignificant part of the whole harvest but its message is the exclusively delivered in one consequent 'direction'. Due to the boosted distribution, the vast 'empty' field outside a hydron has few blind spots that cannot be reached by primal emission that is delivered with overwhelming parental and primal boosts. The embedded differential collective spin coordinates are a minor part of the message.

Any substantial nearby mass will eventually be nudged into a coherent differential orbit around the emitting hydron even though it never held any of its inherent coordinates. As extensive periods allow a build-up of momentum in a neighboring mass, spinfields work diligently to affect or corrupt other movement. The diversity of primal, parental and mattronal spinfields outside a hydron offers little coherent direction and come across to outside mass as Omni directional pressure where only the collective voice of the hydronal spinfield carries through the static.

Given time, opaque outside mass would therefore gradually accelerate in free fall around a hydron (as well as being repelled or attracted by other means). At some far distance, the outside mass is pushed away along a different linear universal line by every exchanged atom but in the same orbital direction by all of them and this makes it harder for a hydron to repel or 'attract' distant mass.

The orbital effect of an inherent spinfield is boosted at greater distances but this is also where the distribution of its virgin emission drops faster and the orbital 'fall' along the inherent (universal) spin field slows due to the fewer communicating virgins while the distant opaque mass is increasingly pushed around a hydron and less away from or towards it.

At far distances, a fading inherent spinfield is (relatively) better at pushing mass sideways as directional push is diluted faster than inherent sweep. In far off orbital bonds, such induced orbital movement is unopposed in the field but considerable distance is needed for an orbital message of collective spin to mature into an effective orbit which causes emission field strength to be extensively exhausted at distances where the orbital message is clearest.

In a hydron grid, unpaired hydrons keep the greatest possible distances apart. The tiny coherent orbital message and its consequent movement has little time to coach faraway hydrons into a decisive free fall around spinning hydrons. As other grid hydrons also restrict such orbits unless the hydrons are individually closer than the average grid distance; collective spinfields are an extremely weak force compared to the other local fields.

Unlike linear push that is neutralized by orbits, the inherent spin message remains consequent at every point. The upper acceleration limit of the free fall speed in orbit through inherent orbital kicks can never outgrow the emitting hydrons inherent rotation speed in its plane of spin. In reality, due to the square root dilution, it remains a marginal fraction of that spin.

Generally speaking, the inherent field of hydron spin is a weak influence, even during extensive periods, but its directional aim is a crucial element in aligning actual orbits that come into play later by more powerful effects.

Local Activation by Intergalactic Flux

I have already briefly looked at how galactic virgins survive across the grid and how they delivered greater long distance pressure from greater swaths of the universe. How local virgins from for example hydron coronas survive in the grid is similar except for the universal bit.

I found that the majority of local virgins survive across 10^{894}ãr between hydrons while only $1,71\times10^{-23}$ of the positive virgins and $1,61\times10^{-22}$ of the negative virgins were activated under way. Galactic flux fields the greatest presence and a virgin moving across a local volume shares that presence in accord with the share of that local volume of the galactic volume.

Although projected local presence disappears as driving inter-galactic flux is harvested by hydrons elsewhere (is no longer locally present), the method of overlapping volumes does not require a local hydron to be alone in the galactic volume to show a correct conflict rate (with $8,38\times10^{68}$ hydrons in the galaxy, it certainly isn't alone). As the method compares actual presence in a given time; it is already divided up in accordance with overlapping volumes and by association, the relevant involved hydron number.

The exposure to galactic flux increases when a virgin in partly overlapping volumes travels with increasing presence across an ever greater volume until it reaches the galactic volume while galactic flux remains constant. Putting it simply; when virgin presence within a given volume and its share of the galactic volume equals the galactic volume, the majority of the virgins will (thereafter) be activated.

In 10^{903}ãt that it takes virgins to escape such a local volume $3,492\times10^{698}$ **positive virgins** are emitted by $3,492\times10^{75}$ negative inter-galactic photons, boosted to $1,79\times10^{463}$¢ across the $3,22\times10^{926}$ãr galaxy and flux harvested by Electrons at $9,13\times10^{228}$¢ during $1,095\times10^{665}$ãt inter-hydron consummation without interference from dark flux. Before harvest, it gains $6,84\times10^{1826}$ã activated presence within the $3,33\times10^{2779}$ã galactic volume. Of this presence in 10^{903}ãt, only 3×10^{-71} or $2,28\times10^{1755}$ã is within the local volume. A virgins $3,16\times10^{952}$ã presence within the smaller volume combines at $7,2\times10^{2707}$ã within a

10^{2709}ã local volume and one in 13,89 positive virgins is activated by galactic flux before leaving this local volume; increase this distance fractionally and the majority of virgins is activated.

Similarly in 10^{903}ãt, some $1,68×10^{701}$ **negative virgins** are emitted by the $1,676×10^{78}$ positive inter-galactic photons and boosted to $1,79×10^{463}$¢ across the $3,22×10^{926}$ãr galaxy to be harvested by Protons at $5,62×10^{230}$¢ (with 5 fold dark matter counted) during $1,78×10^{663}$ãt consummation. It has gained $5,35×10^{1827}$ã negative presence within the $3,33×10^{2779}$ã galactic volume in 10^{903}ãt but of this, only $3×10^{-71}$ or $1,61×10^{1757}$ã is found within the local volume. With the virgins $3,16×10^{952}$ã negative presence combining at $5,09×10^{2709}$ã within this larger 10^{2709}ã volume; one in 5,09 negative virgins is activated by the galactic flux before the virgin leaves this inter-hydronal volume. Fractionally closer, there is no majority activation.

The majority activation by galactic flux is therefore first relevant when a paired hydron distance exceeds 10^{903}ãt which is the norm in inter-stellar grids (10^{904}ãr centimeter) and inter-galactic grids (10^{906}ãr meter) where the activation starts to interfere with standard virgin exchanges (the precursor virgins are activated at less than coronal distances in any grid).

Virgin Activation, Solar Radius

In the 10^{914}ãt time for survivors to escape this local volume $3,492×10^{709}$ **positive virgins** are emitted by $3,492×10^{75}$ negative inter-galactic photons, boosted to $1,79×10^{463}$¢ across the $3,22×10^{926}$ãr galaxy and flux harvested by Electrons at $9,13×10^{228}$¢ during $1,095×10^{665}$ãt inter-hydron consummation without interference from dark flux. Before harvest, it gains $6,84×10^{1837}$ã activated presence within the $3,33×10^{2779}$ã galactic volume but of this presence in 10^{914}ãt, only $3×10^{-38}$ or $2,27×10^{1799}$ã is found within the solar volume. With the virgins 10^{969}ã presence within its volume combining at $2,27×10^{2768}$ã within the 10^{2742}ã local volume, all but $4,41×10^{-27}$ positive virgins are activated by galactic flux before leaving the solar volume.

Similarly in 10^{914}ãt, some $1,68×10^{712}$ **negative virgins** are emitted by the $1,676×10^{78}$ positive inter-galactic photons and boosted to $1,79×10^{463}$¢ across the $3,22×10^{926}$ãr galaxy to be harvested by Protons at $5,62×10^{230}$¢ (with 5 fold dark matter counted) during $1,78×10^{663}$ãt consummation. It has gained $5,35×10^{1838}$ã negative presence within the $3,33×10^{2779}$ã galactic volume in 10^{914}ãt but of this, only $3×10^{-38}$ or $1,61×10^{1801}$ã is found within the solar volume. With the virgins 10^{969}ã negative presence combining at $1,61×10^{2770}$ã within the 10^{2742}ã solar volume; all but $6,21×10^{-29}$ negative virgins are activated by the galactic flux before the remaining virgin leave the solar volume (9,42 times less than is activated of the negative virgins).

Each surviving virgin with 10^{55}¢ orbital sweep has 10^{-124} chance of hitting any Proton (non-aided by its 10^{-7} fractional orbit) and it has a 10^{-67} chance of hitting some of the 10^{57} solar Protons.

If coronal virgin input within a sun is 3×10^{980} virgins during one second, then $10^{57} \times 10^{-271} \times (3 \times 10^{1194}$ãt$)$ and $6,21 \times 10^{-29}$ escapes activation, $1,86 \times 10^{952}$ intact virgins at 10^{55}¢ deliver $1,86 \times 10^{1007}$pu orbital backhand. Only 10^{-67} of them hit one of the 10^{57} hydrons to leave a linear push of $1,86 \times 10^{885}$pu in opposite directions (along with orbital push of $1,86 \times 10^{940}$pu).

The linear virgin shadow is more difficult to assess than that of the stable linear galactic gravity backhand 'G'; $(1,25 \times 10^{-272}$pu per ãnna$)$ due to the nature of uneven grids. Across the sun, the virgins can deliver $1,86 \times 10^{885}$pu potential linear virgin shadows per second or $6,2 \times 10^{-310}$pu per ãnna virgin shadow cast from the surface between local hydrons. This is 2×10^{37} times weaker than 'G' until one looks at the orbital part of this virgin delivery but that must be addressed later.

As inter-galactic flux outside solar grids is focused into suns and galaxies, the space between outside grids also generates lesser activation and stronger virgin gravity. Furthermore, flux gravity shadows are not always distributed along straight lines between local hydron; the last path before blockage can be so convoluted that a flux atom ready for harvest can end up anywhere. This makes flux gravity more of a grid phenomenon and hydrons at closer quarters than the grid structure may register as less predictable than the normal local flux gravity backhand while a larger bigger gathering would see such shadows work far more accurately.

Look at the gravity shadow of Our Solar submist cast by $1,189 \times 10^{57}$ hydrons and $3,736 \times 10^{922}$pus shadow each. The $4,44 \times 10^{979}$pus collective shadow upon a $2,24 \times 10^{1834}$ã Sun-Earth sphere yields a sphere point-push of $1,98 \times 10^{-855}$pus where each of $3,569 \times 10^{51}$ Earth nucleons with 10^{1704}ã point spheres blocks $1,98 \times 10^{849}$pu linear push per second. It takes Earth one year or $3,1536 \times 10^7$ seconds to orbit the Sun during which each of its nucleons blocks linear $6,242 \times 10^{856}$pu push that is neutralized in a one plain orbit by $(2 \times \pi)$ circle constants $(6,283)$. This leaves $9,935 \times 10^{856}$pu linear push in an average balanced solar orbit against each hydrons 10^{1140} atoms; generating an average orbital speed of $9,935 \times 10^{-285}$¢. (Earths actual average measured speed is $1,004$ times faster or $9,974 \times 10^{-285}$¢ or $29,9$ km/sec).

The Collapsing Hydron Field

The likeliest environment that drives oversized hydron embryos in a grid field to collapse into permanent structural entities comes as electromagnetic photons start to pass through a newly formed embryonic galaxy. Until then, all the processes of virgin exchanges are weaker and lengthier. This text shows me attempting at a broad illustration of how all this comes about.

The galaxy has already become slightly granular with up to 10^{12} solar embryos; each gathering roughly 10^{57} hydrons. Oversized galactic hydrons have been slowly consolidating for a long time but are now gaining pressure from the first incoming 10^{752}ãr wide precursored electromagnetic photons arriving from the nearest converting galaxy. These e-photons leave trails of multiple shrinking hydron embryos like a magnetic field around their linear path across consecutive grid distances. At the beginning, the penetration into a galaxy is restricted by oversized embryos this penetration increases as time passes and new e-photons pass further between collapsing hydrons as the galaxy is converted.

A vast gathering 10^{926}ãr wide galaxy becomes transparent almost from the start when a vast average hydron size has reached below 10^{891}αr; without polarization. Incoming e-photons can pass through the interstellar galactic field to emerge on the other side of the galaxy without a majority hitting an embryo (disregarding virgin winds that shift e-photon paths along the way).

The shrinking solar embryos are by nature smaller than the 10^{922}ãr grid distances between nascent suns and when the solar embryos have shrunk below 10^{920}ãr, more e-photons pass through inter-stellar fields than through all solar embryos collectively. As a galaxy and its solar submists shrink hand in hand with hydrons, the galaxy soon turns ever more transparent.

During long early formative periods, all submists are denser toward center as their building units form a tighter grid. In every submist above standard hydrons the gathering units form grids that tighten towards their center and leave inter-unit grid distances in outer layers that are a bit more tenuously dispersed. This intrinsic early formation pattern fits all submist embryos; mattrons and hydrons, suns and galaxies, etc.

Until affected by local electromagnetic splits, solar embryos are initially large but as they are being compressed by virgins, with more compressed outer layers. Greater inter-hydron boosts and external mutually exchanged virgin emission between hydrons causes more density towards center both within suns and free interstellar galactic hydrons. After e-photons start to enter and pass through the galaxy, the diversity of hydron sizes makes any generalization or holistic illustrations that much harder.

The vast forming solar embryos built of oversized Neutrons can (due to lesser external emission) be packed more tightly than repulsive Protons and Electrons eventually allow. As ever more hydrons become small enough to be polarized by-e-photons, they develop into oversized layered Neutrons that have a great affinity to block the mutually exchanged virgins (expansive as well as compressive).

The galaxy has been slowly condensing for a very long time while the vast gathering solar embryos carve globular voids out of its hydron mist but there comes a time when e-photons start to pour out of other converting galaxies into intergalactic space. This gives the galaxy more transparent outer layers as embryos collapse inwards from the galactic surface under the sudden new compression. Before the inter-galactic virgins arrive into another pristine galaxy, their adverse virgins enter ahead of them and provide the first really powerful compression within a pristine galaxy.

While inter-galactic e-photons are fewer than 10^8 mid between the galaxies, their virgins are mostly activated by precursor virgins (flux presence alone is not sufficient to activate a majority). After the 10^8 e-photons reach midways, majority activation puts enough flux in the field to adjust the survival rate of the virgin input. Note that although precursor virgins are boosted to 10^{256}¢, they are for that reason also activated at a higher rate; the average mattron virgin boost applies. The initial hydron compression by flux backhand is at this early stage insignificant.

As e-photons move from a converting galaxy to closer than midways to the next galaxy, their delivered virgins becomes less boosted upon arrival but the drop in compression is outweighed by decreasing activation; and increasing harvests by embryos on galactic outskirts by local hydron embryos. As the conversion of a galaxy takes thousand times longer than for a photon to reach a next galaxy, new photons are, relatively speaking quick in reaching full 10^{78} intergalactic numbers.

At the end of a former galaxies conversion era when a motherload of **10^{78}** positive photons emits its virgins across 10^{928}ãr inter-galactic distances, the flux from 10^{726} inter-galactic virgins gains 10^{464}¢ orbital drive during 10^{698}ãt consummation (10^{900}ãr hydron embryos block flux at around 10^{230}¢) and the 10^{1888}ã flux presence combined with 10^{990}ã virgin presence or 10^{2878}ã in a 10^{2784}ã inter galactic volume allows merely **10^{-94}** of the input or **10^{632}** virgins to avoid activation and to hit the galactic embryos intact at 10^{62}¢.

Even at the beginning with only **10^8** photons midways across the 10^{928}ãr inter galactic distance, the flux from 10^{656} inter-galactic virgins gains 10^{464}¢ orbital drive during 10^{698}ãt consummation; 10^{1888}ã flux presence combines with 10^{990}ã virgin presence or 10^{2808}ã in the 10^{2784}ã inter galactic volume. This allows **10^{-24}** of the 10^{656} negative virgin input; namely the same **10^{632}** virgins to avoid activation and hit galactic embryos at 10^{62}¢. After the first 10^8 photons reach midways, the growing number do not matter as the field delivers the same number of surviving virgins into the new galaxy (until they come closer to the new galaxy than the mid inter-galactic grid).

This purely virgin pressure is not specific to local mass collections such as forming solar embryos in a lumpy galaxy. A linear virgin would hit a 10^{900}ãr wide embryo after penetrating inwards along 10^{-13} galactic radius. Virgins emitted by inter-galactic e-photons arrive into the galaxy with orbital speed of 10^{62} fold linear speed while sweeping repeatedly by vast surface embryos until they hit one.

Look at some aspects involved, as a galaxy of 10^{900}ãr wide surface embryos in a 10^{903}ãr wide grid where roughly 10^{46} hydrons form the outmost galactic layer. A virgin orbit of 10^{928}ãr width takes 10^{866}ãt to complete at 10^{62}¢ and widen its reach into the galaxy across 10^{870}ãr in each of successive non-hit orbits (10^{-4} of those orbits do not lead through the galaxy).

To merely move across the 10^{900}ãr width of a top layer embryo on passing into a galactic 'surface', a virgin must make 10^{30} passes (out of 10^{62} potential passes). The 10^{46} top layer embryos are therefore hit first and consequently shrink first and fastest (10^{900}ãr wide embryos with 10^{1131}primal mass shrink faster than heavier smaller hydrons). As embryos shrink and a grid grows wider between them, it takes more virgin passes and longer periods to hit an embryo; 10^{881}ãr wide embryos in the same 10^{903}ãr hydron grid needs 10^{19} times more sweeps than 10^{900}ãr embryos. Deeper penetration among smaller embryos depends on the virgin orbit (the boost part of a larger orbit cannot make up for the greater time spent outside a galaxy).

After a primeval galaxy sends off its first electromagnetic photons, the massive e-photon effect from inter-galactic space allows one inter-galactic photon to accelerate the ongoing lethargic compression of oversized hydron embryos in neighboring galaxies (by extremely weak grid exchanges). The speed of the collapse is set in the embryos rotation period which is faster in large lighter hydrons, like the 10^{1131}ãt period for a 10^{900}ãr wide embryo.

As the new virgin pressure arrives to compress the light oversized embryos, the harvests by these soon smaller top layer hydrons drop along with orbital gain and results in slowing compression upon their growing primal mass. How many top layer hydrons are hit by a photons virgins can be approached in various ways and I avail myself of the grid exchange method; the virgin of adverse photon emission is aiming through a hydron grid in opening orbits that initially pass repeatedly through the thin outer layers of a galaxy.

Intact virgins from inter-galactic space (whether by 10^8 or 10^{78} photons) counts 10^{632} in 10^{928}ãt or 10^{-296} new virgins per ãnna with 10^{62}¢ orbital sweep delivers a massive 10^{-234}pu per ãnna in collective virgin pressure. Of virgins from the 10^{928}ãr wide inter-galactic void, 10^{-56} would hit a 10^{900}ãr wide embryo but with 10^{34} orbital gain, each virgin has a 10^{-22} chance to hit an embryo within the top layer.

Although it therefore takes merely 10^{22} top layer embryos to stop as many virgins, the delivered 10^{-234}pu collective pressure is shared by all 10^{46} top layer embryos; each embryo harvesting 10^{-280}pu pressure per ãnna. As 10^{-38} works as compression, this 10^{-318}pu compression drive is $1{,}78 \times 10^{31}$ fold the 10^{900}ãr embryos lethargic $5{,}62 \times 10^{-350}$pu expansion drive. The collapse speed generated in the fast 10^{1131}ãt rotation period by 10^{813}pu compression sets its collapse speed at 10^{-318}¢. Although actively shrinking, this slow compression own its own would take 10^{1218}ãt to shrink the embryo across its 10^{900}ãr wide radius (million fold the

$1,58 \times 10^{1212}$ãt galactic conversion itself).

This speed which is exclusively virgin emission from distance inter-galactic photons can continue with successive inputs while diminishing sizes beyond 10^{900}ãr alter the situation. This far off drive is sapped as an embryos shrink and harvests fewer virgins against growing expansion but although the great change is likely to come after the corporeal photons start passing in between embryos in the galactic outer layers; taking 10^{1206}ãt or so to pass through the galaxy, the virgin pressure from outside the galaxy will increase.

One can see how this meager compression changes when the inter-galactic photons advance on the galaxy and affect outermost galactic hydrons prior to entering. Their adverse virgin emission from smaller distances deliver local pressure that is greater in spite of lesser boost (tempered by activation). Look at say 10^{54} e-photons in the mid intergalactic field while other 10^{54} photons are passing or approaching closer to the 10^{926}ãr galaxy at a smaller distance of say 10^{910}ãr.

The 10^{54} midways photons emit 10^{-280} virgins during their 10^{928}αr inter-galactic midways crossing, their flux with 10^{464}¢ boost during 10^{698}ãt consummation gives 10^{1864}ã constant flux presence in that inter galactic space. Of all that flux presence, only 10^{-54} is within the 10^{2730}ã volume of the 10^{54} closer photons or 10^{1810}ã flux presence. Combined with a virgin emitted across the sorter 10^{910}ã and harvested at 10^{53}¢ in the galactic surface after gaining 10^{963}ã presence; the combined 10^{2773}ã presence within the 10^{2730}ã part volume leaves a far greater fraction of **10^{-43}** virgin survivors that can hit the large galactic embryos across shorter distances at slower 10^{53}¢ sweep.

The delivered 10^{-216}pu collective pressure is shared by all the 10^{46} top layer embryos; each embryo harvesting 10^{-262}pu pressure per ãnna. As all of 10^{-38} works as compression, this 10^{-300}pu compression drive harvests 10^{831}pu and generates 10^{-300}¢ collapse speed in 10^{1131}ãt rotation time. Collapse across the embryos 10^{900}αr wide radius takes merely 10^{1200}ãt (roughly twelve days) for the first top layer of 10^{46} out of 10^{68} galactic embryos.

All such numbers are averages spanning from 10^{928}ãr midways and all the way into the galactic surface. The delivered pressure will increase with more approaching photons as smaller joint activation volumes allow more to reach into the galaxy without activation and all these intermingling fields are of course cohabiting at every moment.

This goes to show that initially, the first tier galactic surface embryos lead the shrinkage and embryonic collapses gradually move deeper from there to ultimately embrace all the 10^{69} hydrons as both galaxy and its solar embryos undergo a hydron collapses to ever smaller sizes from the surface inwards.

Exchanged virgin emission pressure between smaller hydrons is growing at the same time but does not come close to photon pressure. The galactic grid is also being slowly condensed towards its center with average grid distances there gradually drawing closer from roughly 10^{903}år. A decreasing exchange boost between central galactic hydrons or between hydrons within denser solar embryos allows larger and lighter embryos than at a galactic surface but this changes when electromagnetic photons actually arrive into the galaxy and the situation really starts to change.

After a more gradual consolidation of the hydron field, a growing number of e-photons enters the galaxy at light speed from all directions and locally rising virgin pressure that easily compresses interstellar and stellar hydrons below the polarization threshold. Inter-stellar hydrons are affected by more numerous passing e-photons from outside a galaxy than in the more tightly packed solar embryos and distances are greater so that the solar embryos remain oversized longer with neutrally compressed mattrons.

A useful generalization is that of all hydrons in a galaxy, before and after the appearance of photons, hydrons within its gathering solar embryos are larger and mutually closer in tighter grids while the interstellar hydrons are smaller and further apart in a wider grid.

Of up to 10^{78} e-photons passing into the galactic surface among 10^{46} top tier hydrons with up to 10^{32} e-photons passing by each embryo, interstellar Neutrons become highly compressed and ripe for a split. When split, they become hydrogen atoms throughout the outer galactic layers.

The chain of events includes an immense number of simultaneous altering situations and an average falters on the fact that this is not meaningful if every link in that chain presents a totally different category of interacting fields. A single snapshot of an entire galaxy can show how far hydrons have shrunk on average at that moment and how tight solar grids have gathered on average but this is treacherous logic as collapses progress inwards from galactic and eventually solar surface layers.

Let me attempt a series of simplified setups to illustrate the principles but the sizes are not to be taken literally. A galaxy can be **10^{926}år** wide and have 10^{12} solar embryos of **10^{914}år** radii with 10^{57} gathered Neutrons each. Snap your fingers and make galactic solar embryos disappear and view the 10^{926}år wide galaxy as a uniformly distributed grid of all remaining hydrons. Assume that interstellar Neutrons and those within solar embryos were at this point of equal numbers; whereupon the purely interstellar field of 10^{69} Neutrons is distributed in an average hydron grid at **10^{903}år** apart.

Snap your fingers again and call back all the solar embryos. Assume that their hydrons are 10^{895}ãr apart throughout the outer layers but more tightly packed at collapsing solar centers where they form 10^{892}ãr wide grids.

Electromagnetic photons cross a 10^{926}ãr wide galaxy in relatively straight lines and stand a 10^{-12} collective chance to hit a sun. There are therefore 10^{12} fold linear photons passing exclusively through the interstellar field than through suns and 10^{12} times more input from passing photons is thus harvested by a interstellar hydrons (compared to what is available to those within solar embryos). As this is more than is made up by closer passages in tighter grids; interstellar embryos are first to be compressed and split.

This average 10^{12} fold virgin input in the interstellar field is also delivered with greater local virgin boost than within solar embryos. The $3,16 \times 10^{49}$¢ between interstellar grid Neutrons, given the above grid parameters, and $3,16 \times 10^{45}$¢ in solar submits shows 10^4 fold interstellar delivery above solar delivery. Although a wider grid allows more virgins to slip by closest hydrons, the pressure becomes divided upon more local hydrons, aided by orbital gain but activation will soon take a bigger bite of such orbiting virgins.

Hydrons can in principle span from 10^{900}ãr sizes down to below standard 10^{850}αr while consolidating Neutron stars could span from 10^{920}ãr to 10^{909}ãr but suns may expand to 10^{914}ãr standard sizes when their central Neutron population starts to split. The interstellar void with smaller hydrons turns transparent faster than solar grids. Photon entrapment is easier with closer passes in solar grids but paths are here more heavily skewed by local virgin winds.

When individual layered Neutrons are temporarily compressed to around 10^{860}αr and drops again, the buffer zone widens and causes a breach that compresses the central Proton core below 10^{855}ãr and facilitates a Neutron split that sends off a massive batch of 2×10^9 electromagnetic photons and jacks up a successively proliferating chain reaction inwards from a galactic surface and later inwards from a solar surface. The inwards moving wave hammers collapsing embryos from oversized embryos down to standard size and moves deeper with an accelerating climax of electromagnetic photons.

Just as in galactic surfaces, e-photons pass into a sun through outermost surface layers and first among a few top layer embryos. A linear photon that crosses a 10^{926}ãr wide galaxy passes between 10^{23} hydron pairs before it exits the galaxy on the other side. A linear photon crossing a 10^{914}ãr wide solar embryo, passes (in principle) between 10^{19} hydron pairs before it exits the sun on the other side. This affords interstellar photons a further 10^4 fold edge in distributing local compression to more hydrons from each photon (and more so with more photons blocked near solar surfaces).

E-photons entering a 10^{895}ãr solar lattice pass on average 10^8 times closer to each hydron embryo that blocks more virgins and their more opaque grids allow less photon penetration into a sun. Long before the photon can pass by 10^{19} hydron pairs on a straight line, each has passed so close to a hydron in the outermost solar layers that its path becomes corrupted which allows more durable local emission.

Selecting an average Neutron size for the galaxy is not easy. From an array of possible candidates, Neutrons shrunk to $\mathbf{10^{876}ãr}$ average sizes can do as well as any. They have started to gather into 10^{914}αr wide solar embryos but there is no reason to complicate the illustration by allowing them to expand from these average sizes within gathering solar embryos and use briefly larger Neutrons at collapsing solar centers.

A standard Neutron has ten million times weaker external main body repulsion than a standard Proton but casts a 'G' shadow equal to a Proton. Even without coronas, recasting Neutrons into repulsive Protons creates immense additional pressure pulse inwards within galactic and solar grids (and later violently outwards on reaching a submist center).

Orbs Gathering Grid Neutrons

While a single disturbed Neutron can in principle go drifting through the evenly dispersed Neutron grid and chance to form a closer bond but in spite of its weak repulsion, this would be rare. Larger Neutron groups at less than an average grid distances cannot form that way and in a grid lattice, locally spinning Neutrons are not appreciably orbiting but held in local positions by exchanged mutual repulsion from all directions.

In the average grid, 'G' gravity influence upon these relatively stationary Neutrons is also countered from opposite directions and neutralized. When local Neutrons begin to split all around the globular solar surface, half of the photons aim deeper into the solar grid which turns denser towards center although I am bound by averages in this illustration.

The rarity of individual Neutron deviations means that initial formation of tightly knit Neutrons must initially be aided by other forces; formation of a baryon pair or group must follow the drive of individual grid Neutrons. Such an individual drive cannot be achieved by the linear virgin winds that touch a great number of grid Neutrons equally but is easily achieved by the narrow powerful orb drives generated by passing electromagnetic photons.

These virgin orbs can under certain circumstances cause oversized but shrinking grid embryos to close ranks far enough for 'G' flux gravity to grab the reins and pull them individually closer; albeit feebly at first. Although the affinity of large embryos in moving closer is aided by weak repulsion and allows easier pairing, the linear push, even during a massive electromagnetic crunch would rarely force hydrons to break rank in the grid.

The nature of a hydron pair or group, once pushed closer than an average grid, is that each can be affected by a local 'G' gravity sway. A couple that has been forced closer by an orb faces neutralization of both repulsion and gravity from the surrounding grid but not mutually from each other. They continue to be repelled or drawn individually closer, which forces them to gradually build up movement in local orbits; albeit feebly at first.

Groups can for natural reasons form slightly denser central areas where the baryon embryos may orbit a center of mass that does not correspond to a position of any corporal embryo within the group. The baryons in the groups less populated outskirts orbit an average center of mass at slower speeds.

In a grid lattice, most locally spinning and mutually repulsive hydrons are not appreciably orbiting each other. This lack of grid motion is an important feature. An advancing e-photon cannot drive embryos more than marginally closer by weak linear push against collective grid repulsion. And no linearly enforced grid shrinkage, even at a collapsing solar core, can bring Neutrons within the stability range where the split of one Neutron can spare the other.

For all of the obvious reasons the similarities to a solar system of groups depending on gravity as the overarching collective influence need not be pointed out. This text aims to illustrate the more focused forces that bring the baryons together into double or multiple systems.

First in line for Neutron pairing or group formation are the electromagnetic photons passing into an inner grid of stable Neutrons from all around the solar surface. The grid Neutrons on the outer layer circumference are the first to split and half of all photons in vast numbers head deeper into a shrinking Neutron grid from all directions around the solar center. Each of these electromagnetic photons affect inner grid Neutrons ahead on a linear path with a powerful virgin orb, long before it upsets the Neutrons in close encounters.

After the electromagnetic implosion from the solar surface starts to eat its way into the solar embryo, an inwards advancing electromagnetic photon in a tightening grid of oversized hydrons emits 10^{-280} per ãnna orbitally boosted virgins that move ahead of the corporeal photon. This is where virgin gravity orbs generated by adverse photon emission become a gathering tool that can bring individual hydrons closer.

Outside of the inwards moving splitting zone, one would find newly split and mostly standard Proton Electron pairs. Inside the splitting zone, the Neutron population remains intact but not for long. The virgin emission rises to a zenith as the noose of the splitting zone draws closer around the inner core. It is the environment closest to the solar center that is most likely to fashion the greatest number of multiple Neutron groups; fashioned by orbs.

As an electromagnetic photon advances into a tightening grid, its emission is projected on globular virgin gravity orb all around the corporeal photon. Backhand dilution on that orb is not by the inverse-square law but rather by it 'un-squared' (how many hydron sized wedding bands areas can be placed on the entire globular orb without any of them coinciding).

The backhand to this powerful virgin orb is a potent candidate in bringing large Neutrons individually together (or asunder). The linear virgin repulsion delivered by the same orb is marginal compared to the sideways push by the same adverse virgins. However, the orbital push is only mutually active while two hydrons are situated on the same virgin orb; the mutual push of its backhand attraction between two hydrons on the same orb lasts only while hydrons both stay on that orb.

The orb width is set by the width of the hydron that blocks and casts the virgin shadow; a variable width down to standard 10^{852}år orb that will seem razor thin at far distances. Globular backhand to a thin orb is for obvious reasons weaker that backhand gathered by a large embryo (which would be shrinking for the same reason). A photon orb between two large Neutrons allows greater harvest of boosted virgins that brings them mutually closer until the linear photon has changed its angle to such a degree that both are no longer on the orbs; or until the photon has passed out of reach.

The great general disturbance surrounding the linear path of the photon will from time to time push two individual Neutrons to mutually approach for a while across part of the globular orb at that distance. The approach along the orb lasts until the orb angle, whether by photon or hydron motion, changes enough for the mutual orb backhand to aim mutually adjacent to a main body; whereupon the entire mutual orb push completely disappears.

For large hydrons, the electromagnetic photon orbs can cause fast mutual approach that can in principle end in a collision if the two hydrons stay long enough on the orb to pick up speed. Although the affected hydrons are at the same time collapsing towards center under harvested virgin pressure, this is generally slower. When the push is lost, a less than accurate approach can continue to brings hydron closer and cause near misses that force them into slower local elliptical 'G' gravity orbits that cut their future staying power on any mutual orb.

I shall now run through a series of backhand orbs generated by a passing e-photon to see how two nearest hydrons are affected by virgin orb pressure during mutual approaches along globular backhands of the same orb. Start with the 10^{903}år wide interstellar grid with illustrative 10^{876}år wide Neutrons and then look at the illustrative average 10^{895}ãr grid distances within a solar grid with similar embryos and lastly within a central core of collapsing solar mass within say a thousand times tighter 10^{892}ãr hydron grid.

Again; if a photon passes between two grid hydrons at a distance closer to one than the other by one hydron radius (a tiny fraction of the distance apart and equal to the orb dilution at that distance), there can be no effective orb backhand at all. The numbers are exponential and the average is a midways passage which highlights the influence of tighter grids. If orb push around a closer Neutron from less than midways is not occupied by another Neutron, the potential backhand cannot be exploited.

A Neutron gathers orb pressure but there is no mutual orbital backhand drive between two embryos after a photon deviates from the average midway distance; the path of greatest likelihood of all possible paths. A less than perfect midways path is to be expected which radically affects the average orb influence while it hardly affects its steady compression at all.

Another point to make is that when multiple electromagnetic photons are passing between hydrons in a local grid at the same time, they can diminish the orb effect upon local hydrons by easily pushing them for a short distance out of other active mutual orbs.

As an electromagnetic photon moves and changes relative angles at light speed, two affected embryos would stay on its orb for a short part of passage at 10^{-280}¢ and, as the angles change, they no longer allow a full backhand shadow. When the common orb gains first contact, its area moves in across an embryos width and a full shadow is first cast if and when they coincide perfectly. A embryo moves into an orb shadow that can encompass only its first tier of mattron only which involves 10^{-76} of its mattron layers. It can also cast a perfect overlay where the orb to hydron areas coincide perfectly. While this is a simple illustration of full orb backhand, there is an average working effect, for those so inclined, that diminishes the average orb during passage.

Whether a couple or a group, the most likely outcome is that the individual Neutrons deep within nascent suns are more often paired as double systems or groups than in interstellar grids; the tighter a grid, the larger presence of bonded units.

The odds of an effective midways path out of all possible paths between two hydrons is small and the chance for two local hydrons to share an active orb is directly tied to a grid distance versus hydron size. If a photon passing between 10^{876}ãr wide hydrons in a 10^{903}ãr interstellar grid but does so closer to one hydron than the other by a fractional **10^{-27}** difference, there can be no mutual orb backhand. This is the average (after 10^{27} passes) while a perfectly placed orb works without any downsizing on this account. Note that photon midways placement does not affect orb delivery of virgin compression.

If the same photon passes between 10^{876}ãr hydrons in 10^{895}ãr solar grids and is closer to one hydron than the other by fractional 10^{-19}, there is no orb backhand. If the same photon passes between 10^{876}ãr embryos in 10^{892}ãr grids at a collapsing solar core; any distance closer to one hydron than the other by a fractional 10^{-18} generates no orb backhand between them.

First; 10^{876}ãr Neutrons in a 10^{903}ãr interstellar grid:

10^{876}ãr÷10^{903}ãr=10^{-54}×(g. $3,16×10^{22}$)=$3,16×10^{-32}×10^{-280}×3,16×10^{49}$¢= 10^{-262}pu×$(10^{-27})^2$ deviation ×orb dilution=10^{-316}pu×10^{1156}ãt=10^{840}pu÷ $3,16× 10^{1135}$mass=$3,16×10^{-296}$¢ across $3,16×10^{860}$ãr in 10^{1156}ãt (if it changes angles across 10^{876} ar).Compression 10^{-262}pu ×10^{-38}=10^{-300}pu and the 10^{876}ãr Neutron is compressed by midways photon vs. $7,48×10^{-326}$pu expansion.

The average 10^{-316}pu orb push is here weak from average e-photons as they pass between two 10^{876}ãr Neutrons and, if the photon angles out of a mutual orb path in 10^{1156}ãt, the $3,16×10^{-296}$¢ temporary attraction speed generated along the orb can bring them mutually closer across $3,16×10^{860}$ãr which is a $3,16×10^{-43}$ fraction of the grid distance during the potentially active orb period.

An orb generated by a perfectly placed midways photon (not sapped by any average cut) delivers 10^{27} fold the above effect and can move the Neutrons closer across $3,16×10^{887}$ãr; even that merely a $3,16×10^{-16}$ fraction of their distance apart and too marginal to play a role in forming double systems.

The 10^{876}ãr Neutron of lighter $3,16×10^{1135}$ primal mass has $7,48×10^{-326}$pu expansion per ãnna that cannot resist 10^{-300}pu compression but during the full 10^{1183}ãt passage, the achieved $3,16×10^{-253}$¢ collapse speed can shrink it (potentially) by $3,16×10^{930}$ãr across its entire radius if it were not losing that same virgin orb harvest due to rapidly shrinking size. Having collapsed from 10^{876}ãr down to somewhere above 10^{870}ãr, its compression harvest comes to equal its increasing expansion and the embryo stabilizes for a while.

When $2×10^9$ photons pass simultaneously between two grid hydrons, the sizes continue to shrink, not least as some of pass closer than others which occasionally leads to massive virgin harvests followed by less active periods. Eventually, a photon passes very close to a single hydron and compresses it to a size that makes electromagnetic splits possible. When the local pressure is released, which must happen, the Neutron expands into a split.

10^{855}ãr÷10^{873}ãr=10^{-36}×(gain $3,16×10^{16}$)=$3,16×10^{-20}×10^{-280}×3,16×10^{34}$¢= 10^{-265}pu×10^{-38}=$3,16×10^{-303}$pu vs. $7,2×10^{-304}$pu.

A 10^{855}ãr Neutron of lighter $2{,}74{\times}10^{1139}$ primal mass has **$7{,}2{\times}10^{-304}$pu** expansion per ãnna. During 10^{1153}ãt photon passage at close 10^{873}ãr away, the 10^{855}ãr Neutron harvests $3{,}16{\times}10^{-303}$ or $3{,}16{\times}10^{850}$pu. This can drive it inwards at $1{,}15{\times}10^{-289}$¢ collapse speed. Such a speed that can in principle bring it to 10^{-9} of its former radius is reversed after it has shrunk merely to within say a tenth of its size when its harvest can no longer beet back its rising expansion drive. The compression seizes and when the photon passes out of reach after 10^{1153}ãt, the ensuing expansion elicits a split of the single Neutron that is positioned 10^{903}ãr away from the next grid Neutron.

Second; 10^{876}ãr Neutrons in 10^{895}ãr outer solar grid:

10^{876}ãr$\div 10^{895}$ãr$=10^{-38}{\times}$(gain $3{,}16{\times}10^{26}$)$=3{,}16{\times}10^{-12}{\times}10^{-280}{\times}3{,}16{\times}10^{45}$¢$=$
10^{-246}pu$\times(10^{-19})^2=10^{-284}pu\times 10^{1156}$ãt $=10^{872}$pu$\div 3{,}16{\times}10^{1135}m=3{,}16{\times}10^{-264}$¢ brings
them closer across $3{,}16{\times}10^{892}$ãr in 10^{1156}ãt (angles across 10^{876}ãr).
10^{-246}pu$\times 10^{-38}=10^{-284}$pu vs. 10^{874}ãr Neutrons $7{,}48{\times}10^{-326}$pu expansion drive.

The average orb push generated by e-photons passing between two **10^{876}ãr** Neutrons (if an electromagnetic photon angles out of contact in 10^{1156}ãt) can bring them closer at $3{,}16{\times}10^{-264}$¢ across $3{,}16{\times}10^{892}$ãr or $3{,}16{\times}10^{-2}$ of their distance apart.

The chance for a perfect midways passing is a rare 10^{-19}. It is not common but when a single photon passes perfectly midways, a 10^{19} fold orb drive can bring two embryos across each other's bows 10^{16} times in orbit (a less than perfect aim prevents most collisions). Note that multiple simultaneously passing photons do not help in bringing them closer in a grid because their orbs become less cohesive (grid exchange orbs have a weaker such effect) while the pressure increases.

Even with a single photon, the massive virgin delivery under way forces the Neutrons to shrink to small sizes before clearing the grid distance. Individual collapses are more the same in all grids, as the effect is purely set by the passing photons and a shrinking embryo. The only thing that changes is the frequency of photons passing at any specific close distance. The odds for an orb that is effective from midways between two 10^{855}ãr photons in a 10^{903}ãr interstellar grid is 10^{-48}, albeit alleviated by 10^{12} times more photons to **10^{-36}**. For 10^{855}ãr photons in 10^{895}ãr solar grids the odds drop to **10^{-40}** and to **10^{-37}** within their 10^{892}ãr central core grids.

The interstellar collapses produce few closely paired Neutrons and even the main outer solar volume does not offer an environment favorable to forming double systems or Neutron groups. It is touch and go whether orbs can bring on occasional close encounters by forcing Neutrons to brake rank from their average solar grid distances. Released local compression during a rapid solar upheaval elicits Neutron splits in the outer solar grid but mostly individually and not in pairs or groups; they are not close enough to allow a Neutron companion to survive under the protective umbrella of a new Proton when the electromagnetic pressure subsides.

Third; 10^{876}ãr Neutrons in 10^{892}ãr central solar grids:

10^{876}ãr÷10^{892}ãr=10^{-32}×(g.10^{28})=10^{-4}×10^{-280}×1044¢=10^{-240}pu×(10^{-16})^2= 10^{-272}pu ×10^{1156}ãt=10^{884}pu÷$3,16$× 10^{1135}m =$3,16$× 10^{-252}¢ brings them closer by $3,16$×10^{904}ãr in 10^{1156}ãt 10^{12} orbits. 10^{-240}pu×10^{-38}=10^{-278}pu÷$3,16$× 10^{1135}m =$3,16$× 10^{837}pu÷$3,16$× 10^{1135}m=10^{-298}¢×10^{1172}ãt across 10^{874}ãr=10^{-2} rad.

During the inwards implosion from the solar surface, a central core grows denser. I illustrate with a grid that has been compressed to from 10^{895}ãr to 10^{892}ãr average lattice occupied by Neutrons that due to lesser boost of local exchanges may have been expanding somewhat if average grid exchange at a solar core cannot hold them but they cannot grow back their mass within the time at hand.

The first photons to pass through this grid are most effective. They avoid the other from each split batch of $2×10^9$ electromagnetic photons crossing deeper into a pristine solar embryo and the local Neutrons are therefore not yet being upset as extensively by multiple photons which alleviates their minute chances for closer passage.

When an average e-photon passes between two such 10^{876} ar hydrons, the orb push in 10^{1156}ãt brings them closer at $3,16$×10^{-281}¢ across $3,16$×10^{904}ãr; a potential distance covering 10^{12} whole mutual orbits. Without averages, a perfectly placed midways orb from a single photon delivers 10^{16} fold push which beds for 10^{28} potential orbits. The chances of bringing two or more embryos across each other's bows in near collisions with less than perfect aim have been significantly raised; as has the harvested pressure.

Across the 10^{892}ãr grid distance, individual 10^{876}ãr wide embryos are now compressed while being brought so close together that interaction by other forces can take the upper hand. As individual hydrons break rank from the average grid and move individually closer within the lattice, all stabilizing contrary forces (linear or orb repulsion or 'G' attraction) are put out of kilter. The new harvests upon the two embryos present a radically different linear drive when not countered by equal drives from opposing directions.

All the weak embryonic orb pushes from surrounding lattice embryos, that were previously countered from opposite directions and distances, are now less favorably positioned to push two closer embryos asunder. Surrounding orb drives are sporadically active with less diluted orb effect between closer embryos (depending on the closeness of the pair) but no longer directly opposed. Orbs between one Neutron of a close pair and distant grid Neutrons are more diluted so that the average local effects will now go unchallenged.

412

Of special interest is the rapidly darkening 'G' gravity pull that is of course active at all times between two allied embryos. Both the 'G' backhand drive and the sporadic orb drives in close encounters can cause a pair to start mutually orbiting at distances far within the average grid distance at speeds that accelerate at closer quarters. Across closer distances, the embryos pick up mutual local orbital speeds which is rarely the case in grids.

this orbital speed will ultimately deter the baryons from drawing closer even after multiple photons join the local grid with great orbs between the close embryos, as these orbs cannot work except for the short periods it takes them to cross an orb the thickness of a small standard hydron.

At a central solar grid where electromagnetic orbs allow group formations, these pairs and groups become lattice units that form average grid distances as the grid adjusts. A pair or group must be considered as a single entity, a grid point among other standalone entities arranged at a similar distance as other standalone units.

Photons continue to pass through the tighter grids where pairs or group embryos are driven closer and variously disturbed by arrays of different close encounters. The local orbital speeds about common points will successively increase; the specifics are for calculators whereas I cannot see anything fundamental that can alter these broad outlines.

Inter-Galactic Flux Distribution

A word is needed on how hydrons within converting galaxies at this early stage of formation block high speed inter-galactic flux. Interstellar and solar embryos harvest of inter-galactic flux that is gradually being focused deeper into a galaxy as it engages with primal particles ripped out of hydrons under way. If this flux would reach larger and mostly unperturbed hydrons deeper within the galaxy (which it does not) beyond the inwards imploding globular splitting milieu, these embryos would block flux at higher speeds. The inner embryos would in other words soak up a greater share of finite inter-galactic flux than standard hydrons left behind outside the inwards moving splitting wave from the galactic surface.

The standard hydrons block negative flux at $5,62 \times 10^{230}$¢ while for example 10^{881}är neutral embryos block all flux at $5,05 \times 10^{244}$¢ which means that the chance of inter-galactic flux being blocked by say equally numerous larger embryos is vastly greater. This however is dependent on the swiftly slowing linear progress of flux into the galaxy; it cannot get ahead of itself and of the inwards moving splitting wave.

Deep interaction by rapidly sweeping but linearly slow moving majority flux will eventually happen but first when the globular inwards moving wall of splitting Neutrons reaches the galactic center from all directions and flux starts to interact directly with inter-galactic flux from the other galactic side. It is first then that its convoluted orbits can cut through the entire galaxy and carry with it the 'G' gravity snippets from all directions towards the galactic center point.

This is just as well as otherwise the standard hydrons left behind outside the inwards moving splitting wave from the galactic surface could not be held to their standard sizes when local pressure outside the wave subsides. Like virgins, flux atoms return again and again in shallow orbit but, unlike virgins, it progresses ever more slowly linearly until blocked by the standard embryos under way and the overwhelming majority of the finite flux input is harvested outside the inwards moving splitting wave.

Bonded Baryon Pairs

The Strong Force of 'G' gravity is here being hampered by rapidly shrinking Neutron sizes; gaining with closeness what is gradually loses by size. There are too many steps down this road of altering exchanges in solar cores from oversized to standard Neutrons and my snapshot is of the ultimate outcome for two standard baryons, in the form of a Neutron or a Proton.

To say that the Strong Force sucks a Neutron and Proton closer with great force in the face of steady main body emission is true, if not the whole truth. It is clear when looking exclusively at the hegemony of 'G' against main body baryon exchanges that both baryons are under the influence of 'G' at great distances; that weak attraction exists at any distance but it cannot outgun repulsion during bouts of coronal activity where baryon pair can be blown apart by trillion times more effective repulsion.

With 'G' flux generated by intergalactic photons eventually crisscrossing the galactic void from a full stock of electromagnetic photons, we can look at two baryons that have broken rank from the now large average 10^{892}ãr grid distances within a solar core. Under the aegis of dark main body orbs some of these have reached closeness below $4,94 \times 10^{868}$ãr maximum distance for future Neutron stability.

Eventually, if one Neutron splits within that distance span, the other will thereafter be exposed to main body emission in excess of $4,08 \times 10^{-286}$pu per ãnna from its Proton companion which suffices to hold a standard Neutrons top layer up to its inner domain. Even full freely escaping coronal emission (which is not in the cards) from a locally orbiting Electron left by the first Neutron split would be unable to balance the second Neutrons top layer.

To have a typical distance to work with and for illustration only, I put the average bonded distance between a Proton and surviving Neutron companion at 10^{860}ãr. This is not a fixed borderline distance; just a distance at which the baryons can coexist in a balanced state. If 10^{860}ãr seems close, this baryon *exchange across 10^8 fold radii is the relative orbiting distance by Our Sun and Proxima Centauri, which puts it nicely in context.*

Standard intact Neutrons cast shadows identical to Protons but they emit $1,78 \times 10^{-307}$pu main body virgins in both signs which is million times weaker repulsion than a Protons. As the two baryons close ranks and 'G' grabs the reins, the Protons main body emission falls short of pushing the two baryons asunder.

A standard Neutron and Proton can serve to set this framework; although as an illustration, this jumps ahead to where electromagnetic photons from Neutron splits in a solar core have left the field after hammering its baryons into standard 10^{852}ãr suits. As a pair or a group of individual hydrons within the 10^{892}ãr solar core grid are forced closer by orbs, the 'G' gravity backhand between them is ever more effectively harnessed.

10^{852}ãr$\div 10^{860}$ãr$=10^{-16}$(orbital gain 10^{20})= all $\mathbf{10^{-300}}$ hit (10^{-304}escape) vs. 'G' $1,25 \times 10^{-272}$pu$\div 10^{16} = 1,25 \times 10^{-288}$pu

The gravity pull across 10^{860}ãr apart places the Strong Force on a secure footing. Without coronal activity to upset its pull, the $1,25 \times 10^{-272}$pu Strong Force always outguns the Protons 10^{-300}pu repulsion; even while all of it is still delivered by way of close orbital gain.

Imagine a pair of 10^{852}ãr wide baryons where one has split into a Proton Electron pair and with a surviving local Neutron companion at standard size orbiting at 10^{860}ãr. See whether the main body emission of either baryon across this distance can limit the effect of potential auxiliary coronas; if recalled.

Upon Neutron$=10^{852}$ãr$\div 10^{860}$ãr$=10^{-16}$(g.10^{20})$=10^{-300}$pu$\times 10^{28}$¢$=10^{-272}$pu. Upon Proton$=10^{852}$ãr$\div 10^{860}$ãr$=10^{-16}$(g.10^{20})$=1,77 \times 10^{-307}$pu$\times 10^{28}$¢$=1,77 \times 10^{-279}$pu

At $\mathbf{10^{860}}$**ãr** the Neutrons main body emission delivers a backhand orb that affects the Proton with $1,77 \times 10^{-279}$pu negative surface pressure which is 10^{28} fold the $1,77 \times 10^{-307}$pu dominant Proton repulsion. The Proton delivers the same $1,77 \times 10^{-307}$pu level of negative pressure into the Neutron accompanied by massive 10^{-272}pu positive virgin pressure that stabilizes it with $2,45 \times 10^{13}$ fold required $4,08 \times 10^{-286}$pu. If a fully inherent auxiliary photon is launched by either baryon, it forms a pod corona with shorter standard $\mathbf{4,21 \times 10^{856}}$**ãr** reach.

At 10^{28} fold the baryons 10^{1140}ãt rotation and launch period, a potential dominant coronas $\mathbf{10^{1168}}$**ãt inactive** period is 10^{28} fold its active period and diminishes the Protons full $7,28 \times 10^{-284}$pu standard free coronal emission to $7,28 \times 10^{-312}$pu free coronal virgins per ãnna at far below a Protons 10^{-300}pu steady main body emission and the Strong Force remains unchallenged by everything except perhaps upheavals after the Neutron splits.

As a unit, this Proton Neutron bond has certain peculiarities. Aside from the ability to hold potential auxiliary launches within a standard pod coronal reach, one in ten thousand of Protons 10^{-300} subservient main body virgins escape clear of the bonded baryons; restricting the pairs free external field emission.

Early Bonding

This brings us to perhaps the main point of this exercise; the fact that the Neutron can remain balanced under the protective umbrella of its Proton but this is not without challenges at the start. The enforced Neutron closeness lies back in time when two baryons in the imploding solar core were brought to within relatively close distances by massive photon orbs in a tight grid; the norm at a collapsing solar center.

The local grid pressure takes time to fade away as photons traverse in all directions within a sun or galaxy. This is a waiting period before the closer Neutrons need be balanced by other means. The general pressure keeps both Neutrons of a pair stable until fewer traversing photons are distant enough for the more expanded Neutron of a local pair to split. That first split, more than anything else, determines whether the bond survives or not.

We can leave the consolidating Electron aside; it will soon have travelled in orbital fashion from the left behind Proton to orbit by the Neutron (with little chance of hitting it) to reach vast distances beyond the baryon pair. For close Neutron companions, the test is whether their closeness allows their bond to survive the repulsion (with subsequent orb attraction) from the split Neutron as 2×10^9 photons between the baryons deliver pour out virgin emission. At first, the remaining Neutron receives this massive effect which affirms its stability in orbital fashion from many directions.

Both baryons receive the virgins from one broad direction albeit downsized in aim by the square root of the distance (their orb width) in baryon radii and repel the baryons mutually apart along that line by non-boosted linear effect. This aim is smeared out and the effect downsized in accord with the mutual 'G' orbital speed already attained by the two baryons.

After a short while, the 2×10^9 negative photons aimed towards the Proton were accepted and the neutrally blended effect can be put aside for the time being while the 2×10^9 positive photons head away from the Proton linearly in that many directions. There is some difference in effect before and after the (closest in between) photons have reached midways between the baryons after which the Neutron receives average roughly half the photon output.

The distance between the two baryons tells us whether the bond can survive the outburst of the first Neutron split. What it really come down to is whether massive adverse e-photon emission from midways between close baryons can blow them apart across the leverage distance before they pass by the Neutron; the photon generated emission orbs after they pass by the Neutron will work partly to push the baryons closer together again.

This push orbital push together again can in theory be with all photons pitching in with virgin orbs (albeit in serious conflict) but as the number of possible orbs between two baryons widely exceeds the photon number, there would be only one 10^{-280}pu orb working to push them together at each moment.

Each virgin represents a one sided totally linear push but the broad lineup of photons heading out at different angles delivers a conflicting linear drive upon a main baryon body. The adverse emission push from a linear photon from a wide corona upon the Neutrons main body (unlike magnetic transfer dilution) is simply downsized in aim by a distance square root; represented by the actual distance measured by the width of the photon sphere.

Baryons that are initially close enough for the 2×10^9 e-photons to push the Neutron away from the Proton at light speed across a leverage distance face different acceleration according to their initial closeness. Harvested emission after passing the $3,96\times10^{861}$âr leverage point distance drops rapidly.

Before the split, the initial baryon distance apart has fashioned relatively robust speed in mutual 'G' orbits. This orbital shift has been downsizing its speed out across the leverage distance by the square root of the difference of the actual initial orbital period and the potential orbital period at the leverage point. But if the accelerated speed when the Neutron passes across the leverage distance is greater than the 'G' orbital speed, the escape speed is merely fractionally slowed.

The question whether the photons can repel a companion Neutron out of its long term stability range is complicated. If a Neutron is repelled away at superluminal speed so that the photon cannot pass beyond it, the Neutron is likely carried beyond its stability range. If, the Neutron cannot be accelerated to superluminal speed, the photons soon pass beyond it and their generated emission orbs are most massive at closest baryon distance where they start to push the baryons closer together again; at rapidly diminishing rate at the larger distances.

Although the Neutron of the pair can survive under the Protons umbrella of main body emission out to a distance of $4,94\times10^{868}$âr, less virgin emission can be harvested at that distance. The stability and survival of the second Neutron is tested relatively long after the first split.

Note also that at very short distances, the mutual virgin harvest delivers less compression by virgin boost. Tangential virgin orbits into the top layer are also less able to deliver orbital compression as more are delivered as linear thrust which heavily downsizes the compressive ability. If a Neutron is too close to a

Proton companion, main body virgins are eventually unable to stabilize the Neutron against the expansion drive upon its top layer.

Two conditions must be met for the companion Neutron to survive after the upheaval. It must not be repelled beyond the $4,94 \times 10^{868}$ãr stability area and it cannot be too close to the Proton as main body Proton virgins that hit a Neutron on shallow sideways aiming paths deliver a diminishing rate of linear compression, the deeper they are within the leverage point which makes them unable to stabilize the Neutron. The closer a Neutron is within its leverage point, less of the virgin boost serves as compression upon its negative outer shell which puts Neutron balance on shakier ground. The Neutron must block $4,08 \times 10^{-286}$pu required pressure for future stability.

The likelihood of being kicked out of a bond increases at both ends of the initial distance apart. We can compare several candidates of initial baryon distances but I have had only hours to think on this and will not be coming back for revisions in the foreseeable future. As always, the quest is to try and come to grips with a framework without getting lost in calculations.

The virgin repulsion and attraction from 2×10^9 electromagnetic photons heading away from the first Neutron that was split is best conceptually explained by repeating a fixed mantra of separate steps.

Proton Neutron at 10^{860}ãr

- **From midways at 10^{860}ãr,** upon the Neutron, the 2×10^{-271} virgins per ãnna linear push from a photon orb all around the Proton is in conflict as to aim. Here I use the inverse r√r where distance 10^8 yields 10^{-12} results in 2×10^{-283}pu repulsion driving the Neutron away at speeds set by its virgin harvest. Note that between baryons repelled by coronal shifts, I let the product of distance dilution 'r' and effective emission part stand for downsizing of repulsive virgin aim from a coronal orb.

- While the Neutron must be accelerated, the photons are at light speed from the start and reach 10^{860}ãr in 10^{1140}ãt during which the Neutron can gather 2×10^{857} virgins that accelerate it to 2×10^{-283}¢ before all the photons pass beyond the Neutron; leaving it more or less in place with the **10^{860}ãr distance** marginally widened by 2×10^{857}ãr.

- **The photon orb 'attraction'.** With the Neutrons continues away in an opening orbit at 2×10^{-283}¢ from the 10^{860}ãr distance, the photons have passed beyond the baryon distance and adverse virgin emission generates massive backhand orbs between the two baryons but the effect is heavily diluted by the greater inter-baryon distance.

10^{852}ãr $\div$ 10^{860}ãr=10^{-16} (gain 10^{20}) all 10^{-280} hit a baryon at 10^{28}¢= 10^{-252}pu orb

- From a distance of 10^{860}ãr, each single electromagnetic photon generates its 10^{-280}pu per ãnna virgin orb that all hit a baryon boosted to 10^{28}¢. There are 2×10^9 photon orbs active between the baryons at the same time and the 2×10^{-243}pu orb orbital shadow backhand from each baryon is diluted to 10^{-16} just for hitting one baryon and by that rate squared for a mutual backhand 10^{-32} between both.

- Across 10^{860}ãr in 10^{1140}ãt, the constant 2×10^{-275}pu active orb works to push the two baryons closer by 2×10^{865}pu at 2×10^{-275}¢ superluminal speed which is here 10^8 times in excess of the 2×10^{-283}¢ needed to break the Neutrons flight speed at that moment.

- With all the 2×10^9 photons outside the short 10^{860}ãr baryon distance, the Neutron is violently kicked back into the fold. The 2×10^{-275}¢ speed would lead to 5×10^{1134}ãt approach period that is 2×10^5 times shorter than the 10^{1140}ãt outwards drive period and working time speed drops by the square root of the difference; by $4,47\times10^2$ to $4,47\times10^{-278}$¢.

$$1,25\times10^{-272}pu \times 1,12\times10^{1129}\tilde{a}t = 1,4\times10^{857} = 1,4\times10^{-283}¢$$

- When the baryons pass for example outside their 10^{852}ãr auras which takes $2,24\times10^{1129}$ãt, the full 'G' force of $1,25\times10^{-272}$pu can affect a corrupted approach with $2,8\times10^{-283}$¢ attraction and in theory forge near miss orbital motion that neutralizes orb attraction; except it now falls woefully short of the $4,47\times10^{-278}$¢ single orb attraction.

- The complicated consequences of a baryon collision leads to upheaval that is likely to leave both in tatters and is not on the agenda here. On the other hand, it is unlikely that two forming Neutrons at vast sizes upwards of 10^{860}ãr can reach such a close 10^{860}ãr side by side distance apart when pushed by the fields available at the time but the scenario shows what would happen in too close encounters. A much further distance apart is a likely scenario but we can first look at say 10^{862}ãr.

- It is not the purview of the Primal Code to study the structures of nuclei; only to provide a conceptual picture of the influences that form them. The treatment at each point is complex with all the changing positions, altered distances and orb parameters to take into account.

Proton Neutron at 10^{862}ãr

- **From midways at 10^{862}ãr**, upon the Neutron, the 2×10^{-271} virgins per ãnna linear push from a photon orb is in conflict to the inverse r√r and downsized to 10^{-15}. The resulting 2×10^{-286}pu repulsion drives the Neutron away at a speed set by its attainable virgin harvest.

- While the Neutron must be accelerated, the photons are at light speed from the start and reach 10^{862}ãr in 10^{1142}ãt during which the Neutron can gather 2×10^{856} virgins that accelerate it to 2×10^{-284}¢ before all the photons pass beyond the Neutron; leaving it more or less in place with the **10^{862}ãr distance** marginally widened by 2×10^{858}ãr.

'G' $1,25\times10^{-272}pu\div10^{20}=1,25\times10^{-292}pu\times8,95\times10^{1146}$ãt$=1,12\times10^{855}$ virgin drive to $1,12\times10^{-285}$¢ during $8,93\times10^{1146}$ãt orbit

---

- The original Neutrons at a distance of 10^{862}ãr apart would already be orbiting under the influence of the $1,25\times10^{-272}$pu 'G' attraction diluted to 10^{-20} at $1,12\times10^{-285}$¢ speed that takes $8,93\times10^{1146}$ãt for each orbit.

- **The photon orb 'attraction'**. With the Neutrons continues away in an opening orbit at 2×10^{-284}¢ out from the 10^{862}ãr baryon distance, the photons have passed beyond that to where adverse virgin emission generates massive backhand orbs between the baryons with an effect heavily diluted by the greater inter-baryon distance.

10^{852}ãr $\div$ 10^{862}ãr$=10^{-20}$(gain 10^{19}) one in ten of 10^{-280} hit the next baryon at 10^{29}¢ casting a 10^{-252}pu backhand orb

---

- From a distance of 10^{862}ãr, each single electromagnetic photon generates its 10^{-280}pu per ãnna virgin orb where one in ten hit the Neutron boosted to 10^{29}¢. There are 2×10^{9} photon orbs active between the baryons at the same time and the 2×10^{-243}pu orb orbital shadow backhand from each baryon is diluted to 10^{-20} for hitting one baryon and by that rate squared for a mutual backhand 10^{-40} between both.

- Across 10^{862}ãr in 10^{1142}ãt, the constant 2×10^{-283}pu active orb has worked to push the baryons closer by 2×10^{859}pu at 2×10^{-281}¢ speed that outguns the 2×10^{-284}¢ required to break the Neutron outwards drive 1000 fold.

- With all the 2×10^{9} photons outside the 10^{862}ãr baryon distance, the Neutron is again kicked violently back into the fold. As 2×10^{-281}¢ leads to 5×10^{1142}ãt period of approach that is 5 fold the photons 10^{1142}ãt outwards drive period and there is no cut to the 2×10^{-281}¢ speed.

$1,25\times10^{-272}pu \times 2\times10^{1133}$ãt $= 2,5\times10^{861} = 2,5\times10^{-279}$¢

- If the baryons first pass is for example outside their 10^{852}ãr auras during 2×10^{1133}ãt, the full 'G' force of $1,25\times10^{-272}$pu can affect any corrupted approach with $2,5\times10^{-279}$¢ attraction that would have forged any near miss into a neutralizing orbital motion by 'G' attraction that side by side outguns the 2×10^{-281}¢ orb drive $1,25\times10^{2}$ times.

10^{852}ãr$\div$ $3,12\times10^{853}$ãr$=3,2\times10^{-2}=1,02\times10^{-3}\times1,25\times10^{-272}pu=1,28\times10^{-275}pu\times1,56\times10^{1134}$ãt $= 2\times10^{859} = 2\times10^{-281}$¢

- A mutual 'G' balance could be found at extremely close $3,12\times10^{853}$ãr distance which is 31,2 fold their radius apart which is too close for comfort because the Neutrons top layer would not find permanent stability at this close distance when the photons are gone.

The Closest Stability Distance

The lower end of the stability span is the compressive effect from main body Proton emission at a close relatively small bonded distance from where the Proton must deliver **$4,08\times10^{-286}$pu** or else the bonded Neutron will split. Orbital virgin pressure that due to closeness arrive in tangential orbits are heavily downsized as compression. The simple method I use to illustrate the compression effect is to take the distance between baryons and divide it by the $3,96\times10^{861}$ãr leverage distance (L) and squaring the result. Five examples at different distances are as follows;

- $(4,21\times10^{856}$ãr $\div$ L$)^2 = 1,13\times10^{-10}$
- $(10^{856}$ãr $\div$ L$)^2 = 6,38\times10^{-12}$
- $(8,4\times10^{855}$ãr $\div$ L$)^2 = 4,5\times10^{-12}$
- $(10^{855}$ãr $\div$ L$)^2 = 6,38\times10^{-14}$
- $(10^{854}$ãr $\div$ L$)^2 = 6,38\times10^{-16}$

10^{852}ãr÷**$4,21\times10^{856}$ãr**$=5,64\times10^{-10}$(gain $4,86\times10^{21}$)=all 10^{-300} sub
$\times(2,05\times10^{26}$¢$\times1,13\times10^{-10})= 2,32\times10^{-284}$pu = **ok**
10^{852}ãr÷**10^{856}ãr**$=10^{-8}$(g10^{23})=10^{-300} sub$\times(10^{26}$¢$\times6,38\times10^{-12})=6,38\times10^{-286}$pu= **ok**
10^{852}ãr÷ **$8,4\times10^{855}$ãr**$=1,35\times10^{-8}$ (gain $1,08\times10^{22}$)= 10^{-300} sub$\times(9,17\times10^{25}$¢$)\times4,5\times$
$10^{-12})=4,12\times10^{-286}$pu = **the last ok**
10^{852}ãr÷**10^{855}ãr**$=10^{-6}$(g. $3,16\times10^{22}$)= 10^{-300} sub$\times(3,16\times10^{25}$¢$\times6,38\times10^{-14})=$
$2,02\times10^{-288}$pu vs. $4,08\times10^{-286}$pu = **the first deficient value**
10^{852}ãr÷**10^{854}ãr**$=10^{-4}$(gain 10^{23})=all 10^{-300} sub$\times(10^{25}$¢$\times6,38\times10^{-16})=$
$6,38\times10^{-291}$pu= **deficient**

The actually delivered pressure across the six distances must meet the required $4,08\times10^{-286}$pu to pacify the Neutron. An orbital bond outside an $8,4\times10^{855}$ãr distance and all the way out to $4,94\times10^{868}$ãr, the main body Proton emission can stabilizes a Neutron. For a distance within $8,4\times10^{855}$ãr and all the way up to a side by side position with the Proton, the delivered main body pressure will prove insufficient for long term stability.

Or, as the Standard Theory tells it; decays of protons and neutrons in nuclear structures occur due to quantum uncertainty in the positions of the quarks that compose them. If the displacement ratio of quantum fluctuation in a bond of shared quarks is too high or low on either side of the bonding distance, the decay rate increases and stable configuration is less likely.

While a direct bond by forcing two single Protons together is extremely hard to achieve as we have seen earlier due to their coronal repulsion and powerful magnetic transfer and one would rarely find two closely paired baryons where one is not a companion Neutron and there is no reason for a companion Neutron to split while it remains perfectly balanced by a Proton in an fast local gravity driven orbit.

If the Neutrons forced to say 10^{855}ãr from a too close starting point, the second Neutron cannot be held stable and will eventually split. But here the rules are different when the same scenario is repeated at close quarters. One might think that the orb drive by photons once outside such a short bonded distance makes an orb driven collision inevitable but this is not the case. At a distance of 10^{855}ãr as the outer Neutron layer has slowly started to expand but barely begun to be accelerated away by the electromagnetic photons that are being split in the widening breach, the consolidating Electron mass orb is without more ado picked up by the close Proton at thousand fold its main body width.

Aided by the powerful orb of 2×10^9 photons between them, the outer layer of this new Neutron is forcefully stabilized by boosted 2×10^{-245}pu harvest of new virgin pressure. The 2×10^{-271}pu linear repulsion by the electromagnetic orb is downsized by r√r from midways between 10^{855}ãr to $3,16\times10^{-4}$. This mutual repulsion would generally last until photons reach outside of the orbiting pair but in this case, the $6,32\times10^{-275}$pu push during 10^{1135}ãt leads to $6,32\times10^{860}$pu repulsion that drives the baryons apart and they orbit away at $6,32\times10^{-280}$¢ superluminal speed.

The 'G' gravity across 10^{855}ãr is driven by $1,25\times10^{-278}$pu and if we look at the $1,58\times10^{1134}$ãt before the baryons start to move apart, the $1,25\times10^{857}$pu flux harvest drives can at best cancel a direct repulsion at $1,252\times10^{-283}$¢. The trailing photons will not catch up with the two baryons heading in opposite directions until after they are kicked out of the central group.

Proton Neutron at 2×10^{862}ãr

From midways at 2×10^{862}ãr, upon the Neutron, the 2×10^{-271} virgins per ãnna linear push from a photon orb is in conflict to the inverse r√r and downsized to $3,54\times10^{-15}$. The resulting $7,08\times10^{-286}$pu repulsion drives the Neutron away at a speed set by its attainable virgin harvest.

- While the Neutron must be accelerated, the photons are at light speed from the start and reach 2×10^{862}ãr in 2×10^{1142}ãt during which the Neutron gathers $1,42\times10^{857}$ virgins that accelerate it to $1,42\times10^{-283}$¢ before all photons pass beyond the Neutron; leaving it more or less in place with **2×10^{862}ãr distance** marginally widened by $2,84\times10^{859}$ãr.

> *'G' $1,25\times10^{-272}$pu$\times2,5\times10^{-21}=3,12\times10^{-293}pu\times2,53\times10^{1147}$ãt$=7,89\times10^{854}$ virgin drive to $7,89\times10^{-286}$¢ in $2,53\times10^{1147}$ãt orbit*

- The original Neutrons at a distance of 2×10^{862}ãr apart would already be orbiting under the influence of $1,25\times10^{-272}$pu 'G' attraction diluted to $2,5\times10^{-21}$ at $7,89\times10^{-286}$¢ speed that takes $2,53\times10^{1147}$ãt in orbit.

- **The photon orb 'attraction'**. With the Neutrons continues away in an opening orbit at $1,42\times10^{-283}$¢ out from the 2×10^{862}ãr baryon distance, the photons have passed beyond that to where adverse virgin emission generates massive backhand orbs between the baryons with an effect heavily diluted by the greater inter-baryon distance.

> 10^{852}ãr $\div$ 2×10^{862}ãr$=2,5\times10^{-21}$ (gain $7,07\times10^{18}$) = $1,77\times10^{-2}$ of 10^{-280} hit a baryon at $1,41\times10^{29}$¢ = $2,5\times10^{-253}$pu orb

- From a distance of 2×10^{862}ãr, each single electromagnetic photon generates its 10^{-280}pu per ãnna virgin orb where $1,77\times10^{-2}$ hit the Neutron boosted to $1,41\times10^{29}$¢ with $2,5\times10^{-253}$pu orb. There are 2×10^{9} photon orbs active between the baryons at the same time and the now 5×10^{-244}pu orbital shadow backhands from all the baryons is diluted to $2,5\times10^{-21}$ just for hitting one baryon and by that rate squared for a mutual backhand $6,25\times10^{-42}$ between both.

- Across 2×10^{862}ãr in 2×10^{1142}ãt, the constant $3,12\times10^{-285}$pu active orb has pushed the baryons closer by $6,24\times10^{857}$pu at $6,24\times10^{-283}$¢ speed. This outguns the $1,42\times10^{-283}$¢ required speed to break the Neutrons outwards drive 4,39 fold at the 2×10^{862}ãr distance.

> 10^{852}ãr $\div$ $8,4\times10^{855}$ãr$=1,42\times10^{-8}\times1,25\times10^{-272}pu=1,78\times10^{-280}pu\times1,35\times10^{1138}$ãt $=2,4\times10^{858}$ = $2,4\times10^{-282}$¢

- With all 2×10^{9} photons outside the 2×10^{862}ãr baryon distance, the companion Neutron is again kicked back into the fold at $6,24\times10^{-283}$¢. A potential $1,35\times10^{1138}$ãt period of approach across the $8,4\times10^{855}$ãr widest exponential distance is not realized because the lower speed allows 'G' already before that distance to heavily affect any inwards path across as its backhand is greater than the inwards orb speed. 'G' is already skewing any approach with an uncertain aim into a closing mutual baryon orbit. At $8,4\times10^{855}$ãr, the 'G' backhand drive is already 3,85 times greater than the inwards orb push.

> 10^{852}ãr $\div$ $3,2\times10^{856}$ãr$=9,76\times10^{-10}\times1,25\times10^{-272}pu=1,22\times10^{-281}pu\times5,13\times10^{1138}$ãt $=6,25\times10^{857}$ = $6,25\times10^{-283}$¢

- Across $3,2\times10^{856}$ãr, the potential $5,13\times10^{1138}$ãt period of approach at $6,24\times10^{-283}$¢ across that widest exponential distance is not realized as 'G' has already at that distance to heavily affect any inwards path across as its backhand is eclipsing its inwards speed of the orb drive and starting to skewing an approach with uncertain aim into a closing mutual baryon orbit at a somewhat closer distance. Any distance on the narrow orbital sphere between $8,4\times10^{855}$ãr and $3,2\times10^{856}$ãr, the Proton supplies enough pressure to keep the Neutrons top layer to permanently stable after the photons are gone and the two baryons continue to fall around each other as any other pair of heavenly bodies under the spell of gravity.

Proton Neutron at 5×10^{862}ãr

From midways at 5×10^{862}ãr, upon the Neutron, the 2×10^{-271} virgins per ãnna linear push from a photon orb is in conflict to the inverse r√r and downsized to $8,94\times10^{-16}$. The resulting $1,79\times10^{-286}$pu repulsion drives the Neutron away at a speed set by its attainable virgin harvest.

- While the Neutron must be accelerated, the photons are at light speed from the start and reach 5×10^{862}ãr in 5×10^{1142}ãt during which the Neutron gathers $8,95\times10^{856}$ virgins that accelerate it to $8,95\times10^{-284}$¢ before all photons pass beyond the Neutron; leaving it more or less in place with **5×10^{862}ãr distance** marginally widened by $4,48\times10^{859}$ãr.

'G' $1,25\times10^{-272}$pu$\times4\times10^{-22}=5\times10^{-294}pu\times10^{1148}$ãt$=5\times10^{854}$ virgin drive to 5×10^{-286}¢ during 10^{1148}ãt orbit

---

- The original Neutrons at a distance of 5×10^{862}ãr apart would already be orbiting under the influence of $1,25\times10^{-272}$pu 'G' attraction diluted to 4×10^{-22} at 5×10^{-286}¢ speed that takes 10^{1148}ãt for each orbit.

- **The photon orb 'attraction'.** With the Neutrons continues away in an opening orbit at $8,95\times10^{-284}$¢ out from the 5×10^{862}ãr baryon distance, the photons have passed beyond that to where adverse virgin emission generates massive backhand orbs between the baryons with an effect heavily diluted by the greater inter-baryon distance.

10^{852}ãr÷ 5×10^{862}ãr$=4\times10^{-22}$ (gain $4,48\times10^{18}$) = $1,79\times10^{-3}$ of 10^{-280} hit a baryon at $2,24\times10^{29}$¢ = $4,01\times10^{-254}$pu orb

---

- From a distance of 5×10^{862}ãr, each single electromagnetic photon generates its 10^{-280}pu per ãnna virgin orb where $1,79\times10^{-3}$ hit the Neutron boosted to $2,24\times10^{29}$¢ with $4,01\times10^{-254}$pu strong orb. There are 2×10^9 photon orbs active between the baryons at the same time and the $8,02\times10^{-245}$pu orbital shadow backhands from all the baryons is diluted to 4×10^{-22} just for hitting one baryon and by that rate squared for a mutual backhand $1,6\times10^{-43}$ between both.

- Across 5×10^{862}ãr in 5×10^{1142}ãt, the constant $1,28\times10^{-287}$pu active orb has worked to push the baryons closer by $6,4\times10^{854}$pu at $6,4\times10^{-286}$¢ speed. This however falls 139,84 times short of $8,95\times10^{-284}$¢ required speed to break the Neutrons outwards drive at 5×10^{862}ãr distance. As the baryons can here generate 5×10^{-286}¢ 'G' orbits at this distance, it cannot prevent the $8,95\times10^{-284}$¢ repulsion. The outwards speed of the Neutron from the Proton continues at 5×10^{862}ãr apart in an opening orbit while orbs from departing photons are rapidly weakening.

- *When one of two bonded Neutrons splits at any starting point beyond 5×10^{862}ãr, the other Neutron will be more firmly repelled and therefore set up for an inevitable split.*

- A Neutron companion survives in orbits from $3,2 \times 10^{856}$år down to $8,4 \times 10^{855}$år around a Proton before it starts to expand into a split. To have some average distance to relate to rather than diverse spans, the **10^{856}år** seems a nice middle ground to use in upcoming illustrations. If you think that two baryons in a stable bond at 10^{856}år keep a small distance apart at ten thousand fold their standard size, keep in mind that if Our Sun was in a double system at this size multiple, the two suns would be mutually orbiting far beyond the distance to Pluto.

Baryon Group Formation by Orbs

Imagine a grid of hydrons at a consolidating solar center. Every way across immense relative distances, single Neutral hydrons form the average lattice which main feature is that they are not affected by attraction and repulsion. Rather, they are affected by both from all directions equally which makes them more or less unable to move one way or the other within the lattice grid. Nevertheless, hydrons are often disturbed and can draw individually closer than the average which makes them vulnerable.

At these grid distances, incoming virgin orbs are far more effective than 'G' gravity in moving vulnerable Neutrons closer together out of grid formation (an orb effect is similar to 'G' gravity but is distributed by virgins on narrow hydron wide orbs instead of holistically from all directions).

In tighter grids with rising field disturbances, Neutron pairs will often form groups at diverse closer distances than the average lattice. The picture is that of a three dimensional solar system where all units are of same size and mass (sun, planet, moon). When this drawing closer has gone on for a long while under the auspices of 'G' and maximum group Neutrons have moved closer, that mutual closeness leads to slow initial orbits at diverse distances from the group center.

It is likely when a solar grid collapses, especially at the later stages, that single Neutrons have started to orbit the group center and those near the group center have become close pairs that have adjusted to 'G' and orbit the center while mutually orbiting each other at higher speeds. The slowest 'G' orbits are those of single maximum Neutrons in slow orbits around a groups far off collective center of gravity.

In later stages of solar collapses, due to the influences of virgin orbs from passing electromagnetic photons forming the group, these stronger orbs affects Neutrons differently. The single maximum Neutrons at the furthest distances within a group are moving in slower collective orbits around the groups center of gravity and are individually pushed along by orbs that cast a backhand along a relatively straight line. The most enduring backhand push effect by common orbs is between two slower single Neutrons that are fastest in approaching each other of all members in the group.

As we have seen, powerful orb backhands from electromagnetic photons, in this case the relatively steady grid distance orbs from passing photons of a solar upheaval are extremely effective in bringing individual baryons closer if they for some reason wander from the average grid lattice distance.

Neutron pairs successively closer to a group center of gravity have already become bonded in smaller individual orbits. These pairs are moving in faster collective orbits around its center of gravity but vastly faster in mutual orbits around each other. Due to local orbits, the staying power on a narrow long distance orb of Neutron width generated by passing photons but common to both Neutrons of the fast orbiting pair is seriously limited. Paired Neutrons, mutually orbiting each other, pass through such a virgin backhand orb and thereafter spend the rest of the local orbit outside that orb; an important character of group formation.

Look at a small forming 'group' of six Neutrons to see what happens when a single photon passes by at say a 10^{892}ãr wide solar core grid distance. For convenience, instead of using 10^{860}ãr maximum Neutrons held to that size by passing local photons; this illustration uses standard Neutrons (from earlier) because although the input is different but the principle is the same.

Say that of the six standard Neutrons, one maximum pair is mutually orbiting at the group center at 2×10^{862}ãr apart in slow $7,89 \times 10^{-286}$¢ 'G' orbit. Another pair could be mutually orbiting at say 10^{866}ãr apart at a distance of 10^{874}ãr, orbiting the groups non-material center of gravity more slowly. The two unpaired single Neutrons orbit the group further away at say 10^{880}ãr and the outermost at 10^{888}ãr (10^4 times closer than the average 10^{892}ãr solar core grid). The group is now an independent point constellation within the general lattice.

The groups central Neutron pair (standard as shown earlier; for illustration only) at 2×10^{862}ãr apart would be mutually orbiting at $7,89 \times 10^{-286}$¢ by the pull of $1,25 \times 10^{-272}$pu 'G' backhand attraction diluted to $2,5 \times 10^{-21}$ speed during $2,53 \times 10^{1147}$ãt in each orbit; also their mutual orbit around the groups nonmaterial center of gravity.

'G' $1,25 \times 10^{-272}$pu÷10^{28}=$1,25 \times 10^{-300}$pu×$8,95 \times 10^{1152}$ãt=$1,12 \times 10^{853}$ *virgin drive to* $1,12 \times 10^{-287}$¢ *during* $8,93 \times 10^{1152}$ãt *orbit*

The standard Neutron pair at 10^{866}ãr apart would be mutually orbiting by the pull of $1,25 \times 10^{-272}$pu 'G' backhand diluted to 10^{-28} at $1,12 \times 10^{-287}$¢ speed that requires $8,93 \times 10^{1152}$ãt for each orbit.

'G' $1,25 \times 10^{-272}$pu÷10^{44}=$1,25 \times 10^{-316}$pu×$8,95 \times 10^{1164}$ãt=$1,12 \times 10^{849}$ *virgin drive to* $1,12 \times 10^{-291}$¢ *during* $8,93 \times 10^{1164}$ãt *orbit*

The same mutually orbiting Neutron pair is simultaneously driven in 10^{874}ãr wide common orbit around the groups center of gravity at $1,12\times10^{-291}$¢ that takes the pair $8,93\times10^{1164}$ãt to complete.

> 'G' $1,25\times10^{-272}pu\div10^{56}=1,25\times10^{-328}pu\times8,95\times10^{1173}$ãt$=1,12\times10^{846}$ *virgin drive to* $1,12\times10^{-294}$¢ *during* $8,93\times10^{1173}$ãt *orbit*

The closest single Neutron, further out at 10^{880}ãr from the group center, is not mutually orbiting any companion but remains under the weaker spell of the groups gravity center in a slow orbit around the group at $1,12\times10^{-294}$¢ which requires $8,93\times10^{1173}$ãt to complete.

> 'G' $1,25\times10^{-272}pu\div10^{72}=1,25\times10^{-344}pu\times8,95\times10^{1185}$ãt$=1,12\times10^{842}$ *virgin drive to* $1,12\times10^{-298}$¢ *during* $8,93\times10^{1185}$ãt *orbit*

The outermost single Neutron, under the weak average central spell in a 10^{888}ãr wide orbit that is slower still around the groups center of gravity that is fuelled by $1,25\times10^{-344}$pu. The 'G' backhand drives its 10^{1140} primal mass at $1,12\times10^{-298}$¢ in its around the groups central pair in $8,93\times10^{1185}$ãt.

Virgins from an electromagnetic photon passing the group at light speed at the 10^{892}ãr grid distance, affects all the local group members individually in accord with the virgin harvest gathered. For the oversized Neutrons, this is a stabilizing effect. For two paired Neutrons during the period of their common simultaneous presence on the orb there is attractive backhand while they pass through the photons virgin orb. A mutually orbiting Neutron pair lines up twice in each differential orbit (also depending on the angle of a passing e-photon) during which it can harvest common orb push.

The angle of a passing electromagnetic photon plays a changing role, so put it at a neutral angle and let it approach the group on a center wise line rather than on a parallel path. The thing that stands out here is that the groups central pair will pass into, pass through and pass out of the baryon thick electromagnetic orb at $7,89\times10^{-286}$¢. This is $7,04\times10^8$ times faster than the inner single Neutron that passes through the same orb at $1,12\times10^{-294}$¢.

All the group Neutrons in local or common center orbits at diverse speeds will from time to time share electromagnetic orbs with the more 'stationary' grid baryons outside the group at 10^{892}ãr distance; before a photon reaches within that distance while it can cast a shadow between both.

One may think that the shorter duration on a common long distance orb makes it likelier that the outermost group Neutrons are carried out from the group than closer into it; which would therefore leave only the closer group pairs. This said, the plain fact is that these long distance orbs provide very diluted backhand across the 10^{892}ãr grid and that resulting backhand is also heavily diluted across the same 10^{892}ãr grid.

From one electromagnetic photon passing at the grid distance, the diluted but boosted backhand from a group baryon is 10^{-312}pu which is on average diluted upon a Neutron in the outside grid by 10^{-80} to an exceedingly weak 10^{-392}pu drive (the drive is greatly enhanced if a photon passes at closer than 10^{892}ãr).

On the other hand, the weak 10^{-312}pu drive is more effective between unpaired Neutrons within the group, here for example the baryon at 10^{888}ãr and the other at 10^{880}ãr where a 10^{-368}pu push forces closer bonds between the slower single group Neutrons over time until they form local orbits that eventually pick up fast enough 'G' speed to avoid further closeness.

Although each grid point within a collapsing solar mass will be subjected to adverse emission from 2×10^9 photons that are continuously followed and replaced by new photons passing out of and through the sun, often at less than grid distances, upholding the active orbs during periods of seconds or even hours during a solar conversion, say 10^{1198}ãt; even the outermost Neutrons are unlikely to join the wider grid while single Neutrons are likely to form pairs or triplets at closer distances where they remain to be tested for the upcoming splitting season.

The few close enough pairs baryon group that can ultimately be stabilized at close enough distances where Neutron companions can enjoy permanent long stability after one of them splits are an exception; the rest are rejected as single Neutrons from far outside the group center in the guise of single remotely tied baryons or as pairs that are just a tad too far apart to allow survival when the photon pressure subsides.

Split into Protons and Electrons, single Protons tend to be repelled by powerful main body and later coronal aversion away from the central group while rejected Electrons from all Neutron splits orbit at greater distances around the central baryon gathering.

Neutron Splits Within Groups

For an illustration of what happens next is that ever more electromagnetic photons are heading inwards from a 'splitting sphere' that is shrinking from the solar surface towards its center; a batch of 2×10^9 photons following the former batch to within that grid distance in a constant stream of 2×10^9 photons where each single photon has a given statistical chance to pass by a Neutron group at a given distance.

For example across 10^{907}ãr (ten meters) across 10^{15} grid domains, one of 2×10^{24} photons would pass at 5×10^{867}ãr from a Neutron group in one of the domains.

10^{852}ãr$\div$ 5×10^{867}ãr$=4\times10^{-32}\times(g.\ 1{,}41\times10^{16})= 5{,}64\times10^{-14}\times\ 10^{-280}=5{,}64\times10^{-294}$pu $\times(7{,}07\times10^{31}¢)=3{,}99\times10^{-262}$pu; *over standard Neutrons* $4{,}23\times10^{-269}$pu *stability pressure. A* 10^{-38}*cut makes it* $3{,}99$ *fold standard* 10^{-300}pu *Proton expansion.*

Under this constant barrage, the large maximum Neutrons in their local groups at a solar core are being compressed to 10^{852}ãr standard sized by the adverse virgin emission from close passing photons. With the *10^{38} shave off for center wise hydron compression, the* 'G' gravity flux has locked all the interior Proton part of the Neutrons into invincible entities.

Better than the statistical chances of close passage are the chances of too far off passages to stabilize a standard Neutron. The expansion of the 10^{852}ãr Neutron Proton is exclusively the departure of its outer negative layer, driven by the underlying Protons negative dominant main body emission; The lesser photon pressure cannot keep its buffer zone from widening into a breach.

There is almost no difference in how orbital virgin pressure from passing photons is harvested by group Neutron; the pressure loss by virgin shadows in even extremely tight pairs is marginal. A greater difference is that of the exchanged boosted main body emission between tighter pairs that affords them slightly more support than wide bonds; the least support is afforded the single Neutrons in the group that are first to be affected and split, which of course temporarily stabilizes the other slower expanding group Neutrons.

Group splits follows the same basic rule; one at the time, individual splits move deeper inwards from the group periphery; each split leaving a standard Proton in its wake and its outgoing photons temporarily stabilizing the rest.

The outer group Protons, usually with their Electrons in orbits that are far larger than an inner group are repulsive upon the rest of the group and they are (relatively speaking) more easily bounced away from the group out to average grid distance. This will later be enhanced when new electromagnetic splits take place nearer to the group center. Photons heading out of a group are instrumental scattering remaining outer Protons into the surrounding grid but unable to bring a stop to ongoing mutual central collapses. This will instead serve to tighten the central pairs before splits occur. This repulsion leaves a steadily dropping number of Protons in the core group population that becomes limited to paired Neutrons within the tighter central group.

10^{852}ãr$\div2{,}2\times10^{883}$ãr$=2{,}07\times10^{-63}\times($*gain* $2{,}13\times10^{8})= 4{,}41\times10^{-55}\times(2\times10^{-271})=$ $8{,}83\times10^{-326}$pu$\times(4{,}69\times10^{39}¢)=4{,}14\times10^{-286}$pu *vs.* $4{,}08\times10^{-286}$pu *stability pressure*

There are 2×10^{9} electromagnetic photons from every split Neutron but even in a group where hundreds of splits can take place, they rarely occur at the same time. No new Neutrons will be split until the 2×10^{9} photons from each split have left across a distance in excess of $2{,}2\times10^{883}$ãr so that all standard group Neutrons can remain stable within that volume for $2,2\times10^{1163}$år before a new burst is allowed from another of the outermost grid Neutrons.

Depending on the distance of a split Neutron on the group outskirts, the initial two billion electromagnetic orbs from the 2×10^9 photons (each casting a backhand orb from all baryons in the group), can either severely overlap at various angles in opposite directions between central baryons or push them along one specific orb as their staying power on any orb is limited when being pushed faster by other orbs.

Orb backhand can expand orbits through myriad steps or diminish them and as the center holds more Protons in a smaller volume, there is a natural reason why the orbs do not disperse the central faster orbiting baryons into wider volumes. When photons reach outside a baryon group, repulsion turns more linear while orb drives fade, especially upon the rapidly orbiting pairs within a small central group volume. Baryons at closer quarters are all orbiting in individual paired relationships at varied speeds according to their distances apart and all of this will notably govern the formation and future structure of heavier atomic nuclei close to a solar center.

And like auxiliary coronal photons crossing a recall orb, Neutrons crossing through an orb in continual orbit gain in linear working time after each passover by square root of the distance in Neutron radii (if the orb persists) and the extra working period brings them closer after each passover. It is however unlikely that two Neutrons line up directionally more than once while a photon heads away at 10^{-280}¢ (which allows the mutual linear push to accumulate over time).

During a Neutron split, the Electron flies away; passing by any companion Neutron while yielding little influence by main body emission. Unlike the electromagnetic photons, the Electrons main body push upon inner baryons is ineffective. Its initially fast flight takes it to least 10^{875}år where eventual exchanges between the single Proton and its single Electron ultimate find some balance, ultimately after the later release of restricted coronas.

Electrons (that harvest weak flux gravity) experience only weak gravity pull from a companion Neutron and hardly none from a Proton. An Electrons is set moving and gathers speed by blocking electromagnetic virgins at a closer distance ahead of the dispersing photons; long before a Neutron companion starts to be accelerated. The Electron contains coordinates that were to large extent one with the Proton while the companion Neutron was never part of the now left behind Proton. The Electron orbits are (to that extent) shaped by inherent coordinates rather than gravity backhand and each will initially (and for a long time) follow earlier coordinates that were common with its Protons inherent coordinates.

Orbiting far away orbiting the central group, there would be equally many Electrons as there are Protons within the central group; all paired to close

*Neutron companions. The **atomic nucleus** holds all the Proton Neutron pairs; encircled by a corresponding number of Electrons in relatively vast orbits.*

In a heavier nuclei, if a baryon pair is too close for the Proton to supply enough pressure to keep the Neutrons top layer permanently stable, its split forces an exchange of the outer negatively polarized layer with the Proton and both baryons are expelled from the central group which eventually lead to a split of the new Neutron. There is generally no single Proton within the atomic nucleus but when it happens, a single Proton orbiting within a tight group is different from the all the other pairs within the baryon group inasmuch as the Protons free main body emission is not blocked by a close companion which allows all of it to become free positive emission that mostly escapes the group.

A free Proton Neutron pair orbiting between the $8,4\times10^{855}$ãr to $3,2\times10^{856}$ãr stability distance is an atomic nucleus of itself; the Deuterium isotope of hydrogen. As noted, a Proton can stabilize a Neutron by $4,08\times10^{-286}$pu main body emission at far greater distance up to $4,94\times10^{868}$ãr and there is nothing wrong with that except such bonds would be unstable and more importantly few as they are an anomaly to the forces that originally set pair distances.

There are various restrictions to group sizes due to internal virgin winds of repulsion and orb attraction where the stabilizing effects of 'G' gravity must coral these effects within a group boundary if it is to remain stable but this is not within the purview of my illustration.

In our present day Milky Way, the formation of a new sun would follow different rules. Any new sun is formed from cloud gatherings of already split hydrogen atoms instead of Neutrons; often interspersed with some heavier nuclei from older dispersed suns. Such solar submists are more resistant by already more powerful internal emission that causes less heavily compressed suns that are unlikely to create as many new heavy nuclei on their own.

Bonded Proton Neutron Corona

The surviving Neutrons of 2×10^{9} positive and $4,17\times10^{6}$ negative photons are in lockdown after a nearby grid Neutron split and so are the Protons 2×10^{9} negative photons but both baryons in the Deuterium atom are yet without the potential auxiliary coronas. An orbital bond outside $8,4\times10^{855}$ãr on a narrow orbital sphere between $8,4\times10^{855}$ãr and $3,2\times10^{856}$ãr, the Proton supplies enough pressure to keep a Neutrons top layer permanently stable after the photons are gone.

At the 10^{856}ãr bonded distance, a Protons main body emission is boosted across that distance where it battens down auxiliary photons on the surface of the rotating and orbiting Neutron. The positive auxiliary photons on the Neutron surface receive the greatest thrust (from subservient 10^{-300}pu versus dominant $1,77\times10^{-307}$pu) and are therefore rarely launched by Neutron surface as the negative dominant exchange between the baryons isa formidable pressure.

> *1,77×10^{-307}dominant× 10^{26}¢ = 1,77×10^{-281}pu vs. 1,77×10^{-307}pu=10^{26}fold the 10^{1140}ãt baryon rotation; the inactive 10^{1166}ãt surface period versus vs. 10^{1176}ãt coordinate adaptation = adapting 10^{-10} of inherent spin!*

Whenever the time comes when a baryon pair can launch their negative auxiliary coronas, all those original auxiliary photons are fully in tune with inherent hydron coordinates (newly blocked photons would in this case carry only 10^{-10} coordinate fraction). The mutual dominant negative pressure effect between baryons at 10^{856}ãr apart delivers 1,77×10^{-281}pu negative orb or 10^{26} fold their 1,77×10^{-307}pu dominant emission.

*The Protons 10^{-300}pu subservient main body emission delivers 10^{-276}pu positive orb upon the Neutron (dropping the leverage point shave-off that does not change enough for the thin top layer). Positive pressure upon the Neutron exceeds its **4,08×10^{-286}pu** requirement and leaves it safely balanced. Being fairly evenly distributed across the Neutron surface, it also keeps all of its potential positive auxiliary photons firmly grounded.*

The inactive period of any eventual negative one jump auxiliary photon emitted into a pod corona would be 10^{26} fold the 10^{1140}ãt period of baryon rotation and after 10^{1164}ãt inactive period coronal launch would cut the standard 7,28×10^{-284}pu free coronal emission to 10^{-26} with 7,28×10^{-310}pu free positive coronal emission supplied by an occasional auxiliary photon. The auxiliary emission across the 4,21×10^{856}ãr trek at 4,21×10^{-284}¢ out to the standard coronal distance takes 10^{1140}ãt which is over 10^{26} times faster than the inactive surface period.

The significant fact is that the 4,21×10^{856}ãr reach of a standard corona is well beyond the two baryons 10^{856}ãr distance apart.

Although it will still be up to the Protons positive main body emission to supply the Neutron with more numerous 10^{-300} free positive virgins per ãnna for stability, this main body exchange orb cannot however recall re-launched auxiliary photons back to either baryon.

> *10^{852}ãr÷10^{856}ãr=10^{-8}(gain 10^{22})=only 10^{-14} of all emitted virgins by the pair slip by the main bodies as free emission.*

As to free field emission; the fact that all surviving Neutrons are this close, the main bodies of each pair block all but 10^{-14} of main body emission with only 10^{-314} positive main body virgins per ãnna slipping into the strayfield along with 3,54×10^{-321}pu negative main body virgins; a 10^{-14} downsizing of free external emission is certainly a massive change.

Since the linear mode of the exchanged main body emission is massive enough to ground negative auxiliary photons at the two 'poles' between the baryons, most auxiliary photons will be launched from the further polar area. Every launched negative auxiliary photon from either baryons 2×10^9 stockpile of singular negative mattrons must on the auxiliary trek out to the coronal distance (along with whatever handful of newly captured photons with less than perfect inherent coordinates) pass through a powerful closer orb that is generated and cast as backhand between the two baryons.

Almost anywhere on the orb sphere, its powerful backhand can cause a mutual one step auxiliary recall; except in the one spot on the polar line out from the further pole where the photon is being attracted equally from all directions on the orb sphere. The orbital backhand between the two baryons cannot recall the outgoing auxiliary photon from the further pole that head unperturbed through that orb on its linear path out to the standard orbital distance. When an auxiliary photon heads out from the further 'pole' at right angles to either baryon and before it forms an orbit, it passes with perfect easy through the backhand of the orb cast by either baryon.

As this extremely stable bond was created with an Electron, albeit further away (due to less free coronal emission), its constant weaker presence places a negative Deuterium corona further from the central baryon pair (due to less magnetic transfer). The joint Proton Neutron corona could soon be fully stocked with 2×10^9 negative fully inherent intermittently recalled auxiliary photons and they may not be able to force the central pair closer by the more diluted orbs; pursuing ever more details will only slow down this illustration. As more coronal emission slips by the two close baryons with a larger corona, this does not put as great a damper on external exchanges.

As a unit, the bonded baryons with their joint corona behaves much like an ordinary Proton with its own corona and although exchanged auxiliary photons from one baryon would often sweep closer by the other baryon and cause faster pod recall but that is a marginal disturbance.

A clearer picture may be needed of the two largely parallel orb planes that form between a bonded Proton Neutron pair and generate mutual recall of auxiliary photons. The emission poles are by nature aligned at right angles to the two parallel recall planes on a straight line through all four launching poles of a hydron pair, two poles between them and two on the far sides.

Mutually exchanged virgins can in principle bombard the emission pole between the hydrons with all the emitted virgins at linear 1¢ drive and hold down negative auxiliary surface photons; driven by the same number of virgins. The Proton has an easier task at keeping down positive auxiliary photons on the near emission pole on the Neutron surface.

The predicament is that a largely equal number of virgins hit the perfect far-side pole where they instead lift a small polar area by the same linear drive. The blocked main body orb virgins are moving linearly away from the baryon that emitted them and therefore also away from the other baryons further pole. For a small number of mattrons at the two perfect poles of a globular surface, their sideways drive does not aim through the top layer but along its; anywhere else, the aim is to some extent compressive.

Even a marginal inwards main body emission aim upon at the diamond smooth surface grid near the further pole does not allow large unhindered polar areas for auxiliary launch. Think of the small further polar area as a low-pressure atmospheric area where a top layer rises to form a bulge on a slightly raised surface as neutral mattrons are pushed exponentially faster sideways and in opposite directions by boosted virgins than they are pushed linearly upwards by the same orb virgins.

In cases of close severe exponential sideways orb pressure, the linear push upon the far polar area fashions a narrow thin and more highly raised polar surface. Without differential rotation, the low-pressure area would create a narrow polar pinnacle built of numerous neutral mattrons close to its top (in the absence of a corona). Rotation is effective in shifting a small polar area and bring its pinnacle under less tangential inwards compression where it starts to subside again.

In a rotating Neutron, the potential pinnacle of neutral surface mattrons broadens in the rotational direction as the sideways compression must work upon more numerous mattrons in the direction of the rotation. At right angle to the rotational direction, the pinnacle does not subside but is reshaped as a raised surface takes a shape of a dolphin breaking the polar surface with a half circle in the mid further polar area; a stationary shape lined up in the rotational direction; periodically flattened by positive pressure from the corona upon neutral mattrons on the polar area but not upon any negative auxiliary photons there.

Forbidden Positions

From the start, there are various orbs formatting the local structure of baryon groups before they stabilize. There are forbidden positions within any system (familiar to us in many guises) that deserve a word. All orbs lend an important hand to this local aspect of a group structuring after 'G' gravity has taken the group in hand.

We are used to obvious 'forbidden' positions such as two planets orbiting a sun on a similar plane and at a similar distance. Jupiter and Saturn would not stay intact under such circumstances; nor will baryons pairs in orbit around a group center. Forbidden positions are by necessity in orbits rather than in stationary positions where two orbiting entities can only avert destruction if both keep permanently and perfectly opposite positions on the same sphere which would neutralize any kind of orb attraction; an unlikely and unstable setup.

A composite atomic nucleus is born as Neutron pairs gather within a globular central region (except in a hydrogen atom of a single baryon). The Helium atom has two baryon pairs within a 'G' domain and two Electrons orbiting at relatively far distance. The two central Protons baryon, each with closely bonded mutually orbiting Neutron companion are within a tight central region and both pairs also orbit more slowly around the groups non-material center of gravity.

Even if this is hardly a group at all, there are forbidden positions. The two Electrons cannot for example orbit on exactly the same sphere. The emission orbs from the two central pairs casts a backhand that would inevitable bring them on a collision course. To permanently escape collision, each Electron sphere must be separated by the possible width across which the inner orbs can strafe it with orbital backhand virgins. The separation on the Electron sphere will therefore not be that significant if it depends on the width of the central group compared to the Electron distance but can vary substantially according to group size.

A group radius can be say 10^{864}ãr with an Electron orbit at least 10^{874}ãr so a separation of the two Electron spheres need only be 10^{-10} of the distance from the center; a minor gap between the active tubular wedding bands from parallel directions. Anything less would be a forbidden position on a given sphere; not least at times when the Electron coronas become grounded and central baryons may form more repulsive coronas.

The orb effect from the center cannot be neutralized by fast mutual local orbits as the Electrons do not form pairs; on the contrary they are mutually repulsive which would eventually widen the gaps on the thin sphere upon which the two Electrons orbit their individual central baryons. Add two hundred Electrons and you have a complicated cloud that still must be governed by forbidden positions.

Individual Electron spheres are only partly set by the positions of their partly inherent baryons within the central group. The Electron from the first split Neutron from an outermost pair within the central group would reach the furthest distance; there is no repulsive Electron orbiting the group that can resist its outwards journey. The next split, from the next outermost Neutron pair would reach a similar distance sphere but it can be affected by the first Electron in orbit. As the first Electron is just as likely to be on the other universal direction from the central group, and more likely a sideways position to the new Electron ejection, a similar distance sphere is likely to accept it on average tad closer in a position that is not a forbidden position. Even after some adjustment by occasional mutual repulsion if the Electrons have close encounters in orbits, these two bonds on the same sphere need not be far apart.

You can see where this is going in subsequent Neutron splits; with each additional orbiting Electron, banded by gaps on that first broader sphere, there is a growing repulsive resistance to allow additional Electrons onto an 'empty slot' on that banded sphere. After that (and partly before) a powerful resistance forces newly split central Electrons to take up a position on a far lower sphere that has fewer available banded slot positions due to a growing overall resistance from the further off first Electron 'shell'. In the end, the lowest shell can carry the fewest Electrons on that banded sphere.

The perfect center of mass need not to be corporal but out of any baryon group, some group baryons would be closer to the gravity center that distributes the baryon pairs into ever shorter orbits closer to the center. For a large group of multiple baryons, we would see pairs positioned at the shortest distances to the group center with all the remaining pairs being positioned out from there at ever wide distances.

The upshot of this is that when you look at a virgin gravity orb generated at the center, it is cast by all other baryon pairs in the group but none of the globular orbs leads through a single other baryon in group orbit (other than the baryon casting it and its mutually orbiting companion). The reason is that a groups baryon pair casting the orb is the only pair that can take up residence at this distance from the group center.

Lay out a two-dimensional illustration of one hundred and one pairs that orbit in a single plane around a baryon groups center of gravity. Each pair on that flat invariable plane also keeps fast mutual local orbits around its close baryon companion. You can imagine this as a strange solar system where all entities are of the same mass and the distances between all pairs widen away from the groups center of gravity (shorten towards its center).

Each of these 101 baryon pairs generates 201 globular three dimensional virgin backhand orbs which virgins are blocked by all the other baryons. Aside from the emitting baryon, this gives rise to 200 tubular backhand orbs of standard thickness that encircle a generating baryon in three dimensional space in all directions from all baryons on that invariable plane. As distances of receiving pairs are different, no globular orbs intersect any two baryons on the invariable plane except intermittently across the local orbit of a mutually orbiting pair.

Each Neutron blocks ten thousand orbs from both directions at right angle to the flat plane but none of these orbs casts an effective backhand between any other two baryons except periodically between closely paired Proton and a Neutron in fast mutual orbits. As the widening distances are different for all the pairs, all forbidden positions are avoided in such a setup.

When you look sideways at the 202 baryons on the flat invariable plane; focusing on the generated orbs (even gravity flux), you notice that the three dimensional orbs reach out of the flat plane to different heights in opposite directions. The distance between a generating hydron and one casting the orb sets the radius of the globular orb. Thus between two hydrons closest to the group center, the orbs are lowest which also means most powerful while the highest and weakest orbs are between across the entire the baryons that occupy the orbits around the flat invariable plane that are furthest from the groups center of gravity (this can be an orbit generated by a central baryon or by an outer baryon of almost twice the group radius).

In between these extremes, orb heights form a salient blend. You can see how the orb effect changes (without changing its nature) if one baryon is allowed to orbit in three dimensions. Not only does the differential orbit of that baryon allow its generated orbs to cast its backhand between baryons one the invariable plane but it also allows orbs generated from that plane to lead a backhand between plane baryons and the right angle baryon.

This will very gradually lower its orbit through every orb as myriad pushes bring it into contact with ever more potent shorter and darker central gravity orbs as its steps towards the system plane grow bigger. As the reach of the central more powerful plane orbs is lowest, it first leads to flattened globular format to the outer edges. The setup shows that a less than spherical three dimensional shape may lead to different forbidden positions in an already slightly flattened globular group.

If a large gathering of baryon is caused to be marginally flattened (not like a rugby ball but more like a deflated soccer ball left on a shelf), the ground may be laid for a less than spherical Neutron group by the initial virgin winds that move ahead of electromagnetic photons crossing into a solar grid. The linear component of these virgins which orbital backhand can pry baryons from their average grid positions and allows them to gather individually closer by the effects of the tubular bands of backhand orbs can also flatter the composition.

Virgins without orbital boost deliver a weak linear push that arrives from one direction and the push upon a forming baryon gathering can set these to move linearly closer as well as orbitally. When outermost group photons draw closer, effects of one sided virgin repulsion can slightly flatten a baryon gathering; an apparently mundane shift that for the above reasons can possibly format such a group structure somewhat more decisively in its later phases.

Again, it is not the purview of the Primal Code to set up precise models; merely to highlight the forces at play. For larger systems such as galaxies, the forces are very different. A spiral galaxy has an invariable plane but its thickness spans the width of myriad suns and grows to a bulge towards its center. A galaxy may be perfectly globular or deformed into an in-between structure. The main difference is that globular virgin orbs spawned by stars and their mutual repulsion are often extremely weak against the mass to be moved while they experience intact gravity pull, such as it is.

Comparisons of orbital speeds

The $\sqrt{m} \div \sqrt{r}$ formula is an exercise in relativity, not actual measurements. It uses a proven benchmark of actual mass and distance to compare the net orbital point movement of any other submist that is driving distant mass in orbits under the influence of 'G' backhand. This could be the point speed of a planet orbiting at any distance around Our Sun or a moon around a planet or a Neutron orbiting a Proton as shown in the following table.

Set Earth speed in orbit as a benchmark with Solar mass calculated in Earth masses with the Sun to Earth distance measured in Solar radii as 'r'. And by using the Earth speed in orbit as the benchmark speed; the formula applies to the point speed generated by any particular submist of solar, planetary or baryon mass when it affects distant exterior mass the orbits that submist.

Using the 577 square root of Our Solar mass (333000 Earth masses) and 14,661 square root of Sun-Earth distance (214,939 solar radii), the $\sqrt{m} \div \sqrt{r}$ is roughly (577÷14,661) and **39,356** is the relative Earth benchmark. This can be coupled to Earths actual 29,78 km/sec speed in Solar orbit; all purely relative without constants.

For larger systems with known orbital distance and mass, the orbital speed generated by gravity pull can be inferred from $\sqrt{m} \div \sqrt{r}$ at the distance where a closed orbit renders perpetual neutralization of gravity backhand in every linear universal direction. As a hydron is the first submist that is able to cast a solid gravity shadow although this may be more disturbed at close distances although a bonded baryon seems to work by comparison. Our Suns mass of $1,989 \times 10^{30}$kg that each consists of $5,976 \times 10^{26}$ baryons casts a backhand by $1,19 \times 10^{57}$ baryons or by $1,19 \times 10^{1197}$ primal mass.

The distance between our Sun and the Earth is one astronomical unit equal to a distance of $1,5 \times 10^{917}$ãr and the $\sqrt{m} \div \sqrt{r}$ yields $8,91 \times 10^{139}$ versus 10^{140} for two baryons in mutual orbit across 10^{860}ãr. As the difference here is 1,12 in favor of the baryon orbit, it exceeds the Earth's 29,78km/s average speed in orbit. The baryon would then orbit at a leisurely 33,41km/s equaling $1,114 \times 10^{-284}$¢.

This is the same speed as came out in an earlier example with 'G' $1,25 \times 10^{-272}$pu gravity input as a central pair attracted by a $1,25 \times 10^{-276}$pu backhand channeling a drive of 10^{1140} primal masses across 10^{860}ãr into mutual orbits at $1,12 \times 10^{-284}$¢ during $8,93 \times 10^{1143}$ãt. The $\sqrt{m} \div \sqrt{r}$ relativity formula may seem a blunt instrument when comparing a nuclei to a solar system, given the roughness of these values but it works as a comparison. For galaxies though, the intergalactic flux drive that forms 'G' will for various reasons be somewhat less consistent.

Sun	$\sqrt{Sm}$	$\sqrt{S\text{-}E}$ distance	Rel. speed	km/sec (actual)
Mercury	577	9,1204	63,26=1,607	47,87(47,36)
Venus	577	12,466	46,286=1,176	35,02(35,02)
Earth	**577**	**14,661**	**39,356=1**	**29,78**
Mars	577	18,099	31,88=0,81	24,12=(24)
Jupiter	577	33,448	17,251=0,4383	13,05(13,06)
Saturn	577	45,382	12,714=0,323	9,62(9,68)
Uranus	577	64,242	8,982=0,2282	6,796(6,8)
Neptune	577	80,363	7,180=0,1824	5,433(5,43)
Europa	17,819 =$\sqrt{Jup_m}$	0,982	18,146=0,4611	13,73(13,74)
Titan	9,75 = $\sqrt{Sat_m}$	1,325	7,359=0,187	5,57(5,57)

The moons of Jupiter and Saturn orbit according to the square root of the planets mass with moon distances divided into the AU distance and the square root used. When divided into the 14,661 Sun Earth benchmark (which holds 1 in relative speed, the result yields orbital speed for to any distance point input in moon orbit.

Even a baryon in local orbit plays by the same rules. Two baryons in mutual orbit at a distance at 10^{860}ãr (10^{-49} km) are affected by the square root of their 10^{1140} primal masses or **10^{570}** which is divided by the square root of the 10^{860}ãr distance where **10^{430}** yields **10^{140}**. Compare this to the Sun Earth benchmark of the square root of the Suns $1,19 \times 10^{1197}$ primal mass ($3,45 \times 10^{598}$) which across the square root of one astronomical unit of $1,5 \times 10^{917}$ãr delivers **$3,87 \times 10^{458}$** with orbital speed by way of **$8,91 \times 10^{139}$**. This gives the baryons a **1,122** fold higher value compared to **10^{140}** and allows the baryon pair to outgun 29,78km/sec Earth speed in orbit and with their own 33,41km/sec in mutual orbits.

Part 12: Photon Properties

New Electromagnetic Splits

After the galactic hydron creation run its initial course in a massive chain reaction of electromagnetic splits and formation of hydron coronas, the field turns relatively tranquil. With the large scale electromagnetic upheavals of a forming galaxy long gone, standard hydrons can split any neutral mattron that enters their aura or approaches too close to auxiliary coronas but the occasions of such electromagnetic splits have become radically restricted.

Aside from Neutrons that escape their Proton bonds, the prerequisite for an electromagnetic split is for a neutral mattron to enter a standard hydron aura, either outwards from its surface or inwards from the surrounding field.

Hydrons in stable grids are rarely destabilized enough for a single neutral mattron of their vast surface populations to lose its footing. Auxiliary photons on a hydron surface cause little local disturbances and are rarely disturbed enough to push neutral mattrons out of top layer contact. In the absence of an orbiting hydron corona that produces inwards coronal pressure, surface splits may be easier but without a serious disturbance to drive individual mattrons out of a top layer into the event horizon, there can be no electromagnetically split.

Auxiliary photons orbiting in hydron coronas deliver robust virgin pressure that compresses the entire hydron surface and prevents neutral mattrons from straying out of the smooth top layer surface. Averages are tricky; but a hydron coronas adverse emission from each auxiliary photon covers merely a fraction of its surface. Each auxiliary photon in an Electron corona aims its linear virgins in tight orbits from the $2,16 \times 10^{1708}$ã coronal sphere of which its body blocks $1,63 \times 10^{1702}$ã; the $1,21 \times 10^{25}$¢ virgin boost limits the spread of virgins per ãnna to a small $1,1 \times 10^{-19}$ patch of main body surface. Emission from a permanent fully stocked Electron corona in perpetual orbit reaches $2,2 \times 10^{-10}$ of its surface from a auxiliary orbit at each time.

In a Proton corona, a single auxiliary photon outside a $1,77 \times 10^{1713}$ã sphere with its main body blocking 10^{1704}ã and with $2,05 \times 10^{26}$¢ boost in less tight orbits has its virgin spread limited to $8,63 \times 10^{-18}$ of the surface per ãnna and in a snapshot of a fully stocked 2×10^9 permanent Proton corona, its virgin emission reaches merely $1,73 \times 10^{-8}$ of the Proton surface at each time.

Parts of a hydron surface are therefore marginally more vulnerable during periods when this spot compression is gone while sweeping elsewhere across the hydron surface. Although averages apply to a greater degree over time as each auxiliary photon orbits about a differentially rotating hydron; both movements distributing adverse emission more evenly across its surface.

There may be auxiliary disturbances that can trigger electromagnetic splits by putting a neutral surface mattron into jeopardy but few situations seem likely to pull that off. The least likely is the effect of precursor emission by a recalled auxiliary photon from within the event horizon; disturbances caused by 10^{124}¢ precursor exit sweep topped up by virgin boost across its distance.

As the precursor that forged a deepening surface crater enters a top layer, it takes its place at the central bottom of the crater; stationed marginally deeper than the average hydron surface. A gargantuan positive Proton of 10^{228} grid mattrons is barely affected by a tiny local surface crater with a negative auxiliary photon at its mid bottom among slightly less disparate top layer mattrons. The shallow surface indentation made by a photon as it joins the hydron surface is short lived; Having joined the top layer surface, any indentation is swiftly adjusted back by the Protons massive underlying collective forces.

The uniformly compressed sideways surface does not seem to rise above the small indentation area sufficiently during its short lived surface footprint to allow an auxiliary photon (sideways below a neutral mattron situated on the craters edge) an effective tangential push out of the top layer.

The incoming dominant virgin wind that brought the auxiliary photon into the surface cannot bury it deeper than its top layer indentation. The photon speed into the surface after shedding a precursor cannot push it far against massive resistance. As the hydron surface turns horizontally smooth again, an auxiliary photon becomes another top layer coordination mattron floating on top of a tiny marginally less disparate entourage. It is hard to see that launched and recalled auxiliary photons can play a part in new electromagnetic surface splits and such a trigger would have to be selectively pulled by external effects.

Outgoing holistic winds across an aura are woven of two constant virgin strands. A hydrons collective subservient and dominant emission are disparate virgin winds remain steady throughout a standard aura where the levels in both signs outgun the electromagnetic threshold and are able to hardwire splits and aim twin photons in opposite relative directions. The problem remains of how to get a neutral mattron to enter the scene where a split can be instigated.

A conceivable trigger for new electromagnetic splits is fractional expansion of a main body after a corona is grounded. While expansion is a small fraction of standard hydron radius, it cannot upset the main hydron as the mattron lockup by inter-galactic flux backhand is barely affected. When a corona is grounded, a lifted compression will cause a patchy expansion of the hydron, layer following layer across a fraction of the hydron radius; a collective expansion set by the relevant fraction of a hydrons expansion drive upon all its standard mattrons.

If a restricted corona with lengthy inactive surface periods is limited to an occasional launch of a single auxiliary photon, this will on average sustain weak compression which in turn allows weak consequent hydron expansion set by the speed and level of any consequent collective subservient auxiliary emission.

When subservient hydron expansion eclipses virgin compression, even by a fraction, the mattron lattice starts to expand and top layers are first out of the gate. For example, a 10^{-60} fractional expansion of a 10^{852}ãr wide hydron does not signal 10^{16} fold widening between two top layers from 10^{776}ãr to 10^{792}ãr; all the surface layers are partaking in the expansion.

The 10^{776}ãr top layer is a mental construction; there is no such thing. That mattrons in a standard lattice keep 10^{776}ãr average distances apart does not allow a top layer to move in rigid unison as there is no separation between layers in a mattron lattice. Across a hydron surface, there are 10^{152} mattrons moving outwards and other mattrons follow suit from behind in an unbroken lattice. An outwards move of a top layer across does not mean 'away from the next layer'. It cannot lead mattrons to form precursors as sideways mattrons still provide flux exchanges with more boosted pressure instead of less.

I have not looked hard at how core release works time-wise versus coronal expansion but assume that core sizes catch up fast if a mattron passes out into the event horizon and swiftly forms a precursor; whether a core release can keep up with coronal mass as a precursor grows to full 10^{752}ãt size in 10^{960}ãt. When large enough to harvest 10^{402}pu during its central mattron rotation, a lone mattron is hardwired into an electromagnetic split that still requires 10^{1027}ãt for internal separation to be consummated. First thereafter is the actual electromagnetic split executed within 10^{912}ãt.

The risk for any outermost mattron to lose surface contact decreases when a fractional speed is slow. With a lower level of collective 10^{-300}pu expansion for a Proton coming up from behind with a pressure drop setting parameters of fractional movement, any uneven expansion movement to wider surface distances can be more effectively evened up before that weak drift can carry a single top layer mattron out of surface contact.

If the expansion push upon individual mattrons is too weak to cause major distress to top layer mattrons across smaller distances at slower speeds, the hydrons surface grid simply expands across a fractional distance and settles without electromagnetic splits. This normal surface expansion can be stabilized by virgin harvests and even large fast surface expansions are unlikely to facilitate new electromagnetic splits due to the steady underlying expansion.

If a full corona is recalled, the reaction of top layer mattrons is to continue out to a specific fractional hydron size and stay there while the corona remains grounded. After auxiliary coronal photons have joined the surface, an expansion thrust remains constant until the correct fractional expansion distance has been reached and it is unlikely that electromagnetic splits occur before that.

The expansion of a hydron is on average set away from its center during one main body rotation in 10^{1140}ãt; the implanted directional drive thereafter receives a continual directional antidote. Expanded by 10^{-300}pu subservient expansion per ãnna by 10^{-528}pu upon each of 10^{228} mattrons, the expansion drive by 10^{612} virgins upon each mattrons 10^{912} primal mass yields 10^{-300}¢ speed and brings it out across 10^{840}ãr in 10^{1140}ãt. This distance is 10^{-12} of a Proton radius, which though small, covers the width of 10^{64} top layers.

It may be helpful to note that if Our Sun expands by 10^{-12} of its radius, it spans 100 millimeters and with the collective expansion driving the neutral mattrons so uniformly outwards, I cannot really see these expansion periods as a trigger for electromagnetic splits.

Fractional Re-Compression

Conversely, a more likely greater potential trigger for new electromagnetic splits comes with the re-collapse of an already expanded main body hydron. The more expanded the hydron surface, the greater is the opportunity for a rapid uneven shrinkage during coronal re-launch that might generate new electromagnetic splits just outside the hydron surface.

A re-launched corona delivers uneven compression upon a hydron surface and this is especially acute over the launching area after the first auxiliary photon rises from a highly expanded hydron after a long time without any corona. Low pressure launching pole area is the easiest place for a neutral mattron to rise within a top layer (although immensely harder than for the auxiliary photon) and any new electromagnetic splits are therefore also likely to aim out from a launching pole.

An auxiliary corona grounded with a weakest possible marginal becomes the optimal electromagnetic environment when launched against a virgin gravity orb with the weakest possible marginal. Even so, it does not seem likely to lead to instigate a split.

If a split occurs, the stronger **subservient** virgin repulsion in a hydron aura aims the subservient mass of a new potential electromagnetic photon; in a sign opposite to the hydrons auxiliary photons. The electromagnetic twin is 'propelled' away by its twin and not by the holistic aural repulsion that set it up for a split. The electromagnetic photon would in this case have taken the inherent standard hydron coordinates completely to heart; unlike electromagnetic photons out of a Neutron breach. This launch into a wider unstable subservient corona follows the rules for such coronas.

With a subservient electromagnetic photon rising swiftly above the hydron surface, its adverse emission starts effectively to beat the expanding surface of neutral mattron contenders back into surface contact. The twins can only briefly emit neutrally blended adverse virgins to pours oil on the waters of other local disturbances and in short order after the split, the potential dominant electromagnetic twin would join the hydron surface to later serve as an ordinary dominant auxiliary photon.

Different faces of Primal Energy

In the present Universe, concentrated emission of primal energy is mainly supplied by already existing photons that deliver their primal virgin energy input locally or universally by different means under various restrictions. All of these virgin winds constitute a delivery of primal energy that becomes ever more evenly blended as the size of the affected volume grows.

In order of increasing power, the virgin winds include constant main body hydron emission, subservient or dominant, and collective locally shifting free adverse emission by auxiliary photons in hydron coronas while the constant adverse emission by electromagnetic photons passing through the grid of Universal systems lends the heaviest hand.

*When it comes to photons, it is not the photon body that signifies its real clout or presence in the field. Immensely more than its actual primal particle mass, what we mean by a 'photon' **is** its singular freely escaping adverse emission.*

The photon mass has constant adverse emission but the collective delivery systems limits its escape into the outside field with auxiliary coronal periods being limited on a scale from all to none.

Another important aspect that will be discussed shortly goes to the heart of the matter; how auxiliary photons harvested by hydrons and re-launched with less than inherent coordinates at boosted orbital speeds affect the way au-photons deliver their constant local adverse emission locally.

Hydrons can encroach on each other territory and alter coronal structures that change the exchanges of this primal energy in different ways. Grounded coronas curtail the delivery of primal energy while re-launched coronas reinstate that power potential. Such shifts and disturbances are a source of heat when hydrons start to move around in an excited state like described by 'Brownian motion'.

Auxiliary photons that are permanently tied to orbit their home hydron deliver emission are in most other aspects exactly the same as the roaming electromagnetic photons but their primal radiation is dosed out with greatly restricted local input.

Globular Electromagnetic Fronts

For present ordinary electromagnetic emission, one can discard the image of a wall of electromagnetic photons moving away from one emission point in a globular fashion around say a collapsing surface of an exploding hydron. Even during great instability, except in Neutron splits, new electromagnetic splits do not send multiple photons in linear directions with a globular pulse.

Nevertheless, it is useful to revisit the globular front because it so clearly shows the nature of a departing photon. Each photon on an expanding front is largely independent and alone (conceptually speaking) even if perception treats it as an integral part of a globular emission front.

The photons adverse virgin emission is vastly more influential in a local field than is the photon body which is merely a side effect; the adverse emission is the photon. Primal emission, even from a photon front, blends with the input of other photons in a local strayfield where singular or neutral virgin pressure moves at 10^{280} fold photon speed and reaches far ahead in the hydron grid; a first delivered pressure is long spent and heavily blended when a photon ultimately arrives with a more intense local pressure.

Every adverse virgin is in possession of its photons collective coordinate system and each is inherently 'oblivious' to any linear universal movement of its photon from another perspective. The photon is a standstill universe to itself while the outside world appears to swoosh by at 10^{-280}¢. In relation to a photons coordinate system, an adverse virgin moves away in any direction at birthspeed and with active birthspin.

One can nevertheless use the image of an expanding globular sphere as a useful conceptual crutch in looking at any photon emission. If more than one photon is emitted in the same burst from an unstable surface area, the span between a first and last photon in one electromagnetic burst (the conceptual thickness of the balloon membrane) would be relatively short.

The multiple photon dots would have the waver thin width of a fraction of a hydron aura. With a sprinkle of photons on an 'imaginary balloon skin'; the principles for one or several photons departing from a troubled hydron are the same and the 'dotted membrane' keeps its width largely unchanged while widening at light speed across the universe.

As the imagined membrane widens, the distance between a couple of close dots on its surface increases sideways with the distance travelled in hydron radii; the original sideways distance of two simultaneous electromagnetic splits on a hydron surface governs the actual distance between the multiple membrane dots when multiplied by the distance traveled in hydron radii. While their sideways distances thus widen, the photon dots themselves are the only part of the membrane that does not expand.

For example, when dots from a simultaneous Proton emission reach one meter (10^{906}ãr) away, a distance of 10^{54} Proton radii, the distance between the dots has grown to 10^{54} fold from their original distance apart during the split. If the two photons were emitted from the same polar spot, let's make their original distance apart less than 10^{794}ãr, the membrane dots would only have moved apart to roughly 10^{848}ãr across the meter distance.

Adverse emission pressure from an electromagnetic photon soon wanes upon its ancestral hydron as more virgins miss it. The rate at which this harvest drop is independent of the photon number; compression upon the home hydron fades at the same rate whether the electromagnetic burst holds one or more photons. The local pressure harvest would be greater with more photons on any expanding 'balloon' patch but if electromagnetic emissions, except in Neutron splits, carry a single photon, it is somewhat superfluous to discuss wave thickness and photon numbers. This brings us to the last important bits of this loose illustration; the reasons for electromagnetic frequencies and wavelengths in our universe.

Electromagnetic Frequency

For the longest time, I have been held back by my unconscious inability to question the definition of frequency and wavelength; an indisputable reality that comes in an exclusively dictated form; so many cars pass this road per hour. I can find no systematic correlation to this type of photon frequency or wavelength in how photons are actually split and emitted outside hydrons. And trust me, I spent years looking.

There are of course many individual electromagnetic splits that are tainted by this sort of frequency where primal backlash from a former split prevents another split by pacifying a hydron surface until a photon has reached so far that it cannot suppress the left behind hydron. This allows a repeat emission but this has nothing to do with electromagnetic frequency or wavelength.

The classical treatment of 'photon waves' is coupled to a frequency where energy delivery over time is a multiple of how many such 'waves' pass during a specified period. In other words, the undefined 'photon wave' is followed and preceded by identical waves at a certain interval (frequency) where the gap between the waves has a corresponding width (wavelength).

Using a wave function, the Standard Theory views each photon as being the whole wave that then realized a single outcome once measured (and no measurement of other possible states can later be made). According to that mathematical view, the photon presents a certain probability to appear to a distant 'observer' and the 'expanding wave vanishes or collapses' when it is measured. The state of the mathematical wave function changes 'instantly' from its full range of possibilities to a single state of the measured particle.

This mathematical method works perfectly for photons. Using a probability wave allows the Standard Theory to sidestep the problem of instant collapse across galactic distances. A probability wave presents a range of possible outcomes and gives a correct statistical answer without tending to what is actually happening. The assumption is that a single wave package of energy arrives with certain periodicity before a next 'photon wave' arrives from the same source delivering energy of a collective 'photon' as its frequency per second multiplied by a constant energy package. The wave function yields the correct outcome but gives no insight into reality on the scene. There is no way to use this method except by turning a mathematically blind eye to the facts on the ground.

I have no recourse in this respect than to hitch the Primal Code to the real material world; to look for an electromagnetic spectrum made up of single photons of variable 'frequency' and 'wavelength' although these are hardly a concept that you take inherently to heart. There is a way to tie a single material photon and its locally delivered adverse emission energy to a Hertz scale where it can be locally deliver a low 1Hz frequency or a yottahertz 10^{24}Hz frequency.

The corporeal photon has a really sophisticated way to deliver its energy with a special frequency or wavelength when using the reality of the Primal Code instead of the probability wave. A large 'wavelength' of solar width can span 3×10^{914}ãr and deliver one Hertz 4,13feV 'energy' per second (3×10^{1194}ãt) while the photon continues linearly at 10^{-280}¢ light speed. The question arises how two identical single photons at the same speed from the same emission point can deliver an effect to an identical local measuring point with such different effect.

The answer is in the way subservient auxiliary photons (at close to linear light speed) are emitted by a hydron that has recently blocked them (wiping their coordinate slate clean). When the auxiliary photon is re-launched, it has rarely collected a full pot of inherent coordinates from the hydron that blocked and is re-launching it. The fraction of inherent coordinates stamps the mostly pristine photon with a precise scale of orbital elements as they head away from the hydron, whether it be a Proton or an Electron.

We have seen a similar phenomenon during a shift an orbital auxiliary photon as an enforcing shift pushes a corporeal photon beyond the center of its own coordinate system. For fully inherent auxiliary photons, this resulted in an orbit set by the distance at which the photon found itself when its orbit passed its own coordinate center (roughly half the original distance).

For a fractionally tainted auxiliary photon, a push away from its coordinate center is only partly inherent. With other majority coordinates involved (the photons private coordinates) the shift of the corporeal photon is mostly governed by enforced coordinates and marginally by inherent coordinates. The effects of these blending coordinate systems is hard to get your head around and may require a special heading to sort out the strings.

Photon Wavelength and Frequency

A less than fully inherent re-launched auxiliary photon is governed by a blend of two coordinate centers which are parts of its new coordinate system as it leaves the hydron aura. The photon continues at the speed gained in the hydron aura and is thereafter no longer accelerated outwards.

For an auxiliary photon (a massive submist), the linear virgins collected in the aura push it 'linearly' away from the gargantuan hydron along a (largely) linear enforced path. These virgins are relatively few compared to the photon mass and the orbital coordinates of these virgins are linearly contradictory and their orbital effect is neutralized in the aura.

The Photons independent (private) coordinate center that remained after its slate of universal coordinates was wiped clear on being blocked by the hydron is at the point of the auxiliary photons departure in control of its own pristine rotation about own central coordinate center. The enforced kick shifts the corporeal photon along with its pristine coordinate center along a linear path away from the hydron but not at the same speed. For a primal particle hit by another primal particle, you recall that the speed away from an emptied coordinate center was halved; for the photon it depends on the push level but I simply allow the same slowdown for illustration).

In the case of the auxiliary photon that has harvested some part of a hydrons collective coordinates, that second coordinate center makes its mark upon the departing auxiliary photon. Some fraction of inherent sideways orbital motion of a hydrons inherent coordinate center (not the photons pristine coordinate center) is also part of the corporeal photon and shapes its outwards path.

If you separate the two coordinate centers, things become clearer. First, its own coordinate center causes the corporeal photon to orbit its independent center in a circular orbit. This fully inherent orbit widens with distance from its center at light speed (or large fraction of that if you like) and orbital speed around its center increases to superluminal speeds. Second, a small part of a hydrons side sweep causes an initially marginal orbital speed of the corporeal photon but this enforced push does not affect the photons linear coordinate center). This side sweep to follow the hydrons rotation accelerates as the distance widens in tune with the inherent share of coordinates harvested within the hydron and carries the corporeal photon orbit sideways in relation to its private coordinate center.

By separating the two coordinate systems, you can see that the pristine but empty photon coordinate center keeps its linear path while the hydrons side sweep shifts the orbit of the corporeal photon. This shifting movement is slow at first in full accordance with the harvested hydron coordinate fraction but the shifting speed is distance dependent and therefore accelerating.

This phenomenon that is similar to coronal shift in magnetic transfer will cause the corporeal photon to be carried beyond its own inherent coordinate center when the speed of the sideways shift comes to equal the speed of light (or any speed at which its perfectly circular coronal orbit widens). A photon will ultimately have its entire orbit moved beyond its empty, fully governing, coordinate center. As the linear distance grows, this inherent sideways shift beyond that coordinate center happens faster because the widening speed remain constant at light speed (or whatever).

A photons central coordinate system holds its dominion over the corporeal photon all the time; the distance of its inherent orbit was been shifted within that system (orbiting slower at the closer side and faster at the further side). Precisely as with magnetic transfer; when the single corporeal photon orbit passes by its own coordinate center, the corporeal photon is again fully at home within its central coordinate system which it continues to orbit (its position has just been shifted within that system).

This new orbit, smaller by roughly half, will now start to widen from that exponential distance by light speed (or whatever) out to the same exponential distance as earlier where it will flip again. The photon again orbits the center of its system and will hereafter swing between midways to greatest possible distance from its coordinate center.

It takes a gradually rising inherent sidesweep a gradually shortening period to bring a circular orbit beyond its central coordinate center by way of rising inherent boost of this component. Although a side shift will relatively soon exceed light speed and the period to shift the orbit across the coronal distance becomes shorter, it does not change much for the orbiting corporeal photon except that the mundane shifts between half and whole orbit become faster. The corporeal photons superluminal orbital speed would also shift by half from the speed reached when the two components became equal at light speed.

After this distance point away from a hydron, the orbit of a corporeal photon around its coordinate center and the distance shift by the fraction of hydron coordinates will not widen. At that point there is a fixed relation between the photons superluminal speed in relatively 'constant' orbit; the larger that orbit, the greater the local orbital speed of the corporeal photon.

Corporeal photons are depicted by identical constant adverse emission. There is no difference in photon emission whether a 'closed' local orbit of a corporeal photon is ten times or ten million times larger. How a photons virgin emission is locally spread is however directly affected by orbital distance in a fixed relation.

The effect from the entire sphere from which a photons adverse emission is spread on its superluminal way is more locally diluted for a large orbit than a small orbit. Obviously, a photon in large orbit, say 10^{914}år solar radius around a coordinate center at that distance, cannot affect a local hydron as much as a photon passing it by in say 10^{888}år orbit. It does not help if the photon in the 10^{914}år orbit can pass close by the local hydron within 10^{888}år, as it passes by at such superluminal speeds that its distribution of local effect is severely diluted in a fixed relationship; it cannot deliver the same local energy.

Naming the fixed orbital distance the photons **'wavelength'** *is a comforting word in this context. The photons orbital speed for a wavelength while it sweeps by a local entity in its orbit, sets the period its adverse emission can affect it and this period of passage can then be called the photons local* **'frequency'**.

After this 'wavelength' distance in accord with inherent hydron coordinates is reached we are seeing photons continue away from a hydron along broad tubular lines where each single corporeal photon in three dimensional differential orbits is constantly shifted between the tube walls in fast superluminal orbits. A larger 'wavelength' yields a lower local 'frequency' and vice versa.

A photon of any wavelength at light speed can be stopped outside or within a hydron aura and be reflected with preserved wavelength and frequency. If it hits a hydron and is absorbed, its inherent Universal coordinate slate is wiped clean.

The Electromagnetic Spectrum

These rough illustration are ill equipped to set a precise electromagnetic spectrum as every aspect of size and influence on the hydrons scale, from the creation constant onwards, is more or less picked out of my hat.

We can start this illustration on the Hertz scale with an extremely low SLF frequency of say 1Hz per second. With one second as 3×10^{1194}ãt according to the Primal Code, the photons vast 3×10^{914}ãr 'wavelength' (which is roughly of solar width) is reached at 10^{-280}¢ in 3×10^{1194}ãt (one second) which gives it a frequency of 1Hz per second.

If a particular **positive** subservient photon is emitted by a Proton soon after being blocked (a Proton is more easily hit by photons due to size and powerful recall orbs) and this photon is re-launched during 10^{1132}ãt with a relatively clean coordinate slate; it would block 10^{632} virgins on its way out of the 10^{852}ãr Proton aura and be pushed to $\mathbf{10^{-280}}$¢ light speed along a linear path during the 10^{1132}ãt period it takes to clear the aura.

The Protons inherent rotation takes 10^{1140}ãt at 10^{-288}¢ surface sweep. The photon must make up for 10^{-8} lacking in surface sweep compared to 10^{-280}¢ linear speed (which it can do at $\mathbf{10^{860}}$ãr with full pot of Proton coordinates).

If a particular photon is swiftly re-emitted and lifts from the Proton surface after 3×10^{-26} part Proton rotation in $3,33 \times 10^{1113}$ãt, its pot of inherent Proton coordinates is $3,33 \times 10^{-63}$ part of the 10^{1176}ãt required for the photon to become fully inherent. The distance from the Proton where 10^{-288}¢ inherent side sweep of a photons orbital drift comes to equal the 10^{-280}¢ linear speed is 3×10^{914}ãr 'wavelength' (10^{852}ãr÷$3,33 \times 10^{-63}$) after one 3×10^{1194}ãt second at 10^{-280}¢ speed.

With all applicable riders for this rough illustration, if a subservient **negative** photon is re-launched by an Electron with a relatively clean coordinate slate by say $6,132 \times 10^{-199}$ of $3,58 \times 10^{-301}$ virgins during $7,63 \times 10^{1130}$ãt, its precursor will harvest $1,68 \times 10^{632}$ virgins that drive out of the $1,277 \times 10^{851}$ãr Electron aura along a linear path and at $\mathbf{1,68 \times 10^{-280}}$¢ speed. If this is correct, this slightly superluminal 1,68 fold light speed takes $7,6 \times 10^{1130}$ãt to clear its aura.

Not only is the linear speed of the outgoing photon slightly faster but an Electrons inherent rotation at $2,08 \times 10^{1137}$ãt is $4,81 \times 10^2$ times faster than a Protons with $6,13 \times 10^{-287}$¢ surface sweep. Compared to its $1,68 \times 10^{-280}$¢ linear speed, the outgoing photon can make up for the $1,03 \times 10^{-6}$ lacking in surface sweep at $1,32 \times 10^{857}$ãr away (if it had a full pot of Electron coordinates).

If a particular negative Electron photon is re-emitted in $8,85 \times 10^{1109}$ãt that is $4,25 \times 10^{-28}$ of the Electrons $2,08 \times 10^{1137}$ãt rotation or $4,25 \times 10^{-64}$ part of the $2,08 \times 10^{1173}$ãt needed for a full inherent pot of Electron coordinates; the rotation speed $6,13 \times 10^{-287}$¢ and linear 10^{-280}¢ turn equal at 3×10^{914}ãr $(1,277 \times 10^{851}$ãr÷ $4,25 \times 10^{-64})$. This is the same 'wavelength' as for the Proton re-emission above but here, for the negative Electron photon, it is set after $1,79 \times 10^{1194}$ãt in 0,6 seconds at a faster $1,68 \times 10^{-280}$¢ linear speed. Many Electron features seem therefore to render a different framework for photon emission and there may even be aspects of coordinate assimilation that are different due to the Electrons internal communications of virgins across its shorter distances.

Staying with the Proton for another example, if a positive photon is re-emitted after 200 Proton rotations in 2×10^{1142}ãt or 2×10^{-34} part of 10^{1176}ãt required for a full pot of Proton coordinates, the distance from the hydron where the 10^{-288}¢ inherent side sweep equals 10^{-280}¢ linear speed is a new 5×10^{885}ãr 'wavelength' $(10^{852}$ãr÷$2 \times 10^{-34})$ after 5×10^{1165}ãt with the photon drive at 10^{-280}¢ speed.

The fixed relationship of wavelength and photon speed in local orbit along a wavelength perimeter from 10^{852}ãr Proton size to 3×10^{914}ãr is a **3×10^{62}** fold increase in wavelength and therefore also 3×10^{62} fold increase in the photon orbital speed along the wavelength perimeter (from 10^{-288}¢ to 3×10^{-226}¢).

After a split of a neutral mattron without any structural coordinates for the adverse twins of the electromagnetic photon, their path is perfectly linear with a wavelength no broader that their precursor paths; this is the smallest possible photon 'wavelength'. The wavelength and orbital speed can of course also be vastly greater than 3×10^{914}αr and even reach universal proportions.

Between the long electromagnetic wavelength of **3×10^{914}ãr** and a mediocre wavelength of **5×10^{885}ãr**, there are 6×10^{28} steps (using a multiple of ten for each step). To a linear electromagnetic 'wavelength' of **10^{852}ãr** there are 3×10^{62} steps but these steps can be divided as one pleases to depict a wavelength that brings some sort of a measurable change in the hydron grid; forming a rather more intricate set of 'wavelengths' within the broader Hertz scale.

There are also slower dominant auxiliary photons with less than full pot of inherent hydron coordinates; all with wavelengths and frequencies of their own set by the same coordinate rules at roughly 10^7 times slower linear speed. This does not remove these photons from of the realm of electromagnetic radiation.

The Photoelectric Effect

Before I begun to grasp frequency according to the Primal Code, I assumed that the photoelectric effect by higher frequency electromagnetic emission would be achieved by delivering multiple photons into individual hydron grid domains. This would put multiple electromagnetic photons close enough to a hydron pair to continue grounding group coronas and allow further jumps in orbit which might ultimately release an Electron across far enough distances to prevent new coronas to materialize and thereby cap its flight.

My early thoughts were that an unbalances at a hydron surface elicited electromagnetic photons and presumed that when these had traveled far enough from the hydron, the adverse emission would become too thinly spread to pacify its surface; whereupon another bout of electromagnetic splits would follow on a globular front. I had thought that a major unbalance would lead to a massive bout of multiple electromagnetic photons which required a short distance and period to pacify an emitting hydron while more marginal unbalances could be pacified by a single photon from a further distance at lower frequency. I never found a cause for this to be so.

As the picture cleared, I came to realize that while surface unbalances and distances of emitted electromagnetic photons are important considerations, these effects could not set precise sets of wavelengths and frequency which then sent me in search for a more logical intricate system.

You have heard this before: *A photon beam delivering energy proportional to its frequency is soaked up by an Electron on material surfaces. The Electron is ejected if the 'absorbed' energy exceeds its 'binding energy'. If the soaked up energy is too weak, the Electron cannot escape and the energy is re-emitted. Increased light intensity raises the photon number in a beam which increases the number of excited electrons but does not increase the energy possessed by each Electron. The energy intake of an emitted Electron depends solely on the frequency of 'one photon' interacting with one outermost electron. This largely follows the all or nothing principle; all the energy from a 'single photon' must be absorbed to liberate a single Electron from its atomic binding. If the photon energy is absorbed, some of it liberates the Electron from its atom while the rest contributes to its 'kinetic energy' as a free particle.*

We have sees how a single Proton Electron pair in a balanced dynamic bond can make a transition jump to a wider orbit during the passage of a distant photon that is able to ground the baryon coronas. A surface of more complex atoms in a metal lattice with a different work function would allow an outermost Electron easier ejection. In a rough illustration based on the Primal Code, look into a tighter metal surface made up of large central baryons groups and their Electrons in wider more vulnerable orbits.

To clarify the general idea without going into any specifics, look at a single largely linear positive au-photon that enters a metal surface grid from one meter (10^{906}ãr) and passes in among atoms in an average 10^{894}ãr grid. Some atoms in this surface are affected as relatively few single photons are divided upon a vast number of atoms. After the individual photons enter and pass between the first top atom layer, they are trapped beyond the top surface and absorbed to turn into just another auxiliary photon deep in the grid.

$$10^{852}\text{ãr} \div 10^{886}\text{ãr} = 10^{-68}\ (gain\ 10^7) = 10^{-61} \times 10^{-280} \times 10^{41}\text{¢} = 10^{-300}pu\ equal\ to\ 10^{-300}pu$$

Let's say that a perfectly linear electromagnetic photon passes at 10^{886}ãr from the metal atom at light speed during 10^{1166}ãt; the odds are that 10^{-6} of such photons pass this close on a path distantly adjacent to the metal atom. A little close and the pod coronas of the many baryons in the metal atom are swiftly grounded or heavily restricted which allows the Electrons greater coronal emission due to dropping inactive surface periods.

A steady local presence of adverse photon emission is needed for considerable periods but the mutual repulsion from a large number of Electrons in a complex and extensive cloud can easily allow a weakly bound outermost Electron to be rejected as it jumps furthest; driven away from the inner Electron cloud by mainly magnetic transfer; this is the 'photoelectric' response.

A linear perfectly linear electromagnetic photon or an auxiliary photon with a medium small 'wavelength' that both move through the grid at light speed and pass adjacently to a metal atom within say **5×10^{885}ãr,** both yield the same local virgin influence as their adverse emission spreads within that space during their 5×10^{1165}ãt period of passage.

If a positive photon is re-emitted by a Proton after say 100 rotations in 10^{1142}ãt or 10^{-34} part of 10^{1176}ãt required for a full pot of Proton coordinates, the distance from the hydron where the 10^{-288}¢ inherent side sweep equals 10^{-280}¢ linear speed yields its 10^{886}ãr 'wavelength' (10^{852}ãr $\div 10^{-34}$).

While a perfectly linear photon and any photon with up to 5×10^{885}ãr wavelength can pass close enough to a local baryon group to ground their coronas, a photon with 10^{886}ãr wavelength and above cannot. The corporeal photon can pass closer but then at such high speed that it cannot deliver its payload locally. If the corporeal photon sweeps close by the metal atom, it passes so quickly that its local energy is still distributed across the entire 10^{886}ãr span of its 'wavelength' whereupon the photon cannot cause a 'photoelectric' effect; no matter how close its sweeps to that metal atom.

It does not help to increase the intensity of this 'light' by increasing the number of such photons, their local effect is too weak to ground coronas and cause

Electrons to be repelled. Only a small number of innumerable atoms on a metal surface are affected by a passing photon in its time of passage.

Photon Absorption

The capture of photons as they pass through various grids among various hydron groups is another relevant aspect of the electromagnetic spectrum; the absorption of photons.

Photon effect can be a temporary local phenomenon as they pass through a hydron grid. There is a chance that some become linearly disoriented as their paths are affected which can increase the rate of photon capture. Many are prevented from leaving a larger grid after being locally affected to take on increasingly curved and convoluted paths. Many are ultimately recalled by a hydron of adverse polarity in incremental orbits to become another auxiliary photon to be recalled and re-launched with the rest of them.

The chance of a perfectly linear photon scoring a direct hit on an 10^{852}ãr wide hydron within an 10^{895}ãr wide average grid in a 10^{914}ãr wide sun of 10^{57} hydrons is miniscule; send a linear electromagnetic photon at a sun and merely one in 10^{67} would hit a hydron if the absorption were exclusively set by hydron sizes.

Linear photons are nevertheless susceptible to hydron interferences (more than direct recalls) because of their slow progress, passing at light speed through myriad powerful local emission fields. A photon is likely to pass by 10^{19} hydrons pairs on its way through a sun and therefore at closer distance than 10^{876}ãr in at least one hydron domain.

$$10^{752}\tilde{a}r \div 10^{880}\tilde{a}r = 10^{-256} \times 2 \times 10^{-271} = 2 \times 10^{-527}pu$$

A negative perfectly linear photon passing through a 10^{895}ãr wide grid at 10^{880}ãr from any hydron with a corona of full free emission that counters its advance with 2×10^{-527} virgins per ãnna. During 10^{1160}ãt the photon collects 2×10^{633} virgins that can reflect it head on at 2×10^{-279}¢ superluminal speed. It would be stopped it in its tracks and returned at its original speed at an angle of reflection equal to its angle of incidence.

Bulls eyes are rare and approaching photons have their linear approach continually rearranged. This scattering and path reversal of electromagnetic photons plays a role when photons on curved and convoluted paths are set to be recalled in incremental steps as they pass through local gravity orbs. For every atom or molecule, the grid character has certain characteristics of different orbs, as is the occasional absorption of passing photons depending on their wavelengths and frequency speed.

However, additionally, a more direct grid absorption of photons is tied to their frequency; the orbital speed the corporeal photon itself keeps within its wavelength distance.

Of course, most photon paths do not lead through opaque suns but rather more transparent clouds or areas of widely different atom constellations. In each sign, each bonded atom has a characteristic rate of photon absorption that depends both on how grid units interact with a passing photon and its wavelength and frequency.

Aside from the recall of photons that pass too close to hydrons through a grid and become disoriented, photons of diverse wavelengths pass by more subdued hydrons where disorientation is not the main cause for entrapment. For occasional direct hits through these material fields, the wavelength plays a more central role (as does the bonded group number of grid atoms).

A perfectly linear photon passing midways through a tenuous 10^{896}år grid where it encounters a single hydrogen atom has a 10^{-88} chance to score a direct linear hit upon it. If this passing photon has a wavelength of **10^{896}år**, its chances of a hit rise to **10^{-52}**; increasing 10^{36} fold by the photons 10^{-244}¢ frequency speed. The photons corporeal path occupies 10^{36} times the volume on crossing this distance than a perfectly linear photon. Its wavelength does not allow the photon to pass closer to any atom or change the 10^{-52} outcome.

When photons of smaller wavelengths pass closer than midways with a certain regularity, their chances of a hit increase. For example, passing at 10^{870}år from a single hydrogen atom renders 10^{-36} chance for a direct linear hit but if that photon has **10^{870}år** wavelength and 10^{-270}¢ frequency speed, its chance of hitting it rises to **10^{-26}**. Its corporeal path, compared to a linear photon, occupies 10^{10} times the volume on crossing this distance. Although its average chance of passing this close is 10^{-26} and leaves the overall chance for at 10^{-52}, Its wavelength does nonetheless allow the photon to pass closer to the grid atom and all such photons will chance to do that and thereby gain a better chance of a direct hit than would the 10^{896}år wavelength photon.

Postscript;

This of course is a thought in progress, an empty template to be filled out by the brilliant minds at work in this field. I am sure that this philosophical search for a foundation to our world is not to everyone's liking but as the Primal Code peeks over the parapet of the Planck constant to view the Universe it has created; for me it has become the end of an interesting journey through a different world.